Aviation Weather Handbook

2022

United States Department of Transportation
FEDERAL AVIATION ADMINISTRATION
Flight Standards Service

Preface

This handbook consolidates the weather information from the following advisory circulars (AC) into one source document. By doing this, the Federal Aviation Administration (FAA) intends to streamline access to the FAA's weather documentation for users of the National Airspace System (NAS). The following ACs will remain in effect, but they will eventually be cancelled at a later date following the publication of this handbook:

- AC 00-6, Aviation Weather.
- AC 00-24, Thunderstorms.
- AC 00-30, Clear Air Turbulence Avoidance.
- AC 00-45, Aviation Weather Services.
- AC 00-54, Pilot Windshear Guide.
- AC 00-57, Hazardous Mountain Winds.

This handbook is designed as a technical reference for all who operate in the NAS. Pilots, dispatchers, and operators will find this handbook a valuable resource for flight planning and decision making.

This handbook conforms to pilot weather training and certification concepts established by the FAA. The discussion and explanations reflect the most commonly used weather products and information.

It is essential for persons using this handbook to also become familiar with and apply the pertinent parts of Title 14 of the Code of Federal Regulations (14 CFR) and the Aeronautical Information Manual (AIM). Title 14 CFR, the AIM, this handbook, current ACs, and other FAA technical references are available via the internet at the FAA home page https://www.faa.gov.

This handbook is available for download in Portable Document Format (PDF) from the FAA's Regulations and Policies web page at https://www.faa.gov/regulations_policies/handbooks_manuals/aviation/.

This handbook is published by the U.S. Department of Transportation (DOT), FAA Flight Technologies and Procedures Division.

The guidance and recommendations in this handbook are not legally binding in their own right and will not be relied upon by the FAA as a separate basis for affirmative enforcement action or other administrative penalty. Conformity with the guidance and recommendations is voluntary only and nonconformity will not affect rights and obligations under existing statutes and regulations.

Comments regarding this publication should be sent, in email form, to the following address: 9-AWA-AFS400-Coord@faa.gov.

Wesley L. Mooty
Acting Deputy Executive Director, Flight Standards Service

11/25/2022

Acknowledgements

This document was prepared under FAA Contract DTFAWA-15-D-00031 with support from AvMet Applications, Inc.

Individual photographic contributions were made by Blake Spry (front cover), Luke Culver (front cover and cover of Chapter 15), and Larry Burch (covers of Chapters 1, 2, 3, 4, 5, 7, 8, 9, 10, 11, 13, 14, 16, 18, 23, 24, 25, 26, 27, 28, Part 1, Part 2 and Part 3). Cover of Chapter 20 courtesy of the National Research Council of Canada.

Notice

The U.S. Government does not endorse products or manufacturers. Trade or manufacturers' names appear herein solely because they are considered essential to the objective of this handbook.

Table of Contents

Preface iii
Acknowledgements iv
Notice v
Table of Contents vi
Chapter 1. Introduction 1-1
Part 1: Overview of the United States Aviation Weather Service Program and Information
Chapter 2. Aviation Weather Service Program 2-1
   2.1   Introduction 2-1
   2.2   National Oceanic and Atmospheric Administration (NOAA) 2-2
       2.2.1   National Environmental Satellite, Data, and Information Service (NESDIS) 2-2
       2.2.2   National Weather Service (NWS) 2-2
   2.3   Federal Aviation Administration (FAA) 2-4
       2.3.1   Air Traffic Control Systems Command Center (ATCSCC) 2-4
       2.3.2   Air Route Traffic Control Center (ARTCC) 2-5
       2.3.3   Airport Traffic Control Tower (ATCT) and Terminal Radar Approach Control (TRACON) 2-5
       2.3.4   Flight Service 2-5
   2.4   Department of Defense (DOD) 2-5
   2.5   Commercial Weather Information Providers 2-5
Chapter 3. Overview of Aviation Weather Information 3-1
   3.1   Introduction 3-1
   3.2   Use of Aviation Weather Information 3-2
       3.2.1   Product Latency 3-2
       3.2.2   Additional Use Information 3-2
   3.3   Obtaining Weather Information 3-2
       3.3.1   Weather Briefings 3-2
       3.3.2   Telephone 3-4
       3.3.3   Self-Briefing 3-5
       3.3.4   In-Flight Updates 3-6
   3.4   Overview of Aviation Weather Products 3-7
       3.4.1   Weather Products versus Weather Elements and Phenomena 3-7
       3.4.2   Summaries of Specific Weather Information Contained in Various Weather Products 3-13

Part 2: Weather Theory and Aviation Hazards

Chapter 4. The Earth's Atmosphere 4-1
    4.1   Introduction 4-1
    4.2   Composition 4-2
        4.2.1   Air Parcel 4-3
    4.3   Vertical Structure 4-3
        4.3.1   Troposphere 4-3
        4.3.2   Stratosphere 4-3
        4.3.3   Mesosphere 4-4
        4.3.4   Thermosphere 4-4
        4.3.5   Exosphere 4-4
    4.4   The Standard Atmosphere 4-5

Chapter 5. Heat and Temperature 5-1
    5.1   Introduction 5-1
    5.2   Matter 5-2
    5.3   Energy 5-2
    5.4   Heat 5-2
    5.5   Temperature 5-2
        5.5.1   Temperature Measurement 5-2
        5.5.2   Temperature Scales 5-2
    5.6   Heat Transfer 5-4
        5.6.1   Radiation 5-4
        5.6.2   Conduction 5-6
        5.6.3   Convection 5-7
    5.7   Thermal Response 5-8
    5.8   Temperature Variations with Altitude 5-11
        5.8.1   Atmospheric Sounding 5-11
        5.8.2   Isothermal Layer 5-11
        5.8.3   Temperature Inversion 5-12

Chapter 6. Water Vapor 6-1
    6.1   Introduction 6-1
    6.2   The Hydrologic Cycle 6-2
        6.2.1   Evaporation 6-2
        6.2.2   Transpiration 6-2
        6.2.3   Sublimation 6-2

	6.2.4	Deposition	6-3
	6.2.5	Condensation	6-3
	6.2.6	Transportation	6-3
	6.2.7	Precipitation	6-3
	6.2.8	Runoff	6-3
	6.2.9	Infiltration	6-3
	6.2.10	Groundwater Flow	6-3
	6.2.11	Plant Uptake	6-3
6.3		Saturation	6-3
6.4		Relative Humidity	6-4
6.5		Dewpoint	6-4
6.6		Temperature-Dewpoint Spread (Dewpoint Depression)	6-4
6.7		Change of Phase	6-5
	6.7.1	Latent Heat	6-6

Chapter 7. Earth-Atmosphere Heat Imbalances 7-1

7.1		Introduction	7-1
7.2		The Earth-Atmosphere Energy Balance	7-2
7.3		Heat Imbalances Between Earth's Surface and the Atmosphere	7-3
	7.3.1	Sensible Heating	7-3
	7.3.2	Latent Heat	7-4
7.4		Heat Imbalance Variations with Latitude	7-5
7.5		Seasons	7-6
7.6		Diurnal Temperature Variation	7-7

Chapter 8. Atmospheric Pressure and Altimetry 8-1

8.1		Introduction	8-1
8.2		Atmospheric Pressure	8-2
	8.2.1	Barometer	8-2
	8.2.2	Atmospheric Pressure Units	8-3
	8.2.3	Station Pressure	8-3
	8.2.4	Pressure Variation	8-4
	8.2.5	Sea Level Pressure	8-6
8.3		Density	8-6
	8.3.1	Volume's Effects on Density	8-7
	8.3.2	Changes in Density	8-8
	8.3.3	Density's Effects on Pressure	8-8

8.3.4		Temperature's Effects on Density	8-9
8.3.5		Water Vapor's Effects on Density	8-9
8.4		Altimetry	8-10
8.4.1		Altitude	8-10

Chapter 9. Global Circulations and Jet Streams 9-1

9.1	Introduction	9-1
9.2	Non-Rotating Earth Circulation System	9-2
9.3	Rotating Earth Circulation System	9-2
9.4	Jet Streams	9-3
9.4.1	Direction of Wind Flow	9-3
9.4.2	Location	9-4

Chapter 10. Wind 10-1

10.1	Introduction	10-1
10.2	Naming of the Wind	10-2
10.3	Forces That Affect the Wind	10-2
10.3.1	Pressure Gradient Force (PGF)	10-2
10.3.2	Coriolis Force	10-3
10.3.3	Friction Force	10-5
10.4	Upper Air Wind	10-6
10.5	Surface Wind	10-7
10.6	Local Winds	10-8
10.6.1	Sea Breeze	10-9
10.6.2	Land Breeze	10-11
10.6.3	Lake Breeze	10-12
10.6.4	Valley Breeze	10-14
10.6.5	Mountain-Plains Wind System	10-14
10.6.6	Mountain Breeze	10-15
10.7	Adverse Winds	10-16
10.7.1	Crosswind	10-16
10.7.2	Gust	10-17
10.7.3	Tailwind	10-17
10.7.4	Variable Wind/Sudden Wind Shift	10-18
10.7.5	Wind Shear	10-18
10.7.6	Adverse Mountain Winds	10-18
10.7.7	Atmospheric Disturbances in Mountainous Areas	10-18

Chapter 11. Air Masses, Fronts, and the Wave Cyclone Model 11-1
11.1 Introduction 11-1
11.2 Air Masses 11-2
11.2.1 Air Mass Classification 11-2
11.2.2 Air Mass Modification 11-3
11.3 Fronts 11-4
11.3.1 Warm Front 11-5
11.3.2 Cold Front 11-6
11.3.3 Stationary Front 11-7
11.3.4 Occluded Front 11-8
11.4 The Wave Cyclone Model 11-9
11.5 Dryline 11-11

Chapter 12. Vertical Motion and Clouds 12-1
12.1 Introduction 12-1
12.2 Vertical Motion Effects on an Unsaturated Air Parcel 12-2
12.3 Vertical Motion Effects on a Saturated Air Parcel 12-3
12.4 Common Sources of Vertical Motion 12-5
12.4.1 Orographic Effects 12-5
12.4.2 Frictional Effects 12-6
12.4.3 Frontal Lift 12-7
12.4.4 Buoyancy 12-7
12.5 Cloud Forms 12-8
12.6 Cloud Levels 12-8

Chapter 13. Atmospheric Stability 13-1
13.1 Introduction 13-1
13.2 Using a Parcel as a Tool to Evaluate Stability 13-2
13.3 Stability Types 13-2
13.3.1 Absolute Stability 13-2
13.3.2 Neutral Stability 13-3
13.3.3 Absolute Instability 13-4
13.3.4 Conditional Instability 13-5
13.3.5 Summary of Stability Types 13-6
13.4 Processes That Change Atmospheric Stability 13-8
13.4.1 Wind Effects on Stability 13-8
13.4.2 Vertical Air Motion Effects on Stability 13-8

	13.4.3 Diurnal Temperature Variation Effects on Stability	13-9
13.5	Measurements of Stability	13-10
	13.5.1 Lifted Index (LI)	13-10
	13.5.2 Convective Available Potential Energy (CAPE)	13-11
13.6	Summary	13-11

Chapter 14. Precipitation 14-1

14.1	Introduction	14-1
14.2	Necessary Ingredients for Formation	14-2
14.3	Growth Process	14-2
14.4	Precipitation Types	14-3
	14.4.1 Snow	14-3
	14.4.2 Ice Pellets	14-4
	14.4.3 Freezing Rain	14-4
	14.4.4 Rain	14-5
	14.4.5 Hail	14-5

Chapter 15. Weather Radar 15-1

15.1	Introduction	15-1
15.2	Principles of Weather Radar	15-2
	15.2.1 Antenna	15-2
	15.2.2 Backscattered Energy	15-2
	15.2.3 Power Output	15-3
	15.2.4 Wavelengths	15-3
	15.2.5 Attenuation	15-4
	15.2.6 Resolution	15-5
	15.2.7 Wave Propagation	15-7
	15.2.8 Radar Beam Overshooting and Undershooting	15-10
	15.2.9 Beam Blockage	15-11
	15.2.10 Ground Clutter	15-11
	15.2.11 Ghost	15-12
	15.2.12 Angels	15-13
	15.2.13 Anomalous Propagation (AP)	15-14
	15.2.14 Other Nonmeteorological Phenomena	15-15
	15.2.15 Precipitation	15-16

Chapter 16. Mountain Weather 16-1

16.1	Introduction	16-1

16.2	Mountain Waves and Adverse Winds		16-2
	16.2.1	Gravity Waves	16-2
	16.2.2	Kelvin-Helmholtz (K-H) Waves	16-3
	16.2.3	Vertically Propagating Mountain Waves	16-5
	16.2.4	Trapped Lee Waves	16-8
	16.2.5	Persistent Horizontal Roll Vortices (Rotors)	16-10
	16.2.6	Smaller-Scale Hazards	16-11
	16.2.7	Visual Indicators of Orographic Wind Fields	16-14
16.3	Mountain/Valley Breezes and Circulation		16-16
16.4	Mountain/Valley Fog		16-16
16.5	Upslope Fog		16-16
16.6	Mountain Obscuration		16-16
16.7	Mountain Turbulence		16-16
16.8	Mountain Icing		16-16
16.9	Density Altitude		16-16
Chapter 17. Tropical Weather			17-1
17.1	Introduction		17-1
17.2	Circulation		17-2
	17.2.1	Subtropical High-Pressure Belts	17-2
	17.2.2	Trade Wind Belts	17-4
	17.2.3	Continental Weather	17-4
	17.2.4	Intertropical Convergence Zone (ITCZ)	17-5
	17.2.5	Monsoon	17-5
17.3	Transitory Systems		17-7
	17.3.1	Remnants of Polar Fronts and Shear Lines	17-7
	17.3.2	Tropical Upper Tropospheric Trough (TUTT)	17-7
	17.3.3	Tropical Wave	17-8
	17.3.4	West African Disturbance Line (WADL)	17-9
	17.3.5	Tropical Cyclones	17-10
Chapter 18. Weather and Obstructions to Visibility			18-1
18.1	Introduction		18-1
	18.1.1	Fog	18-2
	18.1.2	Mist	18-9
	18.1.3	Haze	18-9
	18.1.4	Smoke	18-9

18.1.5	Precipitation	18-9
18.1.6	Blowing Snow	18-10
18.1.7	Dust Storm	18-10
18.1.8	Sandstorm	18-10
18.1.9	Volcanic Ash	18-11
18.2	Low Ceiling and Mountain Obscuration	18-12
18.2.1	Low Ceiling	18-12
18.2.2	Mountain Obscuration	18-12
Chapter 19. Turbulence		19-1
19.1	Introduction	19-1
19.2	Causes of Turbulence	19-2
19.2.1	Convective Turbulence	19-2
19.2.2	Mechanical Turbulence	19-4
19.2.3	Wind Shear Turbulence	19-5
19.2.4	Wind Shear	19-7
19.2.5	The Effects of Orographic Winds and Turbulence on Aviation Operations	19-9
Chapter 20. Icing		20-1
20.1	Introduction	20-1
20.2	Supercooled Water	20-2
20.3	Structural Icing	20-2
20.3.1	Rime Icing	20-2
20.3.2	Clear Icing	20-2
20.3.3	Mixed Icing	20-3
20.3.4	Icing Factors	20-3
20.3.5	Icing in Stratiform Clouds	20-4
20.3.6	Icing in Cumuliform Clouds	20-4
20.3.7	Icing with Fronts	20-4
20.3.8	Icing with Mountains	20-5
20.3.9	Convective Icing	20-6
20.3.10	Icing Hazards	20-6
20.4	Engine Icing	20-7
20.4.1	Carburetor Icing	20-7
20.4.2	High Ice Water Content (HIWC)	20-7
20.5	Additional Information	20-7

Chapter 21. Arctic Weather 21-1

21.1 Introduction 21-1

21.2 Climate, Air Masses, and Fronts 21-2

    21.2.1 Long Days and Nights 21-2

    21.2.2 Land and Water 21-3

    21.2.3 Temperature 21-3

    21.2.4 Clouds and Precipitation 21-3

    21.2.5 Wind 21-3

    21.2.6 Air Masses—Winter 21-3

    21.2.7 Air Masses—Summer 21-4

    21.2.8 Fronts 21-4

21.3 Arctic Peculiarities 21-4

    21.3.1 Effects of Temperature Inversion 21-4

    21.3.2 Light Reflection by Snow-Covered Surfaces 21-4

    21.3.3 Light from Celestial Bodies 21-4

21.4 Arctic Weather Hazards 21-4

    21.4.1 Fog and Ice Fog 21-4

    21.4.2 Blowing and Drifting Snow 21-5

    21.4.3 Frost 21-5

    21.4.4 Whiteout 21-5

Chapter 22. Thunderstorms 22-1

22.1 Introduction 22-1

22.2 Necessary Ingredients for Thunderstorm Cell Formation 22-2

22.3 Thunderstorm Cell Life Cycle 22-2

22.4 Thunderstorm Types 22-3

22.5 Factors that Influence Thunderstorm Motion 22-5

22.6 Thunderstorm Terminology 22-6

22.7 Hazards 22-7

    22.7.1 Low Ceiling and Visibility 22-7

    22.7.2 Lightning 22-7

    22.7.3 Downburst and Microburst 22-7

    22.7.4 Convective Turbulence 22-19

    22.7.5 Convective Icing 22-19

    22.7.6 Hail 22-19

    22.7.7 Rapid Altimeter Changes 22-19

22.7.8	Static Electricity	22-19
22.7.9	Tornado	22-19
22.7.10	Engine Water Ingestion	22-20

22.8 Thunderstorm Avoidance ... 22-20

| 22.8.1 | Airborne Weather Avoidance Radar (Aircraft Radar) | 22-20 |
| 22.8.2 | Thunderstorm Avoidance Guidance | 22-21 |

Chapter 23. Space Weather ... 23-1

23.1 Introduction ... 23-1
23.2 The Sun—Prime Source of Space Weather ... 23-2
23.3 The Sun's Energy Output and Variability ... 23-2
23.4 Sunspots and the Solar Cycle ... 23-2
23.5 Solar Wind ... 23-2
23.6 Solar Eruptive Activity ... 23-3
23.7 Geospace ... 23-3
23.8 Galactic Cosmic Rays (GCR) ... 23-4
23.9 Geomagnetic Storms ... 23-4
23.10 Solar Radiation Storms ... 23-4
23.11 Ionospheric Storms ... 23-5
23.12 Solar Flare Radio Blackouts ... 23-5
23.13 Effects of Space Weather on Aircraft Operations ... 23-6

23.13.1	Communications	23-6
23.13.2	Navigation and GPS	23-6
23.13.3	Radiation Exposure to Flightcrews and Passengers	23-6
23.13.4	Radiation Effects on Avionics	23-6

Part 3: Technical Details Relating to Weather Products and Aviation Weather Tools

Chapter 24. Observations ... 24-1

24.1 Introduction ... 24-1
24.2 Surface Observations ... 24-2

24.2.1	Manual Observation	24-2
24.2.2	Automated Observation	24-2
24.2.3	Augmented Observation	24-2
24.2.4	Recency of Observed Elements at Automated Stations	24-2

24.3 Automated Surface Observing System (ASOS) and Automated Weather Observing System (AWOS) ... 24-2

| 24.3.1 | Automated Surface Observing System (ASOS) | 24-2 |

24.3.2 Automated Weather Observing System (AWOS) 24-3

24.4 Aviation Routine Weather Report (METAR) and Aviation Selected Special Weather Report (SPECI) 24-5

    24.4.1 Aviation Routine Weather Report (METAR) 24-5

    24.4.2 Aviation Selected Special Weather Report (SPECI) 24-6

    24.4.3 METAR/SPECI Format 24-7

24.5 Aircraft Observations and Reports 24-26

    24.5.1 Pilot Weather Reports (PIREP) 24-26

    24.5.2 Aircraft Reports (AIREP) 24-33

    24.5.3 Volcanic Activity Reports (VAR) 24-34

    24.5.4 Turbulence Observations 24-35

24.6 Radar Observations 24-36

    24.6.1 Weather Surveillance Radar—1988 Doppler (WSR-88D) Description 24-36

    24.6.2 Terminal Doppler Weather Radar (TDWR) 24-47

24.7 Satellite Observations 24-48

    24.7.1 Description 24-48

    24.7.2 Imagery Types 24-48

    24.7.3 Polar Operational Environment Satellites (POES) 24-54

24.8 Upper Air Observations 24-54

    24.8.1 Radiosonde Observations (Weather Balloon) 24-54

24.9 Aviation Weather Cameras 24-56

    24.9.1 FAA Aviation Weather Camera Network 24-56

    24.9.2 Visibility Estimation through Image Analytics (VEIA) 24-59

    24.9.3 Visual Weather Observation System (VWOS) 24-60

    24.9.4 Issuance 24-60

Chapter 25. Analysis 25-1

25.1 Introduction 25-1

25.2 Weather Charts 25-2

    25.2.1 Weather Observation Sources 25-2

    25.2.2 Analysis 25-2

    25.2.3 Surface Analysis Chart 25-6

    25.2.4 Unified Surface Analysis Chart 25-15

    25.2.5 AAWU Surface Chart 25-17

25.3 Upper-Air Analysis 25-17

    25.3.1 Issuance 25-19

	25.3.2	Radiosonde Observation (Weather Balloon) Analysis	25-19
25.4		Freezing Level Analysis	25-22
25.5		Icing Analysis (Current Icing Product (CIP))	25-23
25.6		Turbulence (Graphical Turbulence Guidance (GTG)) Analysis	25-25
25.7		Real-Time Mesoscale Analysis (RTMA)	25-26

Chapter 26. Advisories 26-1

26.1		Introduction	26-1
26.2		Significant Meteorological Information (SIGMET)	26-2
	26.2.1	SIGMET Issuance	26-2
	26.2.2	SIGMET Identification	26-4
	26.2.3	SIGMET Standardization	26-5
	26.2.4	Inside the CONUS	26-5
	26.2.5	Outside the CONUS	26-11
26.3		Airmen's Meteorological Information (AIRMET)	26-15
	26.3.1	AIRMET Issuance	26-16
	26.3.2	AIRMET Standardization	26-17
	26.3.3	AIRMET Issuance Times and Valid Periods	26-18
	26.3.4	AIRMET Formats and Examples	26-18
	26.3.5	AIRMET Updates and Amendments	26-21
26.4		Center Weather Advisory (CWA)	26-21
	26.4.1	CWA Issuance	26-21
	26.4.2	CWA Criteria	26-22
	26.4.3	CWA Format and Example	26-23
26.5		Volcanic Ash Advisory (VAA)	26-24
	26.5.1	Volcanic Ash Advisory Center (VAAC)	26-24
26.6		Aviation Tropical Cyclone Advisory (TCA)	26-27
	26.6.1	TCA Issuance	26-27
	26.6.2	TCA Content	26-28
	26.6.3	TCA Example	26-28
	26.6.4	Additional Tropical Cyclone Information Products	26-28
26.7		Space Weather Advisory	26-29
	26.7.1	Space Weather Advisory Issuance	26-29
	26.7.2	Space Weather Advisory Format	26-30
	26.7.3	Space Weather Advisory Examples	26-33
26.8		Low-Level Wind Shear (LLWS)/Microburst Advisories	26-34

26.9	Airport Weather Warning (AWW)		26-34
	26.9.1	AWW Issuance	26-34
	26.9.2	AWW Format and Example	26-35
Chapter 27. Forecasts			27-1
27.1	Introduction		27-1
27.2	Winds and Temperatures Aloft		27-3
	27.2.1	FB Wind and Temperature Aloft Forecast	27-4
27.3	Terminal Aerodrome Forecast (TAF)		27-7
	27.3.1	TAF Responsibility	27-7
	27.3.2	Generic Format of the Forecast Text of an NWS-Prepared TAF	27-7
	27.3.3	TAF Examples	27-18
	27.3.4	Issuance	27-21
27.4	Aviation Surface Forecast and Aviation Clouds Forecast		27-23
27.5	Area Forecasts (FA)		27-25
	27.5.1	FA Standardization	27-25
	27.5.2	FA Issuance Schedule	27-25
	27.5.3	FA Amendments and Corrections	27-26
	27.5.4	FA Issuance	27-26
27.6	Alaska Graphical Forecasts		27-35
	27.6.1	AAWU Flying Weather	27-35
	27.6.2	Alaska Surface Forecast	27-36
	27.6.3	Alaska Icing Forecast	27-37
	27.6.4	Alaska Turbulence Forecast	27-38
	27.6.5	Alaska Convective Outlook	27-39
27.7	World Area Forecast System (WAFS)		27-40
	27.7.1	WAFS Forecasts	27-40
27.8	Significant Weather (SIGWX)		27-43
	27.8.1	Low-Level Significant Weather (SIGWX) Charts	27-43
	27.8.2	Mid-Level Significant Weather (SIGWX) Chart	27-47
	27.8.3	High-Level Significant Weather (SIGWX) Charts	27-49
	27.8.4	Alaska Significant Weather (SIGWX) Charts	27-54
27.9	Short-Range Surface Prognostic (Prog) Charts		27-56
	27.9.1	Content	27-57
27.10	Upper-Air Forecasts		27-58
	27.10.1	Constant Pressure Level Forecasts	27-59

27.11	Freezing Level Forecast Graphics	27-61
27.12	Forecast Icing Product (FIP)	27-62
27.12.1	Icing Probability	27-63
27.12.2	Icing Severity	27-63
27.12.3	Icing Severity—Probability > 25 Percent	27-63
27.12.4	Icing Severity—Probability > 50 Percent	27-63
27.12.5	Icing Severity Plus SLD	27-63
27.13	Graphical Turbulence Guidance (GTG)	27-64
27.14	Cloud Tops	27-66
27.15	Localized Aviation Model Output Statistics (MOS) Program (LAMP)	27-66
27.15.1	Alaska Aviation Guidance (AAG) Weather Product	27-66
27.16	Additional Products for Convection	27-66
27.16.1	Convective Outlook (AC)	27-67
27.16.2	Traffic Flow Management (TFM) Convective Forecast (TCF)	27-67
27.16.3	Extended Convective Forecast Product (ECFP)	27-69
27.16.4	Watch Notification Messages	27-70
27.17	Route Forecast (ROFOR)	27-74
27.18	Aviation Forecast Discussion (AFD)	27-75
27.18.1	Example	27-76
27.19	Meteorological Impact Statement (MIS)	27-76
27.19.1	Example	27-76
27.20	Soaring Forecast	27-76
27.20.1	Example	27-76
27.21	Balloon Forecast	27-77
27.21.1	Example	27-77
Chapter 28. Aviation Weather Tools		**28-1**
28.1	Introduction	28-1
28.2	Graphical Forecasts for Aviation (GFA) Tool	28-2
28.2.1	GFA Static Images	28-3
28.3	FAA Flight Service's Interactive Map	28-3
28.4	Helicopter Emergency Medical Services (HEMS) Tool	28-3
Appendix A. Cloud Types		**A-1**
A.1	High Clouds	A-1
A.1.1	Cirrus (Ci)	A-1
A.1.2	Cirrocumulus (Cc)	A-2

A.1.3	Cirrostratus (Cs)	A-2
A.2	Middle Clouds	A-3
A.2.1	Altocumulus (Ac)	A-3
A.2.2	Altostratus (As)	A-5
A.2.3	Nimbostratus (Ns)	A-6
A.3	Low Clouds	A-7
A.3.1	Cumulus (Cu) and Towering Cumulus (TCu)	A-7
A.3.2	Stratocumulus (Sc)	A-8
A.3.3	Stratus (St)	A-9
A.3.4	Cumulonimbus (Cb)	A-11

Appendix B. Standard Conversion Chart B-1

Appendix C. Density Altitude Calculation C-1

Appendix D. Special Terminal Operation Procedures (STOP) for Operations in a Noncontiguous State D-1

D.1	Introduction	D-1
D.2	Weather Information	D-1
D.2.1	General	D-1
D.2.2	Supplements to NWS or EWINS TAFs	D-1
D.2.3	Manual Surface Observations by Flightcrew Using a Portable Weather Observation Device at STOP Airports	D-3
D.2.4	Documentation and Reporting	D-4
D.3	Airport Data	D-5
D.3.1	Data Maintenance	D-5
D.3.2	Airport Analysis Program	D-5
D.3.3	Civil Twilight Considerations	D-5
D.4	Flight Planning	D-6
D.4.1	General	D-6
D.4.2	IFR Flight Plan	D-6
D.4.3	Composite IFR/VFR Flight Plan	D-6
D.4.4	VFR Flight Plan	D-6
D.4.5	Fuel Requirements	D-6
D.4.6	Weather Minimums at STOP Destination Airports	D-6
D.4.7	Required Destination Primary Alternate	D-7
D.4.8	Designation of Secondary Destination Alternate	D-7
D.4.9	Minimum Equipment List (MEL) Restrictions	D-7

D.5		Preflight Planning and Briefing to STOP Airport	D-8
	D.5.1	Preflight Planning	D-8
	D.5.2	Preflight Briefing	D-8
D.6		En Route, Flight Monitoring, and Communications	D-8
	D.6.1	General	D-8
	D.6.2	Domestic Operations	D-8
	D.6.3	Supplemental Operations	D-9
	D.6.4	Flight Monitoring	D-9
	D.6.5	En Route Terrain Clearance	D-9
	D.6.6	En Route IFR to VFR Transition Point	D-9
D.7		Destination Approach and Arrival	D-10
	D.7.1	General	D-10
	D.7.2	Visual Inspection	D-10
	D.7.3	Barometric Altimeter Validation	D-10
	D.7.4	Radio Altimeter Setting	D-10
D.8		Departure Procedures From a STOP Airport	D-10
	D.8.1	General	D-10
	D.8.2	Departures	D-11
	D.8.3	Departure Data	D-11
D.9		Training for Dispatchers, Flight Followers, and Flightcrew	D-11
	D.9.1	General	D-11
	D.9.2	Introduction for Dispatchers and Flight Followers	D-11
	D.9.3	Introduction for Flightcrews	D-12
D.10		Approval	D-13
D.11		Compliance Monitoring, Quality Assurance (QA), and Auditing	D-13
	D.11.1	Compliance Monitoring	D-13
	D.11.2	Quality Assurance (QA)	D-13
	D.11.3	STOP Audit	D-13

Appendix E. Abbreviations, Acronyms, and Initialisms	E-1
Appendix F. Units of Measurement	F-1
Appendix G. Websites	G-1

1 Introduction

This handbook describes the United States (U.S.) aviation weather program, products, and services. It also documents weather theory and its application to aviation. The objective of this handbook is to help the pilot and operator understand the basics of weather, aviation weather hazards, and aviation weather products. The information included is not prescriptive. Furthermore, the guidance and recommendations in this handbook are not legally binding in their own right and will not be relied upon by the FAA as a separate basis for affirmative enforcement action or other administrative penalty. Conformity with the guidance and recommendations is voluntary only and nonconformity will not affect rights and obligations under existing statutes and regulations.

The handbook is a consolidated source of weather information and, in some cases, best practices to assist with providing safety for flight.

The visualization of the products discussed in this handbook has evolved over the past decade with the use of internet websites. The use of static black and white depictions of aviation forecasts is quickly becoming a thing of the past. Today's websites provide the products in color and offer options to select and overlay multiple pieces of weather information.

Today's aviation weather websites, including those of the National Weather Service (NWS), continue to improve the content and visualization of products. Unfortunately, it is not practical to update this handbook with every change to a weather product.

Examples of weather products in this handbook represent one way of how they can be visualized on a user's viewing device (e.g., computer, tablet, mobile phone, or cockpit display). The examples shown in this handbook are from NWS websites.

This handbook is broken into three parts:

- Part 1: Overview of the United States Aviation Weather Service Program and Information.
- Part 2: Weather Theory and Aviation Hazards.
- Part 3: Technical Details Relating to Weather Products and Aviation Weather Tools.

Part 1:
Overview of the United States Aviation Weather Service Program and Information

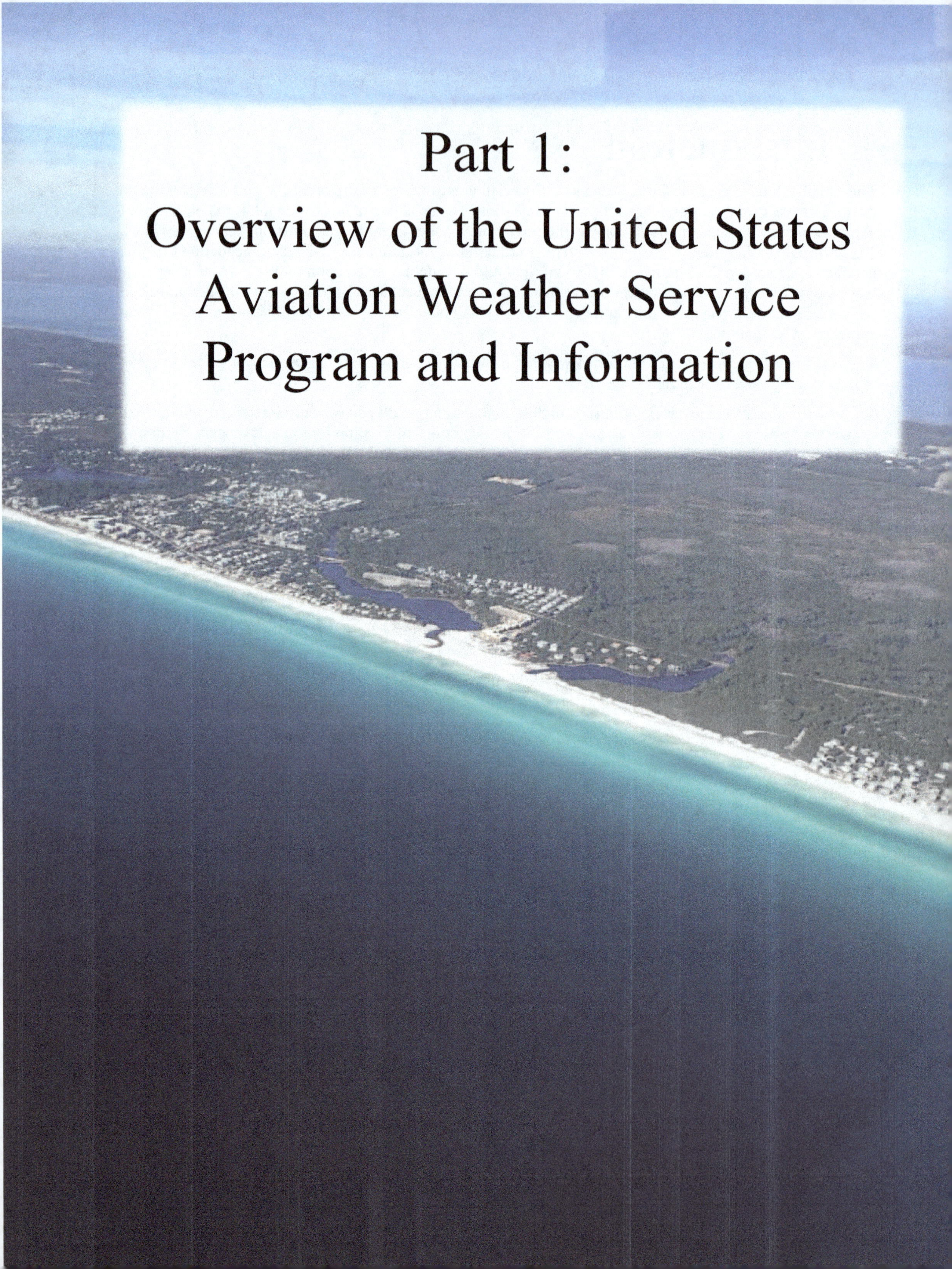

2 Aviation Weather Service Program

2.1 Introduction

The aviation weather service program is a joint effort of the National Oceanic and Atmospheric Administration (NOAA), the Federal Aviation Administration (FAA), the Department of Defense (DOD), and commercial aviation weather providers.

2.2 National Oceanic and Atmospheric Administration (NOAA)

NOAA is an agency of the Department of Commerce (DOC). NOAA conducts research and gathers data about the global oceans, atmosphere, space, and Sun, and applies this knowledge to science and service, which touches the lives of all Americans. Among its six major divisions are the National Environmental Satellite, Data, and Information Service (NESDIS) and the NWS.

2.2.1 National Environmental Satellite, Data, and Information Service (NESDIS)

NESDIS manages the U.S. civil operational remote-sensing satellite systems, as well as other global information for meteorology, oceanography, solid-earth geophysics, and solar-terrestrial sciences. NESDIS provides this data to NWS meteorologists and a wide range of other users for operational weather forecasting.

2.2.1.1 Satellite Analysis Branch (SAB)

NESDIS' Satellite Analysis Branch (SAB) serves as the operational focal point for real-time imagery products and multidisciplinary environmental analyses. The SAB's primary mission is to support disaster mitigation and warning services for the U.S. Federal agencies and the international community. Routine environmental analyses are provided to forecasters and other environmental users, and are used in the numerical models of the NWS. The SAB schedules and distributes real-time satellite imagery products from global geostationary and polar-orbiting satellites to environmental users.

The SAB coordinates the satellite and other information for the NOAA Volcanic Hazards Alert Program, under an agreement with the FAA, and works with the NWS as part of the Washington, DC, Volcanic Ash Advisory Center (VAAC). The Washington, DC, VAAC area of responsibility stretches from 40° W to 130° E and includes the areas of the contiguous United States (CONUS), New York and Oakland Oceanic Flight Information Regions (FIR), and southward through Central America and the Caribbean to 10° S in South America.

2.2.2 National Weather Service (NWS)

The NWS provides weather data, forecasts, and warnings for the United States, its territories, adjacent waters, and ocean areas for the protection of life and property and the enhancement of the national economy. NWS data and products form a national information database and infrastructure that can be used by other government agencies, the private sector, the public, and the global community. The following sections provide a description of NWS offices associated with aviation weather.

2.2.2.1 National Centers for Environmental Prediction (NCEP)

The National Centers for Environmental Prediction (NCEP) is where virtually all global meteorological data is collected and analyzed for the United States. The NCEP then provides a wide variety of national and international weather guidance products to NWS field offices, government agencies, emergency managers, and private sector meteorologists. The NCEP is a critical resource in national and global weather prediction and is the starting point for nearly all weather forecasts in the United States.

The NCEP is comprised of nine distinct centers and the Office of the Director. Each center has its own specific mission. The following NCEP centers provide aviation weather products and services.

2.2.2.1.1 NCEP Central Operations (NCO)

The NCEP Central Operations (NCO) in College Park, MD, sustains and executes the operational suite of the numerical analysis and forecast models and prepares NCEP products for dissemination. It also links all nine of the national centers together via computer and communications-related services.

2.2.2.1.2 Aviation Weather Center (AWC)

The Aviation Weather Center (AWC) in Kansas City, MO, issues a suite of aviation weather forecasts in support of the National Airspace System (NAS) that are described in this handbook, including Airmen's Meteorological Information (AIRMET), Significant Meteorological Information (SIGMET), Convective SIGMETs, and various icing, turbulence, and convective forecast products. The AWC is a Meteorological Watch Office (MWO) for the International Civil Aviation Organization (ICAO). The AWC, in conjunction with the NCO, also serves as one of two ICAO World Area Forecast Centers (WAFC), known as WAFC Washington, issuing flight planning data (winds and temperatures) and Significant Weather (SIGWX) forecasts.

The AWC's website provides the aviation community with textual, digital, and graphical forecasts, analyses, and observations of aviation-related weather variables. Additionally, the website provides information for international flights through the World Area Forecast System (WAFS) Internet File Service (WIFS).

2.2.2.1.3 Weather Prediction Center (WPC)

The Weather Prediction Center (WPC) in College Park, MD, provides analysis and forecast products specializing in multiday, quantitative precipitation forecasts and weather forecast guidance, weather model diagnostics discussions, and surface pressure and frontal analyses.

2.2.2.1.4 Storm Prediction Center (SPC)

The Storm Prediction Center (SPC) in Norman, OK, provides tornado and severe weather watches for the CONUS along with a suite of hazardous weather forecasts.

2.2.2.1.5 National Hurricane Center (NHC)

The National Hurricane Center (NHC) in Miami, FL, provides official NWS forecasts of the movement and strength of tropical weather systems and issues the appropriate watches and warnings for the CONUS and surrounding areas. It also issues a suite of marine products covering the tropical Atlantic, Caribbean, Gulf of Mexico, and tropical eastern Pacific. In support of ICAO, the NHC is also referred to as a Tropical Cyclone Advisory Center (TCAC).

2.2.2.1.6 Space Weather Prediction Center (SWPC)

The Space Weather Prediction Center (SWPC) in Boulder, CO, provides space weather information (e.g., current activity and forecasts) to a wide variety of users. The SWPC is also an ICAO Space Weather Center and is responsible for issuing global advisories for space weather events affecting communication and navigation systems as well as events that pose a potential health hazard to passengers and crew.

2.2.2.1.7 Alaska Aviation Weather Unit (AAWU)

The Alaska Aviation Weather Unit (AAWU), located in Anchorage, AK, is an MWO for ICAO. The AAWU is responsible for the entire Anchorage FIR. They issue a suite of aviation weather products for the airspace over Alaska and adjacent coastal waters, including AIRMETs, SIGMETs, FAs, and SIGWX Prognostic Charts.

The AAWU is also designated as the Anchorage VAAC. The VAAC area of responsibility includes the Anchorage FIR and Far Eastern Russia and is responsible for the issuance of Volcanic Ash Advisories (VAA).

2.2.2.1.8   Center Weather Service Unit (CWSU)

Center Weather Service Units (CWSU) are units of NWS meteorologists under contract with the FAA that are stationed at, and support, the FAA's air route traffic control center (ARTCC).

CWSUs provide timely weather consultation, forecasts, and advice to managers within ARTCCs and to other supported FAA facilities.

Special emphasis is given to those weather conditions that are hazardous to aviation or that could impede the flow of air traffic within the NAS. CWSU meteorologists issue the following products in support of their respective ARTCC: Center Weather Advisories (CWA) and Meteorological Impact Statements (MIS).

2.2.2.1.9   Weather Forecast Office (WFO)

The NWS has 122 Weather Forecast Offices (WFO) across the United States and select territories. An NWS WFO is a multipurpose, local weather forecast center that produces, among its suite of services, aviation-related products. In support of aviation, WFOs issue Terminal Aerodrome Forecasts (TAF) and Aviation Forecast Discussions (AFD), with some offices issuing Airport Weather Warnings (AWW), Soaring Forecasts, and Balloon Forecasts. The WFO also issues weather warnings such as tornado and severe thunderstorm warnings.

The Honolulu WFO is unique among NWS WFOs in that it provides multiple services beyond the typical WFO. WFO Honolulu is also designated as an MWO for ICAO. As a result of this unique designation, WFO Honolulu is the only WFO to issue the following text products: AIRMETs and SIGMETs. WFO Honolulu is co-located with the Central Pacific Hurricane Center (CPHC). The CPHC provides official NWS forecast of the movement and strength of tropical weather systems and issues the appropriate watches and warnings for the central Pacific, including the State of Hawaii. WFO Honolulu also issues a suite of marine products covering a large portion of the Pacific Ocean. In support of ICAO, the CPHC is also referred to as a TCAC.

2.3   Federal Aviation Administration (FAA)

The FAA, a part of the Department of Transportation (DOT), provides a safe, secure, and efficient airspace system for the promotion of U.S. aerospace safety.

Per Title 49 of the United States Code (49 U.S.C.) § 44720, Meteorological Services, the FAA establishes all requirements for aviation weather reports and forecasts.

The FAA provides a wide range of services to the aviation community. The following sections provide a description of those FAA facilities that are involved with aviation weather and pilot services.

2.3.1   Air Traffic Control Systems Command Center (ATCSCC)

The air traffic control systems command center (ATCSCC) has the mission of balancing air traffic demand with system capacity. This ensures maximum safety and efficiency for the NAS, while minimizing delays. The ATCSCC utilizes the Traffic Management System (TMS), aircraft situation display, monitor alert, follow-on functions, and direct contact with ARTCC and Terminal Radar Approach Control (TRACON) facility Traffic Management Units (TMU) to manage flow on a national level.

Because weather is the most common reason for air traffic delays and reroutings, NWS meteorologists support the ATCSCC. These meteorologists, called National Aviation Meteorologists, coordinate NWS operations in support of traffic flow management within the NAS.

2.3.2 Air Route Traffic Control Center (ARTCC)

An ARTCC is a facility established to provide air traffic control (ATC) service to pilots operating on instrument flight rules (IFR) flight plans within controlled airspace, principally during the en route phase of flight. When equipment capabilities and controller workload permit, certain advisory/assistance services may be provided to pilots operating under visual flight rules (VFR).

En route controllers become familiar with pertinent weather information and stay aware of current weather information needed to perform ATC duties. En route controllers advise pilots of hazardous weather that may impact operations within 150 nautical miles (NM) of the controller's assigned sector or jurisdiction and may solicit Pilot Weather Reports (PIREP) from pilots.

2.3.3 Airport Traffic Control Tower (ATCT) and Terminal Radar Approach Control (TRACON)

An airport traffic control tower (ATCT) is a terminal facility that uses air-to-ground communications, visual signaling, and other devices to provide ATC services to aircraft operating in the vicinity of an airport or on the movement area.

Terminal controllers become familiar with pertinent weather information and stay aware of current weather information needed to perform ATC duties. Terminal controllers advise pilots of hazardous weather that may impact operations within 150 NM of the controller's assigned sector or area of jurisdiction and may solicit PIREPs from pilots. ATCTs and TRACONs may opt to broadcast hazardous weather information alerts only when any part of the area described is within 50 NM of the airspace under the ATCT's jurisdiction.

2.3.4 Flight Service

The FAA delivers flight services to pilots in the CONUS, Alaska, Hawaii, and Puerto Rico. Services are provided by phone at 1-800-WX-BRIEF, on the internet through the Flight Service Pilot Web Portal (which can be found in Appendix G, Websites), and in person (Alaska only) at 17 Flight Service Stations (FSS). Services include, but are not limited to: preflight weather briefings, flight planning, and in-flight advisory services.

2.3.4.1 FAA Weather Camera Program

Flight Service also delivers the FAA's Aviation Weather Camera Network. It features an expanding network of camera sites in Alaska, Hawaii, and the CONUS. Weather cameras provide pilots in certain areas with snapshots of current conditions at a specific location, enabling pilots to have additional information for improved decision making. Services may be obtained by visiting the Aviation Weather Cameras website, which can be found in Appendix A. See Section 3.3.3.2 and Section 24.9 for additional information.

2.4 Department of Defense (DOD)

The DOD is a key partner with the FAA and NWS in the next generation weather radar (NEXRAD) program. Various branches of the DOD provide their own weather support, but this is beyond the scope of this handbook.

2.5 Commercial Weather Information Providers

Commercial weather information providers are a major source of weather products for the aviation community. In general, they produce proprietary weather products based on NWS information with formatting and layout modifications, but no material changes to the weather information itself. This is also referred to as "repackaging."

Commercial providers also produce forecasts, analyses, and other proprietary weather products, which may substantially differ from the information contained in NWS-produced products. Operators who desire to use products prepared by a commercial weather provider should consult the Aeronautical Information Manual (AIM), Chapter 7, Section 1, Paragraph 7-1-3, Use of Aviation Weather Products, which provides additional information pertaining to commercial providers, including the Enhanced Weather Information System (EWINS).

3 Overview of Aviation Weather Information

3.1 Introduction

Title 14 of the Code of Federal Regulations (14 CFR) part 91, § 91.103 states the requirements for preflight action for part 91 operations. According to § 91.103, each pilot in command (PIC) shall, before beginning a flight, become familiar with all available information concerning the flight, including weather reports and forecasts.

This handbook describes the weather products primarily produced by the NWS. There is an ever-expanding suite of weather products available on the internet, weather applications (apps) for phones and tablets, as well as commercial Electronic Flight Bags (EFB). Pilots and operators should consult with their Principal Operations Inspector (POI) or their service provider when in doubt about the content and use of new weather products. Note that the FAA does not certify internet providers of aviation weather services.

3.2 Use of Aviation Weather Information

3.2.1 Product Latency

With few exceptions, all weather information and products have latency. Latency is the element of data age. The total latency of weather information and products includes the total time between the actual occurrence of the phenomenon, the data collection, processing, transmittal, and the display or application of the information in the cockpit, on the pilot's EFB, or other publication for use. It is important to be aware of the product time or "valid until" time on the particular data link information displayed in the cockpit or EFB. The amount of latency may limit the use or application of the information or product.

An example of weather information without latency is the wind direction when looking at the windsock along the runway. However, the wind reported in the Automated Weather Observing System (AWOS) or Automated Surface Observing System (ASOS) broadcast has a latency of up to 3 minutes. Why? While the AWOS and ASOS wind (direction and speed) is continuously being recorded by the AWOS/ASOS system processor, the reported wind is the most recent average of the direction and speed over the past 2 minutes. That 2-minute average is then updated once a minute for the radio or telephone broadcast.

Onboard aircraft radar has minimal latency, while NEXRAD data has a latency of 5 to 15 minutes or more with weather apps and data uplink services. This is why NEXRAD data is used for broad strategic avoidance of thunderstorms and never used to navigate through thunderstorms.

3.2.2 Additional Use Information

Details on the use of both government and commercial aviation weather information are discussed in the AIM, Chapter 7, Section 1, Paragraph 7-1-3, Use of Aviation Weather Products. Items discussed include:

- Approved sources for aviation weather information,
- The development of new products through the FAA's Next Generation Air Transportation System (NextGen) Aviation Weather Research Program (AWRP),
- The use of new products to meet regulatory requirements, and
- The use of weather services and products provided by entities other than the FAA, the NWS, or their contractors.

3.3 Obtaining Weather Information

3.3.1 Weather Briefings

Prior to every flight, pilots should gather all information vital to the nature of the flight. This includes a weather briefing obtained by the pilot using online weather resources, a dispatcher, or Flight Service.

Historically, Flight Service has been the primary source for obtaining preflight briefings. Today, increasing numbers of pilots are using online weather resources to obtain weather information through government or commercial providers. Pilots can receive a regulatory compliant briefing without contacting Flight Service. Pilots who prefer to contact Flight Service are encouraged to conduct a self-briefing prior to calling. Conducting a self-briefing before contacting Flight Service provides familiarity of weather conditions applicable to the route of flight and promotes a better understanding of weather information.

To obtain an appropriate weather briefing, pilots need to know which of the three types of briefings is needed for the flight—standard, abbreviated, or outlook. Other necessary information includes whether the flight will be conducted under VFR or IFR, aircraft identification and type, departure point, estimated time of departure, desired flight altitude, route of flight, destination, and estimated time en route. If the briefing updates previously received information, the time of the last briefing is also important.

The information is entered into a flight plan form. When using the route brief feature in the Flight Service Pilot Web Portal (see Appendix G) or speaking to an FSS specialist, the type of weather briefing is recorded. If necessary, the information can be referenced later to file or amend a flight plan. It can also be used when an aircraft is overdue or is reported missing. All briefings provided by Flight Service (online or voice) are time-stamped and archived for 45 days.

Pilots can make a final weather check immediately before departure using online weather apps, when possible.

3.3.1.1 Standard Briefing

A standard briefing provides a complete weather picture and is the most detailed of all briefings. It includes conditions and significant weather information that may influence the pilot in planning, altering, or cancelling a proposed route or flight. A standard briefing provides the following information (if applicable to the route of flight) in sequential order:

- **Adverse Conditions.** This includes significant weather and/or aeronautical information about adverse conditions that may influence a decision to cancel or alter the route of flight (e.g., hazardous weather conditions, airport closures, or air traffic delays). Pilots should also be alert for any reported or forecast icing if the aircraft is not certified for operating in icing conditions. Flying into areas of icing or weather below minimums could have disastrous results.

- **VFR Flight Not Recommended (VNR).** When VFR flight is porposed and sky conditions or visibilities are present or forecast, surface or aloft, that, in the briefer's judgment, would make flight under VFR doubtful, the briefer will describe the conditions, describe the affected locations, and use the phrase "*VFR flight not recommended*." This recommendation is advisory in nature. The final decision as to whether the flight can be conducted safely rests solely with the pilot. Upon receiving a "*VFR flight not recommended*" statement, the non-IFR rated pilot will need to make a "go or no go" decision. This decision should be based on weighing the current and forecast weather conditions against the pilot's experience and ratings. The aircraft's equipment, capabilities, and limitations should also be considered. This advisory is not provided via the internet.

- **Synopsis.** A brief statement describing the type, location, and movement of weather systems and/or air masses that might affect the proposed flight.

- **Current (Latest) Conditions.** This portion of the briefing contains the current (latest reported or received) surface weather summarized from all available resources, including observations, PIREPs, and satellite and radar data along the route of flight. If the departure time is more than 2 hours away, current conditions will not be included in the briefing unless the pilot specifically requests the information.

- **En Route Forecast.** Forecast en route conditions for the proposed route are summarized in logical order (i.e., departure/climbout, en route, and descent). Heights are above mean sea level (MSL), unless the contractions "AGL" or "CIG" are denoted, indicating that heights are above ground.

- **Destination Forecast.** The destination forecast for the planned estimated time of arrival (ETA). Any significant changes within 1 hour before and after the planned arrival are included.

- **Winds and Temperatures Aloft.** Forecast winds aloft will be provided in knots and tens of degrees and referenced to true north. The briefer will interpolate wind directions and speeds between levels and stations as necessary to provide expected conditions at planned altitudes. (Heights are MSL.) Temperature information will be provided on request.

- **Notices to Air Missions (NOTAM), ATC Delays, and Other Information.** Refer to the AIM Chapter 7, Section 1, Paragraph 7-1-5, Preflight Briefing, subparagraphs 8b through 10 for a complete description of this part of the briefing. Also refer to the AIM Chapter 5, Section 1, Paragraph 5-1-3, Notice to Air Missions (NOTAM) System, for a complete description of all NOTAM types.

3.3.1.2 Abbreviated Briefing

An abbreviated briefing is a shortened version of the standard briefing. It can be requested when a departure has been delayed or when specific weather information is needed to update a previous standard briefing. Pilots who prefer to contact Flight Service are encouraged to conduct a self-briefing prior to calling. Conducting a self-briefing before contacting Flight Service provides familiarity of weather conditions applicable to the route of flight and promotes a better understanding of weather information. When contacting Flight Service by phone, the FSS specialist will ask for the time and source of the previous briefing so the specialist does not inadvertently omit the necessary weather information.

3.3.1.3 Outlook Briefing

An outlook briefing can be requested when a planned departure is 6 or more hours away. It provides initial forecast information that is limited in scope due to the timeframe of the planned flight. This type of briefing is a good source of flight planning information that can influence decisions regarding route of flight, altitude, and ultimately the "go, no-go" decision. A followup standard or abbreviated briefing prior to departure is advisable, since an outlook briefing generally only contains information based on weather trends and existing weather in geographical areas at or near the departure airport.

3.3.2 Telephone

3.3.2.1 Flight Service 1-800-WX-BRIEF

For flights within the CONUS, Alaska, Hawaii, and Puerto Rico, call 1-800-WX-BRIEF (1-800-992-7433).

Additionally, for flights within Alaska, individual FSS telephone numbers are listed in the FAA Chart Supplement for Alaska, Section 4, Associated Data.

FSS specialists are qualified and certified as Pilot Weather Briefers by the FAA. They are not authorized to make original forecasts, but are certified to translate and interpret available weather information directly into terms describing the weather conditions that can be expected at the departure, along the route of flight, and at the destination.

The FSS' purpose is to serve the aviation community. Pilots can ask questions and discuss factors they do not fully understand. The briefing is considered complete when the pilot has a clear picture of what weather to expect.

3.3.2.2 Airport Weather

The latest weather reports from airport automated observing systems (e.g., AWOS and ASOS) can be accessed from the phone. Phone numbers can be found in FAA Chart Supplements.

3.3.3 Self-Briefing

Preflight decision making using online weather information continues to offer more options for the pilot. The advent of interactive online aviation weather has allowed pilots to assemble aviation weather information into a better decision making process.

Pilots can receive a regulatory compliant briefing through online weather resources. Pilots that prefer to contact an FSS are encouraged to use the online weather resources prior to calling. Some online weather sources do not provide Flight Information Services (FIS), such as NOTAMs and Temporary Flight Restrictions (TFR). However, this information can also be found online through other websites.

3.3.3.1 Flight Service 1800wxbrief.com

The FAA contract provider for flight services provides a website (https://www.1800wxbrief.com) that allows pilots to review weather information, receive online preflight briefings, file flight plans, and receive automatic notifications and alerts. The website also offers an interactive map to allow pilots to view a variety of weather products and access to a variety of aeronautical information that can be tailored to their planned flight route. See Chapter 28, Aviation Weather Tools, for additional information.

3.3.3.2 Aviation Weather Cameras

The FAA's Aviation Weather Cameras website provides access to current weather camera images from the FAA's Aviation Weather Camera Network. On the website, pilots can compare the images to clear day views or play back a loop of past images to establish weather trends. Weather camera images are a supplementary product and may only be used to improve situational awareness.

The website also delivers a variety of safety of flight information including adverse conditions (e.g., AIRMETs and SIGMETs), current conditions (e.g., Aviation Routine Weather Reports (METAR), radar, satellite imagery, and weather trends), TAFs, PIREPs, and other aeronautical information (e.g., remote communications outlets (RCO), TFRs, and charts).

The FAA's Aviation Weather Cameras website can be found in Appendix G. The website uses a set of progressive web application standards that enables an application-level experience on certain mobile devices. See Section 24.9 for additional information.

3.3.3.3 AviationWeather.gov

The website https://aviationweather.gov is operated by the NWS AWC in Kansas City, MO. It is a major aviation weather website for obtaining text and graphical preflight weather information and products.

3.3.3.3.1 Graphical Forecasts for Aviation (GFA) Tool

The Graphical Forecasts for Aviation (GFA) Tool is a set of web-based displays that provide the necessary aviation weather information to give users a complete picture of the weather that may impact flights in the CONUS, Gulf of Mexico, the Caribbean, portions of the Atlantic Ocean, and portions of the Pacific Ocean, including the Hawaiian Islands and Alaska. See Chapter 28 for additional information.

3.3.3.4 Flight Information Service-Broadcast (FIS-B)

Pilots can receive a regulatory compliant briefing through online weather resources that can be used in conjunction with the Flight Information Service-Broadcast (FIS-B) products. See Section 3.3.4.1 for information on FIS-B.

3.3.3.5 Commercial Services

There are several commercial aviation weather providers that offer aviation weather and flight information suitable for tablets and EFBs. These have a subscription fee for the service.

3.3.3.6 Weather Applications (Apps)

There are an increasing number of weather apps that allow pilots to access a wide range of weather reports and forecasts from their phone, tablet, and computer. Some apps include flight planning services, in-flight updates, NOTAMs, and TFRs. Details on these weather apps and other online weather services can be found in the information from the service provider.

3.3.4 In-Flight Updates

3.3.4.1 Flight Information Service-Broadcast (FIS-B)

FIS-B over Universal Access Transceiver (UAT) datalink service provides aeronautical information and meteorological information to the flight deck for aircraft operating in the NAS. These products are broadcast over the Automatic Dependent Surveillance-Broadcast (ADS-B) UAT link so that pilots have timely information of regional weather and NAS status/changes that might affect flight. FIS-B aeronautical information and meteorological information products provide strategic information to the flight deck that enhances a preflight briefing. FIS-B products do not include all NOTAMs.

Advisory Circular (AC) 00-63, Use of Flight Deck Displays of Digital Weather and Aeronautical Information, contains detailed information concerning FIS-B meteorological products. The AIM Chapter 3, Airspace; Chapter 4, Air Traffic Control; and Chapter 5, Air Traffic Procedures, contain information on Special Use Airspace (SUA), TFR, and NOTAM products.

FIS-B update intervals are defined as the rate at which the product data is available from the source. Transmission intervals are defined as the amount of time within which a new or updated product transmission must be completed and the rate or repetition interval at which the product is rebroadcast. Refer to the AIM, Chapter 7, Section 1, Table 7-1-2, FIS-B Over UAT Product Update and Transmission Intervals, for update and transmission intervals for each FIS-B product.

Where applicable, FIS-B products include a look-ahead range expressed in nautical miles for three service domains: Airport Surface, Terminal Airspace, and En Route/Gulf of Mexico. The AIM, Chapter 7, Section 1, Table 7-1-3, Product Parameters for Low/Medium/High Altitude Tier Radios, provides service domain availability and look-ahead ranging for each FIS-B product.

Details on the content, format, and symbology of individual FIS-B products can be obtained from the manufacturer of the avionics equipment used to receive and display them.

3.3.4.2 Automated Surface Observing System (ASOS) and Automated Weather Observing System (AWOS)

ASOS and AWOS information can be retrieved in flight. Typically, the range of an ASOS/AWOS broadcast is 25 NM. ASOS and AWOS frequencies are printed on Sectional Charts and listed in FAA Chart Supplements. The majority of ASOS and AWOS are on airports, but there are a few located off-airport, such as in a mountain pass. See Section 24.3 for more information on ASOS and AWOS.

3.3.4.3 Automatic Terminal Information Service (ATIS)

The Automatic Terminal Information Service (ATIS) is a continuous broadcast on an assigned frequency of recorded information in selected terminal areas to provide pilots with necessary airport and local area information prior to arrival or departure.

ATIS frequencies can be found on Sectional Charts and Chart Supplements, as well as on instrument approach charts.

The ATIS broadcast is updated upon the receipt of new hourly weather, special weather, or when there is a change in other pertinent data, such as runway change, instrument approach in use, etc.

3.3.4.4 Flight Service

FSS specialists provide in-flight weather updates as well as collect PIREPs. FSS frequencies are listed on Sectional Charts and Chart Supplements.

3.3.4.5 Automatic Flight Information Service (AFIS) – Alaska Only

The Automatic Flight Information Service (AFIS) provides a continuous broadcast of recorded non-control information at airports in Alaska where Flight Service provides local airport advisory (LAA) service. The AFIS broadcast automates the repetitive transmission of essential but routine information, such as weather, wind, altimeter, favored runway, braking action, airport NOTAMs, and other applicable information. The information is continuously broadcast over a discrete very high frequency (VHF) radio frequency (usually the ASOS frequency). When rapidly changing conditions exist, the ceiling, visibility, altimeter, wind, or other conditions may be omitted from the AFIS and will be issued by the FSS specialist on the appropriate radio frequency. AFIS frequencies can be found on Sectional Charts and the Alaska Chart Supplement.

3.4 Overview of Aviation Weather Products

There are many aviation weather products available to the pilot through the internet and mobile phone apps. Each product has a specific purpose that provides the user with reported or forecast weather conditions either at an airport or aloft. Table 3-1 and Table 3-2 are matrices of aviation weather products versus the weather elements and the weather phenomena applicable to aviation. These tables are a high-level overview and do not attempt to capture all products and elements.

A brief summary of the weather products that contain information pertaining to each weather element and weather phenomenon is provided in individual tables in the next section. Technical specifications (e.g., codes and contents) pertaining to the products in Table 3-1 and Table 3-2, as well as others, are provided in Chapters 24, 25, 26, 27, and 28.

3.4.1 Weather Products versus Weather Elements and Phenomena

Table 3-1. High-Level Overview of Select Weather Products and Select Weather Elements and Phenomena that are of Interest to Aviation Users (See Table 3-3 for acronyms and abbreviations.)

An 'X' indicates that the weather product contains information pertaining to the weather element or phenomenon.

Product	T-Storm	Wind	Wind Shear	Visibility	Precip	Fog	In-Flight Icing	Freezing Level	Turb	Ash	TC	SS/DS	Mtn Obsc
Observations													
METAR, SPECI (ASOS, AWOS, ATIS)	X	X	X	X	X	X			X			X	
PIREP/AIREP /VAR	X		X	X	X	X	X	X	X	X		X	X
Radar	X				X				X	X	X	X	
Lightning Data	X												
Satellite	X								X	X	X	X	
Weather Camera	X			X	X	X				X			X

Product	T-Storm	Wind	Wind Shear	Visibility	Precip	Fog	In-Flight Icing	Freezing Level	Turb	Ash	TC	SS/DS	Mtn Obsc
Analysis													
Surface Analysis Charts		X		X	X	X					X		
Upper-Air Analysis													
Freezing Level Analysis								X					
CIP							X						
GTG									X				
Advisories													
Convective-SIGMET	X		X				X		X		X		
SIGMET	X						X		X	X	X	X	
AIRMET		X	X	X	X	X	X	X	X				X
CWA	X	X		X	X		X		X	X		X	
VAA										X			
TCA		X									X		
Space Weather Advisory													
LLWAS			X										
AWW	X	X			X								
Forecasts													
Wind and Temp Aloft		X											
TAF	X	X	X	X	X	X			X	X		X	
Aviation Surface Forecast	X	X		X	X	X						X	
Aviation Clouds Forecast							X						X
AK FA	X	X		X	X	X	X	X	X	X			X
HI FA	X	X		X	X	X			X				
Gulf of Mexico and Caribbean FA	X	X		X	X	X	X	X	X	X			
Low-Level SIGWX							X	X				X	
Med-Level SIGWX	X	X					X		X	X	X		
High-Level SIGWX	X	X							X	X	X		
AK SIGWX	X						X	X					
AK Surface Chart	X				X	X							
AK Convective Outlook	X												
Surface Prog Charts	X				X						X		
WAFS	X	X					X		X				
Upper Air Forecasts		X						X					
FIP							X						
Cloud Top													
Convective Outlook	X												
TCF	X												
ECFP	X												

Product	T-Storm	Wind	Wind Shear	Visibility	Precip	Fog	In-Flight Icing	Freezing Level	Turb	Ash	TC	SS/DS	Mtn Obsc
SAW	X												
AFD	X	X		X	X	X				X	X	X	X
MIS	X			X	X	X	X		X	X	X		
Tools													
GFA Tool	X	X	X	X	X	X	X	X	X	X	X	X	X
Flight Service Interactive Map	X	X	X	X	X	X	X	X	X	X	X	X	X

Table 3-2. Continuation of Table 3-1 (See Table 3-3 for acronyms and abbreviations.)

An 'X' indicates that the weather product contains information pertaining to the weather element or phenomenon.

Product	IFR	MVFR	Cloud Cover	Cloud Base	Cloud Tops	Cloud Layers	Pressure	Fronts	Temp Aloft	Space Weather	Density Altitude
Observations											
METAR, SPECI (ASOS, AWOS, ATIS)	X	X	X	X		X	X	X	X		X
PIREP/AIREP			X	X	X	X			X		
Radar											
Lightning Data											
Satellite			X								
Weather Camera			X	X							
Analysis											
Surface Analysis Charts			X	X			X	X	X		
Upper-Air Analysis									X		
Freezing Level Analysis									X		
CIP											
GTG											
RTMA									X		
Advisories											
Convective SIGMET					X						
SIGMET					X						
CWA	X			X							
VAA											
TCA											
Space Weather Advisory										X	
LLWAS										X	
AWW											
Forecasts											
Wind and Temp Aloft									X		
TAF	X	X	X	X			X		X		
Aviation Surface Forecast	X										
Aviation Clouds Forecast			X	X	X	X					
AK FA	X	X	X	X	X	X			X		
HI FA	X	X	X	X	X	X			X		
Gulf of Mexico and Caribbean FA	X	X	X	X	X	X			X		

Chapter 3, Overview of Aviation Weather Information

Product	IFR	MVFR	Cloud Cover	Cloud Base	Cloud Tops	Cloud Layers	Pressure	Fronts	Temp Aloft	Space Weather	Density Altitude
Low-Level SIGWX	X	X	X	X					X		
Med-Level SIGWX											
High-Level SIGWX											
AK SIGWX	X	X	X	X				X			
AK Surface Chart								X			
Surface Prog Charts							X	X			
WAFS									X		
Upper Air Forecasts									X		
FIP											
GTG											
Cloud Top			X		X						
Convective Outlook											
TCF					X						
ECFP											
SAW											
AFD	X	X	X	X				X			
MIS	X			X							
Tools											
GFA Tool	X	X	X	X	X	X					
Flight Service Interactive Map	X	X	X	X		X					

Table 3-3. Acronyms and Abbreviations Used in Product/Element/Phenomenon Tables

Acronym/Abbreviation	Definition
AFD	Aviation Forecast Discussion
AGL	Above Ground Level
AIREP	Aircraft Report
AIRMET	Airmen's Meteorological Information
AK	Alaska
ARTCC	Air Route Traffic Control Center
Ash	Volcanic Ash
ASOS	Automated Surface Observing System
ATIS	Automatic Terminal Information Service
AWOS	Automated Weather Observing System
AWW	Airport Weather Warning
BCFG	Patchy Fog
BKN	Broken
BLDU	Blowing Dust

Acronym/Abbreviation	Definition
BLSA	Blowing Sand
BR	Mist
CB	Cumulonimbus
CIP	Current Icing Product
CONUS	Contiguous United States
CWA	Center Weather Advisory
DRDU	Drifting Dust
DRSA	Drifting Sand
DS	Dust Storm
ECFP	Extended Convective Forecast Product
FA	Area Forecast
FG	Fog
FIP	Forecast Icing Product
FL	Flight Level
FROPA	Frontal Passage
ft	Feet
FZFG	Freezing Fog
GFA	Graphical Forecasts for Aviation
GTG	Graphical Turbulence Guidance
HI	Hawaii
HZ	Haze
ICAO	International Civil Aviation Organization
IFR	Instrument Flight Rules
kt	knot
LIFR	Low Instrument Flight Rules
LLWAS	Low-Level Wind Shear Alert System
LLWS	Low-Level Wind Shear
METAR	Aviation Routine Weather Report
MIFG	Shallow Fog
MIS	Meteorological Impact Statement
MSL	Mean Sea Level
Mtn Obsc	Mountain Obscuration
MVFR	Marginal Visual Flight Rules
NIL SIG	No Significant

Acronym/Abbreviation	Definition
NOAA	National Oceanic and Atmospheric Administration
Obsc	Obscuration
OVC	Overcast
PIREP	Pilot Weather Report
Precip	Precipitation
PRFG	Partial Fog
Prog	Prognostic
RMK	Remarks
RTMA	Real-Time Mesoscale Analysis
SAW	Aviation Watch Notification Message
SIGMET	Significant Meteorological Information
SIGWX	Significant Weather
SLD	Supercooled Large Drop
sm	statute mile
SPECI	Aviation Selected Special Weather Report
SS	Sandstorm
TAF	Terminal Aerodrome Forecast
TC	Tropical Cyclone
TCA	Tropical Cyclone Advisory
TCF	Traffic Flow Management Convective Forecast
Temp	Temperature
TS	Thunderstorm
TSRA	Thunderstorm with Rain
T-Storm	Thunderstorm
Turb	Turbulence
VA	Volcanic Ash
VAA	Volcanic Ash Advisory
VAR	Volcanic Activity Report
VC	Vicinity
VCFG	Fog in the Vicinity
VCTS	Thunderstorm in the Vicinity
VFR	Visual Flight Rules
WAFS	World Area Forecast System
WDSPR DS	Widespread Dust Storms

Acronym/Abbreviation	Definition
WDSPR SS	Widespread Sandstorms
Wx or WX	Weather
Z	Zulu

3.4.2 Summaries of Specific Weather Information Contained in Various Weather Products

The following set of tables identify specific weather elements (e.g., thunderstorms, turbulence) and list the products that contain information relating to those elements. A brief summary description is also provided. These summaries are not intended to provide all of the details. More detailed information can be found in Chapters 24, 25, 26, 27, and 28.

3.4.2.1 Products with Thunderstorm Information

Table 3-4. Weather Element: Thunderstorm

Type: text (T), graphic (G), image (I), voice (V)

Thunderstorm Information Contained In:	Type	Summary
Observations		
METAR and SPECI (ASOS, AWOS, ATIS)	T, G, V	Thunderstorms are reported in the body section of the METAR/SPECI when observed or detected by lightning networks or observers. Lightning information is provided in the RMK section of the METAR/SPECI.
PIREP	T, G, V	Included when reported.
Radar	I	Radar provides intensity levels of precipitation. Thunderstorms can typically be distinguished based on intensity, but not always. Certain radar limitations can be misleading.
Lightning data	G	Used in the METAR, SPECI, ASOS, AWOS, ATIS, as well as incorporated into other product overlays (e.g., satellite pictures, radar display).
Satellite	I	Thunderstorms can be inferred by a meteorologist or trained specialist, unless they are embedded in other cloud layers.
Weather Camera	I	CB clouds may be seen on the camera.
Analysis		
Advisories		
Convective SIGMET	T, G	Active area of thunderstorms. Only issued for the CONUS instead of a SIGMET for thunderstorms.
SIGMET	T, G	Active area of thunderstorms. Issued for areas outside the CONUS.
CWA	T, G	Active area of thunderstorms. Issued as a supplement to a Convective SIGMET or when Convective SIGMET criteria have not been met.
AWW	T	Intended for ground operations at select airports. Criteria may vary depending on user needs. For example, it can be issued for cloud to ground lightning within 5 miles of the airport.

Thunderstorm Information Contained In:	Type	Summary
Forecasts		
TAF	T, G	Thunderstorm included using various descriptors. For example: - TS: thunderstorm without precipitation (means dry thunderstorms). - TSRA: thunderstorm with precipitation (in this case, rain). - VCTS: thunderstorms in the vicinity (i.e., within 5–10 sm of the center of the airport).
Aviation Surface Forecast	G	Derived from the GFA and includes forecasts of areas of thunderstorms at specified valid times.
Alaska Area Forecast	T	A description of significant clouds and weather including thunderstorms and CB clouds.
Hawaii Area Forecast	T	A description of significant clouds and weather including thunderstorms and CB clouds.
Gulf of Mexico and Caribbean Area Forecast	T	A description of significant clouds and weather including thunderstorms and CB clouds.
Med-Level SIGWX	G	Forecast of significant weather including areas of CB clouds at specified valid times for limited geographic areas around the globe.
High-Level SIGWX	G	Global forecast of significant weather including areas of CB clouds at specified valid times.
Alaska SIGWX Chart	G	Forecast of significant weather including areas of thunderstorms at specified valid times.
Alaska Surface Chart	G	Forecast of surface weather features including areas of thunderstorms at specified valid times.
Alaska Convective Outlook	G	Seasonal product that provides forecasts that indicate where conditions are favorable for towering cumulus and thunderstorms at specified valid times.
Surface Prog Charts	G	CONUS forecast that includes areas of thunderstorms at specified valid times.
WAFS	G	Global forecasts that include areas of CB clouds.
TCF	G	The TCF depicts areas of convection meeting select criteria.
ECFP	G	The ECFP is a planning forecast mainly intended for air traffic managers.
Convective Outlook	T, G	Convective Outlooks provide the potential for severe (tornado, wind gusts 50 knots or greater, or hail 1-in diameter or greater) and non-severe (general) convection and specific severe weather threats during the following 8 days.
SAW	T, G	Formerly known as the AWW, the SAW provides an approximation of the area covered by a Severe Thunderstorm Watch or Tornado Watch.
Severe Thunderstorm Watch	T, G	A watch is when conditions are favorable for severe thunderstorms, which is hail 1-in diameter or greater and/or damaging winds of 50 knots or greater. Not to be confused with a Severe Thunderstorm Warning, which is issued when severe thunderstorms are occurring.
Tornado Watch	T, G	A watch is when conditions are favorable for tornadoes. Not to be confused with a Tornado Warning, which is issued when tornadoes are occurring.
AFD	T	Describes the weather conditions as they relate to a specific TAF or group of TAFs, and may provide additional aviation weather-related issues that cannot be encoded into the TAF, such as the reasoning behind the forecast.
MIS	T	A discussion of meteorological events (including significant convection) causing or expecting to impact the flow of air traffic across an ARTCC.

Thunderstorm Information Contained In:	Type	Summary
Tools		
GFA Tool	G	Interactive website that includes forecast thunderstorm areas with different coverage levels. Also overlays radar Convective SIGMETs, CWAs for thunderstorms and Severe Thunderstorm/Tornado Warnings.
Flight Service Interactive Map	G	Interactive website that includes overlays of radar, Convective SIGMETs, CWAs for thunderstorms, and Severe Thunderstorms/Tornado Watches and Warnings. Also includes an overlay for radar echo tops.

3.4.2.2 Products with Wind Information

Table 3-5. Weather Element: Wind

Type: text (T), graphic (G), image (I), voice (V)

Wind Information Contained In:	Type	Summary
Observations		
METAR and SPECI (ASOS, AWOS, ATIS)	T, G, V	Surface wind speed and direction are included. Wind direction is reported relative to magnetic north in ATIS as well as ASOS and AWOS radio (voice) broadcasts. Otherwise reported relative to true north. Wind speed is reported in knots.
PIREP/AIREP	T, G, V	Wind direction is reported in tens of degrees magnetic north and wind speed in knots.
Analysis		
Surface Analysis Chart	G	Surface wind speed and direction (true north) are depicted with standard symbols on station plot models.
Upper Air Analysis	G	At select pressure levels, the wind direction aloft is displayed in tens of degrees true north and wind speed aloft is displayed in knots.
Advisories		
AIRMET	T, G	AIRMET Tango issued when sustained surface winds greater than 30 knots are occurring or expected to occur.
CWA	T, G	May be issued if surface wind gusts are at or above 30 knots.
TCA	T	Includes maximum sustained surface winds.
AWW	T	Intended for ground operations at select airports. Criteria may vary depending on user needs.
Forecasts		
Winds and Temperature Aloft	T, G	Wind direction aloft is indicated in text format as tens of degrees with reference to true north, and wind speed aloft in knots. The graphical forecast uses standard wind barb display for wind speed/direction (true north).
TAF	T, G	Surface wind forecasts of direction rounded to the nearest 10 degrees (true north) and the surface mean wind speed in knots are included in the wind group.
Aviation Surface Forecast	G	Derived from the GFA and includes forecasts of surface winds at specified valid times. Also includes AIRMET for surface winds.
Alaska Area Forecast	T	Surface winds greater than 20 knots are included in the description of significant clouds and weather.

Wind Information Contained In:	Type	Summary
Hawaii Area Forecast	T	Sustained surface winds of 20 knots or greater are included in the description of significant clouds and weather.
Gulf of Mexico and Caribbean Area Forecast	T	Sustained surface winds greater than or equal to 20 knots are included in the description of significant clouds and weather.
Medium-Level SIGWX	G	A jet stream axis with a wind speed aloft of more than 80 knots is depicted with double hatched lines identifying 20-knot speed changes.
High-Level SIGWX	G	Global forecast of a jet stream axis with a wind speed aloft of more than 80 knots is depicted with double hatched lines identifying 20-knot speed changes.
WAFS	G	Winds aloft are issued at select FLs and are provided in chart and grid point formats. Wind speed and direction (true north) are displayed with wind barbs.
Upper Air Forecasts	G	Model outputs for winds aloft are shown at select pressure levels. Wind speed and direction (true north) are displayed with wind barbs.
AFD	T	Describes the weather conditions as they relate to a specific TAF or group of TAFs, including surface winds.
Tools		
GFA Tool	G	Interactive website that includes surface winds (true north) and winds aloft at select altitudes/FLs.
Flight Service Interactive Map	G	Interactive website that includes overlap of AIRMET for surface winds and winds aloft (true north) at select altitudes/FLs.

3.4.2.3 Products with Wind Shear Information

Table 3-6. Weather Element: Wind Shear

Type: text (T), graphic (G), image (I), voice (V)

Wind Shear Information Contained In:	Type	Summary
Observations		
ATIS	V	LLWS and/or microburst is included in ATIS broadcast for 20 minutes following last report.
PIREP/AIREP	T, G, V	Wind shear is included in the RMK section when reported.
Analysis		
Advisories		
Convective SIGMET	T, G	Possible LLWS is implied within the convective SIGMET area.
AIRMET	T, G	AIRMET Tango issued when nonconvective LLWS potential below 2,000 ft AGL is occurring or expected to occur.
LLWAS	T	A wind shear alert occurs when wind shear ±15 kt is detected.
Forecasts		
TAF	T, G	Included in the nonconvective LLWS group.
AFD	T	Describes the weather conditions as they relate to a specific TAF or group of TAFs, including nonconvective LLWS.

Wind Shear Information Contained In:	Type	Summary
Tools		
GFA Tool	G	Interactive website that includes overlay of AIRMET for nonconvective LLWS.
Flight Service Interactive Map	G	Interactive website that includes overlay of AIRMET for nonconvective LLWS.

3.4.2.4 Products with Visibility Information

Table 3-7. Weather Element: Visibility

Type: text (T), graphic (G), image (I), voice (V)

Visibility Information Contained In:	Type	Summary
Observations		
METAR and SPECI (ASOS, AWOS, ATIS)	T, G, V	Visibility is reported in the body section of the METAR/SPECI. The visibility is reported in statute miles.
PIREP/AIREP	T, G, V	Distance for visibility is reported by the pilot in statute miles.
Weather Camera	I	A rough estimate could be inferred when known distance features are seen on the camera.
Analysis		
Surface Analysis Chart	G	Certain surface analysis charts may include the reported visibility in the station plot model.
Advisories		
AIRMET	T, G	AIRMET Sierra depict areas of surface-based IFR (ceilings less than 1,000 ft and/or visibilities less than 3 sm) that is occurring or expected to occur.
CWA	T, G	Issued if conditions are at or approaching low IFR (ceiling less than 500 ft and/or visibilities less than 1 sm).
Forecasts		
TAF	T, G	Included under the visibility group when prevailing visibility is forecast to be less than or equal to 6 sm.
Aviation Surface Forecast	G	Derived from the GFA and includes forecasts of visibility at specified valid times.
Alaska Area Forecast	T	Visibilities of 6 sm or less and obscurations to visibility are included.
Hawaii Area Forecast	T	Visibilities of 6 sm or less and obscurations to visibility are included.
Gulf of Mexico and Caribbean Area Forecast	T	Visibilities of 6 sm and obscurations to visibility are included.
AFD	T	Describes the weather conditions as they relate to a specific TAF or group of TAFs, including visibilities.

Visibility Information Contained In:	Type	Summary
Tools		
GFA Tool	G	Interactive website that includes visibility.
Flight Service Interactive Map	G	Interactive website that includes overlay of AIRMET Sierra and CWAs.

3.4.2.5 Products with Precipitation Information

Table 3-8. Weather Element: Precipitation

Type: text (T), graphic (G), image (I), voice (V)

Precipitation Information Contained In:	Type	Summary
Observations		
METAR and SPECI (ASOS, AWOS, ATIS)	T, G, V	Precipitation is reported in the body section of the METAR/SPECI when observed or detected. Some AWOS systems do *not* report precipitation.
PIREP	T, G, V	Pilot reports may include precipitation, which in places in the "/WX" section of the PIREP.
Radar	I	Radar provides intensity levels of precipitation.
Weather Camera	I	Precipitation may be seen on the camera.
Analysis		
Surface Analysis Chart	G	Precipitation is included within the station plot models or when charts are combined with radar imagery.
Advisories		
AIRMET	G	AIRMETs are issued when weather phenomena such as precipitation restricts visibility (less than 3 sm).
CWA	T, G	May be issued for heavy, extreme, or frozen precipitation.
AWW	T	Intended for ground operations at select airports. Criteria may vary depending on user needs. Examples of criteria that could issue this warning are heavy snow or freezing rain.
Forecasts		
TAF	T, G	Included when conditions are expected and indicated by various descriptors based on the type of precipitation. Intensity is also coded with precipitation types.
Aviation Surface Forecast	G	Derived from the GFA and includes forecasts of precipitation at specified valid times includes weather phenomena.
Alaska Area Forecast	T	A description of significant clouds and weather, including precipitation.
Hawaii Area Forecast	T	A description of significant clouds and weather, including precipitation.
Gulf of Mexico and Caribbean Area Forecast	T	A description of significant clouds and weather, including precipitation.

Precipitation Information Contained In:	Type	Summary
Alaska Surface Chart	G	Includes forecast of precipitation.
Surface Prog Charts	G	Depicts the type of precipitation and the forecast percent probability of measurable precipitation.
AFD	T	Describes the weather conditions as they relate to a specific TAF or group of TAFs, including precipitation.
MIS	T	A nontechnical discussion of meteorological events (including precipitation) causing or expecting to impact the flow of air traffic across an ARTCC.
Tools		
GFA Tool	G	Interactive website that includes precipitation information and overlay of radar.
Flight Service Interactive Map	G	Interactive website that includes overlay of radar.

3.4.2.6 Products with Fog Information

Table 3-9. Weather Element: Fog

Type: text (T), graphic (G), image (I), voice (V)

Fog Information Contained In:	Type	Summary
Observations		
METAR and SPECI (ASOS, AWOS, ATIS)	T, G, V	Fog (including mist and haze) information is reported in METAR/SPECI from manual and ASOS stations, and included in ASOS and ATIS broadcasts. Fog/mist/haze is not included in METARs/SPECIs from most AWOS, nor most AWOS broadcasts, depending on the type of AWOS. FG is reported when visibility is less than 5/8 sm. FZFG is reported when temperature is below 00 °C. BR or HZ is reported for visibilities from 5/8 sm to less than 7 sm, depending on the difference between the temperature and dewpoint. If the difference is 40 °F (~20 °C) or less, then FG is reported. Otherwise HZ is reported.
PIREP	T, G, V	Included if reported.
Weather Camera	I	Fog may be seen on the camera.
Analysis		
Surface Analysis Chart	G	Noted on the chart or in the station plots.
Advisories		
AIRMET	T, G	Issued when weather phenomena such as fog/mist occurs or is expected to occur that could restrict visibility (less than 3 sm).

Fog Information Contained In:	Type	Summary
Forecasts		
TAF	T, G	A visibility threshold must be met for fog to be included in the TAF (visibility of less than 5/8 sm). The fog code is "FG," with the following additional terms: - Freezing fog (FZFG). - Shallow fog (MIFG). - Patchy Fog (BCFG). - Partial Fog (PRFG). - Fog in the vicinity (VCFG). Vicinity (VC) is defined as area 5 to 10 sm from the center of the airport. - BR is included for visibilities from 5/8 to 6 sm.
Aviation Surface Forecast	G	Derived from NOAA's GFA and includes forecasts of areas of thunderstorms at specified valid times includes obscurations such as fog/mist.
Alaska Area Forecast	T	Visibilities of 6 sm or less and obstruction(s) to visibility are included.
Hawaii Area Forecast	T	Visibilities of 6 sm or less with obstruction(s) to visibility are included.
Gulf of Mexico and Caribbean Area Forecast	T	Visibilities of 6 sm or less and obstruction(s) to visibility are included.
Alaska Surface Chart	G	Forecast of surface weather features including areas of fog/mist at specified valid times.
AFD	T	Describes the weather conditions as they relate to a specific TAF or group of TAFs, including fog/mist.
MIS	T	A discussion of meteorological events (including fog/mist) causing or expecting to impact the flow of air traffic across an ARTCC.
Tools		
GFA Tool	G	Interactive website that includes fog/mist. Also includes AIRMET Sierra.
Flight Service Interactive Map	G	Interactive website that includes overlay of AIRMET Sierra.

3.4.2.7 Products with In-Flight Icing Information

Table 3-10. Weather Element: In-Flight Icing

Type: text (T), graphic (G), image (I), voice (V)

In-Flight Icing Information Contained In:	Type	Summary
Observations		
PIREP/AIREP	T, G, V	Icing intensity, type, and altitude reported. Noted as an Urgent PIREP or Special AIREP when severe.
Analysis		
CIP	G	Computer model's 0-hour forecast (referred to on the product as an analysis) depicting five icing analysis products: - Icing Probability. - Icing Severity. - Icing Severity – Probability > 25%. - Icing Severity – Probability > 50%. - Icing Severity plus SLD.
Advisories		
Convective SIGMET	T, G	Possible severe icing is implied within the convective SIGMET area.
SIGMET	T, G	Nonconvective SIGMETs are issued for severe icing.
AIRMET	T, G	AIRMET Zulu depicts areas of active or expected moderate icing.
CWA	T, G	May be issued for moderate or greater icing.
Forecasts		
Aviation Clouds Forecast	G	Includes a depiction of icing AIRMETs.
Alaska Area Forecast	T	May includes a description of icing not meeting AIRMET criteria otherwise "NIL SIG" is noted if no significant icing is forecast.
Gulf of Mexico and Caribbean Area Forecast	T	A description of moderate or severe icing conditions.
Medium-Level SIGWX	G	Areas moderate or severe icing are depicted.
WAFS	G	Global icing forecasts provided in a grid point format.
FIP	G	Five icing forecast graphics derived from computer model data for the following: - Icing Probability. - Icing Severity. - Icing Severity – Probability > 25%. - Icing Severity – Probability > 50%. - Icing Severity plus SLD.
MIS	T	A discussion of meteorological events (including inflight icing information) causing or expecting to impact the flow of air traffic across an ARTCC.

In-Flight Icing Information Contained In:	Type	Summary
Tools		
GFA Tool	G	Interactive website that includes in-flight icing (PIREPs, FIPs, AIRMETs, SIGMETs, and CWAs).
Flight Service Interactive Map	G	Interactive website that includes icing.

3.4.2.8 Products with Freezing Level Information

Table 3-11. Weather Element: Freezing Level

Type: text (T), graphic (G), image (I), voice (V)

Freezing Level Information Contained In:	Type	Summary
Observations		
PIREP/AIREP	T, G, V	Included when reported.
Analysis		
Freezing Level Analysis	G	Depicts the height (MSL) of the lowest freezing level.
Advisories		
AIRMET	T, G	Contains one or more of the following: - Areas of multiple freezing levels. - Range of freezing levels over the area. - Lowest freezing levels at intervals of 4,000 ft MSL.
Forecasts		
Alaska Area Forecast	T	A description of significant clouds and weather that includes freezing levels.
Gulf and Caribbean Area Forecast	T	A description of significant clouds and weather that includes freezing levels.
Low-Level SIGWX	G	Freezing level at the surface are depicted.
Alaska SIGWX	G	Freezing levels are shown for the surface and at 2,000-ft intervals.
Upper Air Forecasts	G	Computer model outputs for freezing levels at select height levels (available levels vary depending on the model).
Tools		
GFA Tool	G	Interactive website that includes freezing levels.
Flight Service Interactive Map	G	Interactive website that includes freezing levels.

3.4.2.9 Products with Turbulence Information

Table 3-12. Weather Element: Turbulence

Type: text (T), graphic (G), image (I), voice (V)

Turbulence Information Contained In:	Type	Summary
Observations		
PIREP/AIREP	T, G, V	Turbulence reports include location, altitude, and aircraft type. The pilot determines the degree of turbulence, intensity, and duration. Severe or extreme turbulence is reported as an Urgent PIREP or Special AIREP. The vast majority of AIREPs are automated and include turbulence reports derived from the aircraft's motion.
Radar	I	Convective weather on radar could indicate potential areas of severe turbulence.
Satellite	I	Potential turbulence areas may be inferred from certain cloud patterns by a meteorologist or trained specialist. CB always implies severe turbulence.
Analysis		
GTG	G	Product provides a computer analysis of turbulence based on multiple computer algorithms. Graphic also depicts any turbulence PIREPs.
Advisories		
Convective SIGMET	T, G	Possible severe turbulence is implied within the convective SIGMET area.
SIGMET	T, G	Issued for severe turbulence not associated with thunderstorms.
AIRMET	T, G	AIRMET Tango depicts areas of active or expected moderate turbulence. The product is divided into high and low altitude at 18,000 ft.
CWA	T, G	May be issued for moderate or greater turbulence not covered by an existing AIRMET or SIGMET.
Forecasts		
TAF	T, G	Low-level mechanical turbulence could be inferred when strong and gusty surface winds are forecast
Alaska Area Forecast	T	Expected turbulence conditions are included along with the following information: - AIRMET information for turbulence or LLWS. - Turbulence not meeting SIGMET/AIRMET criteria during the 6 to 12-hour period. - "NIL SIG" is noted if no significant turbulence in forecast.
Gulf of Mexico and Caribbean Area Forecast	T	Moderate or greater turbulence is noted at the end of forecast bulletin.
Low-Level SIGWX	G	Moderate or greater turbulence is depicted.
Medium-Level SIGWX	G	Areas of nonconvective clouds with moderate or severe turbulence are depicted.
High-Level SIGWX	G	Global forecasts of moderate or severe turbulence associated with wind shear zones and mountain waves are depicted.
Alaska SIGWX	G	Areas of forecast moderate or greater nonconvective low-level turbulence are depicted.
WAFS	G	Global forecasts of turbulence provided in a grid point format primarily intended for use in flight-planning systems.

Turbulence Information Contained In:	Type	Summary
MIS	T	A discussion of meteorological events (including turbulence information) causing or expecting to impact the flow of air traffic across an ARTCC.
Tools		
GFA Tool	G	Interactive website that includes turbulence (PIREP/AIREP, GTG, AIRMET, SIGMET, and CWA).
Flight Service Interactive Map	G	Interactive website that includes turbulence (PIREP/AIREP, GTG, AIRMET, SIGMET, and CWA).

3.4.2.10 Products with Volcanic Ash Information

Table 3-13. Weather Element: Volcanic Ash

Type: text (T), graphic (G), image (I), voice (V)

Volcanic Ash Information Contained In:	Type	Summary
Observations		
METAR, SPECI (ATIS)	T, G, V	Reported in manual and select augmented METAR/SPECIs as "VA." Included in ATIS as appropriate.
PIREP/AIREP/VAR	T, G, V	Reported by pilot and noted as an Urgent PIREP or Special AIREP. First part of VAR reported immediately to ATC, second part of VAR submitted postflight.
Radar	I	Potentially visible on radar.
Satellite	I	May be visible on satellite if not obscured by cloud cover.
Weather Camera	I	Volcanic ash cloud may be seen on the camera.
Analysis		
		See SIGMET and VAA
Advisories		
SIGMET	T, G	One of the conditions that triggers the issuance of a SIGMET. Provides an analysis and 6-hour forecast location of the ash cloud.
VAA	T, G	Provides an analysis, 6-, 12-, 18-, and 24-hour forecast location of the ash cloud. Issued every 6 hours until the volcanic ash is no longer discernible from satellite and no further reports of volcanic ash are received from the area.
CWA	T, G	One of the conditions that triggers the issuance of a CWA.
Forecasts		
TAF	T	Included when conditions are expected and indicated in the forecast as "VA" under the significant weather group.
Alaska Area Forecast	T	Obstructions to visibility such as volcanic ash are noted in the description of significant clouds and weather for the region during the first 12 hours of the forecast period.
Hawaii Area Forecast	T	Obstructions to visibility such as volcanic ash are noted in the description of significant clouds and weather for the region during the first 12 hours of the forecast period.
Gulf of Mexico and Caribbean Area Forecast	T	Obstructions to visibility such as volcanic ash are noted in the description of significant clouds and weather for the region during the first 12 hours of the forecast period.

Volcanic Ash Information Contained In:	Type	Summary
Medium-Level SIGWX	G	Erupting volcano is identified with a trapezoidal symbol along with the name, latitude, and longitude. Location of ash cloud is not depicted.
High-Level SIGWX	G	Erupting volcano is identified with a trapezoidal symbol along with the name, latitude, and longitude. Location of ash cloud is not depicted.
AFD	T	May include a discussion on volcanic ash when applicable.
MIS	T	A discussion of meteorological events (including volcanic ash cloud) causing or expecting to impact the flow of air traffic across an ARTCC.
Tools		
GFA Tool	G	Interactive website that includes SIGMETs and CWAs for volcanic ash cloud when they are in effect.
Flight Service Interactive Map	G	Interactive website that includes SIGMETs and CWAs for volcanic ash cloud when they are in effect.

3.4.2.11 Products with Tropical Cyclone Information

Table 3-14. Weather Element: Tropical Cyclone (e.g., Hurricane)

Type: text (T), graphic (G), image (I), voice (V)

Tropical Cyclone Information Contained In:	Type	Summary
Observations		
Radar	I	Tropical cyclones are visible on radar when in range.
Satellite	I	Tropical cyclones are visible on satellite.
Analysis		
Surface Analysis Chart	G	Tropical cyclones are included using standard symbols.
Advisories		
Convective SIGMET	T, G	Issued for thunderstorms associated with tropical cyclones.
SIGMET	T, G	SIGMETs are issued for tropical cyclones except over the CONUS and associated coastal waters.
TCA	T, G	TCAs are issued four times daily and report the current and forecast tropical cyclone position and intensity.
Forecasts		
Medium-Level SIGWX	G	Tropical cyclones are included using standard symbols.
High-Level SIGWX	G	Global forecasts that include tropical cyclones using standard symbols.
Surface Prog Charts	G	Tropical depressions, tropical storms, and hurricanes are included using symbols.
AFD	T	May include a discussion on tropical cyclones when applicable.
MIS	T	A nontechnical discussion of meteorological events (including tropical cyclones) causing or expecting to impact the flow of air traffic across an ARTCC.

Tropical Cyclone Information Contained In:	Type	Summary
Tools		
GFA Tool	G	Interactive website that includes tropical cyclones, if applicable.
Flight Service Interactive Map	G	Interactive website that includes tropical cyclones, if applicable.

3.4.2.12 Products with Sandstorm/Dust Storm Information

Table 3-15. Weather Element: Sandstorm/Dust Storm

Type: text (T), graphic (G), image (I), voice (V)

Sandstorm/Dust Storm Information Contained In:	Type	Summary
Observations		
METAR and SPECI (ASOS, AWOS, ATIS)	T, G, V	SS, DS, BLDU and BLSA are reported in manual observations, at some augmented observations, but not automated observations. Automated observations may report these as HZ.
PIREP	T, G, V	Included if reported.
Radar	I	Potentially visible on radar.
Satellite	I	Sandstorms/dust storms may be visible on satellite if not obscured by cloud cover.
Advisories		
SIGMET	T, G	Widespread sandstorms/dust storms (WDSPR DS, WDSPR SS) are conditions that trigger the issuance of a SIGMET.
CWA	T, G	One of the conditions that triggers the issuance of a CWA.
Forecasts		
TAF	T, G	Included as BLDU, BLSA, DRDU and DRSA.
Aviation Surface Forecast	G	Derived from the GFA and includes forecasts of obscurations such as sandstorms/dust storms.
Low-Level SIGWX	G	Could be inferred when IFR or MVFR conditions are depicted in desert areas and supported by other products (e.g., METAR, TAF, SIGMET).
AFD	T	May include a discussion on current or forecast sandstorms/dust storms when applicable.
Tools		
Graphical Forecast of Aviation (GFA) Tool	G	Interactive website that includes sandstorms/dust storms, if applicable.
Flight Service Interactive Map	G	Interactive website that includes sandstorms/dust storms, if applicable.

3.4.2.13 Products with Mountain Obscuration Information

Table 3-16. Weather Element: Mountain Obstruction

Type: text (T), graphic (G), image (I), voice (V)

Mountain Obscuration Information Contained In:	Type	Summary
Observations		
METAR, SPECI (ASOS, AWOS, ATIS)	T, V	May be included in RMK section at some mountain airports. At times it may be inferred or implied by the reporting of clouds at ASOS/AWOS located in mountain passes.
PIREP	T, G, V	Included when reported.
Weather Camera	I	Mountain obscuration may be visible on the camera.
Advisories		
AIRMET	T, G	AIRMET Sierra are issued when widespread mountain obstruction is occurring or expected to occur.
Forecasts		
Aviation Clouds Forecast	G	Part of the derived forecasts from the GFA and includes overlays of mountain obscuration when applicable.
Alaska Area Forecast	T	A description of significant clouds and weather including mountain obscuration.
AFD	T	May include a discussion on current or forecast mountain obscuration.
Tools		
GFA Tool	G	Interactive website that includes mountain obscuration, if applicable.
Flight Service Interactive Map	G	Interactive website that includes mountain obscuration, if applicable.

3.4.2.14 Products with Surface-Based IFR Information

For most aviation weather products, IFR refers to ceilings less than 1,000 feet (ft) (i.e., above ground level (AGL)) and/or surface visibilities less than 3 statute miles (sm). But many aviation weather websites provide a graphical depiction of METARs or Aviation Selected Special Weather Reports (SPECI), and sometimes TAFs, using color-coded station plots for various Weather Flight Categories. These Weather Flight Categories and color codes serve as a means to help pilots visually assess ceilings and visibilities on a map of METARs. Most aviation weather websites use the following color codes and definitions for their display of METARs/SPECIs (and sometimes TAFs) on their website:

- Purple: LIFR = Low IFR, ceilings less than 500 ft and/or visibilities less than 1 sm.
- Red: IFR = Ceiling 500 ft to less than 1,000 ft and/or visibility 1 sm to less than 3 sm.
- Blue: MVFR = Ceiling 1,000 to 3,000 ft and/or visibility 3 to 5 sm.
- Green: VFR = Ceiling greater than 3,000 ft and visibility greater than 5 sm.

Note: The VFR Weather Flight Category is not to be confused with the basic VFR weather minimums given in § 91.155. Weather Flight Categories are only intended for situational awareness.

Table 3-17. Weather Element: Surface IFR

Type: text (T), graphic (G), image (I), voice (V)

Surface IFR Information Contained In:	Type	Summary
Observations		
METAR and SPECI (ASOS, AWOS, ATIS)	T, G, V	Ceiling less than 1,000 ft and/or visibility less than 3 sm. *Note: Websites may graphically depict METAR/SPECI as IFR when ceilings are from 500 ft to less than 1,000 ft and/or visibility 1 sm to less than 3 sm.*
Analysis		
Advisories		
AIRMET	T, G	AIRMET Sierra is issued when surface-based IFR conditions are occurring or expected to occur.
CWA	T, G	May be issued if conditions are at or approaching LIFR conditions (ceilings less than 500 ft and/or visibilities less than 1 sm).
Forecasts		
TAF	T, G	Ceiling less than 1,000 and/or visibility less than 3 sm. *Note: Websites may graphically depict TAFs as IFR when ceilings are from 500 ft to less than 1,000 ft and/or visibility 1 sm to less than 3 sm.*
Aviation Surface Forecast	G	Derived from NOAA's GFA and includes forecasts of surface IFR at specified valid times.
Alaska Area Forecast	T	A 12 to 18-hour categorical outlook for IFR is included in the description of significant clouds and weather.
Hawaii Area Forecast	T	A 12 to 18-hour categorical outlook for IFR is included in the description of significant clouds and weather.
Gulf of Mexico and Caribbean Area Forecast	T	A 12 to 24-hour categorical outlook for IFR is included in the description of significant clouds and weather.
Low-Level SIGWX	G	Areas of forecast IFR conditions are depicted.
Alaska SIGWX	G	Areas of forecast IFR conditions are depicted.
AFD	T	Describes weather conditions such as surface IFR as they relate to the TAF and provide additional aviation weather-related issues.
MIS	T	A nontechnical discussion of meteorological events (including surface IFR) causing or expecting to impact the flow of air traffic across an ARTCC.
Tools		
GFA Tool	G	Interactive website that includes observed and forecast IFR, including AIRMET for IFR.
Flight Service Interactive Map	G	Interactive website that includes METAR and TAF station plots depicting surface-based IFR conditions. AIRMETs for IFR are also shown.

3.4.2.15 Products with Surface-Based MVFR Information

MVFR is a Weather Flight Category. While the "R" in the acronym means "rules," there are no part 91 MVFR weather minimums. The MVFR weather category is defined as ceilings from 1,000 ft to and including 3,000 ft (AGL), and/or surface visibilities from 3 sm to and including 5 sm.

Table 3-18. Surface MVFR

Type: text (T), graphic (G), image (I), voice (V)

Surface MVFR Information Contained In:	Type	Summary
Observations		
METAR and SPECI (ASOS, AWOS, ATIS)	T, G, V	Ceiling 1,000 to 3,000 ft (inclusive) and/or visibility 3 to 5 sm (inclusive). Graphical depictions may provide color-coded flight categories, including MVFR.
Analysis		
Advisories		
Forecasts		
TAF	T, G	Ceiling 1,000 to 3,000 ft (inclusive) and/or visibility 3 to 5 sm (inclusive). Graphical depictions may provide color-coded MVFR flight categories for the TAF.
Alaska Area Forecast	T	A 12 to 18-hour categorical outlook for MVFR is included in the description of significant clouds and weather.
Hawaii Area Forecast	T	A 12 to 18-hour categorical outlook for MVFR is included in the description of significant clouds and weather.
Gulf of Mexico and Caribbean Area Forecast	T	A 12 to 24-hour categorical outlook for MVFR is included in the description of significant clouds and weather.
Low-Level SIGWX	G	Areas of forecast MVFR conditions are depicted.
Alaska SIGWX	G	Areas of forecast MVFR conditions are depicted.
AFD	T	Describes weather conditions such as surface MVFR as they relate to the TAF and provide additional aviation weather-related issues.
Tools		
GFA Tool	G	Interactive website that includes observed and forecast MVFR.
Flight Service Interactive Map	G	Interactive website that includes METAR and TAF station plots depicting MVFR conditions.

3.4.2.16 Products with Cloud Coverage Information

Table 3-19. Weather Element: Cloud Coverage

Type: text (T), graphic (G), image (I), voice (V)

Cloud Coverage Information Contained In:	Type	Summary
Observations		
METAR, SPECI (ASOS, AWOS, ATIS)	T, G, V	Cloud coverage is included.
PIREP	T, G, V	Includes sky condition such as cloud coverage when reported.

Cloud Coverage Information Contained In:	Type	Summary
Satellite	I	A general cloud coverage can be inferred from satellite.
Weather Camera	I	Clouds over the horizon can be seen if in view of the camera and there is sufficient daylight. Overhead clouds cannot be seen.
Analysis		
Surface Analysis Charts	G	Cloud coverage amounts are included within the station plot models and can be inferred when charts are combined with satellite imagery.
Advisories		
AIRMET	T, G	Issued for areas of IFR ceilings and/or visibilities (i.e., BKN or OVC less than 1,000 ft and/or 3 sm).
Forecasts		
TAF	T, G	Cloud coverage is included.
Aviation Clouds Forecast	G	Part of the derived forecasts from the GFA and includes cloud coverage.
Alaska Area Forecast	T	Cloud coverage is included.
Hawaii Area Forecast	T	Cloud coverage is included.
Gulf of Mexico and Caribbean Area Forecast	T	Cloud coverage is included.
Low-Level SIGWX	G	Depicts forecasts areas of IFR ceilings and/or visibilities (i.e., BKN or OVC less than 1,000 ft and/or less than 3 sm) as well as MVFR (i.e., BKN or OVC from 1,000 ft to 3,000 ft and/or 3 to 5 sm).
Alaska SIGWX	G	Areas of forecast IFR and MVFR conditions are depicted.
Cloud Top Forecast	G	Computer model cloud amount and height of cloud tops.
AFD	T	May include a discussion on current or forecast cloud cover.
Tools		
GFA Tool	G, I	Interactive website that includes cloud coverage overlay, METAR station plots, and satellite imagery.
Flight Service Interactive Map	G, I	Interactive website that includes METAR station plots and satellite imagery.

3.4.2.17 Products with Cloud Base Information

Table 3-20. Weather Element: Cloud Base

Type: text (T), graphic (G), image (I), voice (V)

Cloud Base Information Contained In:	Type	Summary
Observations		
METAR, SPECI (ASOS, AWOS, ATIS)	T, G, V	Cloud bases included when reported.
PIREP	T, G, V	Includes cloud bases when reported.
Weather Camera	I	Cloud bases are difficult to estimate or impossible to determine from a camera.
Analysis		
Surface Analysis Charts	G	Some surface analysis charts may plot the height of a BKN or OVC ceiling.
Advisories		
AIRMET	G	Issued for areas of IFR ceilings and/or visibilities (i.e., BKN or OVC less than 1,000 ft and/or less than 3 sm).
CWA	T, G	May be issued for ceilings less than 500 ft.
Forecasts		
TAF	T, G	Cloud bases included.
Aviation Cloud Forecast	G	Part of the derived forecasts from GFA and includes cloud bases.
Alaska Area Forecast	T	Cloud bases included.
Hawaii Area Forecast	T	Cloud bases included.
Gulf of Mexico and Caribbean Area Forecast	T	Cloud bases included.
Low-Level SIGWX	G	Depicts forecast areas of IFR ceilings and/or visibilities (i.e., BKN or OVC less than 1,000 ft and/or less than 3 sm) as well as MVFR (i.e., BKN or OVC from 1,000 ft to 3,000 ft and/or 3 to 5 sm).
Alaska SIGWX	G	Areas of forecast IFR and MVFR conditions are depicted.
AFD	T	May include a discussion on current or forecast cloud bases and/or IFR, MVFR, etc. conditions.
MIS	T	May include a discussion on current or forecast IFR conditions.
Tools		
GFA Tool	G	Interactive website that includes overlay of forecast cloud bases as well as METAR station plots.
Flight Service Interactive Map	G	Interactive website that includes METAR station plots.

3.4.2.18 Products with Cloud Tops Information

Table 3-21. Weather Element: Cloud Tops

Type: text (T), graphic (G), image (I), voice (V)

Cloud Tops Information Contained In:	Type	Summary
Observations		
PIREP	T, G, V	Cloud tops included when reported.
Analysis		
Advisories		
Convective SIGMET	T, G	CB cloud tops are included in Convective SIGMETs.
SIGMET	T, G	CB cloud tops are included in SIGMETs for thunderstorms (outside the CONUS).
Forecasts		
Aviation Cloud Forecast	G	Part of the derived forecasts from the GFA and includes cloud tops. This product provides a forecast of cloud coverage and height (in hundreds of feet MSL).
Alaska Area Forecast	T	A description of significant clouds and weather for the first 12 hours includes cloud tops.
Hawaii Area Forecast	T	A description of significant clouds and weather for the first 12 hours includes cloud tops.
Gulf of Mexico and Caribbean Area Forecast	T	A description of significant clouds and weather for the first 12 hours includes cloud tops.
Medium-Level SIGWX	G	Forecast height of CB tops included.
High-Level SIGWX	G	Global forecasts that include forecast CB tops.
Cloud Top Forecast	I	Computer model cloud amount and height of cloud tops.
TCF	G	The TCF includes forecast CB tops.
Tools		
GFA Tool	G	Interactive website that includes overlay of forecast cloud tops.

3.4.2.19 Products with Cloud Layers Information

Table 3-22. Weather Element: Cloud Layers

Type: text (T), graphic (G), image (I), voice (V)

Cloud Layers Information Contained In:	Type	Summary
Observations		
METAR, SPECI, (ASOS, AWOS, ATIS)	T, G, V	Reports the sky condition including cloud layers.
PIREP/AIREP	T, G, V	PIREPs may include cloud layers.
Analysis		
Advisories		
Forecasts		
TAF	T, G	Cloud layers are included.
Aviation Cloud Forecast	G	Part of the derived forecasts from GFA and includes cloud layers.
Alaska Area Forecast	T	A description of cloud layers is included.
Hawaii Area Forecast	T	A description of cloud layers is included.
Gulf of Mexico and Caribbean Area Forecast	T	A description of cloud layers is included.
Tools		
GFA Tool	G	Interactive website that includes overlay of forecast cloud layers as well as METAR station plots.
Flight Service Interactive Map	G	Interactive website that includes METAR station plots.

3.4.2.20 Products with Pressure Information

Table 3-23. Weather Element: Pressure

Type: text (T), graphic (G), image (I), voice (V)

Pressure Information Contained In:	Type	Summary
Observations		
METAR, SPECI (ASOS, AWOS, ATIS)	T, G, V	Altimeter setting is included. Sea level pressure included in the RMK section of the METAR.
Analysis		
Surface Analysis Chart	G	Sea level pressure is depicted as isobars and within the station plots. High and low pressure centers are also shown.

Pressure Information Contained In:	Type	Summary
Advisories		
Forecasts		
Surface Prog Charts	G	Sea level pressure is depicted as isobars. High and low pressure centers depicted.
Tools		

3.4.2.21 Products with Fronts Information

Table 3-24. Weather Element: Fronts

Type: text (T), graphic (G), image (I), voice (V)

Fronts Information Contained In:	Type	Summary
Observations		
METAR, SPECI (ASOS, AWOS, ATIS)	T, G, V	Manually produced METARs and SPECIs may report "FROPA" in the RMK portion.
Analysis		
Surface Chart Analysis	G	Depicts the location of fronts as well as the high and low pressure systems.
Advisories		
Forecasts		
TAF	T	Fronts are not explicitly mentioned under weather phenomena, but a significant change in wind direction coupled with changes in other elements can imply a forecast frontal passage.
Alaska Area Forecast	T	There is a brief discussion of the synoptic weather affecting the region during the first 18-hour valid period, which could include frontal boundaries.
Hawaii Area Forecast	T	There is a brief discussion of the synoptic weather affecting the region during the first 18-hour valid period, which could include frontal boundaries.
Gulf of Mexico and Caribbean Area Forecast	T	There is a brief discussion of the synoptic weather affecting the region during the entire 24-hour valid period, which could include frontal boundaries.
Alaska SIGWX	G	Pressure systems and fronts are included using standard symbols.
Alaska Surface Chart	G	Fronts are depicted using standard symbols for this chart. This product is issued every 6 hours with forecasts valid for 00Z, 06Z, 12Z, and 18Z.
Surface Prog Charts	G	Fronts are included using standard symbols.
AFD	T	May include a discussion on fronts when applicable.
Tools		

3.4.2.22 Products with Temperature Information

Table 3-25. Weather Element: Temperature

Type: text (T), graphic (G), image (I), voice (V)

Temperature Aloft Information Contained In:	Type	Summary *Note: Temperatures are in degrees Celsius (°C)*
Observations		
METAR, SPECI (ASOS, AWOS, ATIS)	T, G, V	Surface temperature is included.
PIREP/AIREP	T, G, V	Temperature aloft is included if reported.
Analysis		
Upper-Air Analysis	T, G	At select pressure levels, the wind direction and speed are displayed in tens of degrees and wind speed is in knots.
Freezing Level Analysis	G	Temperatures aloft can be indicated with limitations. Depicts the freezing level at the lowest altitude in the atmosphere over a given location at which the air temperature reaches 0 °C.
RTMA	T	Surface temperature is included.
Advisories		
Forecasts		
Winds and Temperature Aloft	T, G	The text format provides the temperature aloft in a coded format for select height levels and locations. Graphical format provides contours of temperatures aloft.
Low-Level SIGWX	G	Temperatures aloft can be indicated with limitations. Multiple freezing levels can be forecast when temperatures are 0 °C at more than one altitude aloft.
WAFS	G	Global forecasts include temperatures aloft, at selected height levels, from model data in a grid point format.
Upper Air Forecasts	G	Computer model outputs for temperatures aloft at select height levels (available levels vary depending on the model). Depicted using contours (isotherms).
Tools		

3.4.2.23 Products with Space Weather Information

Table 3-26. Weather Element: Space Weather

Type: text (T), graphic (G), image (I), voice (V)

Space Weather Information Contained In:	Type	Summary
Observations		
		(See Space Weather Advisory)
Analysis		

Space Weather Information Contained In:	Type	Summary
Advisories		
Space Weather Advisory	T	Issued whenever space weather conditions exceed predefined ICAO thresholds for both moderate and severe impacts. Provides an observed or expected location for the impact and 6-, 12-, 18-, and 24-hour forecasts.
Forecasts		
		(See Space Weather Advisory)
Tools		

3.4.2.24 Products with Density Altitude Information

Table 3-27. Weather Element: Density Altitude

Type: text (T), graphic (G), image (I), voice (V)

Density Altitude Information Contained In:	Type	Summary
Observations		
ASOS, AWOS and ATIS	V	Included in the ASOS and AWOS broadcasts (phone and radio) when density altitude exceeds the field elevation by more than 1,000 ft. A density altitude advisory (i.e., "check density altitude") is broadcast on ATIS when appropriate.
Analysis		
Advisories		
Forecasts		
Tools		

Part 2: Weather Theory and Aviation Hazards

Photo Credit: UCAR

4 The Earth's Atmosphere

4.1 Introduction

The Earth's atmosphere is a cloud of gas and suspended solids extending from the surface out many thousands of miles, becoming thinner with distance, but always held by the Earth's gravitational pull. The atmosphere is made up of layers surrounding the Earth that holds the air that people breathe, protects us from outer space, and holds moisture (e.g., vapor, clouds, and precipitation), gases, and tiny particles. In short, the atmosphere is the protective bubble that people live in.

This chapter covers the atmosphere's composition and vertical structure and the standard atmosphere.

4.2 Composition

The Earth's atmosphere consists of numerous gases (see Table 4-1) with nitrogen, oxygen, argon, and carbon dioxide making up 99.998 percent of all gases. Nitrogen, by far the most common, dilutes oxygen and prevents rapid burning at the Earth's surface. Living things need it to make proteins. Oxygen is used by all living things and is essential for respiration. Plants use carbon dioxide to make oxygen. Carbon dioxide also acts as a blanket and prevents the escape of heat to outer space.

Table 4-1. Composition of a Dry Earth's Atmosphere

Gas	Symbol	Content (by Volume)
Nitrogen	N_2	78.084%
Oxygen	O_2	20.947%
Argon	Ar	0.934%
Carbon Dioxide	CO_2	0.033%
Neon	Ne	18.20 parts per million
Helium	He	5.20 parts per million
Methane	CH_4	1.75 parts per million
Krypton	Kr	1.10 parts per million
Sulfur dioxide	SO_2	1.00 parts per million
Hydrogen	H_2	0.50 parts per million
Nitrous Oxide	N_2O	0.50 parts per million
Xenon	Xe	0.09 parts per million
Ozone	O_3	0.07 parts per million
Nitrogen Dioxide	NO_2	0.02 parts per million
Iodine	I_2	0.01 parts per million
Carbon Monoxide	CO	trace
Ammonia	NH_3	trace

Note: The atmosphere always contains some water vapor in amounts varying from trace to about 4 percent by volume. As water vapor content increases, the other gases decrease proportionately.

Weather (the state of the atmosphere at any given time and place) strongly influences daily routine as well as general life patterns. Virtually all of our activities are affected by weather, but, of all of our endeavors, perhaps none more so than aviation.

4.2.1 Air Parcel

An air parcel is an imaginary volume of air to which any or all of the basic properties of atmospheric air may be assigned. A parcel is large enough to contain a very large number of molecules, but small enough so that the properties assigned to it are approximately uniform. It is not given precise numerical definition, but a cubic centimeter of air might fit well into most contexts where air parcels are discussed. In meteorology, an air parcel is used as a tool to describe certain atmospheric processes, and air parcels will be referred to in this handbook.

4.3 Vertical Structure

The Earth's atmosphere is subdivided into five concentric layers (see Figure 4-1) based on the vertical profile of average air temperature changes, chemical composition, movement, and density. Each of the five layers is topped by a pause, where the maximum changes in thermal characteristics, chemical composition, movement, and density occur.

4.3.1 Troposphere

The troposphere begins at the Earth's surface and extends up to about 11 kilometers (km) (36,000 ft) high. This is where people live. As the gases in this layer decrease with height, the air becomes thinner. Therefore, the temperature in the troposphere also decreases with height. Climbing higher, the temperature drops from about 15 degrees Celsius (°C) (59 degrees Fahrenheit (°F)) to -56.5 °C (-70 °F). Almost all weather occurs in this region.

The vertical depth of the troposphere varies due to temperature variations, which are closely associated with latitude and season. It decreases from the Equator to the poles, and is higher during summer than in winter. At the Equator, it is around 18–20 km (11–12 miles (mi)) high, at 50° N and 50° S latitude, 9 km (5.6 mi), and at the poles, 6 km (3.7 mi) high.

The lowest portion of the troposphere is known as the planetary boundary layer. The height of the boundary layer varies depending on terrain and time of day, and is directly affected by surface heating and cooling. It has an important role in transporting heat and moisture into the atmosphere.

The transition boundary between the troposphere and the layer above is called the tropopause. Both the tropopause and the troposphere are known as the lower atmosphere.

4.3.2 Stratosphere

The stratosphere extends from the tropopause up to 50 km (31 mi) above the Earth's surface. This layer holds 19 percent of the atmosphere's gases, but very little water vapor.

Temperature increases with height as radiation is increasingly absorbed by oxygen molecules, leading to the formation of ozone. The temperature rises from an average -56.6 °C (-70 °F) at the tropopause to a maximum of about -3 °C (27 °F) at the stratopause due to this absorption of ultraviolet radiation. The increasing temperature also makes it a calm layer, with movements of the gases being slow.

Commercial aircraft often cruise in the lower stratosphere to avoid atmospheric turbulence and convection in the troposphere. Severe turbulence during the cruise phase of flight can be caused by the convective overshoot of thunderstorms from the troposphere below. The disadvantages of flying in the stratosphere can include increased fuel consumption due to warmer temperatures, increased levels of radiation, and increased concentration of ozone.

4.3.3 Mesosphere

The mesosphere extends from the stratopause to about 85 km (53 mi) above the Earth. The gases, including the number of oxygen molecules, continue to become thinner and thinner with height. As such, the effect of the warming by ultraviolet radiation also becomes less and less pronounced, leading to a decrease in temperature with height. On average, temperature decreases from about -3 °C (27 °F) to as low as -100 °C (-148 °F) at the mesopause. However, the gases in the mesosphere are thick enough to slow down meteorites hurtling into the atmosphere, where they burn up, leaving fiery trails in the night sky.

4.3.4 Thermosphere

The thermosphere extends from the mesopause to 690 km (430 mi) above the Earth. This layer is known as the upper atmosphere.

The gases of the thermosphere become increasingly thin compared to the mesosphere. As such, only the higher energy ultraviolet and x ray radiation from the Sun is absorbed. But because of this absorption, the temperature increases with height and can reach as high as 2,000 °C (3,600 °F) near the top of this layer.

Despite the high temperature, this layer of the atmosphere would still feel very cold to our skin because of the extremely thin air. The total amount of energy from the very few molecules in this layer is not sufficient enough to heat our skin.

4.3.5 Exosphere

The exosphere is the outermost layer of the atmosphere and extends from the thermopause to 10,000 km (6,200 mi) above the Earth. In this layer, atoms and molecules escape into space and satellites orbit the Earth. The transition boundary that separates the exosphere from the thermosphere is called the thermopause.

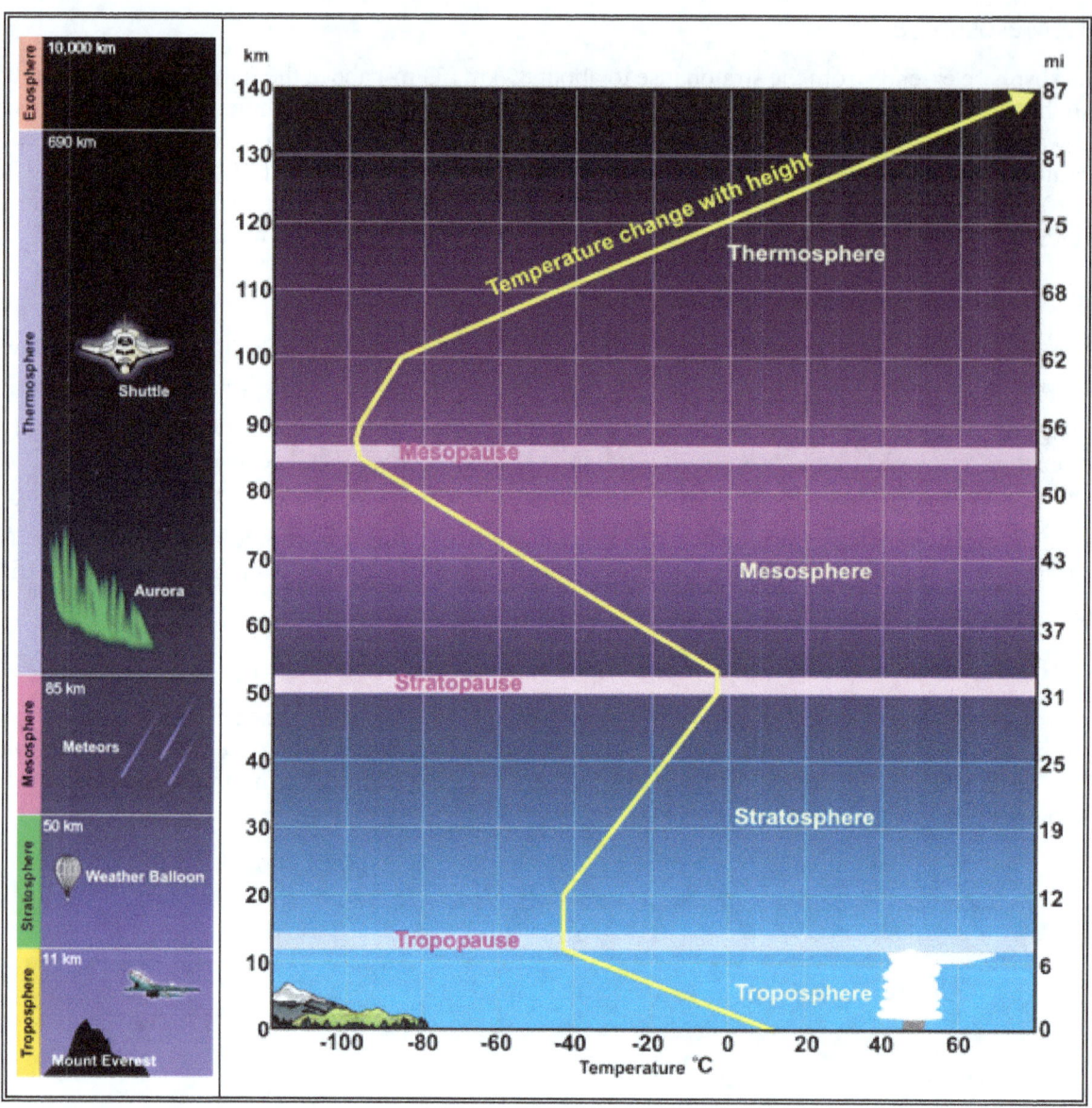

The regions of the stratosphere and the mesosphere, along with the stratopause and mesopause, are called the middle atmosphere. The transition boundary that separates the stratosphere from the mesosphere is called the stratopause.

Figure 4-1. Vertical Structure of the Atmosphere

4.4 The Standard Atmosphere

Continuous fluctuations of atmospheric properties create problems for engineers and meteorologists who need a fixed standard for reference. To solve this problem, they defined a standard atmosphere, which represents an average of conditions throughout the atmosphere for all latitudes, seasons, and altitudes.

Standard atmosphere is a hypothetical vertical distribution of atmospheric temperature, pressure, and density that, by international agreement, is taken to be representative of the atmosphere for purposes of pressure altimeter calibrations, aircraft performance calculations, aircraft and missile design, ballistic tables, etc. (see Table 4-2 and Figure 4-2). Weather-related processes are generally referenced to the standard atmosphere, as are the examples in this handbook.

Table 4-2. Selected Properties of the Standard Atmosphere

Property	Metric Units	English Units
Sea level pressure	1013.25 hectopascals (hPa)	29.92 inches of mercury (inHg)
Sea level temperature	15 °C	59 °F
Lapse rate of temperature in the troposphere	6.5 °C/1,000 m	3.57 °F/1,000 ft
Pressure altitude of the tropopause	11,000 m	36,089 ft
Temperature at the tropopause	-56.5 °C	-69.7 °F

Note: 1 hectopascal = 1 millibar.

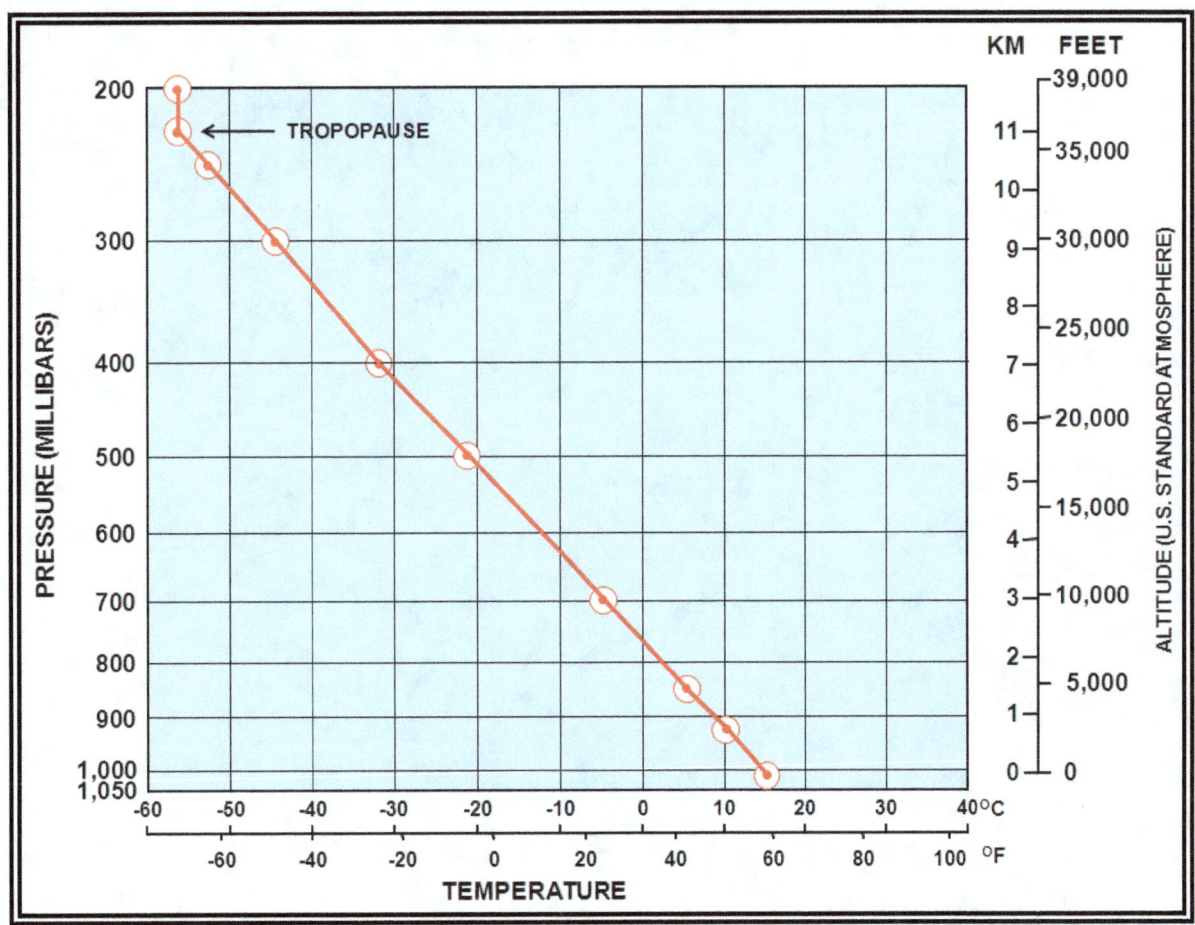

Figure 4-2. U.S. Standard Atmosphere Within the Troposphere

5 Heat and Temperature

5.1 Introduction

Temperature is one of the most basic variables used to describe the state of the atmosphere. Air temperature varies with time from one season to the next, between day and night, and even from one hour to the next. Air temperature also varies from one location to another, from high altitudes and latitudes to low altitudes and latitudes. Temperature can be critical to some flight operations. As a foundation for the study of temperature effects on aviation and weather, this chapter describes temperature, temperature measurement, and heat transfer and imbalances.

5.2 Matter

Matter is the substance of which all physical objects are composed. Matter is composed of atoms and molecules, both of which occupy space and have mass. The Earth's gravity acting on the mass of matter produces weight.

5.3 Energy

Energy is the ability to do work. It can exist in many forms and can be converted from one form to another. For example, if a ball is located at the edge of a slide, it contains some amount of potential energy (energy of position). This potential energy is converted to kinetic energy (energy of motion) when the ball rolls down the slide. Atoms and molecules produce kinetic energy because they are in constant motion. Higher speeds of motion indicate higher levels of kinetic energy.

5.4 Heat

Heat is the total kinetic energy of the atoms and molecules composing a substance. The atoms and molecules in a substance do not all move at the same velocity. Thus, there is actually a range of kinetic energy among the atoms and molecules.

5.5 Temperature

Temperature is a numerical value representing the average kinetic energy of the atoms and molecules within matter. Temperature depends directly on the energy of molecular motion. Higher (warmer) temperatures indicate a higher average kinetic energy of molecular motion due to faster molecular speeds. Lower (colder) temperatures indicate a lower average kinetic energy of molecular motion due to slower molecular speeds. Temperature is an indicator of the internal energy of air.

5.5.1 Temperature Measurement

A thermometer is an instrument used to measure temperature. Higher temperatures correspond to higher molecular energies, while lower temperatures correspond to lower molecular energies.

5.5.2 Temperature Scales

Many scientists use the Kelvin (K) scale, which is a thermodynamic (absolute) temperature scale, where absolute zero, the theoretical absence of all thermal energy, is 0 K. Thus, the Kelvin scale is a direct measure of the average kinetic molecular activity. Because nothing can be colder than absolute zero, the Kelvin scale contains no negative numbers.

The Celsius (°C) scale is the most commonly used temperature scale worldwide and in meteorology. The scale is approximately based on the freezing point (0 °C) and boiling point (100 °C) of water under a pressure of one standard atmosphere (approximately sea level). Each degree on the Celsius scale is exactly the same size as a degree on the Kelvin scale.

In the early 1990s, the United States aligned with ICAO standards by moving to the metric system for aviation weather reports. While some websites and applications provide temperature from the METAR in degrees Fahrenheit, these are done by the conversion software, as the temperature in the METAR is strictly reported in degrees Celsius. The United States uses the Fahrenheit (°F) scale for everyday temperature measurements for non-aviation purposes. In this scale, the freezing point of water is 32 °F and the boiling point is 212 °F.

See Table 5-1 and Table 5-2 for conversion information between temperature scales.

Table 5-1. Celsius Temperature Conversion Formulae

	From Celsius	To Celsius
Fahrenheit	[°F] = ([°C] × 9/5) + 32	[°C] = ([°F] − 32) × 5/9
Kelvin	[K] = [°C] + 273.15	[°C] = [K] − 273.15
For temperature *intervals* rather than specific temperatures: ±1 °C = ±1 K = ±1.8 °F		

Table 5-2. Fahrenheit Temperature Conversion Formulae

	From Fahrenheit	To Fahrenheit
Celsius	[°C] = ([°F] − 32) × 5/9	[°F] = ([°C] × 9/5) + 32
Kelvin	[K] = ([°F] + 459.67) × 5/9	[°F] = ([K] × 9/5) − 459.67
For temperature *intervals* rather than specific temperatures: ±1 °F = ±.56 °C = ±.56 K		

A thermometer changes readings due to the addition or subtraction of heat. Heat and temperature are not the same, but they are related.

Figure 5-1 gives a comparison of Kelvin, Celsius, and Fahrenheit temperature scales.

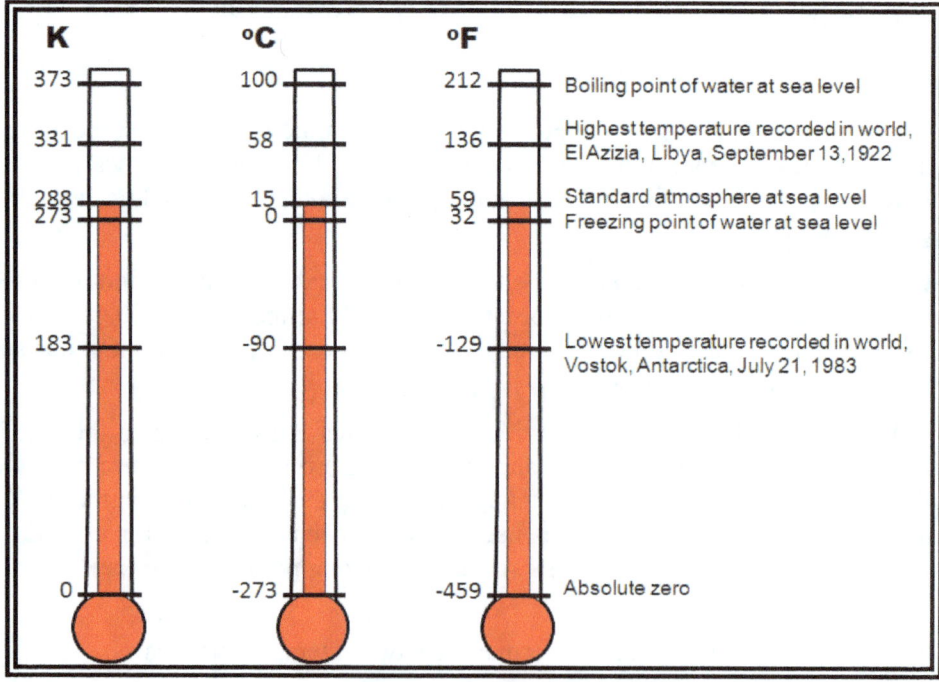

Figure 5-1. Comparison of Kelvin, Celsius, and Fahrenheit Temperature Scales

Chapter 5, Heat and Temperature

5.6 Heat Transfer

Heat transfer is energy transfer as a consequence of temperature difference. When a physical body (e.g., an object or fluid) is at a different temperature than its surroundings or another body, transfer of thermal energy, also known as heat transfer (or heat exchange), occurs in such a way that the body and the surroundings reach thermal equilibrium (balance). Heat transfer always occurs from a hot body to a cold body. Where there is a temperature difference between objects in proximity, heat transfer between them can never be stopped; it can only be slowed down.

The heat source for the surface of Earth is the Sun. Energy from the Sun is transferred through space and through the Earth's atmosphere to the Earth's surface. Since this energy warms the surface and atmosphere, some of it becomes heat energy. There are three ways heat is transferred into and through the atmosphere: radiation, conduction, convection, or any combination of these. Heat transfer associated with the heat change of water from one phase to another (i.e., liquid water absorbs heat when changed to a vapor and liquid water releases heat when it changes to ice) can be fundamentally treated as a variation of convective heat transfer. The heat transfer associated with water will be discussed in Chapter 6, Water Vapor.

5.6.1 Radiation

If a person has ever stood in front of a fireplace or near a campfire, then they have felt the heat transfer known as radiation (see Figure 5-2). The side of the body nearest the fire warms, while the other side remains unaffected by the heat. Although people are surrounded by air, the air has nothing to do with this type of heat transfer. Heat lamps that keep food warm work in the same way.

Radiation is the transfer of heat energy through space by electromagnetic radiation. These electromagnetic waves travel at the speed of light and are usually described in terms of wavelength or frequency. Frequencies range from gamma rays on the high end to radio waves on the low end. Also contained in the spectrum are x ray, ultraviolet, visible, infrared, and microwave.

Figure 5-2. Radiation Example

All objects emit (radiate) energy as the heat energy within the object is converted to radiation energy. This transmitted radiation passes through entities such as air, water, or space. Along the way, the radiation can be reflected, which occurs when the wave energy changes direction when encountering an object. Eventually, the radiation is absorbed and the electromagnetic wave energy is converted to heat energy by the absorbing object. The emitting object loses heat energy, and the absorbing object gains heat energy during this process.

5.6.1.1 Solar and Terrestrial Radiation

All objects emit radiation energy, including the Sun (solar radiation) and the Earth (terrestrial radiation). An object's wavelength of maximum radiation is inversely related to its temperature; the hotter (colder) the object, the shorter (longer) the wavelength. The Sun's wavelength of maximum radiation is relatively short and is centered in the visible spectrum. The Earth's wavelength of maximum radiation is relatively long and is centered in the infrared spectrum.

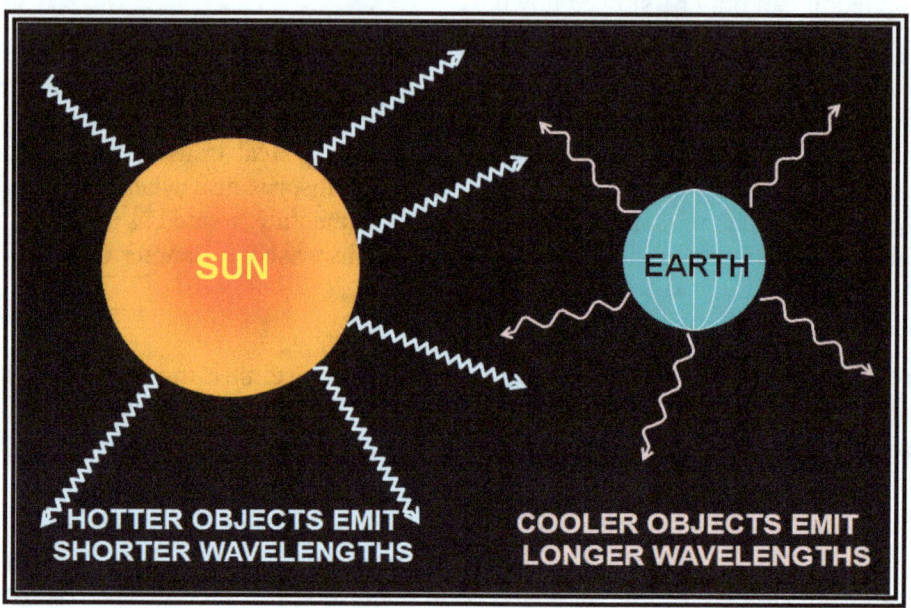

Figure 5-3. Temperature's Effect on Radiation Wavelength

Some of the solar radiation that reaches the Earth's surface is radiated back into the atmosphere to become heat energy. Dark-colored objects such as asphalt absorb more of the radiant energy and warm faster than light-colored objects. Dark objects also radiate their energy faster than light-colored objects.

5.6.1.2 Solar Zenith Angle

The intensity of incoming solar radiation that strikes the Earth's surface (insolation) varies with solar zenith angle. Solar zenith angle is the angle measured from the Earth's surface between the Sun and the zenith (i.e., directly overhead). Solar zenith angle varies with latitude, season, and the diurnal cycle (sunrise/sunset).

Figure 5-4 illustrates this concept. Insolation is maximized when the solar zenith angle is zero degrees (0°), which means the Sun is directly overhead. With increasing solar zenith angle, the insolation is spread over an increasingly larger surface area (y is greater than x) so that the insolation becomes less intense. Also, with increasing solar zenith angle, the Sun's rays must pass through more of the Earth's atmosphere, where they can be scattered and absorbed before reaching the Earth's surface. Thus, the Sun can heat the surface to a much higher temperature when it is high in the sky, rather than low on the horizon.

Figure 5-4. Solar Zenith Angle

5.6.2 Conduction

Conduction is the transfer of energy (including heat) by molecular activity from one substance to another in contact with or through a substance. Heat always flows from the warmer substance to the colder substance. The rate of heat transfer is greater with larger temperature differences and depends directly on the ability of the substance(s) to conduct heat. During conduction, the warmer substance cools and loses heat energy, while the cooler substance warms and gains heat energy.

Heat (thermal) conductivity is the property of a substance that indicates its ability to conduct heat as a consequence of molecular motion. Units are watts per meter-kelvin (W m^{-1} K^{-1}). Table 5-3 provides the heat (thermal) conductivity of various substances. Note that air is a poor thermal conductor.

Table 5-3. Heat (Thermal) Conductivity of Various Substances

Material	Phase	Heat (Thermal) Conductivity (W m^{-1} K^{-1})
Silver	Solid	429
Copper	Solid	401
Aluminum	Solid	250
Iron	Solid	80
Sand (saturated)	Solid	2.7
Water (ice)	Solid (0 °C)	2.18
Sandstone	Solid	1.7
Limestone	Solid	1.26–1.33
Glass	Solid	1.05
Water (liquid)	Liquid	0.58
Sand (dry)	Solid	0.35
Soil	Solid	0.17–1.13
Wood (oak)	Solid	0.17
Wood (balsa)	Solid	0.055
Snow	Solid (<0 °C)	0.05–0.25
Air	Gas	0.024
Water (steam)	Gas (125 °C)	0.016

All measurements are at 25 °C unless otherwise noted.
Note: 1 K equals -272.15 °C.

5.6.3 Convection

Convection is the transport of heat within a fluid, such as air or water, via motions of the fluid itself. This type of heat flow takes place in liquids and gases because they can move freely and it is possible to set up currents within them. Water boiling in a pot is an example of convection. Because air is a poor thermal conductor, convection plays a vital role in the Earth's atmospheric heat transfer process. Figure 5-5 illustrates examples of various heat transfer processes.

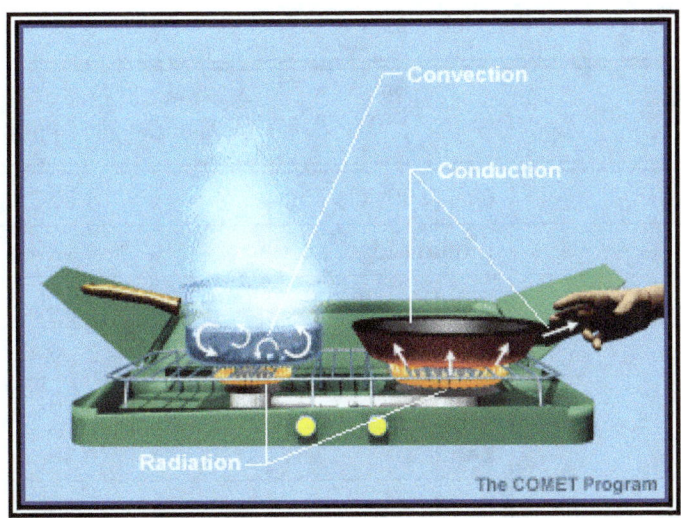

Figure 5-5. Heat Transfer Examples[1]

5.7 Thermal Response

Whether by radiation, conduction, convection, or a combination of these, the temperature response to the input (or output) of some specified quantity of heat varies from one substance to another. Specific heat capacity, also known simply as specific heat, is defined as the measure of heat energy required to increase the temperature of a unit quantity of a substance by a certain temperature interval. Specific heat capacity is typically expressed in units of joules per gram-kelvin ($J\ g^{-1}\ K^{-1}$). Thus, two different substances with identical temperature measurements do not necessarily possess the same amount of heat energy. When exposed to the same amount of heat energy, a substance with a low specific heat capacity warms up more than a substance with a higher specific heat capacity. Table 5-4 lists the specific heat capacity of various substances.

---

[1] The source of this and other material labeled COMET® is the COMET® website at https://meted.ucar.edu/ of the University Corporation for Atmospheric Research (UCAR), sponsored in part through cooperative agreement(s) with NOAA, U.S. DOC. ©1997-2017 University Corporation for Atmospheric Research. All Rights Reserved.

Table 5-4. Specific Heat Capacity of Various Substances

Substance	Phase	Specific Heat Capacity (J g^{-1} K^{-1})
Water (steam)	Gas (100 °C)	4.22
Water	Liquid (25 °C)	4.18
Wood (balsa)	Solid	2.90
Water (ice)	Solid (0 °C)	2.05
Wood (oak)	Solid	2.00
Soil (wet)	Solid	1.48
Sandy clay	Solid	1.38
Air (sea level, dry)	Gas	1.01
Asphalt	Solid	0.92
Clay	Solid	0.92
Aluminum	Solid	0.91
Brick (common)	Solid	0.90
Concrete	Solid	0.88
Glass	Solid	0.84
Limestone	Solid	0.84
Sand (quartz)	Solid	0.83
Soil (dry)	Solid	0.80
Granite	Solid	0.79
Iron	Solid	0.46
Copper	Solid	0.39
Mercury	Liquid	0.14
Lead	Solid	0.13

All measurements are at 25 °C unless otherwise noted.
Note: 1 K equals -272.15 °C.

Water has the highest specific heat capacity of any naturally occurring substance. That means it has a much higher capacity for storing heat energy than other substances, such as soil, sand, rock, or air. Water can store large amounts of heat energy while only experiencing a small temperature change.

Figure 5-6 compares the specific heat capacity of water and sand. The specific heat capacity of water is more than five times that of quartz sand. Thus, 4.18 J of heat are required to raise the temperature of 1 gram (g) of water by 1 °C, while only 0.83 J are required to raise the temperature of 1 g of quartz sand by 1 °C. This is one reason why beach sand is hotter than water on a sunny, summer afternoon.

Figure 5-6. Specific Heat Capacity: Water versus Sand

The difference in specific heat capacities is one of the primary reasons why the temperature of a body of water, such as a lake or the ocean, is less variable with time than the surface temperature of land. Water heats up more slowly than land during the day and during the summer, and cools down more slowly at night and during the winter. Thus, a body of water exhibits greater resistance to temperature change (called thermal inertia) than does a land mass.

Heat flow differences are another reason why water bodies warm up and cool down more slowly than land. Incoming solar radiation penetrates water to significant depths, but can only heat the top skin layer of soil and rock. Also, since water is a fluid, its heat energy can be circulated through great volumes and depths via convection. Water temperature changes occur to depths of 6 meters (m) (20 ft) or more on a daily basis, and 200 to 600 m (650 to 1950 ft) annually. The process is more problematic over land since heat must be transferred via the slow process of conduction. Land temperature changes occur to depths of only 10 centimeters (cm) (4 inches (in)) on a daily basis and 15 m (50 ft) or less annually.

Water is much more resistant to temperature changes than land. It warms up and cools down more slowly than land and helps to moderate nearby air temperature. This is why islands and localities located immediately downwind from the ocean or a large lake (maritime locations) exhibit smaller diurnal and seasonal temperature variations than localities well inland (continental locations). Figure 5-7 illustrates this effect. Although both cities are at approximately the same latitude, the temperature is much less variable in San Francisco (maritime) than St. Louis (continental).

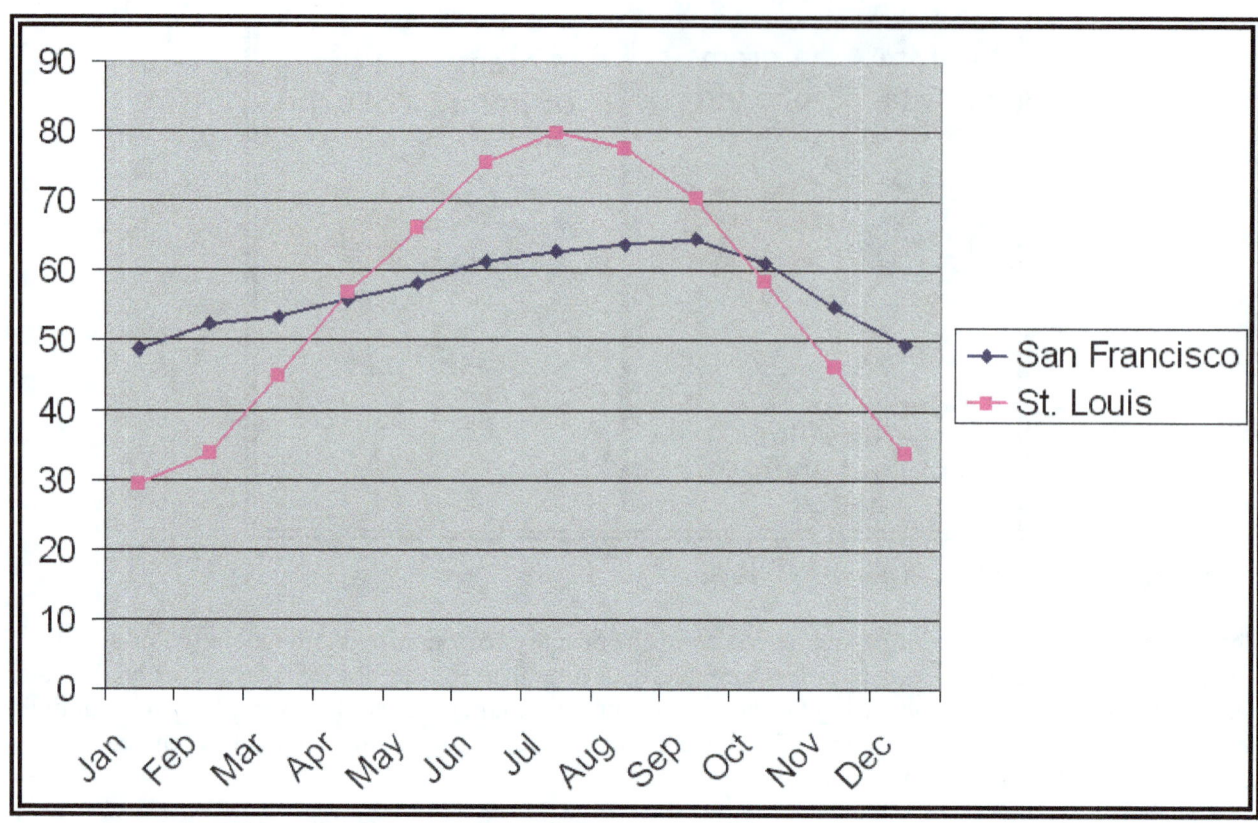

Figure 5-7. Variation of Mean Daily Temperatures for San Francisco (Maritime) and St. Louis (Continental)

5.8 Temperature Variations with Altitude

A lapse rate of temperature is defined as a decrease in temperature with height. In Figure 4-2, it was stated that the temperature decreases 6.5 °C/km (3.57 °F/1,000 ft) in the standard atmosphere. But since this is an average, the exact value seldom exists. In fact, temperature in the troposphere sometimes remains constant or even increases with height. Caution should be taken when using the standard lapse rate to estimate the freezing level. Quite often the boundary layer is dry adiabatic and the estimate of freezing level could be in error.

5.8.1 Atmospheric Sounding

An atmospheric sounding, or simply sounding, is a plot of the vertical profile of one or more atmospheric parameters, such as temperature, dewpoint, or wind above a fixed location. Soundings are used extensively by meteorologists to determine the state of the atmosphere.

5.8.2 Isothermal Layer

An isothermal layer is a layer within the atmosphere where the temperature remains constant with height (see Figure 5-8).

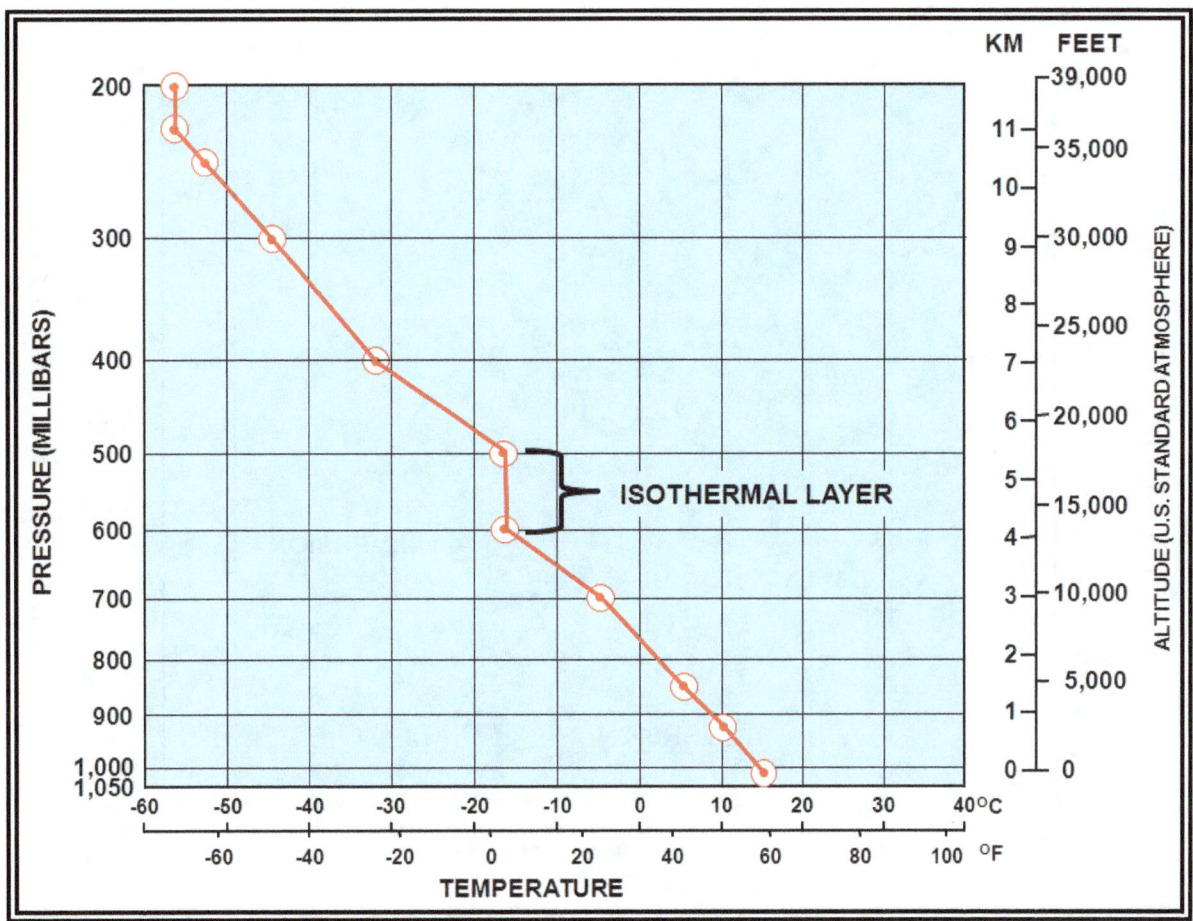

Figure 5-8. Sounding with an Isothermal Layer

5.8.3 Temperature Inversion

A temperature inversion, or simply inversion, is a layer in which the temperature increases with altitude. If the base of the inversion is at the surface, it is termed a surface-based inversion. If the base of the inversion is not at the surface, it is termed an inversion aloft (see Figure 5-9).

A surface-based inversion typically develops over land on clear nights when wind is light. The ground radiates and cools much faster than the overlying air. Air in contact with the ground becomes cool, while the temperature a few hundred feet above changes very little. Thus, temperature increases with height.

An inversion may also occur at any altitude when conditions are favorable. For example, a current of warm air aloft overrunning cold air near the surface produces an inversion aloft. Inversions are common in the stratosphere.

The principal characteristic of an inversion layer is its marked stability, so that very little turbulence can occur within it. Turbulence will be discussed at length in Chapter 19, Turbulence.

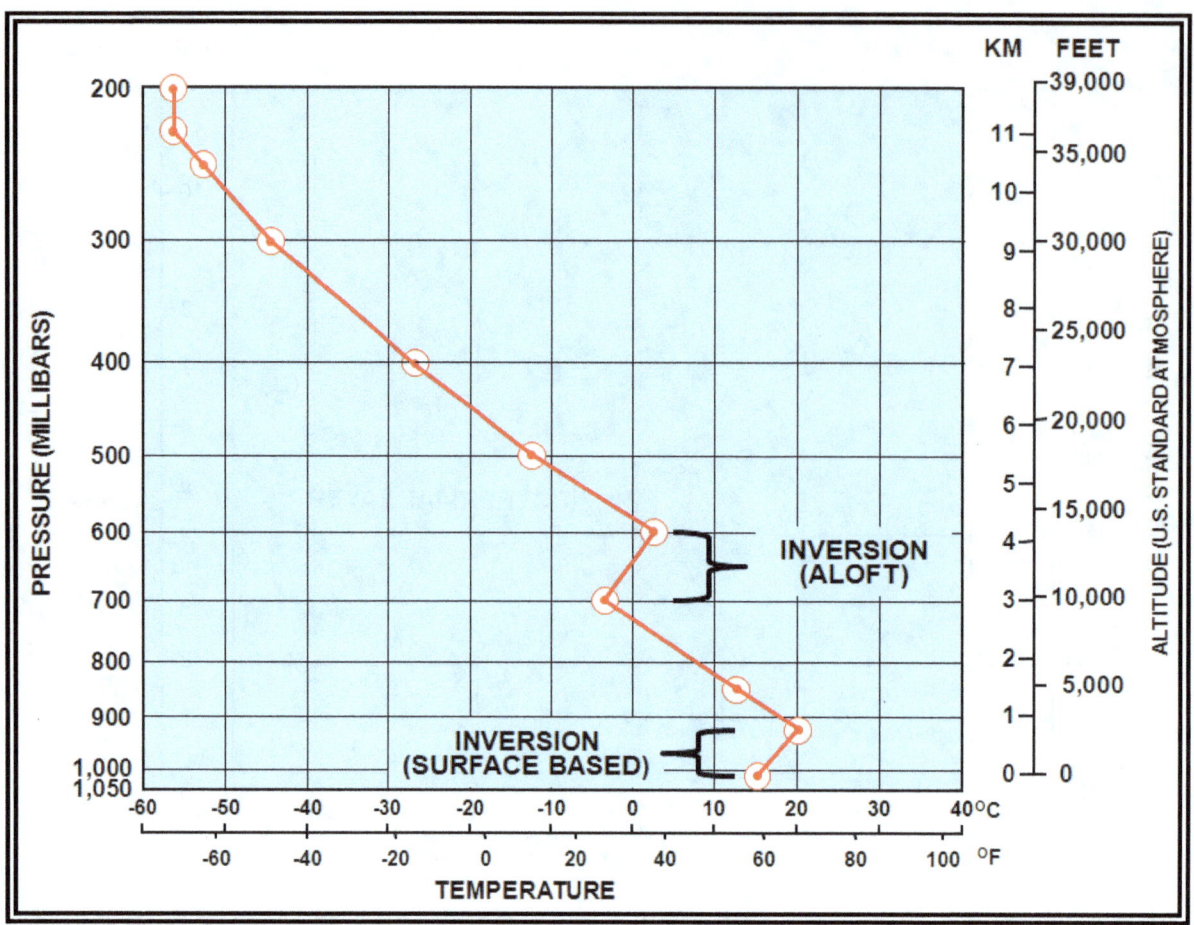

Figure 5-9. Sounding with a Temperature Inversion

6 Water Vapor

6.1 Introduction

Water vapor is the gaseous form of water and one of the most important of all constituents of the atmosphere. It constitutes only a small percentage of the Earth's atmosphere, varying from only trace amounts to 4 percent by volume, and its amount varies widely in space and time. Approximately half of all of the atmospheric water vapor is found below 2 km (6,500 ft) altitude, and only a minute fraction of the total occurs above the tropopause.

The development of clouds and precipitation can have potential impacts on flight operations. Water vapor is important, not only as the raw material for clouds and precipitation (e.g., rain and snow), but also as a vehicle for the transfer of heat energy and as a regulator of the Earth's temperatures through absorption and emission of radiation, most significantly in the thermal infrared (i.e., the greenhouse effect). The amount of water vapor present in a given air sample may be measured in a number of different ways, involving such concepts as relative humidity and dewpoint. Before talking about these subjects, how water cycles through the Earth-atmosphere system will be discussed.

6.2 The Hydrologic Cycle

The hydrologic cycle (see Figure 6-1) involves the continuous circulation of water in the Earth-atmosphere system. Water vapor plays a critical role in the cycle.

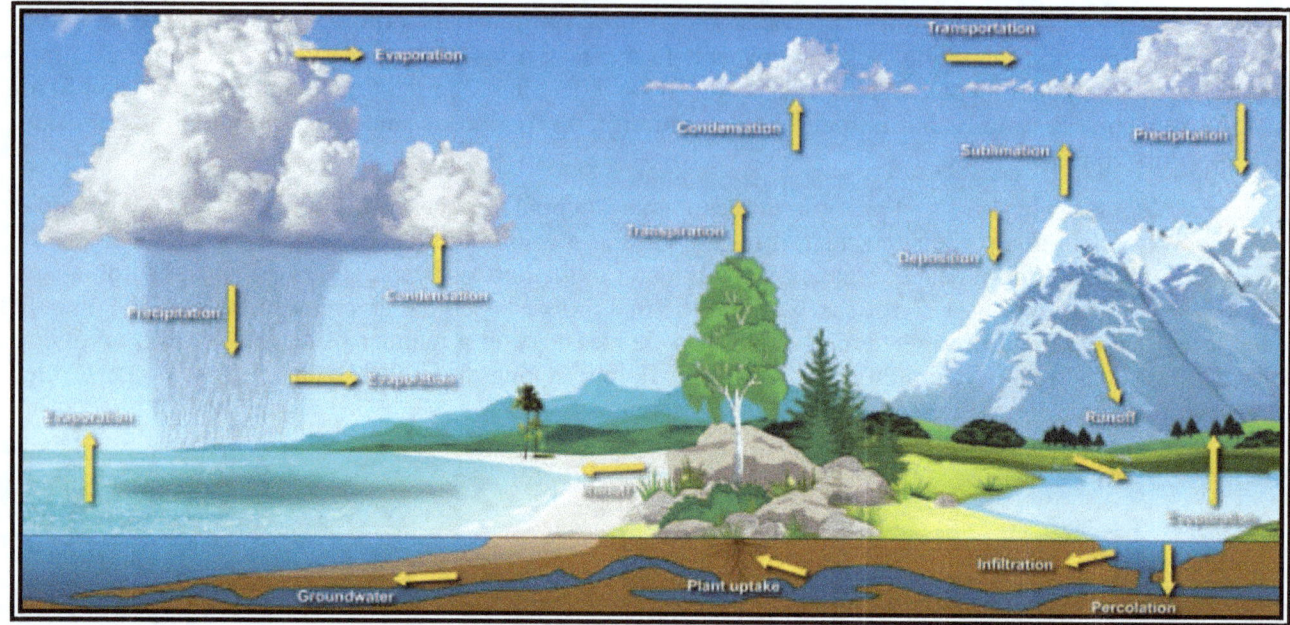

Figure 6-1. The Hydrologic Cycle

6.2.1 Evaporation

Evaporation is the phase transition by which a liquid is changed to a vapor (gas). In meteorology, the substance meteorologists are concerned about the most is water, and the primary source is the ocean. On average, about 120 cm (47 in) is evaporated into the atmosphere from the ocean each year. For evaporation to take place, energy is required. The energy can come from any source: the Sun, the atmosphere, the Earth, or objects on the Earth, such as humans.

Everyone has experienced evaporation personally. When the body heats up due to the air temperature or through exercise, the body sweats, secreting water onto the skin. The purpose is to cause the body to use its heat to evaporate the liquid, thereby removing heat and cooling the body. The same effect can be seen when stepping out of a shower or swimming pool. The coolness felt is from the removal of bodily heat used to evaporate the water on the skin.

6.2.2 Transpiration

Transpiration is the evaporation of water from plants. In most plants, transpiration is a passive process largely controlled by the humidity of the atmosphere and the moisture content of the soil. Of the transpired water passing through a plant, only 1 percent is used in the growth process of the plant. The remaining 99 percent is passed into the atmosphere.

6.2.3 Sublimation

Sublimation is the phase transition by which a solid is changed into vapor (a gas) without passing through the liquid phase. In the atmosphere, sublimation of water occurs when ice and snow (solids) change into water vapor (a gas).

6.2.4 Deposition

Deposition is the phase transition by which vapor (a gas) is changed into a solid without passing through the liquid phase. In the atmosphere, deposition of water occurs when vapor (a gas) in a sub-freezing cloud changes into ice crystals (solid).

6.2.5 Condensation

Condensation is the phase transition by which vapor (a gas) is changed into a liquid. In the atmosphere, condensation may appear as clouds, fog, mist, dew, or frost, depending upon the physical conditions of the atmosphere.

6.2.6 Transportation

Transportation is the movement of solid, liquid, and gaseous water through the atmosphere. Without this movement, the water evaporated over the ocean would not precipitate over land.

6.2.7 Precipitation

Precipitation results when tiny condensation particles grow in the atmosphere through collision and coalescence and then fall to the Earth's surface.

6.2.8 Runoff

Runoff occurs when there is excessive precipitation and the ground is saturated (i.e., cannot absorb any more water). This runoff flows into streams and rivers and eventually back into the sea.

Evaporation of this runoff into the atmosphere begins the hydrologic cycle over again. Some of the water percolates into the soil and into the ground water, only to be drawn into plants again for transpiration to take place.

6.2.9 Infiltration

Infiltration is the movement of water into the ground from the surface.

6.2.10 Groundwater Flow

Groundwater flow is the flow of water underground in aquifers. The water may return to the surface in springs or eventually seep into the oceans.

6.2.11 Plant Uptake

Plant uptake is water taken from the groundwater flow and soil moisture.

6.3 Saturation

Saturation is the maximum possible quantity of water vapor that an air parcel can hold at any given temperature and pressure. The term "saturated air" means an air parcel has all of the water vapor it can hold, while "unsaturated air" means an air parcel can hold more water vapor.

6.4 Relative Humidity

Relative humidity is the ratio, usually expressed as a percentage, of water vapor actually in the air parcel compared to the amount of water vapor the air parcel could hold at a particular temperature and pressure.

$$Relative\ Humidity = \frac{Water\ vapor\ content}{Water\ vapor\ capacity}$$

While relative humidity is the most common method of describing atmospheric moisture, it is also the most misunderstood. Relative humidity can be confusing because it does not indicate the actual water vapor content of the air, but rather how close the air is to saturation. An air parcel with 100 percent relative humidity is saturated, while an air parcel with relative humidity less than 100 percent is unsaturated.

An air parcel's capacity to hold water vapor (at a constant pressure) is directly related to its temperature. It is possible to change an air parcel's relative humidity without changing its water vapor content. Figure 6-2 illustrates this concept. An air parcel at sea level at a temperature of 30 °C has the capacity to hold 27 g of water vapor. If it actually held 8 g, its relative humidity would be 30 percent, and it would be unsaturated. However, if the air parcel's temperature decreases to 20 °C, its water vapor storage capacity decreases to 15 g and its relative humidity rises to 53 percent. At 10 °C, the air parcel's water vapor storage capacity decreases to equal the amount of water vapor it actually holds (8 g), its relative humidity increases to 100 percent, and it becomes saturated. During this cooling process, the air parcel's actual water vapor content remained constant, but relative humidity increased with decreasing temperature.

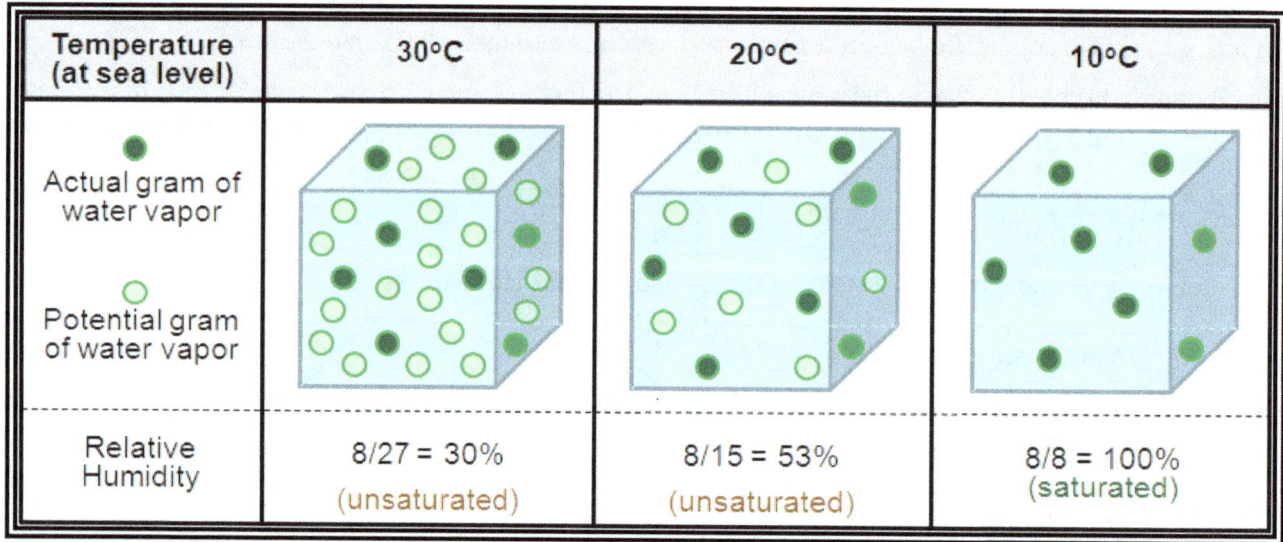

Figure 6-2. Temperature Effects on Relative Humidity

6.5 Dewpoint

Dewpoint is the temperature an air parcel must be cooled at constant pressure and constant water vapor pressure to allow the water vapor in the parcel to condense into water (dew). When this temperature is below 0 °C (32 °F), it is sometimes called the frost point. Lowering an air parcel's temperature reduces its capacity to hold water vapor.

6.6 Temperature-Dewpoint Spread (Dewpoint Depression)

The difference between an air parcel's temperature and its dewpoint is the dewpoint depression, or commonly referred to as the spread. Surface aviation weather reports (e.g., METARs/SPECIs) provide observations of both temperature and dewpoint. The temperature greatly affects the air parcel's ability to hold water vapor, while the dewpoint indicates the actual quantity of water vapor in the parcel. As the spread decreases, relative humidity increases. When the spread decreases to zero, relative humidity is 100 percent, and the air parcel is saturated. Figure 6-3 illustrates the relationship between temperature-dewpoint spread and relative humidity.

Surface temperature-dewpoint spread is important in anticipating fog, but has little bearing on precipitation. To support precipitation, air must be saturated through thick layers aloft.

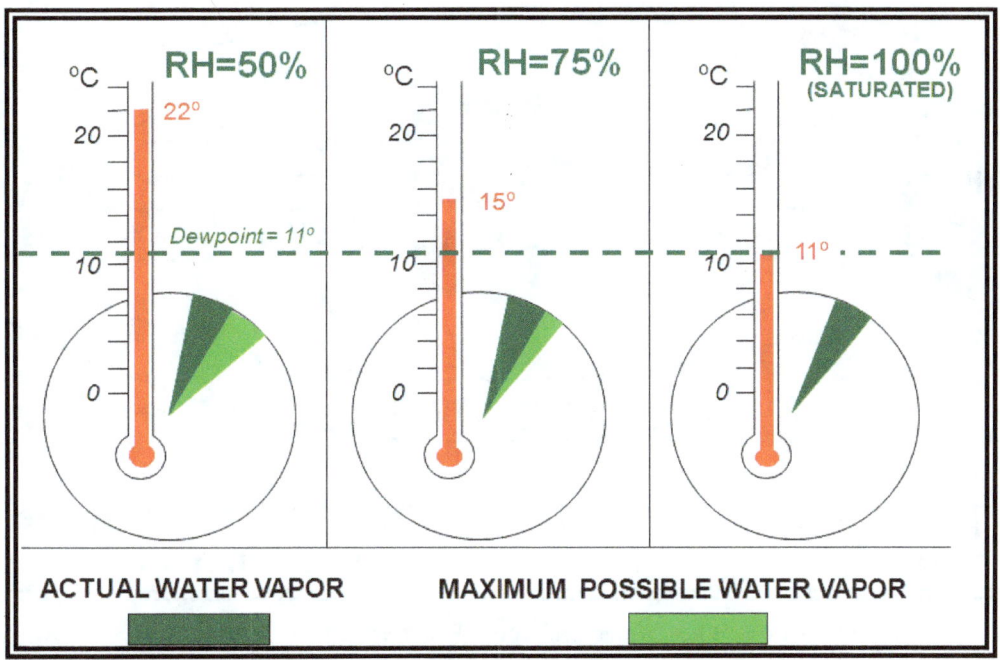

Figure 6-3. Temperature-Dewpoint Spread Effect on Relative Humidity

Relative humidity depends on the temperature-dewpoint spread. In Figure 6-3, dewpoint is constant but temperature decreases from left to right. On the left panel, relative humidity is 50 percent, which indicates the air parcel could hold twice as much water vapor as is actually present. As the air parcel cools, the temperature-dewpoint spread decreases while relative humidity increases. When the air parcel's temperature cools to equal its dewpoint (11 °C), its capacity to hold water vapor is reduced to the amount actually present. The temperature-dewpoint spread is zero, relative humidity is 100 percent, and the air parcel is now saturated.

6.7 Change of Phase

Water changes from one state of matter (e.g., solid, liquid, or vapor) to another at the temperatures and pressures experienced near the surface of the Earth. Interestingly, water is the only substance on Earth that exists naturally in all three phases: as water droplets, ice crystals (visible as clouds), and water vapor.

Water has some unique thermal properties, which make it a powerful heat transport mechanism. It has the highest specific heat capacity of any naturally occurring substance (see Table 5-4). That means water has a much higher capacity for storing heat energy (with little resulting temperature change) than other substances. These properties make water an ideal heat transport mechanism and have important implications on weather and climate.

6.7.1 Latent Heat

Latent heat is the quantity of heat energy either released or absorbed by a unit mass of a substance when it undergoes a phase transition (change of state). Units are typically expressed in terms of joules per gram (J/g). Figure 6-4 illustrates the latent heat transactions that occur when water undergoes phase transition.

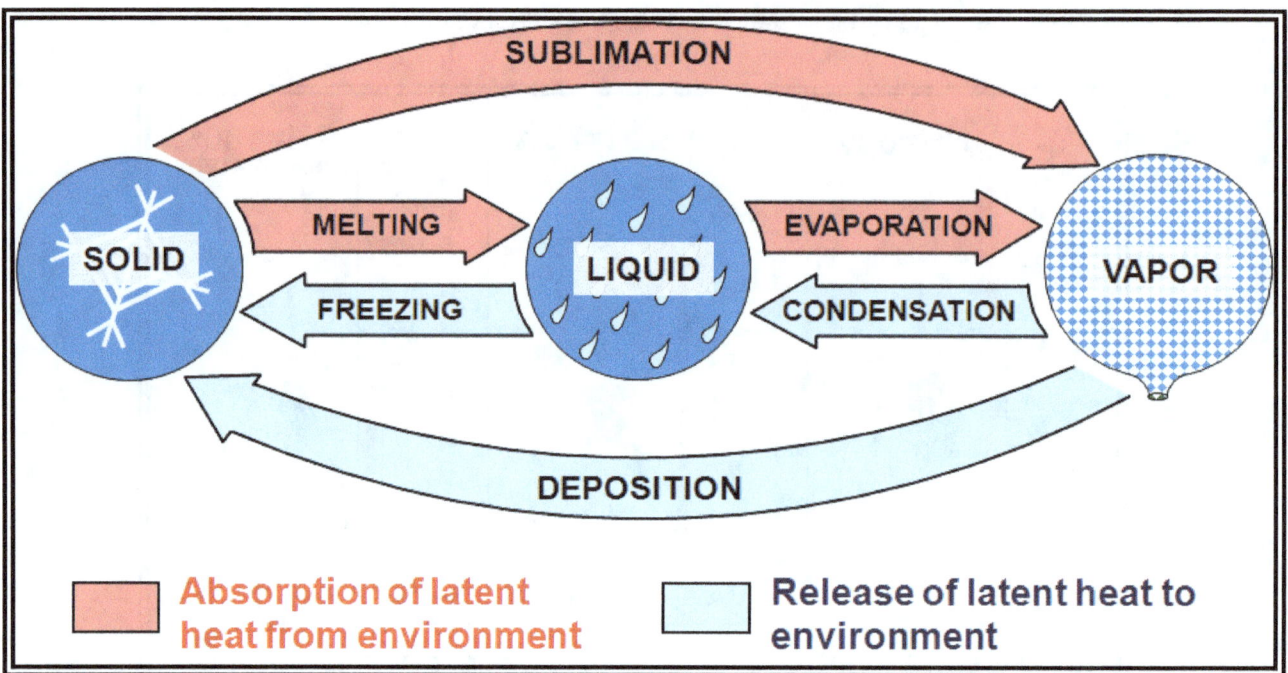

Figure 6-4. Latent Heat Transactions When Water Undergoes Phase Transition

Heat is exchanged between water and its environment during phase transition. Although the temperature of the environment changes in response, the temperature of the water undergoing the phase transition remains constant until the phase change is complete; that is, the available heat, latent heat, is involved exclusively in changing the phase of water and not in changing its temperature. There are six phase transitions, three of which are associated with the absorption of latent heat by water from the environment (melting, evaporation, and sublimation), and three of which are associated with the release of heat energy by water to the environment (freezing, condensation, and deposition).

Melting is the phase transition by which a solid is changed to a liquid. During melting, water absorbs 334 J/g due to the latent heat of fusion. Freezing, the reverse process, releases 334 J/g back to the environment.

Evaporation is the phase transition by which a liquid is changed to a vapor. During evaporation, water absorbs 2,501 J/g due to the latent heat of vaporization. Condensation, the reverse process, releases 2,501 J/g back to the environment.

Sublimation is the phase transition by which a solid is changed to a vapor. During sublimation, water absorbs 2,834 J/g due to the latent heat of sublimation. Deposition, the reverse process, releases 2,834 J/g back to the environment. Table 6-1 lists the latent heat exchanges of water.

Table 6-1. Latent Heat of Water at 0 °C

Latent Heat Type	Energy Exchange (J/g)
Latent heat of sublimation	2,834
Latent heat of vaporization	2,501
Latent heat of fusion	334

The amount of energy associated with latent heat exchange should not be understated. An average hurricane releases 52 million trillion (5.2×10^{19}) joules per day as water vapor condenses into clouds and precipitation. This is equivalent to about 40 times the total worldwide energy consumption per day in 2005!

7 Earth-Atmosphere Heat Imbalances

7.1 Introduction

Weather is not a capricious act of nature, but rather the atmosphere's response to unequal rates of radiational heating and cooling across the surface of the Earth and within its atmosphere. The absorption of incoming solar radiation causes heating, while the emission of outgoing terrestrial radiation causes cooling. However, imbalances in the rate of heating and cooling create temperature gradients.[2] Atmospheric circulations and weather are the atmosphere's never-ending attempt to redistribute this heat and achieve equilibrium. This chapter provides a background on the interaction between the Earth and the atmosphere for a better understanding of the weather that will affect flight operations.

---
[2] A gradient describes the rate of change of a feature (in this case, temperature) per unit of distance.

7.2 The Earth-Atmosphere Energy Balance

The Earth-atmosphere energy balance is the balance between incoming energy from the Sun (solar radiation) and outgoing energy from the Earth (terrestrial radiation), as seen in Figure 7-1. When solar radiation reaches the Earth, some is reflected back to space by air (8 percent), clouds (17 percent), or the surface (6 percent). Some is absorbed by water vapor/dust/ozone (19 percent) or by clouds (4 percent). The remainder is absorbed by the Earth's surface (46 percent).

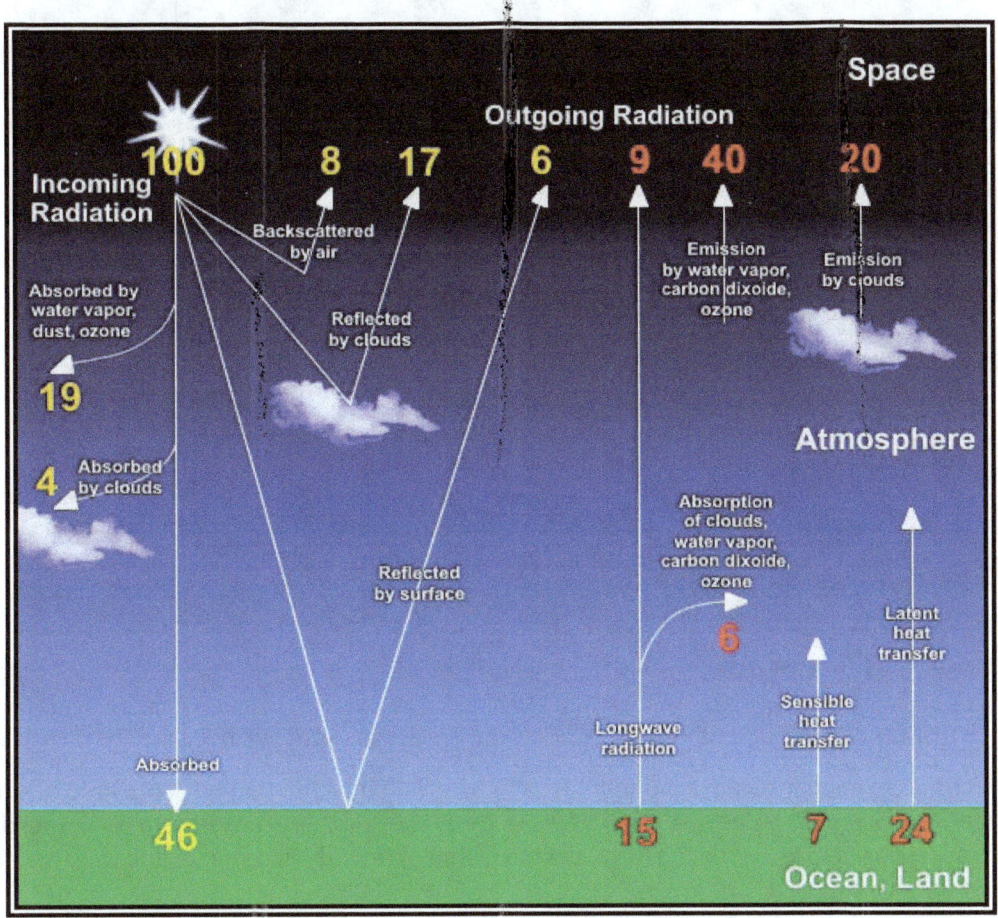

Figure 7-1. Earth-Atmosphere Energy Balance

In Figure 7-1, 100 units of incoming radiation from the Sun is balanced by 100 units of outgoing radiation from the Earth.

However, since the Earth is much cooler than the Sun, its radiating energy is much weaker (long wavelength) infrared energy. Indirectly, this energy can be seen radiating into the atmosphere as heat (e.g., rising from a hot road, creating shimmers on hot sunny days). The Earth-atmosphere energy balance is achieved as the energy received from the Sun (solar radiation) balances the energy lost by the Earth back into space (terrestrial radiation). In this way, the Earth maintains a stable average temperature.

The absorption of infrared radiation trying to escape from the Earth back to space is particularly important to the global energy balance. Energy absorption by the atmosphere stores more energy near its surface than it would if there was no atmosphere. The average surface temperature of the Moon, which has no atmosphere, is -18 °C (0 °F). By contrast, the average surface temperature of the Earth is 15 °C (59 °F). This heating effect is called the greenhouse effect.

Greenhouse warming is enhanced during nights when the sky is overcast (see Figure 7-2). Heat energy from the Earth can be trapped by clouds, leading to higher temperatures as compared to nights with clear skies. The air is not allowed to cool as much with overcast skies. Under partly cloudy skies, some heat is allowed to escape, and some remains trapped. Clear skies allow for the most cooling to take place.

Figure 7-2. Greenhouse Effect on Nighttime Radiational Cooling

7.3 Heat Imbalances Between Earth's Surface and the Atmosphere

The Earth-atmosphere energy balance numbers (see Figure 7-1) indicate that both sensible heat (7 percent) and latent heat (24 percent) processes transfer heat from the Earth's surface into its atmosphere. Both processes are necessary to prevent the Earth's surface from continually heating up and the atmosphere from continually cooling down.

7.3.1 Sensible Heating

Sensible heating involves both conduction and convection. It occurs due to differences in air density. Warm air is less dense than cool air.

On warm sunny days, the Earth's surface is heated by incoming solar radiation or insolation. However, the heating is somewhat uneven because certain areas of the Earth's surface absorb more heat from the Sun than others. Heat is conducted from the relatively warm ground to the cooler overlying air, which warms a shallow layer of air near the ground. The heated air expands, becomes less dense than the surrounding cooler air, and rises. Through this process, a large bubble of warm air called a thermal rises and transfers heat energy upwards (see Figure 7-3). Cooler, denser air sinks toward the ground to replace the rising air. This cooler air becomes heated in turn, rises, and repeats the cycle.

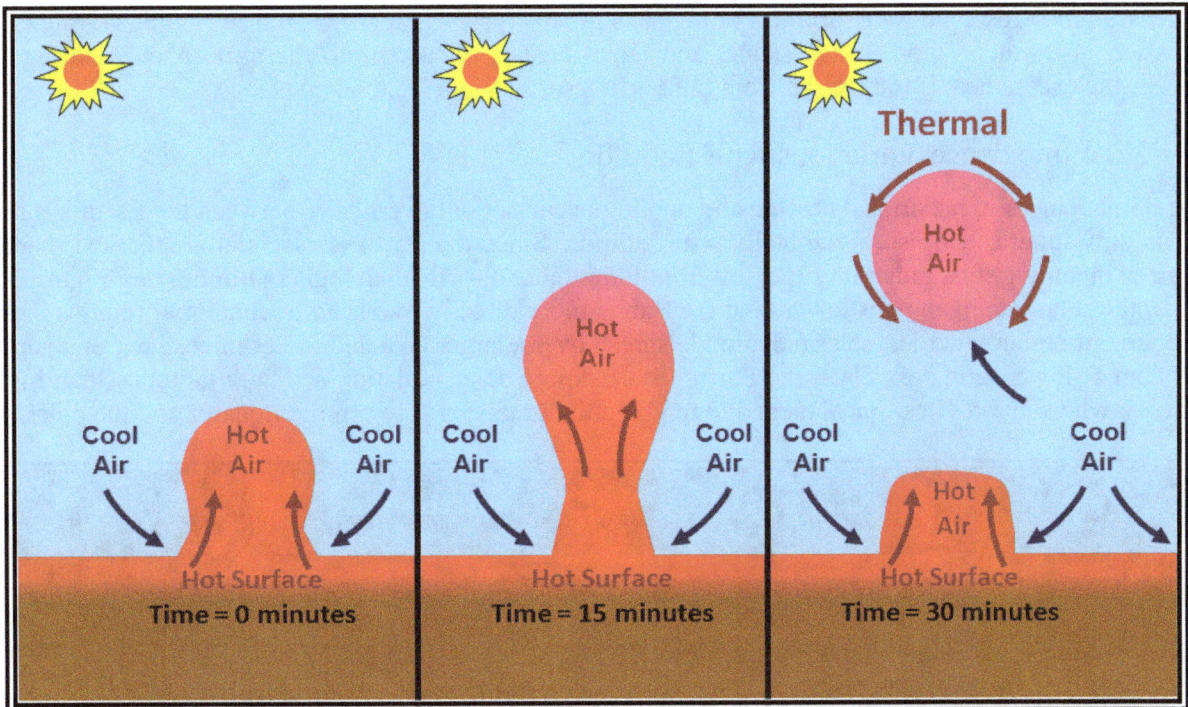

Figure 7-3. Development of a Thermal

In this manner, convection transports heat from the Earth's surface into the atmosphere. Because air is a poor conductor of heat (see Table 5-3), convection is much more important than conduction as a heat transport mechanism within the atmosphere.

Figure 7-4. Example of Convection in the Atmosphere

7.3.2 Latent Heat

The phase transition of water and associated latent heat exchanges are largely responsible for transferring the excess heat from the surface of the Earth into its atmosphere. As the Earth's surface absorbs radiation, some of the heat produced is used to evaporate (vaporize) water from oceans, lakes, rivers, soil, and vegetation. The water absorbs heat energy due to the latent heat of vaporization. Some of this water vapor condenses to microscopic water droplets or deposits as ice crystals that are visible as clouds. During cloud formation, the water vapor changes state, and latent heat is released into the atmosphere. During this process, the excess heat is transferred from the Earth's surface into its atmosphere.

7.4 Heat Imbalance Variations with Latitude

Global imbalances in radiational heating and cooling occur not only vertically between the Earth's surface and its atmosphere, but also horizontally with latitude. Since the Earth is essentially spherical, parallel beams of incoming solar radiation strike lower latitudes more directly than higher latitudes (see Figure 7-5); that is, the solar zenith angle is lower, and the Sun is more directly overhead in equatorial regions than at the poles. At higher latitudes, solar radiation is spread over a larger area and is less intense per unit surface area than at lower latitudes. Thus, the Earth absorbs more solar radiation at lower latitudes than higher latitudes, which creates heat imbalances and temperature gradients between the Equator and the poles.

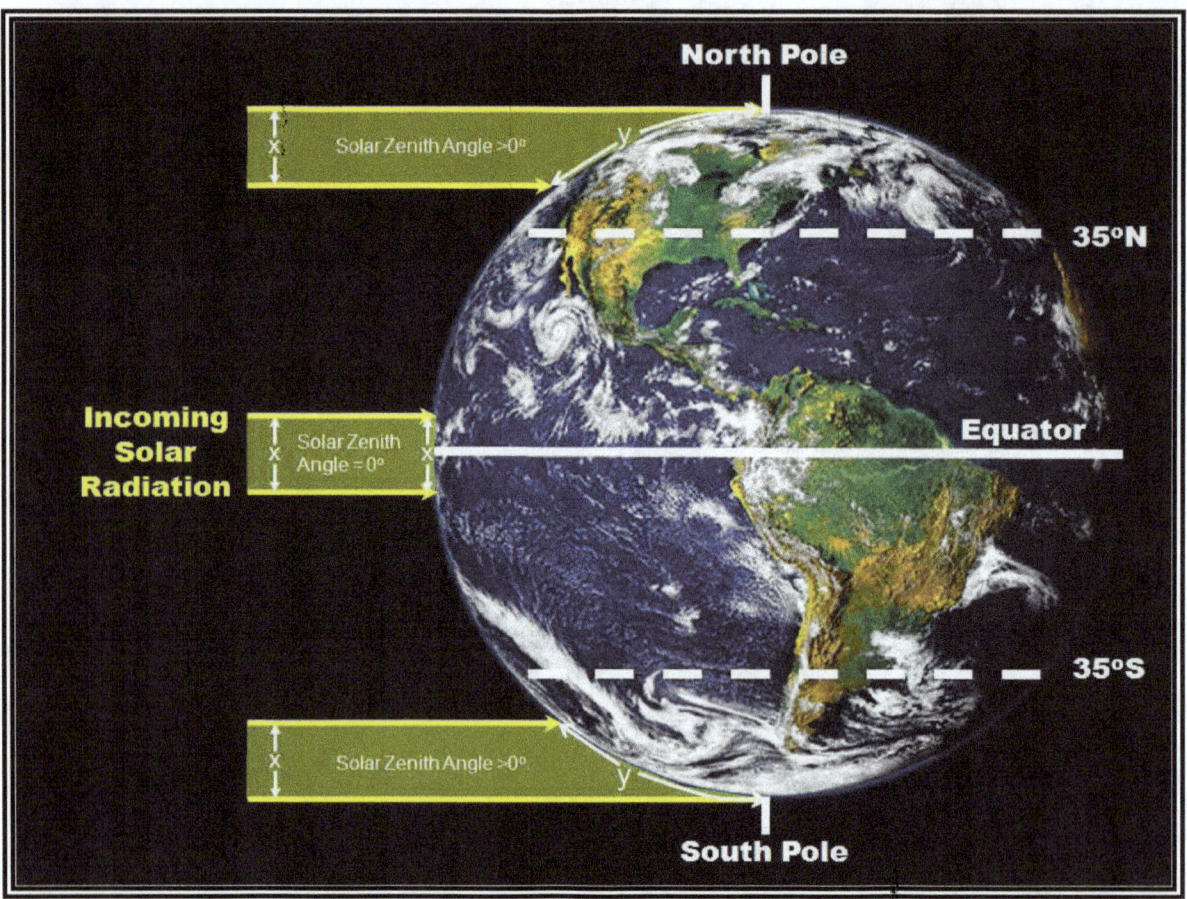

Figure 7-5. Solar Zenith Angle Variations with Latitude

The emission of terrestrial radiation also varies by latitude, but less so than the absorption of solar radiation. Terrestrial radiation emission decreases with increasing latitude due to a drop in temperature with latitude. Thus, at higher latitudes, the annual rate of cooling exceeds the rate of warming, while the reverse is true at lower latitudes.

Averaged over all latitudes, incoming solar radiation must equal outgoing terrestrial radiation. Otherwise, the Earth would be unable to maintain a constant average temperature. About 35° latitude in both hemispheres is where incoming and outgoing radiation is equal. This implies there is annual net cooling at higher latitudes and net warming at lower latitudes; however, this is untrue. The excess heat in the tropics must be transported polar by some mechanism(s). This poleward heat transport is accomplished by atmospheric circulations, weather, and ocean currents.

7.5 Seasons

Seasons are caused by the tilt of the Earth's rotational axis as the Earth orbits the Sun (see Figure 7-6). The Earth's rotational axis is tilted by 23½° from the perpendicular drawn to the plane of the Earth's orbit about the Sun and points the same direction in space all year long. The North Pole is tilted most directly toward the Sun during the summer solstice. Thus, in the Northern Hemisphere, the longest day of the year (lowest solar zenith angle) occurs on the summer solstice (approximately June 22), while the shortest day of the year (highest solar zenith angle) occurs on the winter solstice (approximately December 22). Day and night are of equal length (12 hours) worldwide on the vernal equinox (approximately March 21) and the autumnal equinox (approximately September 23).

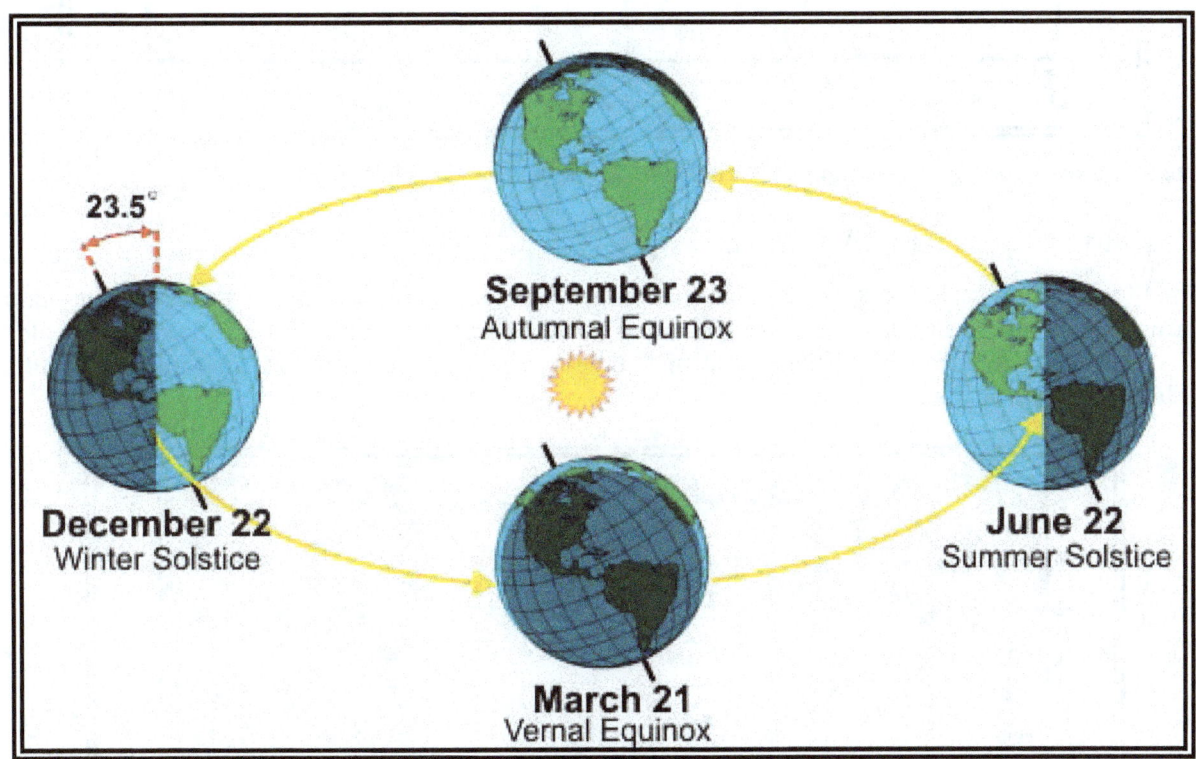

Figure 7-6. Solar Zenith Angle Variations with Northern Hemisphere Seasons

Figure 7-7 illustrates the average seasonal temperature variation in the Northern Hemisphere. Note that the warmest (coldest days) of the year occur after the summer (winter) solstice. This is due to the time lag necessary for heat flow processes to fully heat (cool) the surface of the Earth.

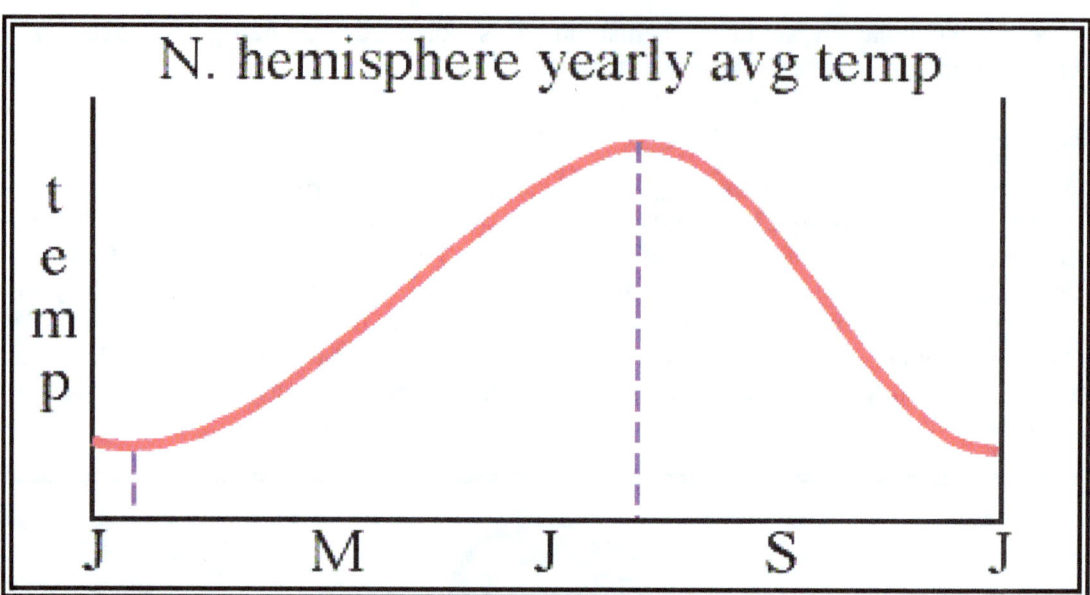

Figure 7-7. Average Seasonal Temperature Variation in the Northern Hemisphere

7.6 Diurnal Temperature Variation

Diurnal temperature variation is the daytime maximum and nighttime minimum of air temperature due to variations of insolation caused by the rising and setting of the Sun (i.e., variations of solar zenith angle) as the Earth rotates around its axis. Figure 7-8 depicts the typical diurnal temperature and radiation variations over land when the sky is clear.

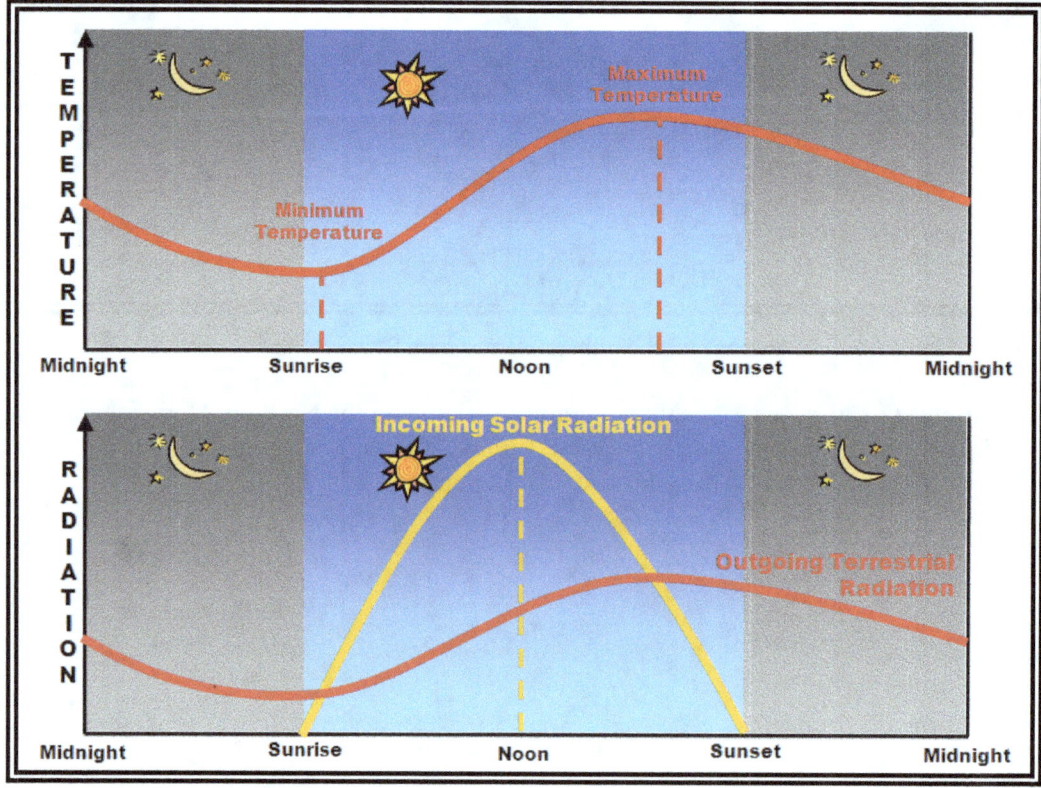

Figure 7-8. Clear Sky Diurnal Temperature and Radiation Variations Over Land

Warming and cooling of the Earth depend on an imbalance between solar and terrestrial radiation. The Earth receives heat during the day through incoming solar radiation. It loses heat to space by outgoing terrestrial radiation both day and night.

Shortly after sunrise, incoming solar radiation received at the Earth's surface (insolation) becomes greater than outgoing terrestrial radiation and the Earth's surface warms. Peak insolation occurs around noon, but maximum surface air temperature usually occurs during the midafternoon. This lag is necessary for the air near the ground to heat up due to conduction and convection with the surface. The Earth begins to cool once the rate of outgoing terrestrial radiation exceeds the rate of insolation.

At night, insolation is absent, but outgoing terrestrial radiation continues and the Earth's surface continues to cool. Cooling continues until shortly after sunrise, when incoming solar radiation once again exceeds outgoing terrestrial radiation. Minimum surface air temperature usually occurs shortly after sunrise.

The magnitude of diurnal temperature variation is primarily influenced by surface type, latitude, sky cover (e.g., clouds or pollutants), water vapor content of the air, and wind speed. Temperature variation is maximized over land, at low latitudes, with a clear sky, dry air, and light wind. Conversely, temperature variation is minimized over water, at high latitudes, with a cloudy sky, moist air, and strong wind.

8 Atmospheric Pressure and Altimetry

8.1 Introduction

Atmospheric pressure is one of the most basic variables used to describe the state of the atmosphere and is commonly reported in weather observations. Unlike temperature and relative humidity, changes in atmospheric pressure are not as readily sensed by people. However, variations of pressure across the Earth are associated with pressure centers (either high-pressure centers or low-pressure centers) that cause the wind to blow and can bring important weather changes. Density, which is directly related to pressure, is a property of the atmosphere, which can be used by pilots to help determine how their aircraft will perform at various altitudes.

This chapter discusses atmospheric pressure, how it is measured, and how it varies across the Earth. This chapter also covers the altimeter, which is a pressure sensor used by pilots to determine altitude. Finally, density will be discussed, along with its relationship to density altitude.

8.2 Atmospheric Pressure

The atoms and molecules that make up the various layers in the atmosphere are always moving in random directions. Despite their tiny size, when they strike a surface, they exert pressure.

Each molecule is too small to feel and only exerts a tiny bit of pressure. However, when adding up all the pressures from the large number of molecules that strike a surface each moment, the total pressure is considerable. This is air pressure. As the density of the air increases, the number of strikes per unit of time and area also increases.

Since molecules move in all directions, they even exert air pressure upwards as they smash into objects from underneath. Air pressure is exerted in all directions.

Atmospheric pressure is the force per unit area exerted by the weight of the atmosphere. Since air is not solid, it cannot be weighed with conventional scales. Yet, three centuries ago, Evangelista Torricelli proved he could weigh the atmosphere by balancing it against a column of mercury. He actually measured pressure, converting it directly to weight.

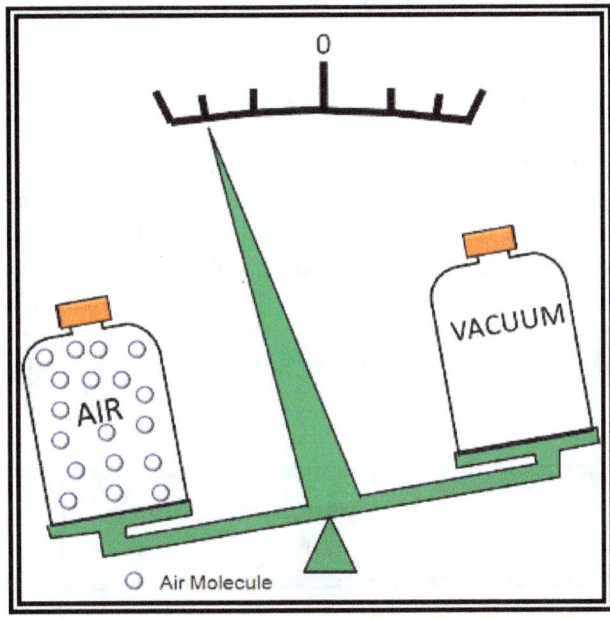

Air is composed of matter and, thus, has weight due to the pull of Earth's gravity.

Figure 8-1. Air Has Weight

8.2.1 Barometer

The instrument Torricelli designed to measure pressure was called a barometer. The aneroid barometer is the type most commonly used by meteorologists and the aviation community.

Essential features of an aneroid barometer (see Figure 8-2) are a flexible metal cell and the registering mechanism. Air is taken out of the cell to create a partial vacuum. The cell contracts or expands as pressure changes. One end of the cell is fixed, while the other end moves the registering mechanism. The coupling mechanism magnifies the movement of the cell driving an indicator hand along a scale graduated in pressure units.

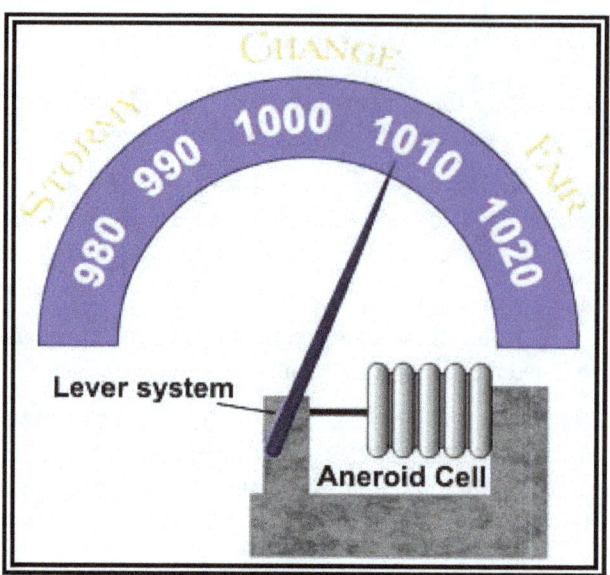

Figure 8-2. Aneroid Barometer

8.2.2 Atmospheric Pressure Units

Atmospheric pressure is expressed in many ways throughout the world (see Table 8-1). Meteorologists worldwide have long measured atmospheric pressure in millibars (mb or mbar), which denote pressure as a force per square centimeter. However, after the introduction of the International System of Units (SI) in 1960, the hectopascal (hPa) was adopted by most countries and is used in the METAR/SPECI code first developed in 1968. Many meteorologists prefer to use the term they learned during their education and work experience. Therefore, some continue to use the term "millibars," while others use "hectopascal" (which are equivalent). The unit inch of mercury (inHg or Hg) is still used in the United States for altimetry.

Table 8-1. Units of Pressure

Units of Pressure	Standard Atmosphere Value at Sea Level	Common Use
Hectopascals (hPa)	1013.2 hPa	METAR/SPECI
Millibars (mb or mbar)	1013.2 mb	U.S. Weather Charts
Inches of mercury (inHg or Hg)	29.92 inHg	U.S. Aviation
Pounds per square inch (psi)	14.7 psi	U.S. Engineering

8.2.3 Station Pressure

The pressure measured at an airport is called station pressure, or the actual pressure at field elevation. Pressure is lower at higher altitudes. Therefore, airports with higher field elevations usually have lower pressure than airports with lower field elevations. For instance, station pressure at Denver is less than at New Orleans (see Figure 8-3).

Figure 8-3. Station Pressure

The next few sections will examine some factors that influence pressure.

8.2.4 Pressure Variation

Atmospheric pressure varies with altitude and the temperature of the air, as well as with other minor influences, such as water vapor.

8.2.4.1 Pressure Changes with Altitude

As a person moves upward through the atmosphere, the weight of the air above the person decreases. If a person carries a barometer, then they can measure a decrease in pressure as the weight of the air above them decreases. Figure 8-4 shows the pressure decrease with height in the standard atmosphere.

The standard altitudes in Figure 8-4 are based on standard temperatures. In the real atmosphere, temperatures are seldom standard, so temperature's effects on pressure will be explored in the following section.

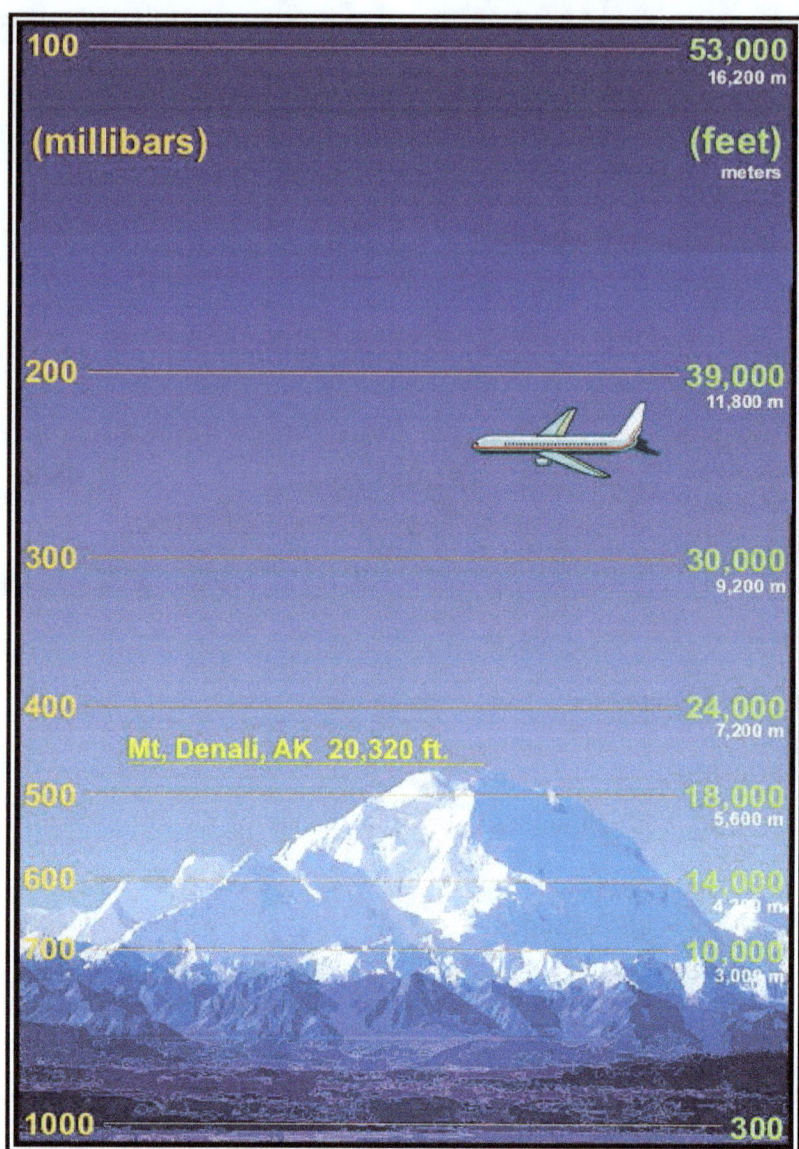

Figure 8-4. Air Pressure in the Standard Atmosphere

8.2.4.2 Temperature's Effects on Pressure

Like most substances, air expands as it becomes warmer and contracts as it cools. Figure 8-5 shows three columns of air: one colder than standard, one with standard temperature, and one warmer than standard. Pressure is equal at the bottom and top of each column. Vertical expansion of the warm column has made it taller than the column at standard temperature. Contraction of the cold column has made it shorter than the standard column. Since the total pressure decrease is the same in each column, the rate of decrease of pressure with height in warm air is less than standard, while the rate of decrease in pressure with height in cold air is greater than standard.

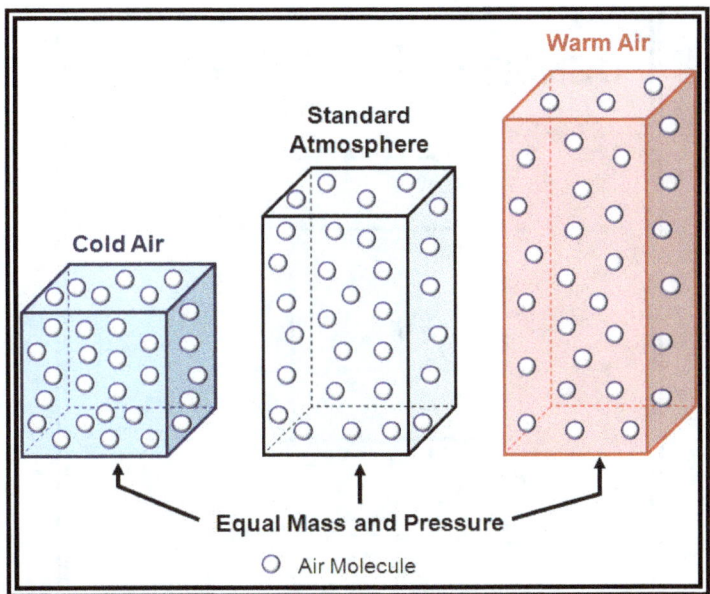

Figure 8-5. Temperature's Effect on Pressure

8.2.5 Sea Level Pressure

Since pressure varies greatly with altitude, people cannot readily compare station pressures between stations at different altitudes. To make them comparable, they are adjusted to some common level. Mean sea level (MSL) is the most useful common reference. In Figure 8-6, pressure measured at a station at a 5,000-ft elevation is 25 inHg; pressure increases about 1 inHg for each 1,000 ft, or a total of 5 inHg. Sea level pressure is approximately 25 + 5, or 30 inHg.

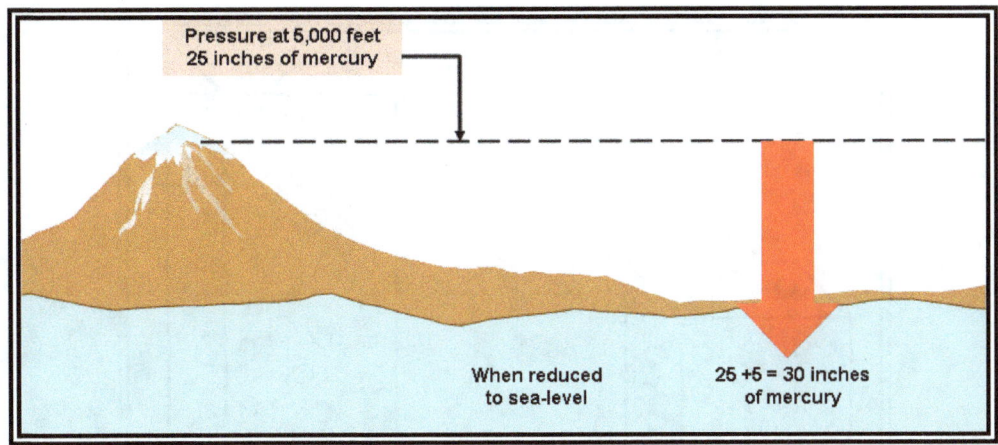

Figure 8-6. Reduction of Station Pressure to Sea Level

Sea level pressure is typically displayed on surface weather charts. Pressure continually changes across the Earth, so a sequence of surface charts must be viewed to follow these changing pressures.

8.3 Density

Density is the ratio of any quantity to the volume or area it occupies. Atmospheric density is defined as ratio of the mass (or weight) of the air to the volume occupied by it, usually expressed in kilograms per cubic meter (see Figure 8-7).

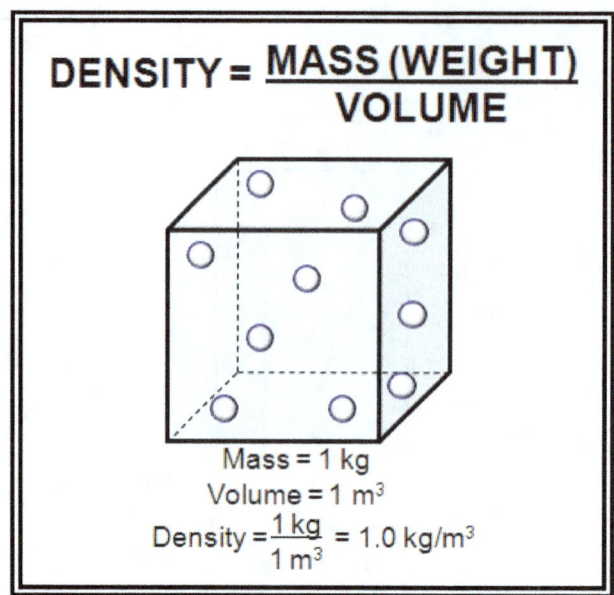

Figure 8-7. Density is Mass (Weight) per Volume

8.3.1 Volume's Effects on Density

The density of an air parcel varies inversely with its volume. Assuming equal mass, an air parcel with a higher density has a smaller volume than an air parcel with a lower density (see Figure 8-8).

The shorter parcel (i.e., the parcel with the smaller volume) has a higher density than the taller parcel, which contains the larger volume. This is due to the fact that the air molecules within the shorter parcel must be compressed within the smaller volume.

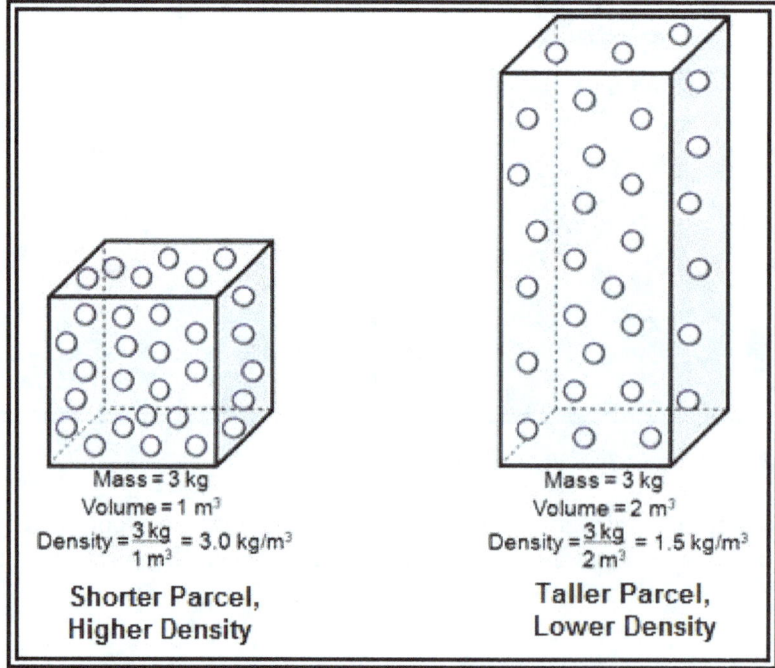

Figure 8-8. Volume's Effects on Density

8.3.2 Changes in Density

In general, the density of an air parcel can be changed by changing its mass, pressure, or temperature. Boyle's law says that the density of an ideal gas (ρ, the Greek letter rho) is given by:

$$\rho = \frac{MP}{RT}$$

Where M is the molar mass, P is the pressure, R is the universal gas constant, and T is the absolute temperature.

8.3.3 Density's Effects on Pressure

Density is directly related to pressure. Assuming constant mass and temperature, an air parcel with a higher pressure is denser than an air parcel with a lower pressure.

As previously discussed, air pressure decreases with height in the atmosphere. Therefore, the density also decreases with height (see Figure 8-9). In the atmosphere, pressure has the greatest effect on density in the vertical direction.

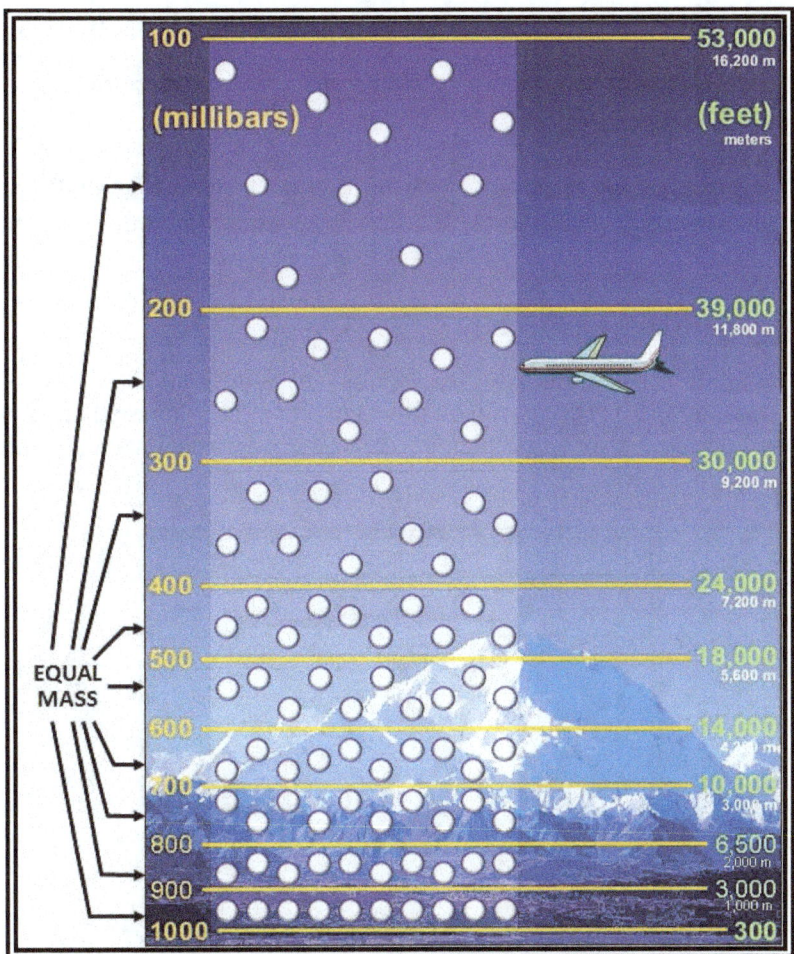

Figure 8-9. Pressure's Effects on Density in the Atmosphere

8.3.4 Temperature's Effects on Density

Density is inversely related to temperature. Assuming constant mass and pressure, an air parcel with a higher temperature is less dense than an air parcel with a lower temperature (see Figure 8-10). This is because the warmer air occupies a large volume.

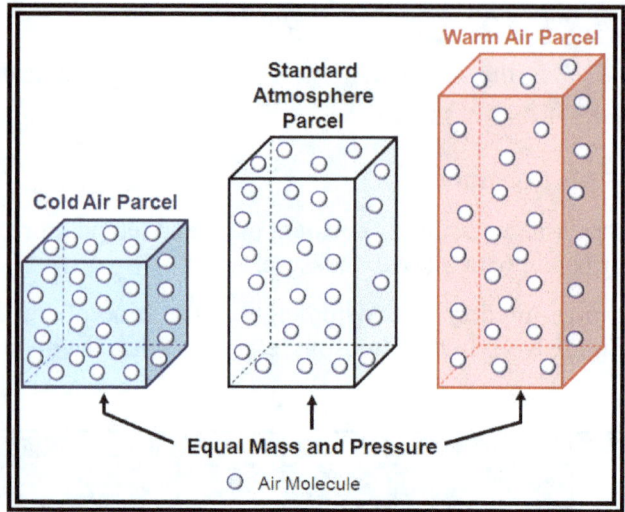

Figure 8-10. Temperature's Effects on Density

In the atmosphere, temperature has the most effect on density in the horizontal direction; that is, with horizontal changes of location (e.g., New York City versus Miami), temperature has the greatest effect on density.

8.3.5 Water Vapor's Effects on Density

Density of an air parcel is inversely related to its quantity of water vapor. Assuming constant pressure, temperature, and volume, air with a greater amount of water vapor is less dense than air with a lesser amount of water vapor. This is because dry air molecules have a larger mass (weight) than water vapor molecules, and density is directly related to mass (see Figure 8-11).

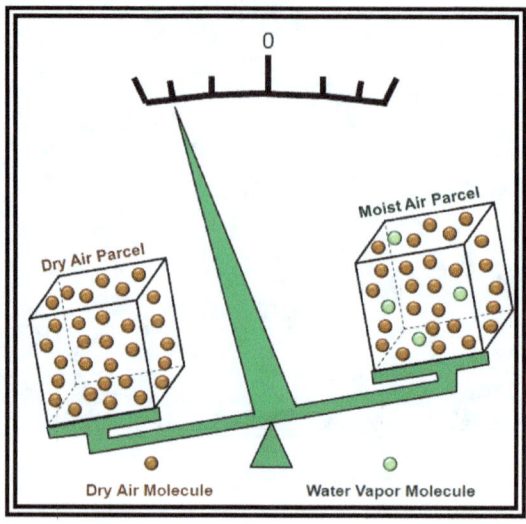

Figure 8-11. Water Vapor's Effects on Density

8.4 Altimetry

The altimeter is essentially an aneroid barometer. The difference is the scale. The altimeter is graduated to read increments of altitude rather than units of pressure. The standard for graduating the altimeter is the standard atmosphere.

8.4.1 Altitude

Altitude seems like a simple term; it means the vertical elevation of an object above the surface of the Earth. However, in aviation, it can have many meanings.

8.4.1.1 True Altitude

Since existing conditions in a real atmosphere are seldom standard, altitude indications on the altimeter are seldom actual or true altitudes. True altitude is the actual vertical distance above MSL. If an altimeter does not indicate true altitude, what does it indicate?

8.4.1.2 Indicated Altitude

Figure 8-10 shows the effect of mean temperature on the thickness of three columns of air. Pressures are equal at the bottoms and tops of the three layers. Since an altimeter is essentially an aneroid barometer, altitude indicated by the altimeter at the top of each column would be the same. To see this effect more clearly, see Figure 8-12. In the warm air column, a pilot would fly at an altitude that is higher than the indicated altitude. In the cold air column, the pilot would fly at an altitude lower than the indicated altitude.

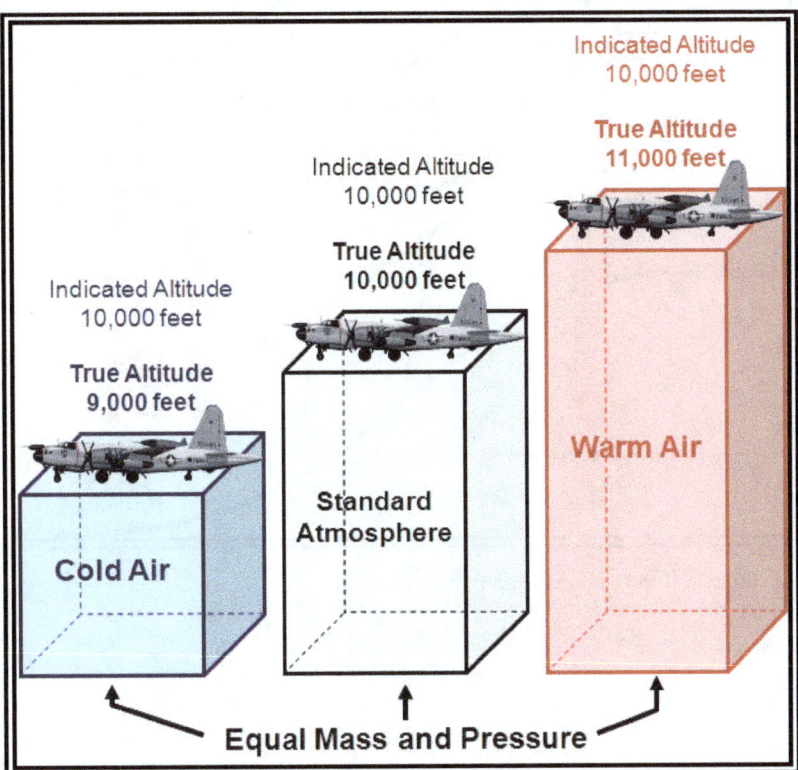

Figure 8-12. True Versus Indicated Altitude

Chapter 8, Atmospheric Pressure and Altimetry

Height indicated on the altimeter also changes with changes in surface pressure. A movable scale on the altimeter permits the pilot to adjust for surface pressure, but the pilot has no means of adjusting the altimeter for mean temperature of the column of air below. Indicated altitude is the altitude above MSL indicated on the altimeter when set at the local altimeter setting. But what is altimeter setting?

8.4.1.2.1 Altimeter Setting

Since the altitude scale is adjustable, a pilot can set the altimeter to read true altitude at some specified height. Takeoff and landing are the most critical phases of flight; therefore, airport elevation is the most desirable altitude for a true reading of the altimeter. The altimeter setting is the value to which the scale of the pressure altimeter is set so the altimeter indicates true altitude at field elevation.

To ensure the altimeter reading is compatible with altimeter readings of other aircraft in the vicinity, a pilot should ensure the altimeter setting is current. The pilot should adjust it frequently while in flight, according to the nearest surface weather reporting station. Figure 8-13 shows the trouble a pilot can encounter if not vigilant in adjusting the altimeter during flight. As the pilot flies from high pressure to low pressure, the plane is lower than the altimeter indicates.

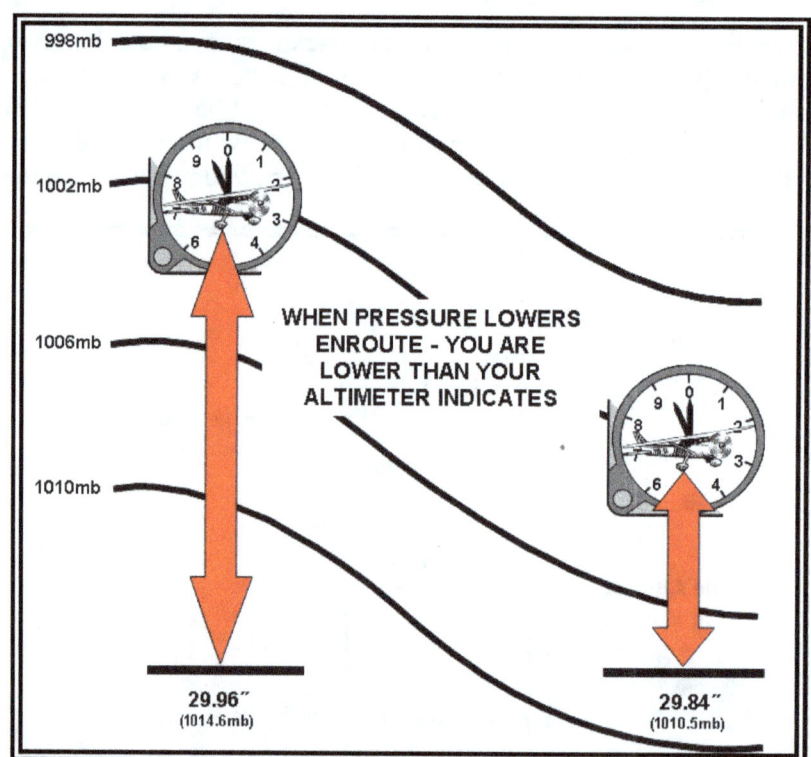

Figure 8-13. Pressure Change's Effects on Altimeter Readings

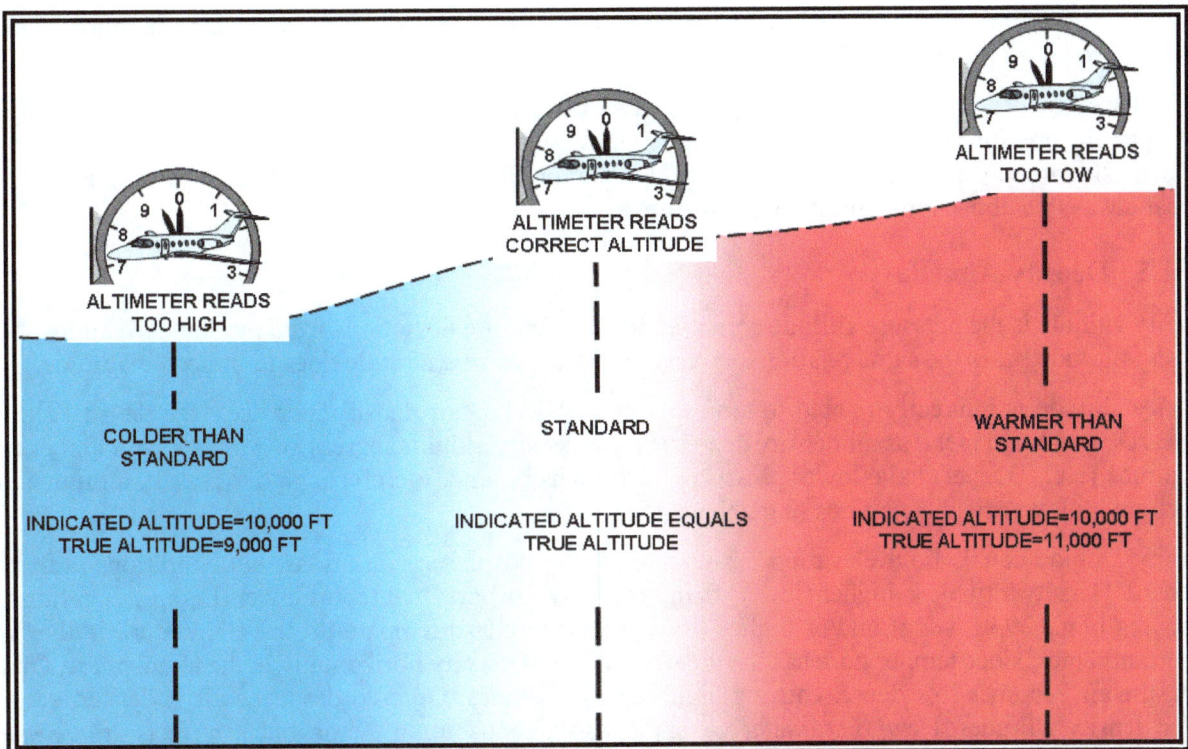

Figure 8-14. Temperature Change's Effects on Altimeter Readings

Figure 8-14 shows that as a pilot flies from warm to cold air, the altimeter reads too high—the pilot is lower than the altimeter indicates. Over flat terrain, this lower-than-true reading is no great problem; other aircraft in the vicinity are also flying indicated rather than true altitude, and everyone's altimeter readings are compatible. If flying in cold weather over mountainous areas, however, a pilot needs to take this difference between indicated and true altitude into account. The pilot needs to know that the true altitude assures clearance of terrain, and compute a correction to indicated altitude.

8.4.1.3 Corrected (Approximately True) Altitude

If a pilot could always determine the mean temperature of a column of air between the aircraft and the surface, flight computers would be designed to use this mean temperature in computing true altitude. However, the only guide a pilot has to temperature below is free air temperature at the pilot's altitude. Therefore, the flight computer uses outside air temperature (OAT) to correct indicated altitude to approximately true altitude. The corrected (approximately true) altitude is indicated altitude corrected for the temperature of the air column below the aircraft, the correction being based on the estimated deviation of the existing temperature from standard atmosphere temperature. It is a close approximation to true altitude and is labeled true altitude on flight computers. It is close enough to true altitude to be used for terrain clearance, provided the pilot has the altimeter set to the value reported from a nearby reporting station.

8.4.1.4 Pressure Altitude

In the standard atmosphere, sea level pressure is 29.92 inHg (1013.2 mb). Pressure decreases at a fixed rate upward through the standard atmosphere. Therefore, in the standard atmosphere, a given pressure exists at any specified altitude. Pressure altitude is the altitude (above MSL) shown by the altimeter when set to 29.92 inHg. In other words, it is the altitude associated with a specific pressure measured by the static port when the altimeter is set to 29.92. Since pressure is the same everywhere regardless of the specific pressure altitude, a constant-pressure surface defines a constant-pressure altitude. When a pilot flies a constant-pressure altitude, the pilot is flying a constant-pressure surface.

As discussed earlier, constant-pressure surfaces have different heights across them. Therefore, when flying at a specific pressure altitude (i.e., constant-pressure surface) a pilot's true altitude will change with distance. However, since pressure altitudes are flown at or above FL180 (in the United States), a pilot will almost always be above the highest terrain features.

8.4.1.5 Density Altitude

Density altitude is the pressure altitude corrected for temperature deviations from the standard atmosphere. Density altitude bears the same relation to pressure altitude as true altitude does to indicated altitude.

Density altitude is indirectly related to atmospheric density; as air density increases, the density altitude decreases, and conversely, as air density decreases, the density altitude increases. Airports with higher field elevations (e.g., Denver) have lower pressure, lower density, and, therefore, higher density altitudes than airports with lower field elevations (e.g., New Orleans).

Density altitude equals field elevation during standard atmospheric conditions, but conditions are rarely standard. Density altitude is higher (lower) than standard at airports that report lower (higher) than standard pressures (e.g., 29.92 inHg) and/or higher (lower)-than-standard temperatures. Temperature is the most important factor, since temperature has the greatest effect on density horizontally in the atmosphere. On hot days, the air becomes less dense, causing high density altitudes. On cold days, the air is denser, causing lower density altitudes. Dewpoint (water vapor) is also a contributing factor, but its effects are generally negligible.

Density altitude is an index to aircraft performance. Higher density altitude decreases aircraft performance. Lower density altitude increases aircraft performance. High density altitude is a hazard, since it reduces aircraft performance in the following three ways:

1. It reduces power because the engine takes in less air to support combustion.
2. It reduces thrust because there is less air for the propeller to work with, or a jet has less mass of gases to force out of the exhaust.
3. It reduces lift because the light air exerts less force on the airfoils.

A pilot cannot detect the effect of high density altitude on the airspeed indicator. The aircraft lifts off, climbs, cruises, glides, and lands at the prescribed indicated airspeeds; but at a specified indicated airspeed, the pilot's true airspeed and groundspeed increase proportionally as density altitude becomes higher.

The net results are that high density altitude lengthens a pilot's takeoff and landing rolls and reduces the rate of climb. Before lift-off, the plane must attain a faster groundspeed, and, therefore, needs more runway; and the reduced power and thrust add a need for still more runway. The plane lands at a faster groundspeed and, therefore, needs more room to stop. At a prescribed indicated airspeed, it is flying at a faster true airspeed, and, therefore, covers more distance in a given time, which means climbing at a shallower angle. Adding to this are the problems of reduced power and rate of climb. Figure 8-15 shows the effect of density altitude on takeoff distance and rate of climb.

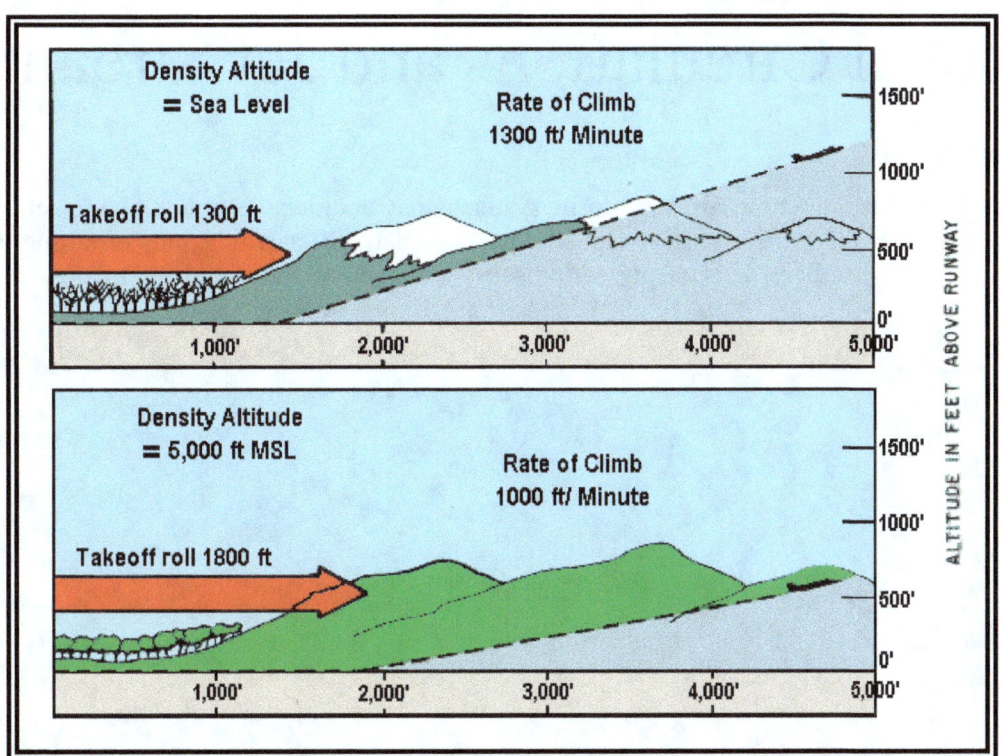

Figure 8-15. High Density Altitude's Effects on Flight

High density altitude also can be a problem at cruising altitude. When air temperature is higher (warmer) than standard atmosphere, the higher density altitude lowers the service ceiling. For example, if temperature at a pressure altitude of 10,000 ft is 20 °C, density altitude is 12,700 ft. A pilot's aircraft will perform as though it were at 12,700 ft indicated with a normal temperature of -8 °C.

To compute density altitude, a pilot can set the altimeter to 29.92 in (1,013.2 mb), read the pressure altitude from the altimeter, obtain the OAT, and then use a flight computer to compute density altitude.

9 Global Circulations and Jet Streams

9.1 Introduction

Global circulations explain how air and storm systems that could have potential impacts on aircraft operations travel over the Earth's surface. Jet streams are relatively narrow bands of strong wind in the upper levels of the atmosphere. This chapter will discuss global circulations and jet streams.

9.2 Non-Rotating Earth Circulation System

The global circulation would be simple if the Earth did not rotate, the rotation was not tilted relative to the Sun, and had no water.

Without those factors, the ground and atmosphere directly beneath the Sun would be subject to more of the Sun's heat than anywhere else on the planet. The result would be the Equator becoming very hot, with the hot air rising into the upper atmosphere.

That hot air would then move toward the poles, where it would become very cold and sink, returning to the Equator (see Figure 9-1). One large area of high pressure would be at each of the poles, with a large belt of low pressure around the Equator.

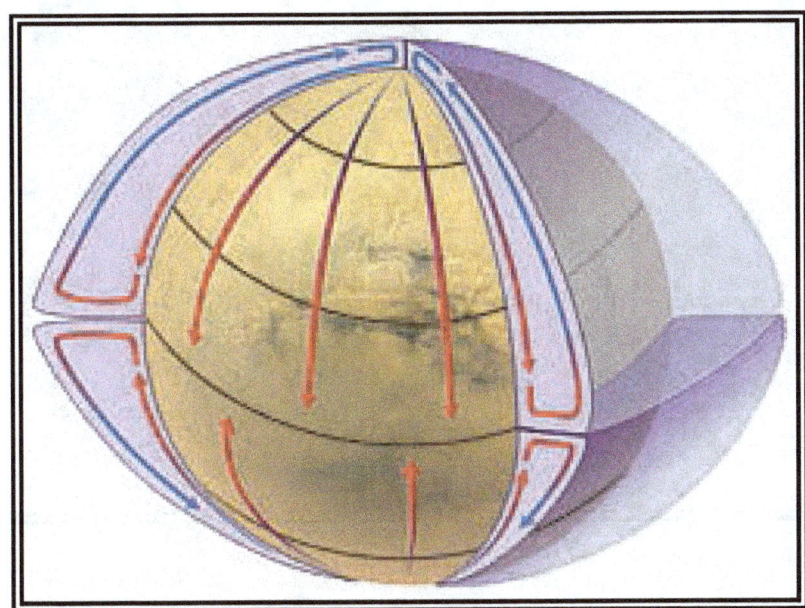

Figure 9-1. Non-Rotating, Non-Tilted, Waterless, Earth Circulation System

9.3 Rotating Earth Circulation System

However, since the Earth rotates, the axis is tilted, and there is more land mass in the Northern Hemisphere than in the Southern Hemisphere, the actual global pattern is much more complicated.

Instead of one large circulation between the poles and the Equator, there are three circulations (see Figure 9-2):

- Hadley cell—Low-latitude air movement toward the Equator that, with heating, rises vertically with poleward movement in the upper atmosphere. This forms a convection cell that dominates tropical and subtropical climates.

- Ferrel cell—A mid-latitude mean atmospheric circulation cell for weather, named by William Ferrel in the 19th century. In this cell, the air flows poleward and eastward near the surface, and equatorward and westward at higher levels.

- Polar cell—Air rises, diverges, and travels toward the poles. Once over the poles, the air sinks, forming the polar highs. At the surface, air diverges outward from the polar highs. Surface winds in the polar cell are easterly (polar easterlies).

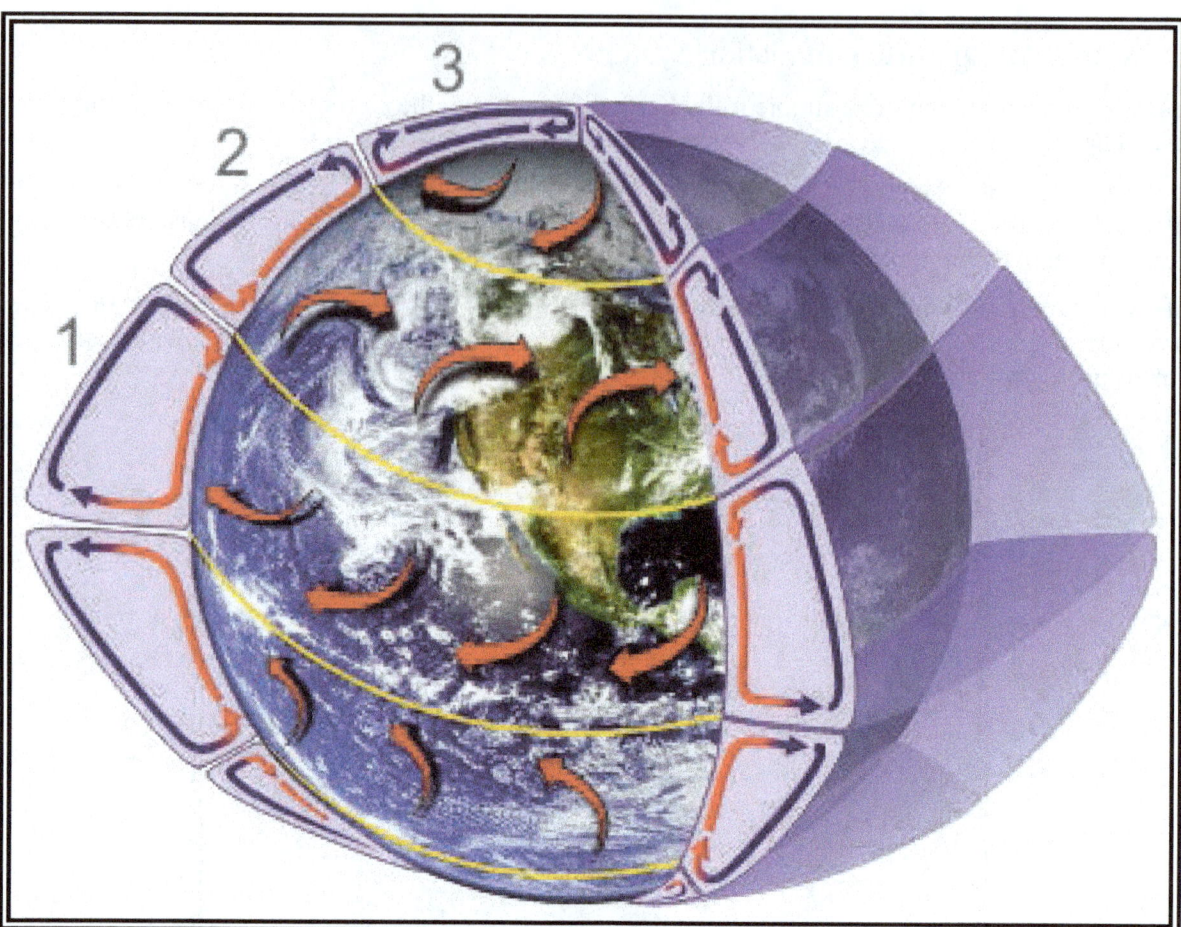

Figure 9-2. Earth Circulation System

Between each of these circulation cells are bands of high and low pressure at the surface. The high-pressure band is located about 30° N/S latitude and at each pole. Low-pressure bands are found at the Equator and 50° to 60° N/S latitude.

Usually, fair and dry/hot weather is associated with high pressure, and rainy and stormy weather is associated with low pressure. The results of these circulations become evident on a globe. Consider the number of deserts located along the 30° N/S latitude around the world compared to the region between 50°–60° N/S latitude. The higher latitudes, especially near the west coast of continents, tend to have more precipitation due to more storms moving around the Earth at these latitudes.

9.4 Jet Streams

Jet streams are relatively narrow bands of strong wind in the upper levels of the atmosphere. The winds blow from west to east in jet streams, but the flow often meanders southward and northward in waves. Jet streams follow the boundaries between hot and cold air. Since these hot and cold air boundaries are most pronounced in winter, jet streams are the strongest for both the Northern and Southern Hemisphere winters.

9.4.1 Direction of Wind Flow

Why do the jet stream winds blow from west to east? As stated in the previous section, if the Earth was not rotating, the warm air would rise at the Equator and move toward both the poles. The Earth's rotation divides this circulation into three cells. Likewise, the Earth's rotation is responsible for the jet stream.

The motion of the air is not directly north and south, but rather is affected by the momentum the air has as it moves away from the Equator and how fast a location on or above the Earth moves relative to the Earth's axis.

An object's speed relative to the Earth's axis depends on its location. Someone standing on the Equator is moving much faster than someone standing on a 45° latitude line. In Figure 9-3, the person at the position on the Equator arrives at the yellow line sooner than the other two. Someone standing on a pole is not moving at all (except that person would be slowly spinning). The speed of the rotation is great enough to cause a person to weigh 1 pound (lb) less at the Equator than they would at the North or South Pole.

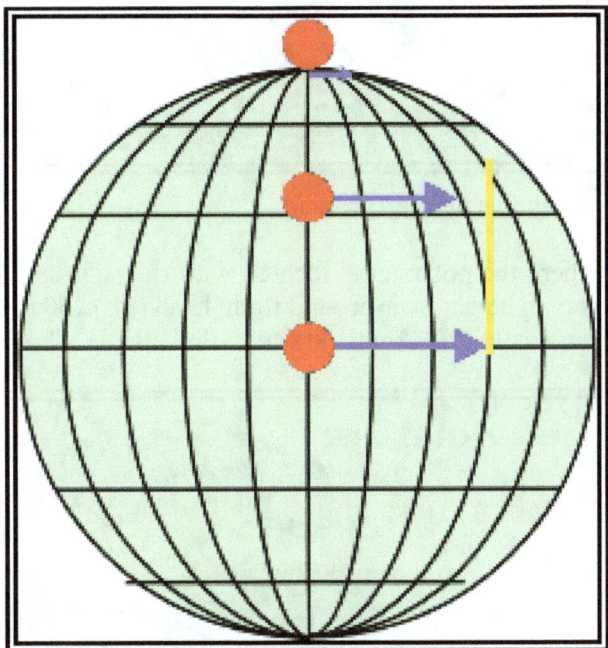

Figure 9-3. Speed Relative to the Earth's Axis Versus Latitude

The momentum of air as it travels around the Earth is conserved, which means as the air that is over the Equator starts moving toward one of the poles, it keeps its eastward motion constant. The Earth below the air, however, moves slower, as that air travels toward the poles. The result is that the air moves faster and faster in an easterly direction (relative to the Earth's surface below) the farther it moves from the Equator.

9.4.2 Location

In addition, with the three cell circulations mentioned previously, the regions around 30° N/S and 50°–60° N/S are areas where temperature changes are the greatest (see Figure 9-4). As the difference in temperature between the two locations increases, the strength of the wind increases. Therefore, the regions around 30° N/S and 50°–60° N/S are also regions where the wind in the upper atmosphere is the strongest.

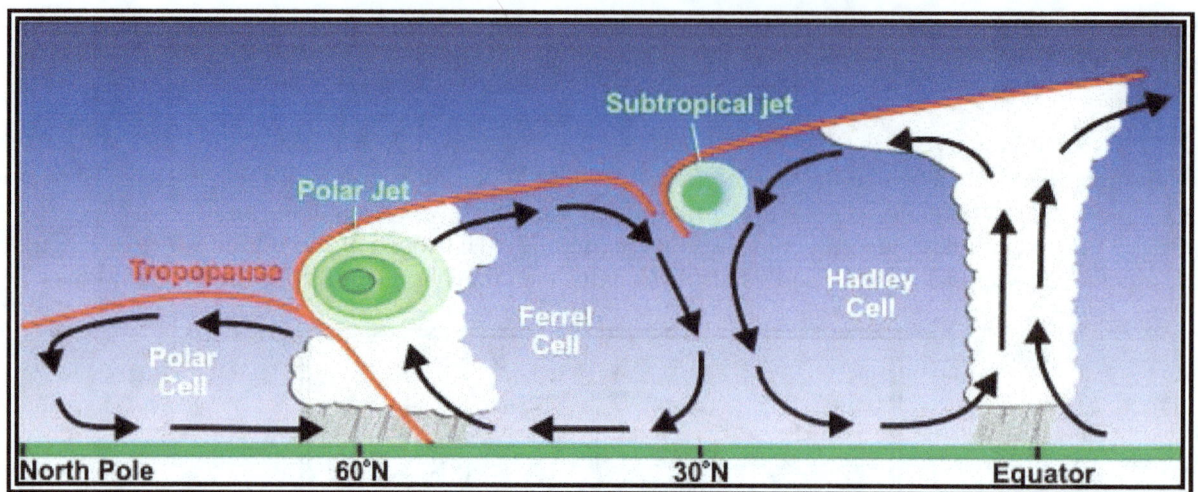

Figure 9-4. Three Cell Circulations and Jet Stream Location

The 50°–60° N/S region is where the polar jet is located with the subtropical jet located around 30° N (see Figure 9-5). Jet streams vary in height from around flight level (FL) 200 to FL450 and can reach speeds of more than 275 miles per hour (mph) (239 knots (kt)/442 kilometers per hour (km/h)).

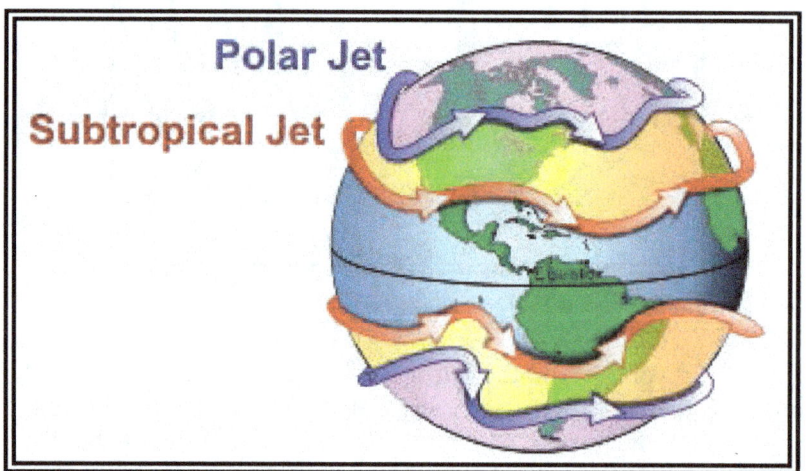

Figure 9-5. Illustration of Polar and Subtropical Jet Streams and Their Relative Locations Around the Globe

The actual appearance of jet streams results from the complex interaction between many variables, such as the location of high- and low-pressure systems, warm and cold air, and seasonal changes. They meander around the globe, dipping and rising in altitude/latitude, splitting at times and forming eddies, and even disappearing altogether to appear somewhere else.

Jet streams also follow the Sun, in that as the Sun's elevation increases each day in the spring, the jet streams shift north, moving into Canada by summer. As autumn approaches and the Sun's elevation decreases, the jet stream moves south into the United States, helping to bring cooler air to the country.

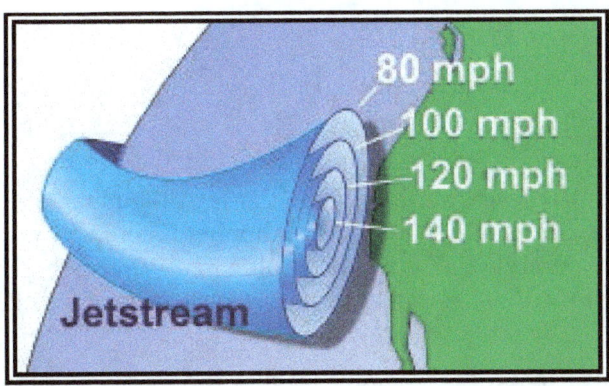

Figure 9-6. Jet Stream Wind Speeds

Also, the jet stream is often indicated by a line on maps, and shown by television meteorologists. The line generally points to the location of the strongest wind (see Figure 9-6). In reality, jet streams are typically much wider. They are less a distinct location, and more a region where winds increase toward a core of highest speed.

One way of visualizing this is to consider a river. The river's current is generally the strongest in the center, with decreasing strength as one approaches the river's bank. It can be said that jet streams are rivers of air.

10 Wind

10.1 Introduction

Wind is the air in motion relative to the surface of the Earth. Although people cannot actually see the air moving, it can be measured by its motion of force that it applies on objects. For example, leaves rustling or trees swaying on a windy day indicate that the wind is blowing. Winds are a major factor to both weather and aircraft. Winds cause the formation, dissipation, and redistribution of weather. Winds also affect aircraft during all phases of flight.

Adverse wind is a category of hazardous aviation weather that is responsible for many weather-related accidents. Adverse winds include: crosswinds, gusts, tailwind, variable wind, sudden wind shift, wind shear, and mountain wind hazards. Takeoff and landing are the most critical periods of any flight and are most susceptible to the effects of adverse wind. The most at-risk group is pilots flying aircraft with lower crosswind and tailwind threshold values.

This chapter discusses the origin of wind as well as adverse winds.

10.2 Naming of the Wind

Wind is named according to the direction from which it is blowing. For example, a west wind indicates the wind is blowing from the west to the east. There are 36 specific azimuth degrees expressed in intervals of 10 degrees. In aviation, the points of the compass are normally used to represent the direction from which the wind is blowing. For example, north winds come from 360°, east from 90°, south from 180°, and west from 270°.

There are also 16 cardinal compass directions relative to wind. The four primary cardinal directions are north (N), south (S), east (E), and west (W). There are also four intermediate directions, such as northeast (NE), northwest (NW), southeast (SE), and southwest (SW). Additionally, there are eight subdivisions, including north-northeast (NNE), north-northwest (NNW), south-southeast (SSE), south-southwest (SSW), east-northeast (ENE), east-southeast (ESE), west-southwest (WSW), and west-northwest (WNW).

10.3 Forces That Affect the Wind

Three primary forces affect the flow of wind: Pressure Gradient Force (PGF), Coriolis force, and friction.

10.3.1 Pressure Gradient Force (PGF)

Wind is driven by pressure differences, which create a force called the PGF. Whenever a pressure difference develops over an area, the PGF makes the wind blow in an attempt to equalize pressure differences. This force is identified by height contour gradients on constant-pressure charts and by isobar gradients on surface charts.

PGF is directed from higher height/pressure to lower height/pressure and is perpendicular to contours/isobars. Whenever a pressure difference develops over an area, the PGF begins moving the air directly across the contours/isobars. See Figure 10-1.

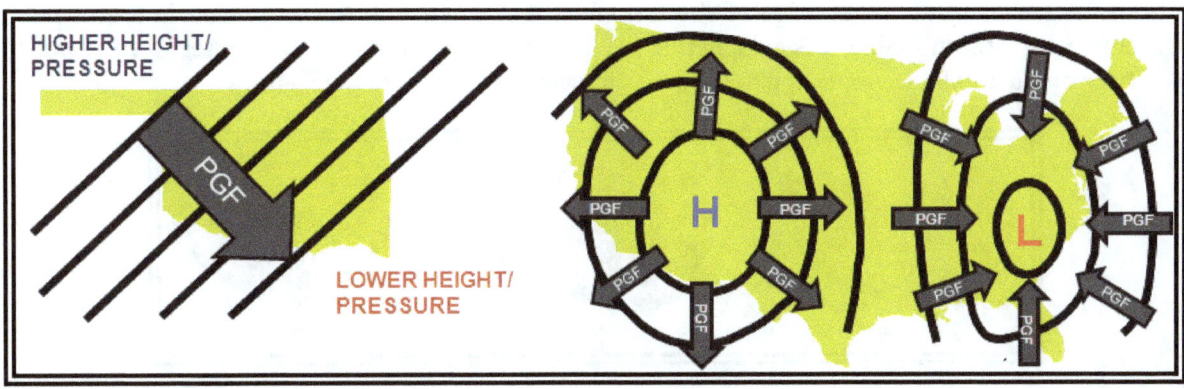

PGF is directed across contours/isobars towards lower height/pressure.

Figure 10-1. Direction of Pressure Gradient Force

Wind speed is directly proportional to the PGF, which itself is directly proportional to the contour/isobar gradient. Closely spaced contours/isobars indicate strong winds, while widely spaced contours/isobars mean lighter wind. From a pressure analysis, users can get a general idea of wind speed from contour/isobar spacing.

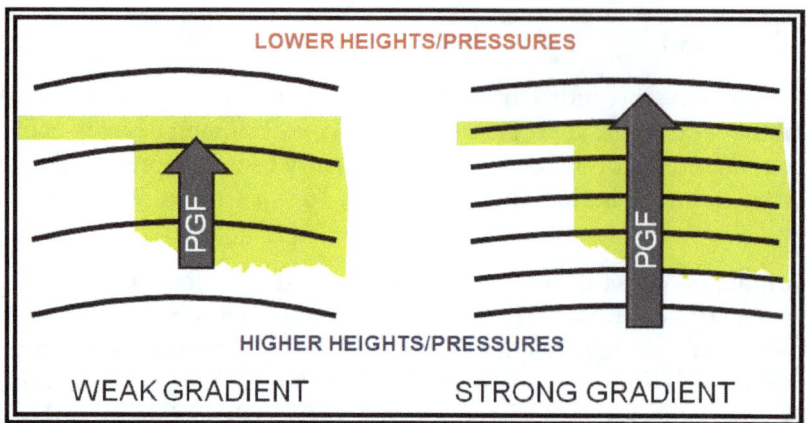

Figure 10-2. Magnitude of Pressure Gradient Force

In Figure 10-2, on the left panel, the contours/isobars are widely spaced apart, PGF is weak, and the wind speed is weak. On the right panel, the contours/isobars are more closely spaced, the PGF is stronger, and the wind speed is stronger.

The wind would flow from high to low pressure if the PGF was the only force acting on it. However, because of the Earth's rotation, there is a second force called the Coriolis force that affects the direction of wind flow.

10.3.2 Coriolis Force

A moving mass travels in a straight line until acted on by some outside force. However, if one views the moving mass from a rotating platform, the path of the moving mass relative to his platform appears to be deflected or curved. To illustrate, consider a turntable. If one used a pencil and a ruler to draw a straight line from the center to the outer edge of the turntable, the pencil will have traveled in a straight line. However, stopping the turntable, it is evident that the line spirals outward from the center (see Figure 10-3). To a viewer on the turntable, some apparent force deflected the pencil to the right.

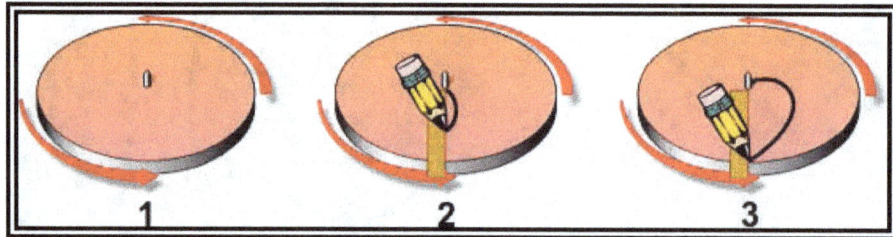

Figure 10-3. Illustration of Coriolis Force

A similar apparent force deflects moving particles on the Earth. Because the Earth is spherical, the deflective force is much more complex than the simple turntable example. The principle was first explained by Gaspard-Gustave de Coriolis, and now carries his name—the Coriolis force.

Coriolis force is an apparent force that affects all moving objects. The force deflects air to the right in the Northern Hemisphere and to the left in the Southern Hemisphere.

Coriolis force is at a right angle to wind direction and directly proportional to wind speed; that is, as wind speed increases, Coriolis force increases. At a given latitude, double the wind speed and the Coriolis force is doubled. Why at a given latitude?

Coriolis force varies with latitude from zero at the Equator to a maximum at the poles. It influences wind direction everywhere except immediately at the Equator, but the effects are more pronounced in middle and high latitudes.

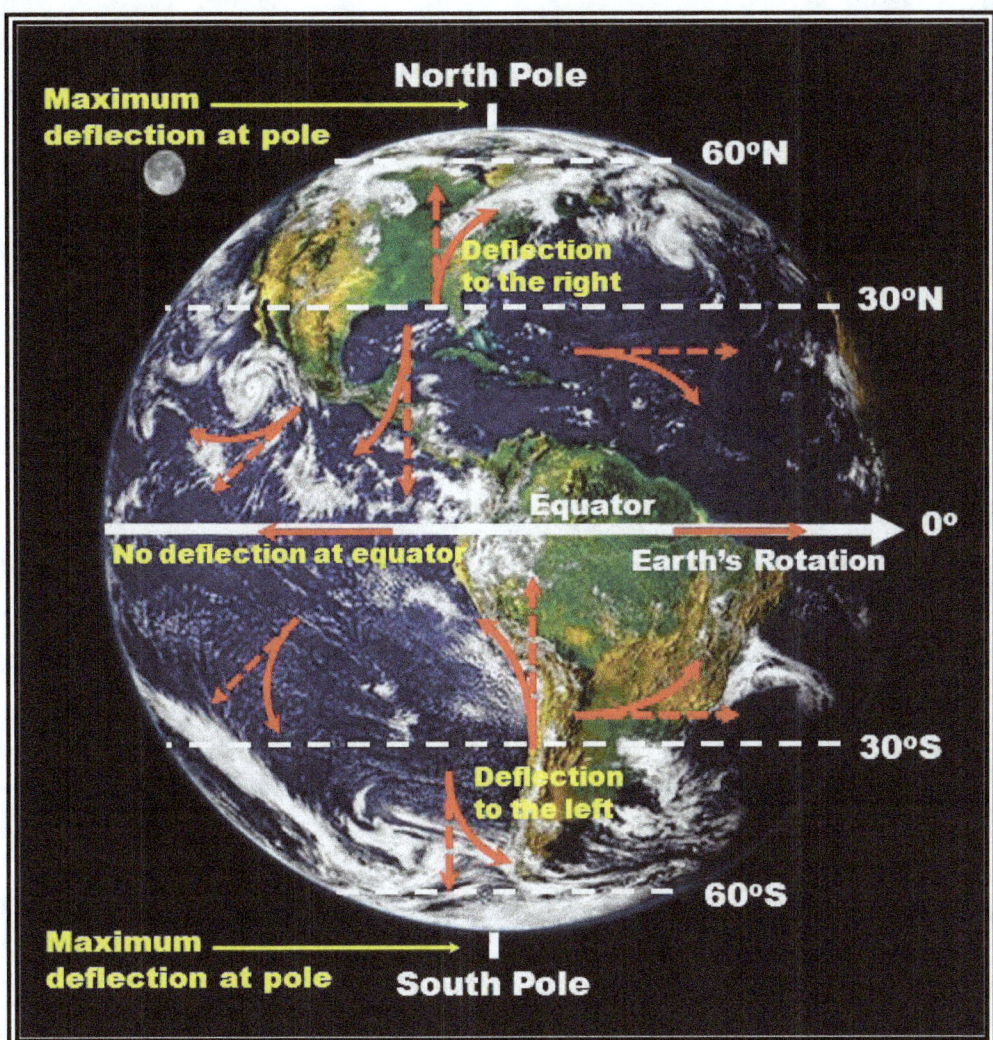

Figure 10-4. Coriolis Force Variations Across the Earth

Coriolis force deflects moving objects to the right of their path in the Northern Hemisphere and to the left of their path in the Southern Hemisphere. Coriolis deflection is maximized at the poles and zero at the Equator.

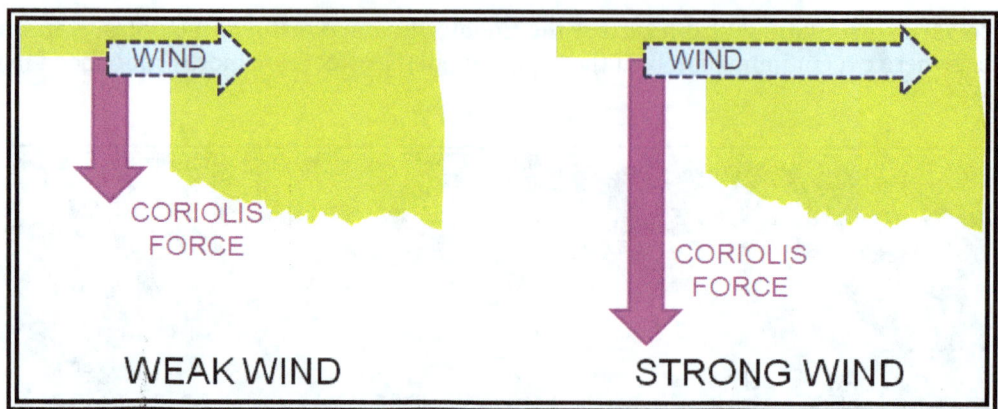

Figure 10-5. Coriolis Force Magnitude Variations with Wind Speed

Coriolis force magnitude is directly proportional to wind speed. In Figure 10-5, wind speed is twice as strong in the right panel; thus, the Coriolis force is doubled.

10.3.3 Friction Force

Friction between the wind and the terrain surface slows the wind. The rougher the terrain, the greater the frictional effect. Also, the stronger the wind speed, the greater the friction. One may not think of friction as a force, but it is a very real and effective force always acting opposite to wind direction.

Figure 10-6. Friction Force Magnitude Variations with Terrain Roughness

Friction force magnitude is directly proportional to terrain roughness. Even though the wind speed is the same in both panels in Figure 10-6, the terrain is rougher in the right panel; thus, the friction force is stronger.

Figure 10-7. Friction Force Magnitude Variations with Wind Speed

Friction force magnitude is directly proportional to wind speed. In Figure 10-7, wind speed is twice as strong in the right panel; thus, the friction force is doubled.

The frictional drag of the ground normally decreases with height and becomes insignificant above the lowest few thousand feet. However, this may vary somewhat, since both strong winds and rough terrain extend the friction layer to higher altitudes.

10.4 Upper Air Wind

In the atmosphere above the friction layer (lowest few thousand feet), only PGF and Coriolis force affect the horizontal motion of air. Remember that the PGF drives the wind and is oriented perpendicular to height contours. When a PGF is first established, wind begins to blow from higher to lower heights directly across the height contours. However, the instant air begins moving, Coriolis force deflects it to the right. Soon the wind is deflected a full 90° and is parallel to the height contours. At this time, Coriolis force exactly balances PGF, as shown in Figure 10-8 on a 500 mb constant-pressure chart (see Section 25.3 for information on constant-pressure charts). With the forces in balance, wind will remain parallel to height contours as shown in Figure 10-9. This is called the geostrophic wind.

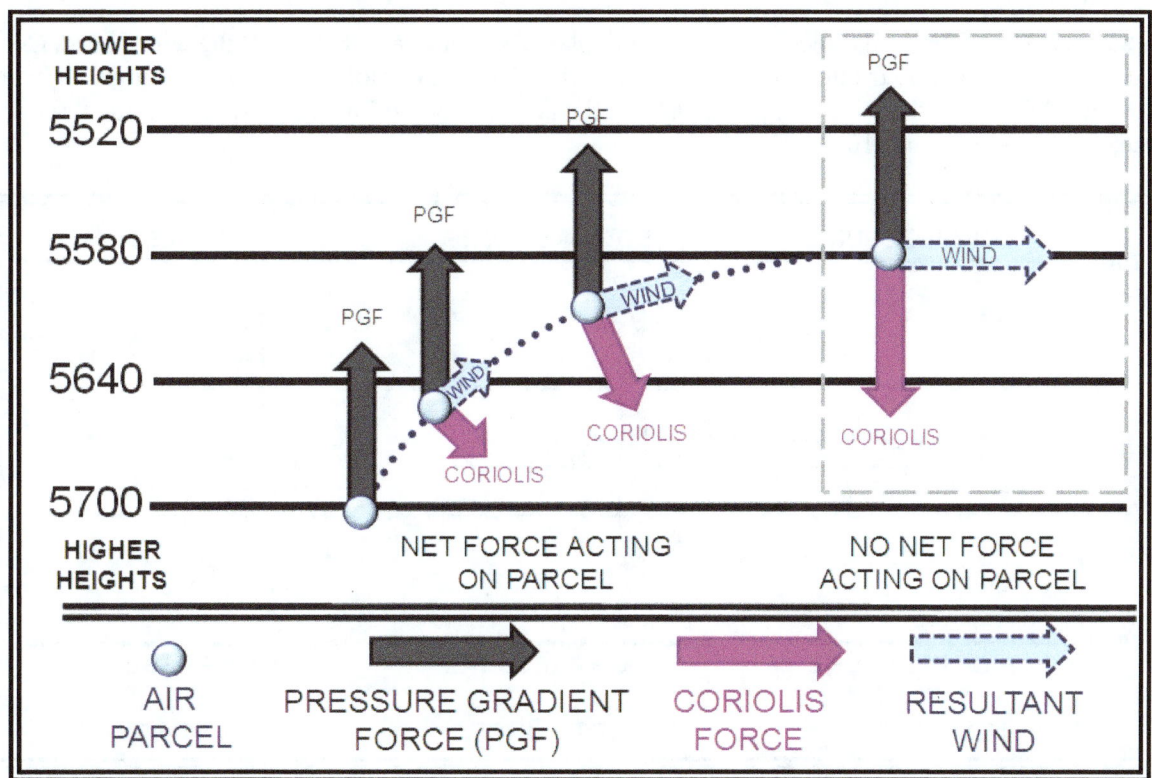

Figure 10-8. Geostrophic Wind

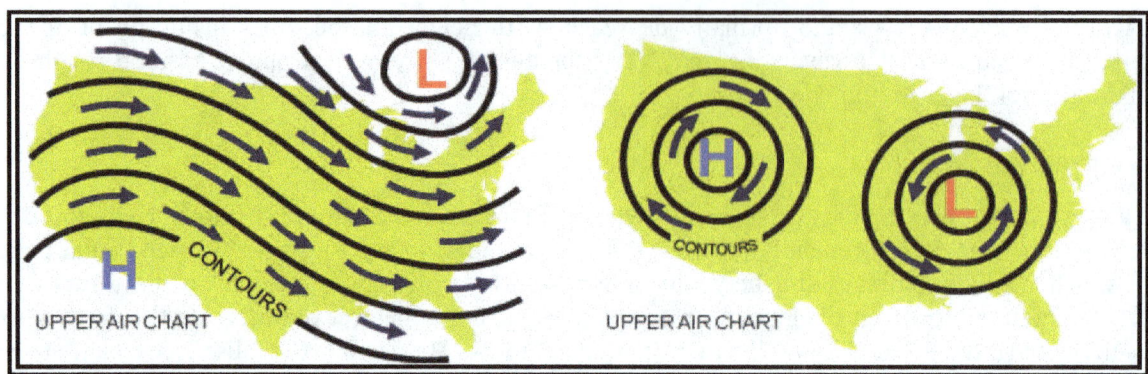

Figure 10-9. Upper Air Wind Flow

10.5 Surface Wind

At the surface of the Earth, all three forces come into play. As frictional force slows the wind speed, Coriolis force decreases. However, friction does not affect PGF. PGF and Coriolis force are no longer in balance. The stronger PGF turns the wind at an angle across the isobars toward lower pressure until the three forces balance, as shown in Figure 10-10.

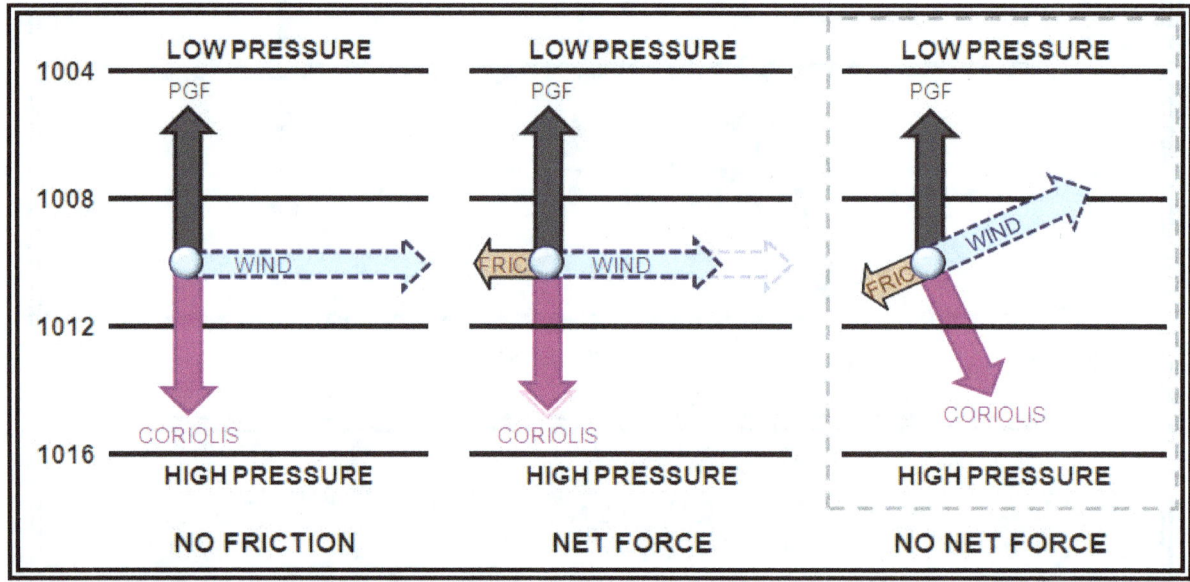

Figure 10-10. Surface Wind Forces

The angle of surface wind to isobars is about 10° over water, increasing to as high as 45° over rugged terrain. The end result is, in the Northern Hemisphere, the surface wind spirals clockwise and outward from high pressure, and counterclockwise and inward into low pressure (see Figure 10-11). In mountainous regions, one often has difficulty relating surface wind to pressure gradient because of immense friction and because of local terrain effects on pressure.

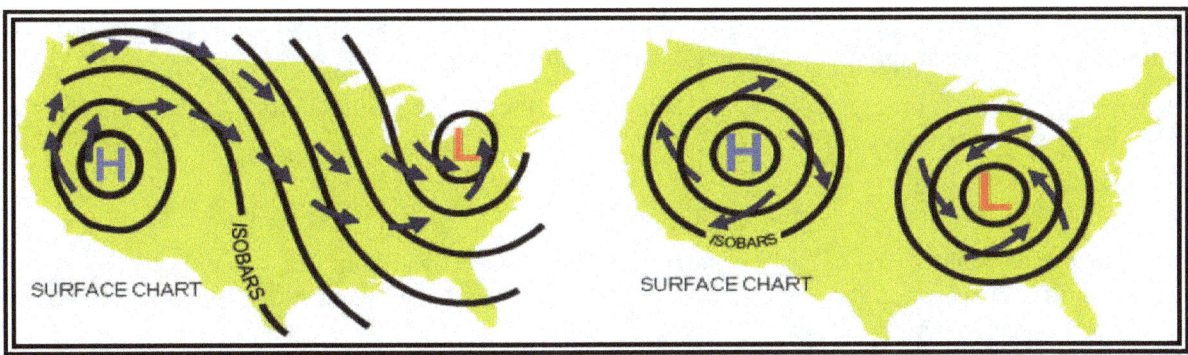

Figure 10-11. Surface Wind Flow

10.6 Local Winds

Local winds are small-scale wind field systems driven by diurnal heating or cooling of the ground. Air temperature differences develop over adjacent surfaces. Air in contact with the ground heats during the day and cools at night. Low-level pressure gradients develop with higher pressure over the cooler, denser air, and lower pressure over the warmer, less dense air (see Figure 10-12).

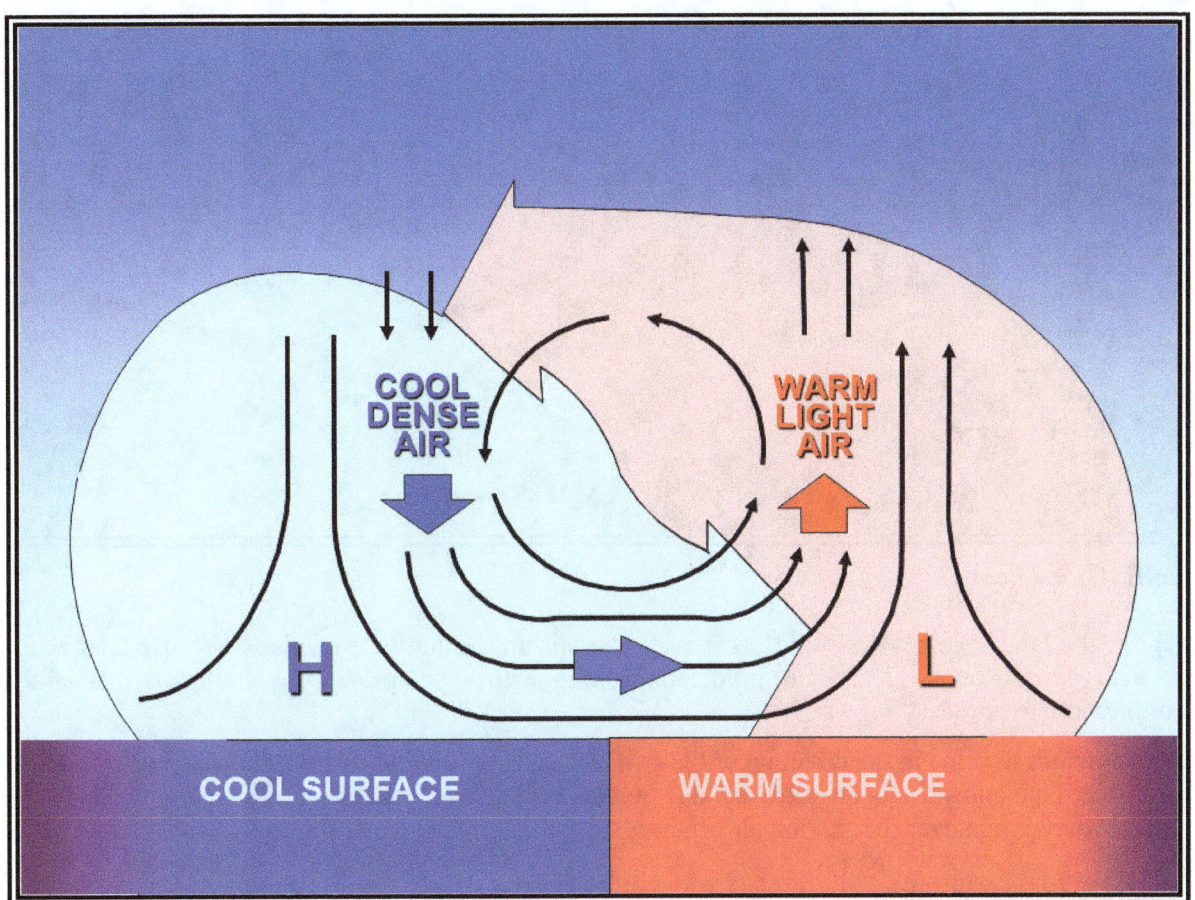

Figure 10-12. Local Wind Circulation

Low-level winds develop in the direction of the PGF. Coriolis force is insignificant, because the circulation's dimension (less than 100 mi) and lifespan (less than 12 hours) are too short for significant Coriolis deflection. Thus, the wind generally blows from a high-pressure cool surface to a low-pressure warm surface. Air rises over the warmer surface and sinks over the cooler surface. A local wind circulation is easiest to identify when synoptic-scale wind patterns are weak.

Local winds include: sea breeze, land breeze, lake breeze, lake effect, valley breeze, mountain-plains wind circulation, and mountain breeze.

10.6.1 Sea Breeze

A sea breeze (see Figure 10-13) is a coastal local wind that blows from sea to land and is caused by temperature differences when the sea surface is colder than the adjacent land. Sea breezes usually blow on relatively calm, sunny, summer days.

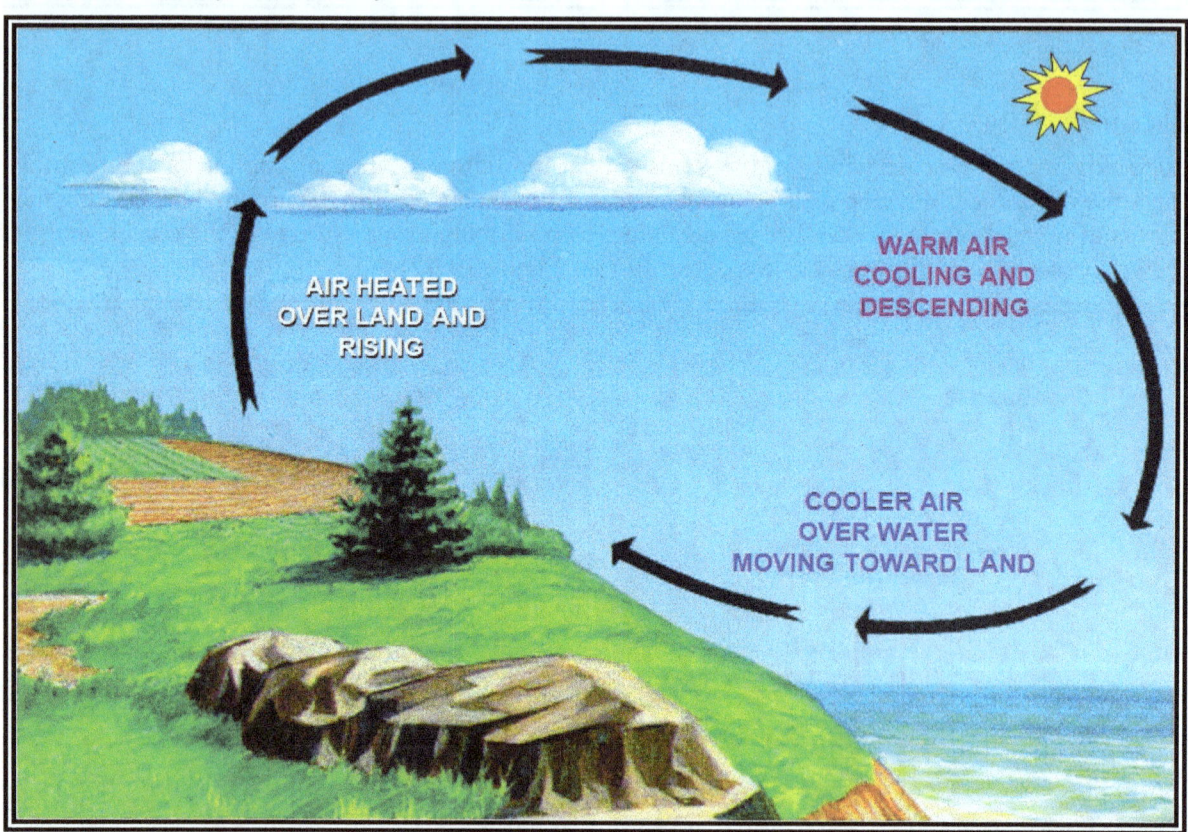

Figure 10-13. Sea Breeze

Air above the land becomes warmer (less dense) than air above the water. This is because land heats up faster than water. Low-level pressure gradients develop with lower pressure over the warmer land and higher pressure over the cooler water.

Low-level winds develop in the direction of the PGF. Thus, the wind blows from the water to the land. The air rises over land and sinks over water. Clouds (and precipitation) may develop in the rising air over land with cloud dissipation over the sinking air offshore.

10.6.1.1 Sea Breeze Front

A sea breeze front (see Figure 10-14) is the horizontal discontinuity in temperature and humidity that marks the leading edge of the intrusion of cooler, moister marine air associated with a sea breeze. It often produces a wind shift and enhanced cumulus clouds along its leading edge. Cumuliform clouds may be absent if the air mass being lifted over land is dry or stable.

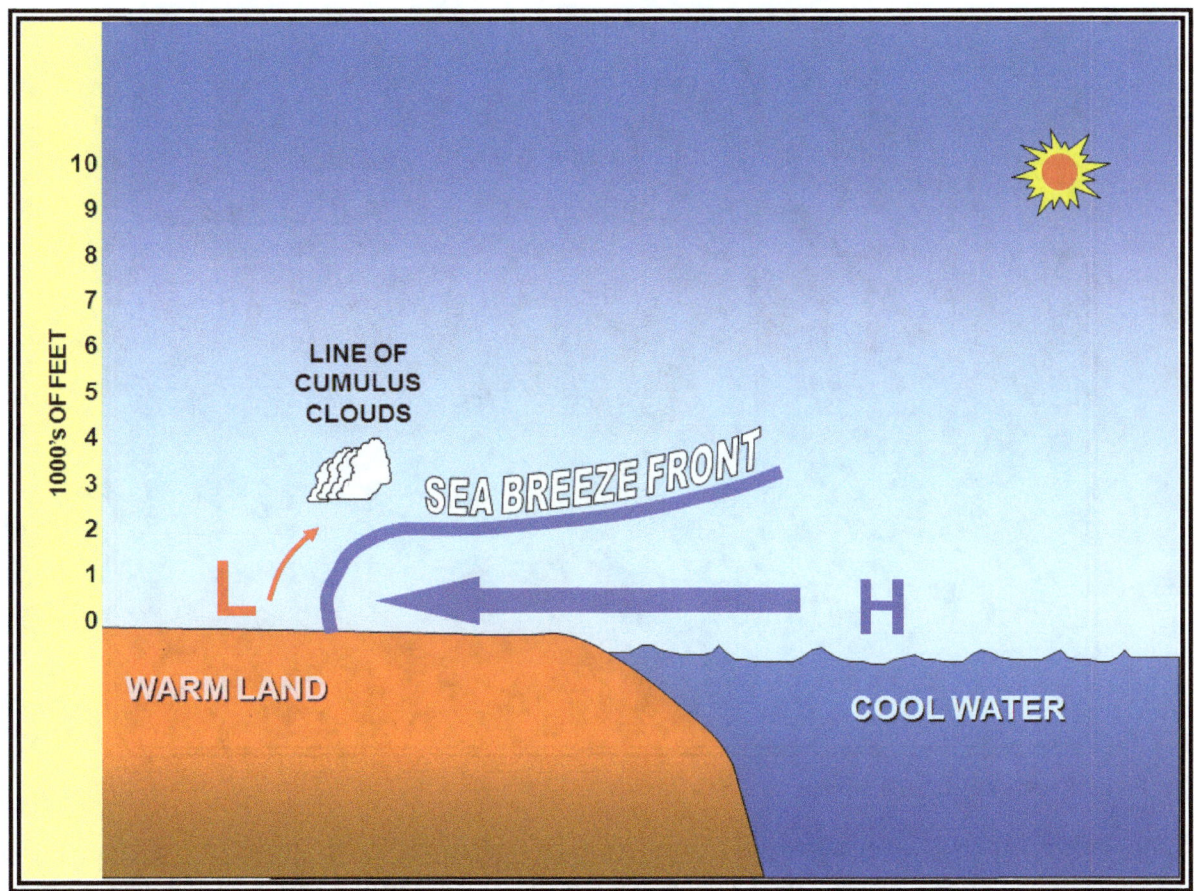

Figure 10-14. Sea Breeze Front

A sea breeze front's position and movement are influenced by coastline shape, low-level wind direction and speed, and temperature difference between land and sea surface. This temperature difference can be affected by the presence of cloud cover over land and the diurnal cycle. The depth of convection is usually too shallow for precipitation to develop. However, sea breeze fronts can be a lifting mechanism for shower and thunderstorm development.

10.6.1.2 Effects of Coastline Shape

Locally, the shape of the coastline plays an important role in the development of convection along sea breezes (see Figure 10-15). A narrow peninsula or island is generally an area of strong convective development during the late morning or early afternoon. This is because the sea breezes formed along opposing shores merge near the peninsula's or island's center.

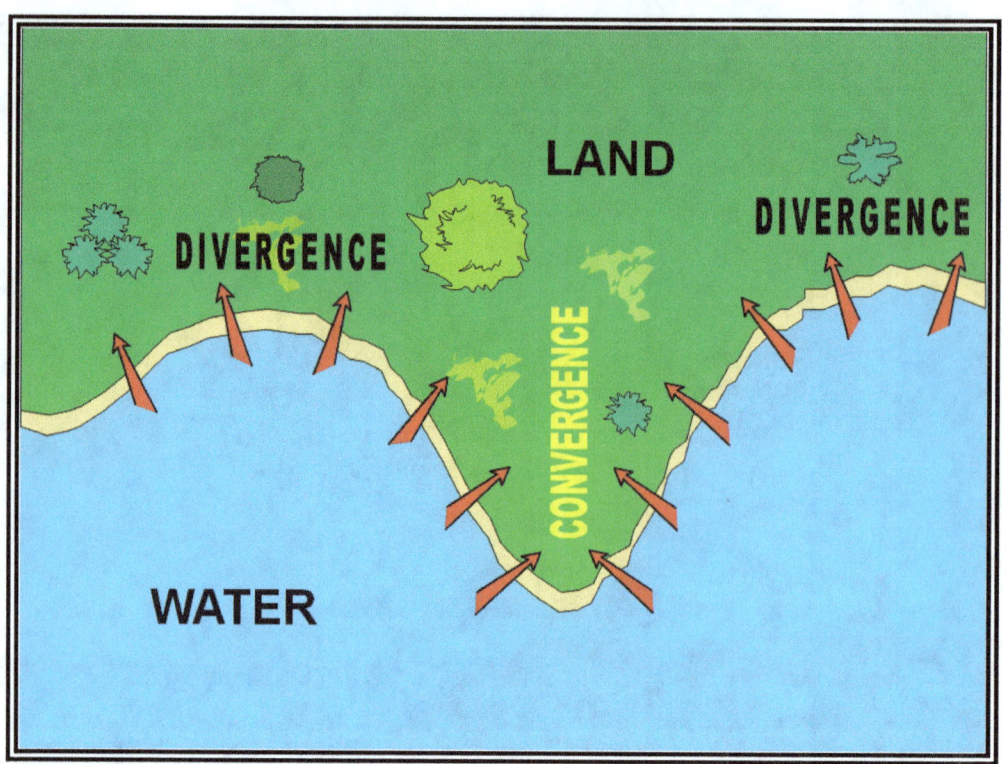

Figure 10-15. Effects of Coastline Shape on a Sea Breeze

In Figure 10-15, convergence occurs where sea breezes merge from opposite directions. Stronger lift may be sufficient to initiate showers and thunderstorms if the air mass is sufficiently moist and unstable.

10.6.2 Land Breeze

A land breeze (see Figure 10-16) is a coastal breeze blowing from land to sea caused by the temperature difference when the sea surface is warmer than the adjacent land. Land breezes usually occur at night and during early morning.

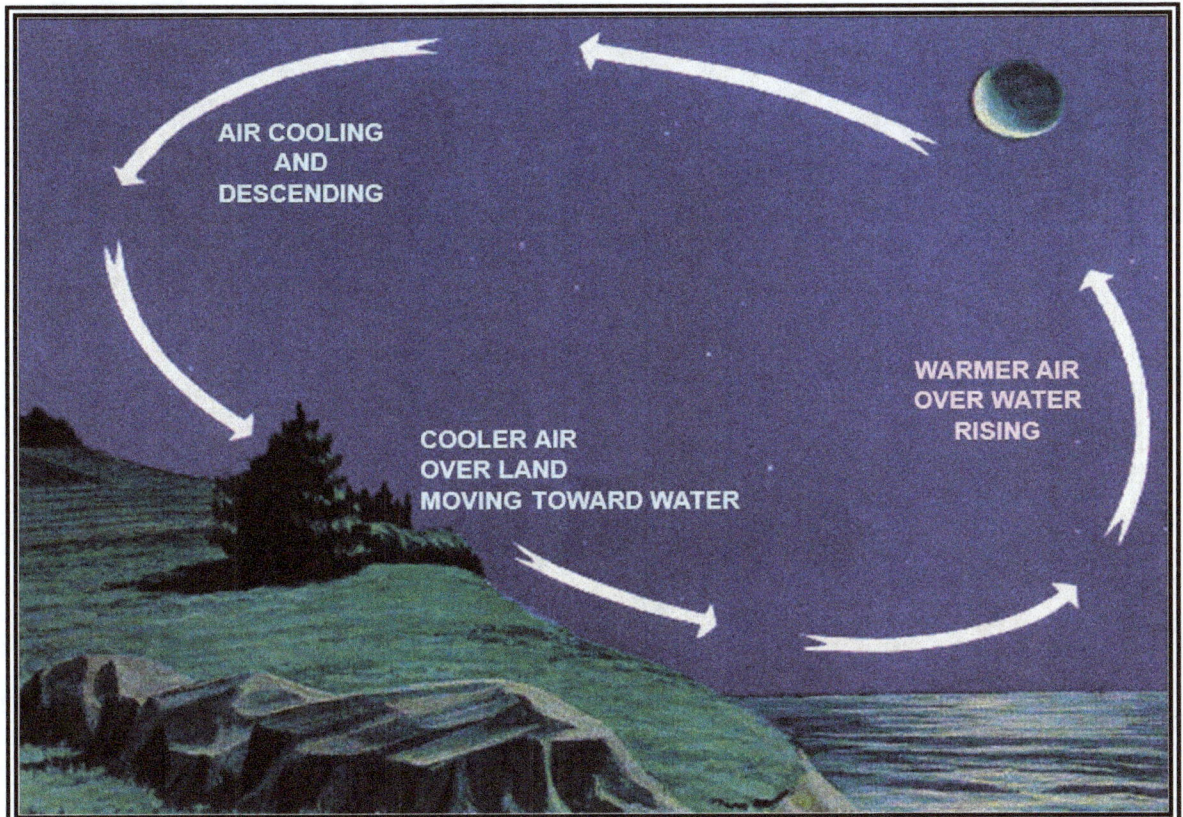

Figure 10-16. Land Breeze

Air above the land becomes cooler (denser) than air above the water due to conduction. This is because land cools faster than water. Low-level pressure gradients develop with higher pressure over the cooler land and lower pressure over the warmer water.

Low-level winds develop in the direction of the PGF. Thus, the wind blows from the land to the water. The land breeze is usually weaker than the sea breeze. The air rises over water and sinks over land. Clouds and precipitation may develop in the rising air over the water.

10.6.3 Lake Breeze

A lake breeze (see Figure 10-17) is a local wind that blows from the surface of a large lake onto the shores during the afternoon and is caused by the temperature difference when the lake surface is colder than the adjacent land. The lake breeze is similar in origin to the sea breeze, and is common in the Great Lakes. Both occur during the warm season, primarily spring and summer. Both are easiest to detect in light synoptic wind conditions.

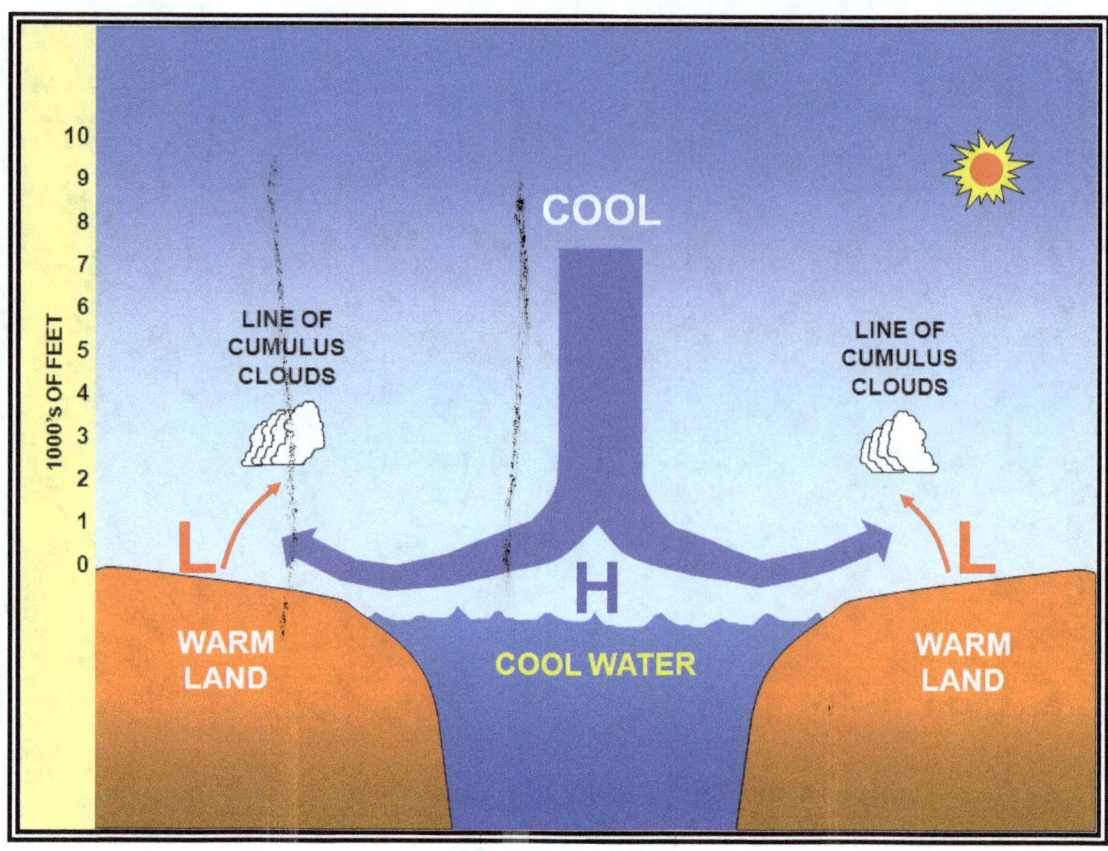

Figure 10-17. Lake Breeze

As with sea breezes, thunderstorms are favored in the upward motion branch of the lake breeze circulation. This is especially true where breezes from adjacent lakes collide.

The strength of the lake breeze circulation is affected by a lake's depth. A shallow lake (e.g., Lake Erie and Lake St. Clair) warms up rapidly and is less effective as the source of a lake breeze in summer than a deep lake (e.g., the other Great Lakes).

Figure 10-18. Sea Breeze/Lake Breeze Example (National Aeronautics and Space Administration (NASA))

In Figure 10-18, the sinking air behind the lake breeze inhibits clouds over Lake Ontario and Lake Erie and for miles inland.

10.6.4 Valley Breeze

A valley breeze (see Figure 10-19) is a wind that ascends a mountain valley during the day. Air in contact with the sloping terrain becomes warmer (less dense) than air above the valley. This is because the air in contact with the sloping terrain heats up faster than air above the valley.

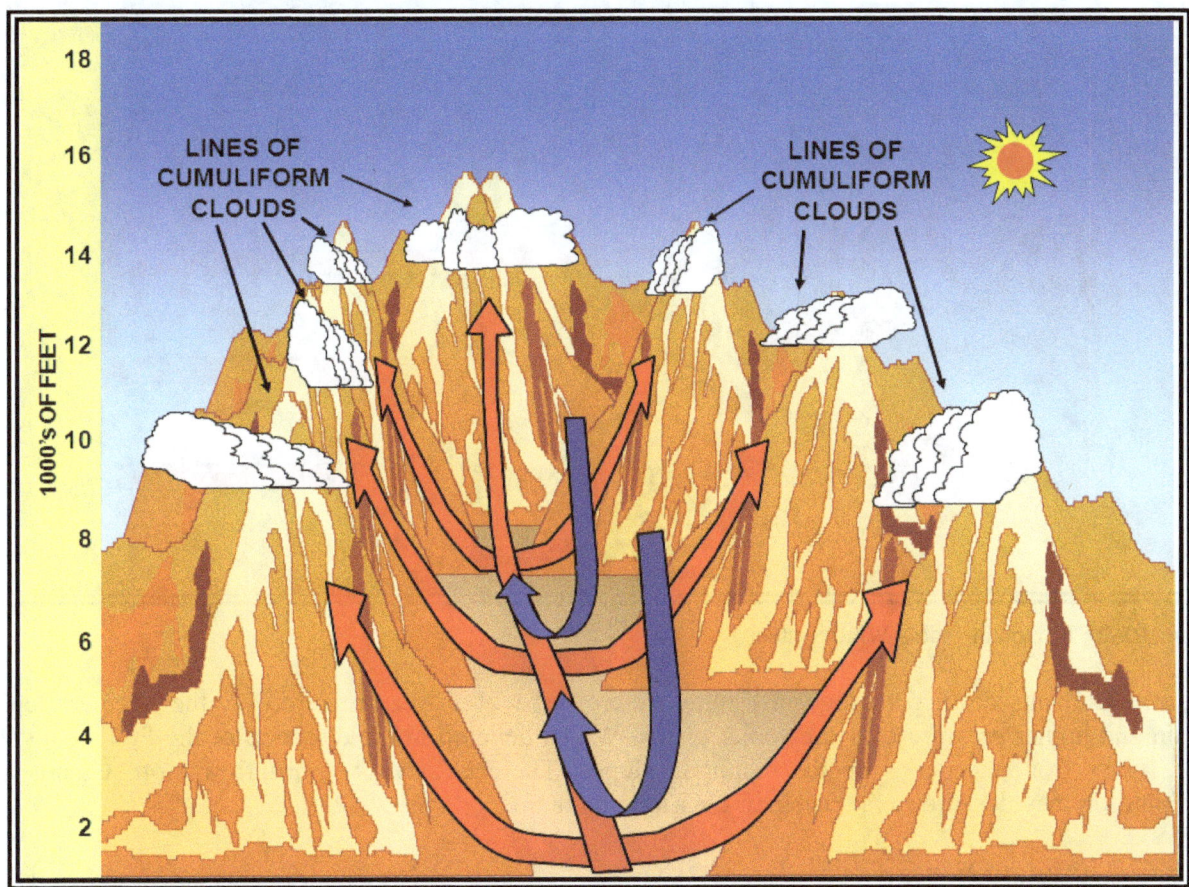

Figure 10-19. Valley Breeze

Pressure gradients develop (along a horizontal reference) with lower pressure over the warmer sloping terrain and higher pressure over the cooler valley. Winds develop in the direction of the PGF. Thus, the wind blows from the valley up the mountain slopes. Air rises over sloping terrain and sinks over the valley. Clouds and precipitation may develop over mountain slopes.

10.6.5 Mountain-Plains Wind System

A mountain-plains wind system (see Figure 10-20) is the diurnal cycle of local winds between a mountain or mountain range and the adjacent plains. During the daytime, this wind system is the equivalent of one-half of a valley breeze. Air in contact with the sloping terrain becomes warmer (less dense) than air above the plains. This is because the air in contact with the sloping terrain heats up faster than the air above the plains.

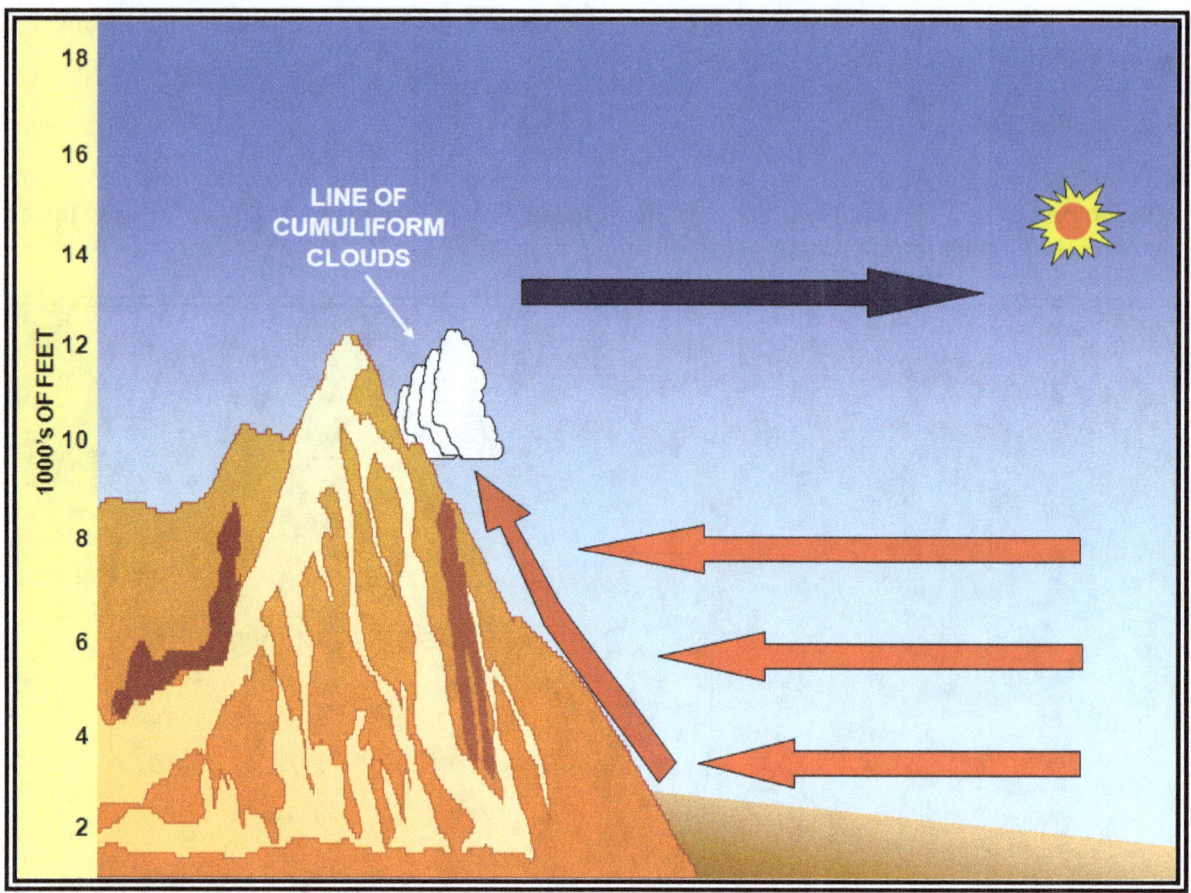

Figure 10-20. Mountain-Plains Wind System

Pressure gradients develop (along a horizontal reference) with lower pressure over the warmer sloping terrain and higher pressure over the cooler plains. Winds develop in the direction of the PGF. Thus, the wind blows from the plains up the mountain slopes. There is a weak return flow aloft. Clouds and precipitation may develop in the rising air over the mountain.

10.6.6 Mountain Breeze

A mountain breeze (see Figure 10-21) is the nightly downslope winds commonly encountered in mountain valleys. Air in contact with the sloping terrain cools faster than air above the valley. Pressure over the sloping terrain is higher than over the valleys (along a horizontal reference). Cooler air over the sloping terrain is denser than warmer air over the valley.

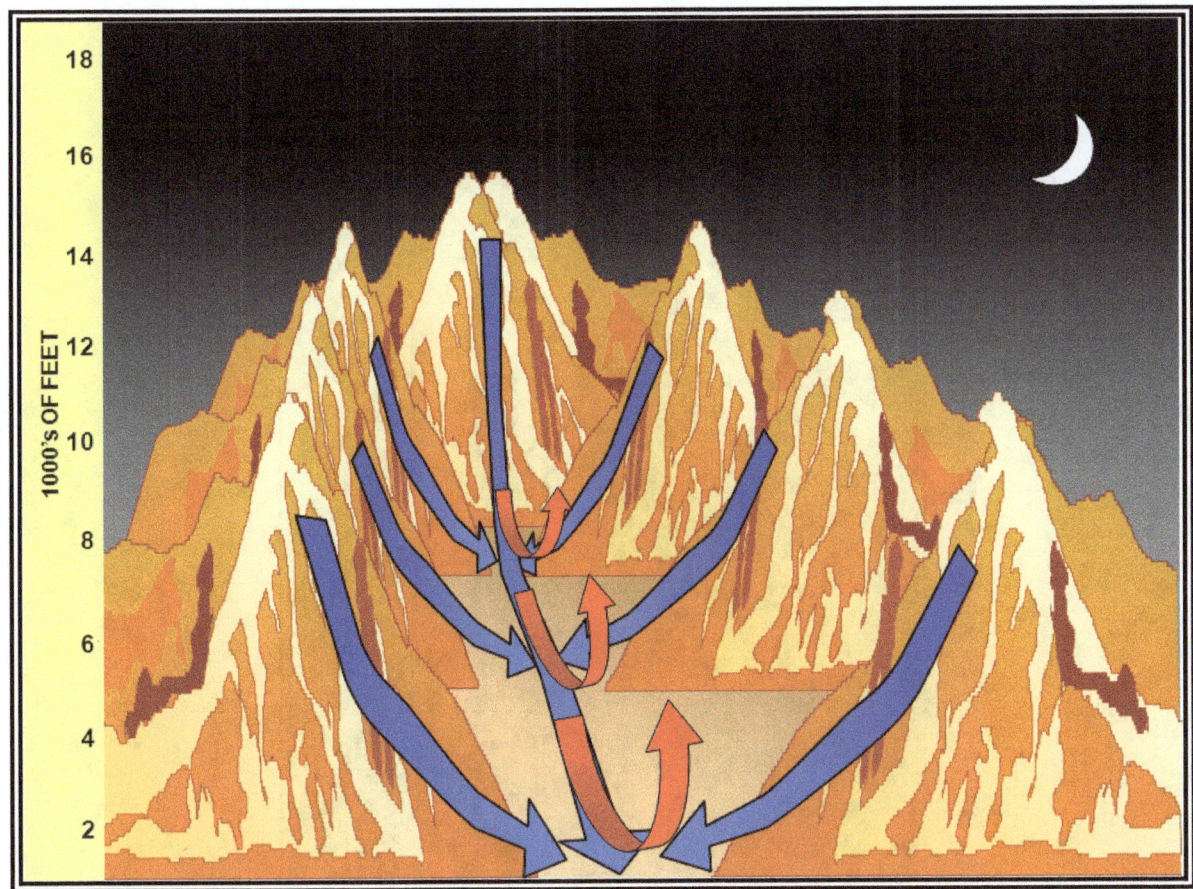

Figure 10-21. Mountain Breeze

Surface wind flows from the mountain down the sloping terrain into the valley. Air rises over the valley and sinks over the sloping terrain.

10.7 Adverse Winds

10.7.1 Crosswind

A crosswind is a wind that has a component directed perpendicularly to the heading of an aircraft (see Figure 10-22). The potential of drift produced by crosswind is critical to air navigation and can have its biggest impact during takeoff and landing. Airplanes take off and land more efficiently when oriented into the wind. The aircraft's groundspeed is minimized, a shorter runway is required to achieve lift-off, and the pilot has more time to make adjustments necessary for a smooth landing. As the wind turns more perpendicular to the runway to become a crosswind, the airplane's directional control is affected. If a pilot does not correctly compensate for the crosswind, the aircraft may drift off the side of the runway or side load on landing gear might occur. In extreme cases, the landing gear may collapse.

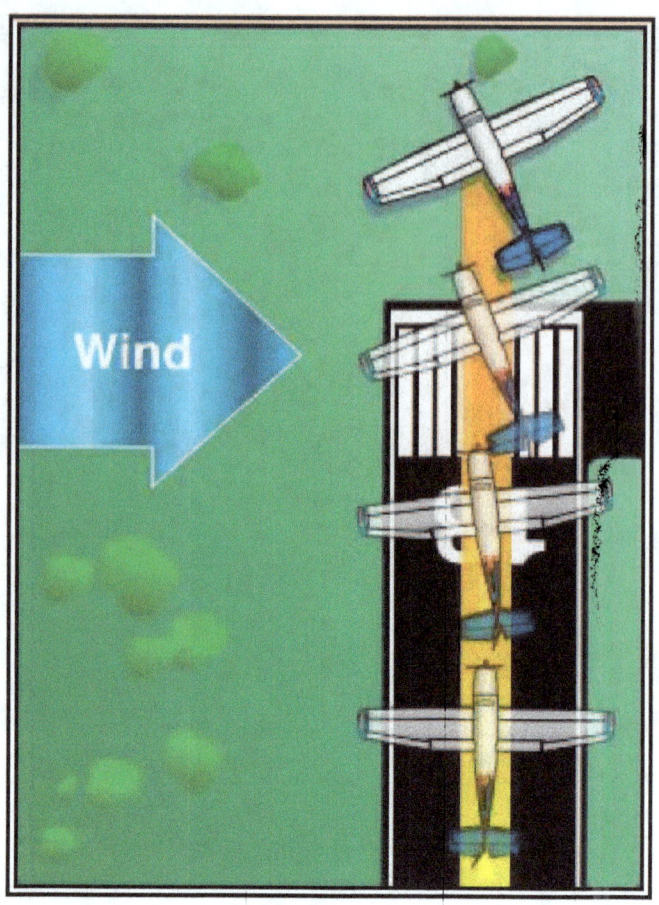

Figure 10-22. Crosswind Climb Flightpath

10.7.2 Gust

A gust is a fluctuation of wind speed with variations of 10 knots (kt) or more between peaks and lulls.

Even if the airplane is oriented into the wind, gusts during takeoff and landing cause airspeed fluctuations that can cause problems for pilots. A gust increases airspeed, which increases lift, and may cause an aircraft to briefly balloon up. Once the gust ends, a sudden decrease of airspeed occurs, which decreases lift and causes the aircraft to sink. Gusty winds at the point of touchdown provide significant challenges to a safe landing.

10.7.3 Tailwind

A tailwind is a wind with a component of motion from behind the aircraft.

A tailwind can be hazardous during both takeoff and landing. A longer takeoff roll is necessary because a higher groundspeed is needed to generate sufficient lift, and the aircraft may roll off the end of the runway before lift-off. Also, a smaller initial climb gradient occurs during takeoff, which may be insufficient to clear obstacles at the end of the runway. During a landing, a longer landing roll is needed because the aircraft will touch down at a higher groundspeed. Wind should always be considered in takeoff performance planning.

10.7.4 Variable Wind/Sudden Wind Shift

A variable wind is a wind that changes direction frequently, while a sudden wind shift is a line or narrow zone along which there is an abrupt change of wind direction. Both, even at low wind speeds, can make takeoffs and landings difficult. A headwind can quickly become a crosswind or tailwind.

10.7.5 Wind Shear

See Chapter 19, Turbulence, for information on wind shear.

10.7.6 Adverse Mountain Winds

See Chapter 16, Mountain Weather, for information on adverse mountain winds.

10.7.7 Atmospheric Disturbances in Mountainous Areas

See Chapter 19, Turbulence, for information on mountain-related turbulence.

11 Air Masses, Fronts, and the Wave Cyclone Model

11.1 Introduction

An air mass is a large body of air with generally uniform temperature and humidity.

A wave cyclone is a low-pressure circulation that forms and moves along a front.

This chapter will discuss air masses, fronts, and the wave cyclone model.

11.2 Air Masses

The area from which an air mass originates is called a source region.

Air mass source regions range from extensive snow-covered polar areas to deserts to tropical oceans. The United States is not a favorable source region because of the relatively frequent passage of weather disturbances that disrupt any opportunity for an air mass to stagnate and take on the properties of the underlying region. The longer the air mass stays over its source region, the more likely it will acquire the properties of the surface below.

11.2.1 Air Mass Classification

Air masses are classified according to the temperature and moisture properties of their source regions (see Figure 11-1).

11.2.1.1 Temperature Properties

- Arctic (A)—An extremely deep, cold air mass that develops mostly in winter over arctic surfaces of ice and snow.
- Polar (P)—A relatively shallow, cool to cold air mass that develops over high latitudes.
- Tropical (T)—A warm to hot air mass that develops over low latitudes.

11.2.1.2 Moisture Properties

- Continental (c)—A dry air mass that develops over land.
- Maritime (m)—A moist air mass that develops over water.

11.2.1.3 Five Air Masses

When this classification scheme is applied, the following five air masses (see Figure 11-1) may be identified:

- Continental Arctic (cA)—Cold, dry.
- Continental Polar (cP)—Cold, dry.
- Continental Tropical (cT)—Hot, dry.
- Maritime Polar (mP)—Cool, moist.
- Maritime Tropical (mT)—Warm, moist.

Note: Maritime Arctic (mA) is not listed, since it seldom (if ever) forms.

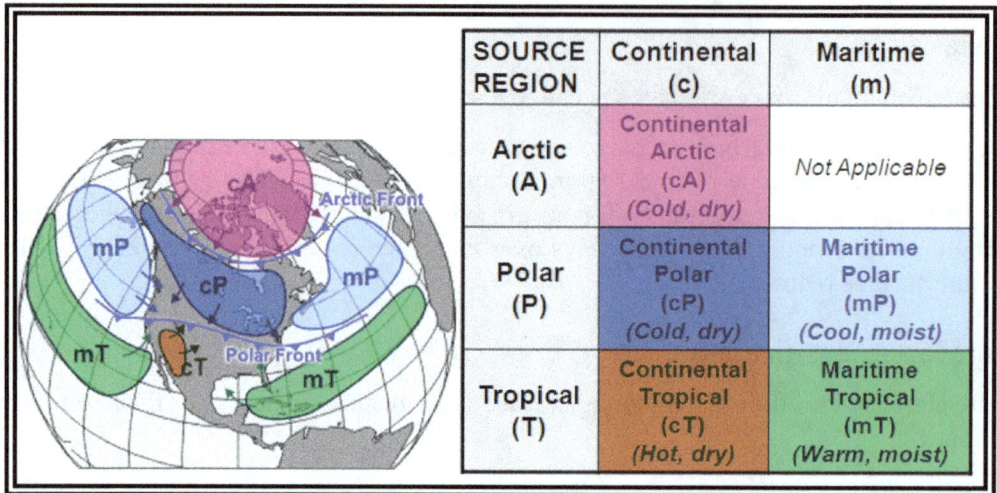

Figure 11-1. Air Mass Classification

11.2.2 Air Mass Modification

As these air masses move around the Earth, they can begin to acquire different attributes. For example, in winter, an arctic air mass (very cold and dry air) can move over the ocean, picking up some warmth and moisture from the warmer ocean and becoming a maritime polar (mP) air mass—one that is still fairly cold but contains moisture. If that same polar air mass moves south from Canada into the southern United States, it will pick up some of the warmth of the ground, but due to lack of moisture, it remains very dry. This is called a continental polar (cP) air mass.

The Gulf Coast states and the eastern third of the country commonly experience the tropical air mass in the summer. Continental tropical (cT) air is dry air pumped north, off of the Mexican Plateau. If it becomes stagnant over the Midwest, a drought may result. Maritime tropical (mT) air is air from the tropics that has moved north over cooler water.

A warm, moist air mass moving over a cold surface (see Figure 11-2) produces stable air associated with stratiform clouds, fog, and drizzle.

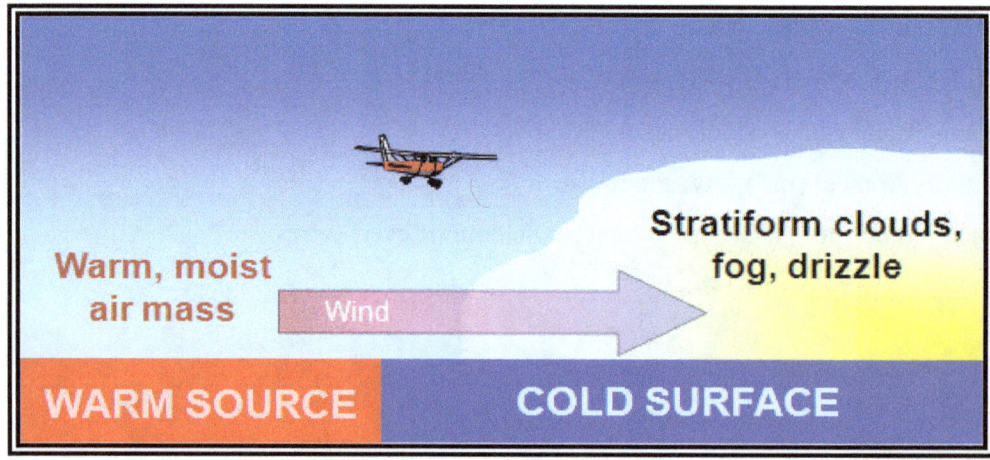

Figure 11-2. Air Mass Modification—Warm, Moist Air Mass Moving Over a Cold Surface

11.2.2.1 Lake Effect

Lake effect is the effect of any lake in modifying the weather near its shore and for some distance downwind. In the United States, the term is applied specifically to the region around the Great Lakes, and sometimes the Great Salt Lake in Utah. A lake effect can sometimes generate spectacular snowfall amounts to the lee side of the Great Lakes. This phenomenon is termed "lake effect snow."

In autumn and winter, cumuliform clouds and showers often develop in bands over, and to the lee of, large, ice-free lakes (see Figure 11-3). As initially cold, dry, stable polar air over land flows over the relatively warm water, the air is heated and moistened, and stability decreases. Shallow cumuliform clouds develop with low tops. The strength of the convection increases with increasing temperature differences between warm water and cold air, increasing wind speeds, and decreasing relative humidity within the cold, dry air.

Figure 11-3. Lake Effect

11.3 Fronts

Air masses can control the weather for a relatively long time period ranging from days to months. Most weather occurs along the periphery of these air masses at boundaries called fronts. A front is a boundary or transition zone between two air masses. Fronts are classified by which type of air mass (cold or warm) is replacing the other (see Figure 11-4).

FRONT	CHART SYMBOL	DEFINITION
Cold Front	▼▼	A front that moves in such a way that colder air replaces warmer air.
Warm Front	●●	A front that moves in such a way that warmer air replaces colder air.
Stationary Front	▼●	A front which is stationary or nearly so.
Occluded Front	▲●	A composite of two fronts as a cold front overtakes a warm front or stationary front.

Note: Frontal symbols point in the direction of frontal movement.

Figure 11-4. Fronts

Fronts are usually detectable at the surface in a number of ways: significant temperature gradients, or differences, exist along fronts (especially on the cold air side); winds usually converge, or come together, at fronts; and pressure typically decreases as a front approaches and increases after it passes.

Fronts do not exist only at the surface of the Earth; they have a vertical structure in which the front slopes over the colder (denser) air mass.

11.3.1 Warm Front

A warm front occurs when a warm mass of air advances and replaces a body of colder air. Warm fronts move slowly, typically 10 to 25 mph. The slope of the advancing front slides over the top of the cooler air and gradually pushes it out of the area. Warm fronts typically have a gentle slope, so the warm air rising along the frontal surface is gradual (see Figure 11-5). This favors the development of widespread layered or stratiform cloudiness and precipitation along, and ahead of, the front if the warm rising air is stable.

Generally, prior to the passage of a warm front, cirriform or stratiform clouds, along with fog, can be expected to form along the frontal boundary. In the summer months, cumulonimbus clouds (thunderstorms) are likely to develop.

Light to moderate precipitation is probable, usually in the form of rain, sleet, snow, or drizzle, accentuated by poor visibility. The wind blows from the south-southeast, and the outside temperature is cool or cold with an increasing dewpoint. Finally, as the warm front approaches, the barometric pressure continues to fall until the front passes completely.

During the passage of a warm front, stratiform clouds are visible and drizzle may be falling. The visibility is generally poor, but improves with variable winds. The temperature rises steadily from the inflow of relatively warmer air. For the most part, the dewpoint remains steady and the pressure levels off. After the passage of a warm front, stratocumulus clouds predominate and rain showers are possible. The visibility eventually improves, but hazy conditions may exist for a short period after passage. The wind blows from the south-southwest. With warming temperatures, the dewpoint rises and then levels off. There is generally a slight rise in barometric pressure, followed by a decrease of barometric pressure.

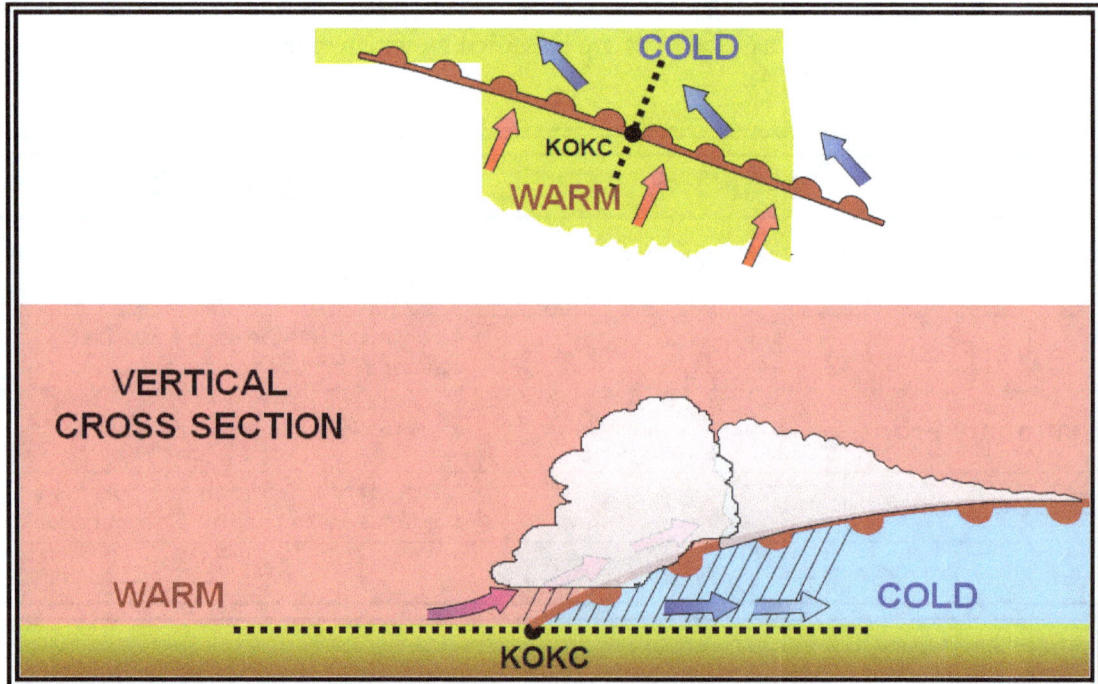

Figure 11-5. Warm Front

11.3.2 Cold Front

A cold front occurs when a mass of cold, dense, and stable air advances and replaces a body of warmer air. It is so dense, it stays close to the ground and acts like a snowplow, sliding under the warmer air and forcing the less dense air aloft. Cold fronts have a steep slope, and the warm air is forced upward abruptly (see Figure 11-6). This often leads to a narrow band of showers and thunderstorms along, or just ahead of, the front if the warm rising air is unstable. Cold fronts move more rapidly than warm fronts, progressing at a rate of 25 to 30 mph. However, extreme cold fronts have been recorded moving at speeds of up to 60 mph.

The rapidly ascending air causes the temperature to decrease suddenly, forcing the creation of clouds. The type of clouds that form depends on the stability of the warmer air mass. A cold front in the Northern Hemisphere is normally oriented in a northeast to southwest manner and can be several hundred miles long, encompassing a large area of land.

Prior to the passage of a typical cold front, cirriform or towering cumulus clouds are present, and cumulonimbus clouds may develop. Rain showers may also develop due to the rapid development of clouds. A high dewpoint and falling barometric pressure are indicative of an imminent cold front passage.

As the cold front passes, towering cumulus or cumulonimbus clouds continue to dominate the sky. Depending on the intensity of the cold front, heavy rain showers form and may be accompanied by lightning, thunder, and/or hail. More severe cold fronts can also produce tornadoes. During cold front passage, the visibility is poor with winds variable and gusty, and the temperature and dewpoint drop rapidly. A quickly falling barometric pressure bottoms out during frontal passage, and then begins a gradual increase. After frontal passage, the towering cumulus and cumulonimbus clouds begin to dissipate to cumulus clouds with a corresponding decrease in the precipitation. Good visibility eventually prevails with the winds from the west-northwest. Temperatures remain cooler and the barometric pressure continues to rise.

Fast-moving cold fronts are pushed by intense pressure systems far behind the actual front. The friction between the ground and the cold front retards the movement of the front and creates a steeper frontal surface. This results in a very narrow band of weather, concentrated along the leading edge of the front. If the warm air being overtaken by the cold front is relatively stable, overcast skies and rain may occur for some distance behind the front. If the warm air is unstable, scattered thunderstorms and rain showers may form. A continuous line of thunderstorms, or squall line, may form along or ahead of the front. Squall lines present a serious hazard to pilots as squall-type thunderstorms are intense and move quickly. Behind a fast-moving cold front, the skies usually clear rapidly, and the front leaves behind gusty, turbulent winds and colder temperatures.

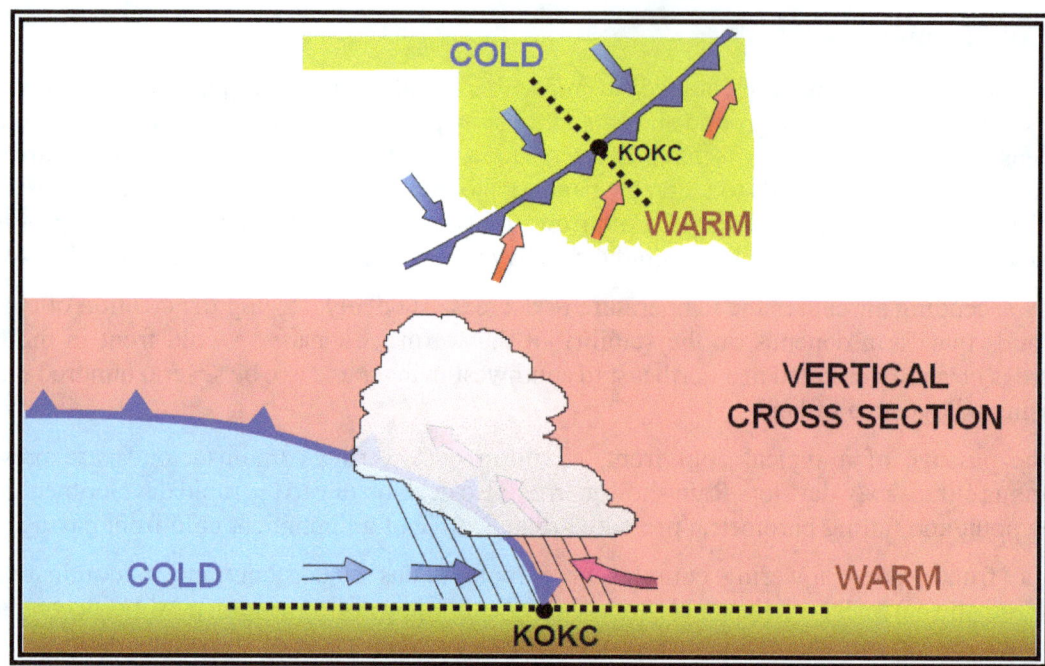

Figure 11-6. Cold Front

11.3.3 Stationary Front

When the forces of two air masses are relatively equal, the boundary or front that separates them remains stationary and influences the local weather for days. This front is called a stationary front. Stationary frontal slope can vary, but clouds and precipitation would still form in the warm rising air along the front (see Figure 11-7). The weather associated with a stationary front is typically a mixture that can be found in both warm and cold fronts.

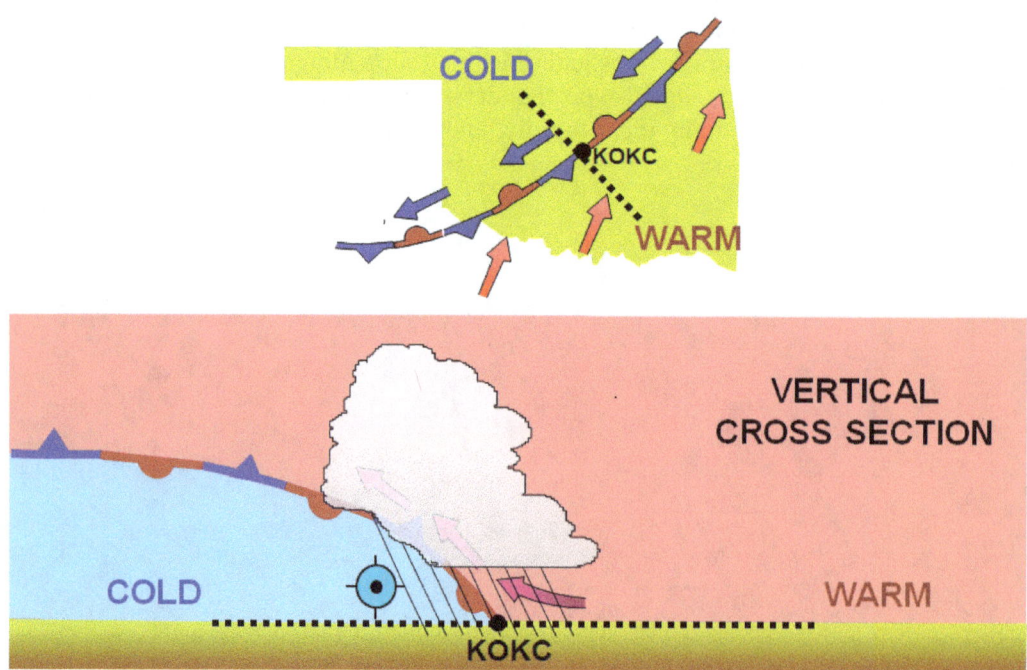

Figure 11-7. Stationary Front

Chapter 11, Air Masses, Fronts, and the Wave Cyclone Model

11.3.4 Occluded Front

Cold fronts typically move faster than warm fronts, so in time they catch up to warm fronts. As the two fronts merge, an occluded front forms (see Figure 11-8). At the occluded front, the cold air undercuts the retreating cooler air mass associated with the warm front, further lifting the already rising warm air. Clouds and precipitation can occur in the areas of frontal lift along, ahead of, and behind the surface position of an occluded front. As the occluded front approaches, warm front weather prevails, but is immediately followed by cold front weather.

There are two types of occluded fronts that can occur, and the temperatures of the colliding frontal systems play a large part in defining the type of front and the resulting weather.

A cold front occlusion occurs when a fast-moving cold front is colder than the air ahead of the slow-moving warm front. When this occurs, the cold air replaces the cool air and forces the warm front aloft into the atmosphere. Typically, the cold front occlusion creates a mixture of weather found in both warm and cold fronts, providing the air is relatively stable.

A warm front occlusion occurs when the air ahead of the warm front is colder than the air of the cold front. When this is the case, the cold front rides up and over the warm front. If the air forced aloft by the warm front occlusion is unstable, the weather is more severe than the weather found in a cold front occlusion. Embedded thunderstorms, rain, and fog are likely to occur.

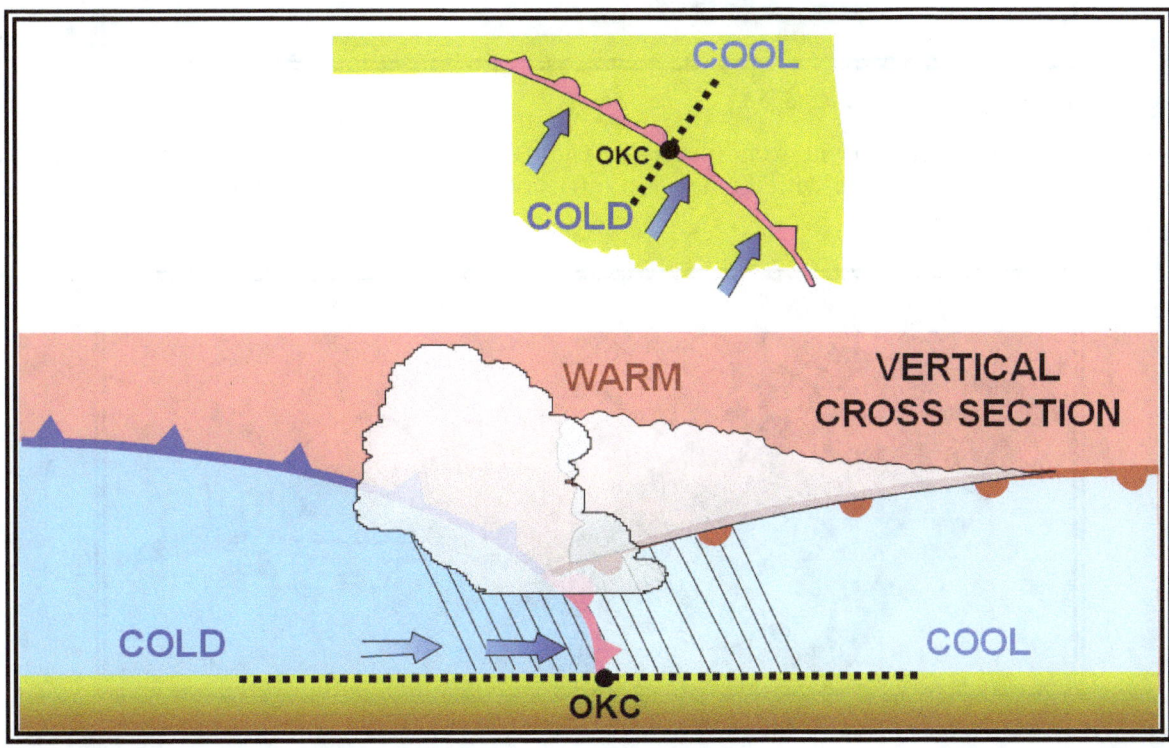

Figure 11-8. Occluded Front

11.4 The Wave Cyclone Model

A wave cyclone[3] is a low-pressure circulation that forms and moves along a front. The circulation about the cyclone center tends to produce a wavelike kink along the front. Wave cyclones are the primary weather producers in the mid-latitudes. They are large lows that generally travel from west to east along a front. They last from a few days to more than a week.

A wave cyclone typically follows a predictable evolution. Initially, there is a stationary front separating warm air from cold air (see Figure 11-9).

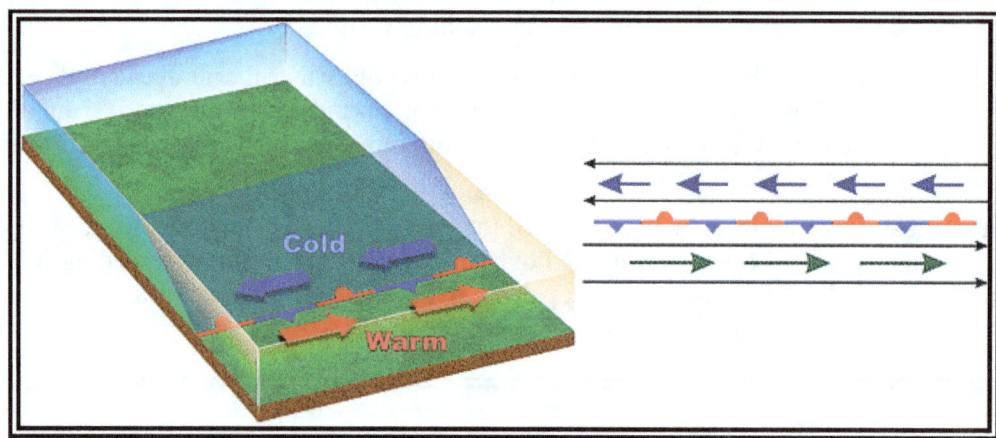

Figure 11-9. Wave Cyclone Model—Stage 1

A low-pressure wave forms on the front (see Figure 11-10). The front develops a kink where the wave develops. Precipitation develops with the heaviest intensity (dark green) located in the zone of lift along the front.

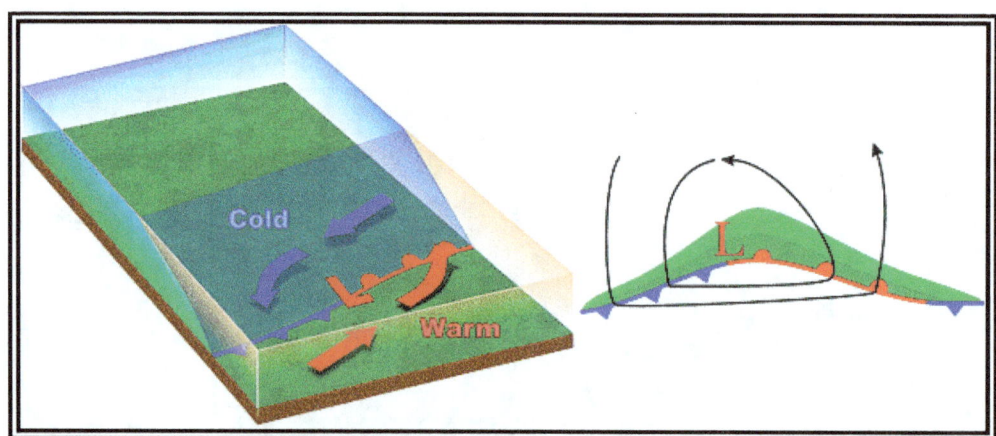

Figure 11-10. Wave Cyclone Model—Stage 2

As the wave intensifies, both the cold and warm fronts become better organized (see Figure 11-11).

---

[3] A wave cyclone should not be confused with the alternative name for a tornado. They are quite different.

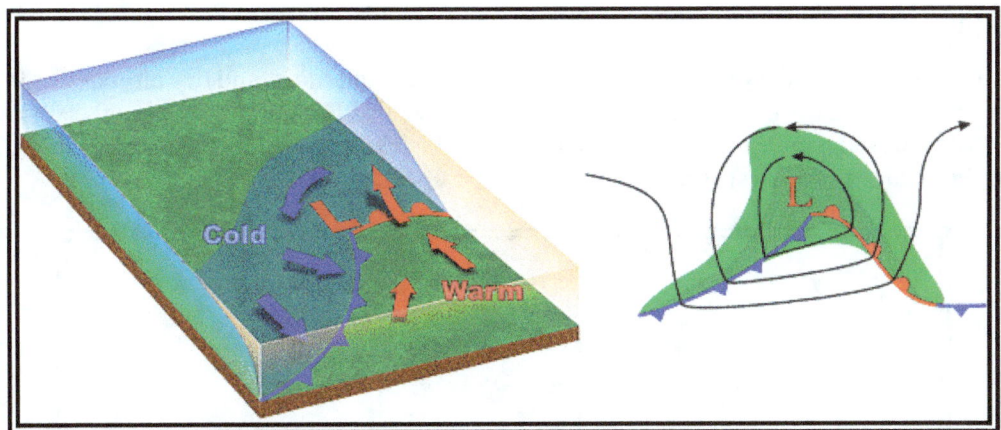

Figure 11-11. Wave Cyclone Model—Stage 3

In the fourth stage, the wave becomes a mature low (see Figure 11-12). The occluded front forms as the cold front overtakes the warm front.

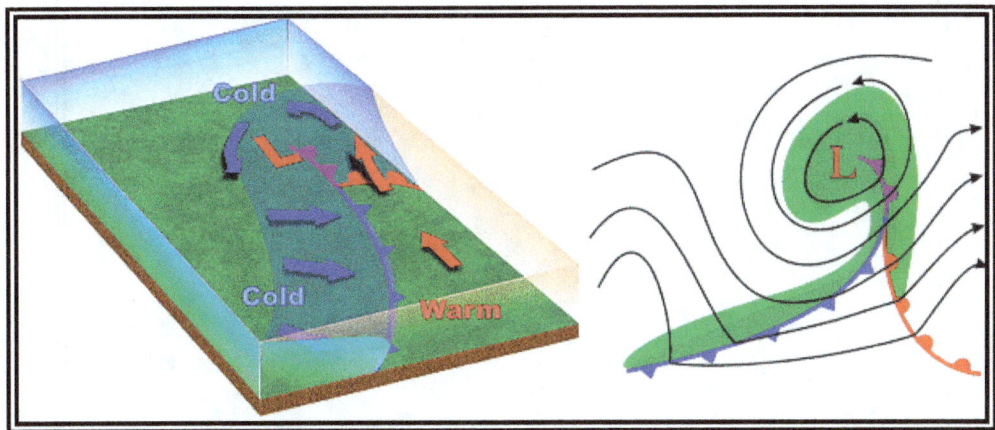

Figure 11-12. Wave Cyclone Model—Stage 4

As the cold front continues advancing on the warm front, the occlusion increases and eventually cuts off the supply of warm moist air (see Figure 11-13). This causes the low to gradually dissipate.

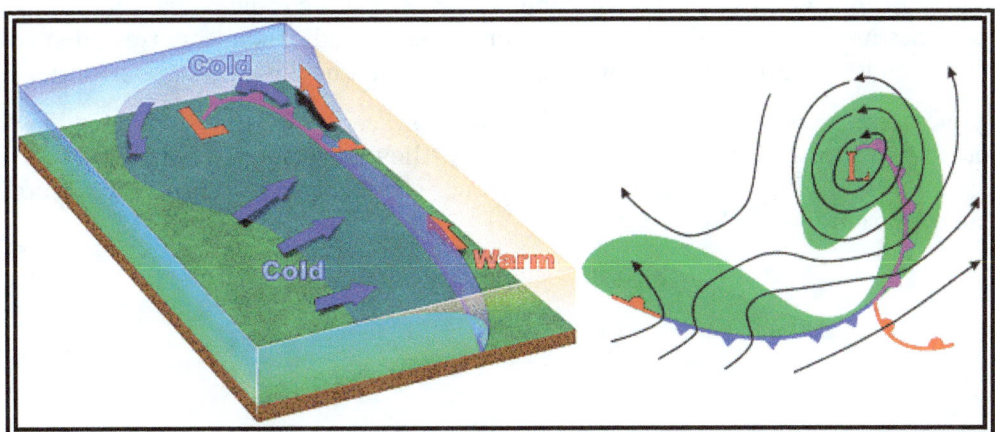

Figure 11-13. Wave Cyclone Model—Stage 5

11.5 Dryline

A dryline is a low-level boundary hundreds of miles long separating moist and dry air masses. In the United States, it typically lies north-south across the southern and central High Plains during the spring and early summer, where it separates moist (mT) air from the Gulf of Mexico to the east and dry desert (cT) air from the southwestern states to the west (see Figure 11-14).

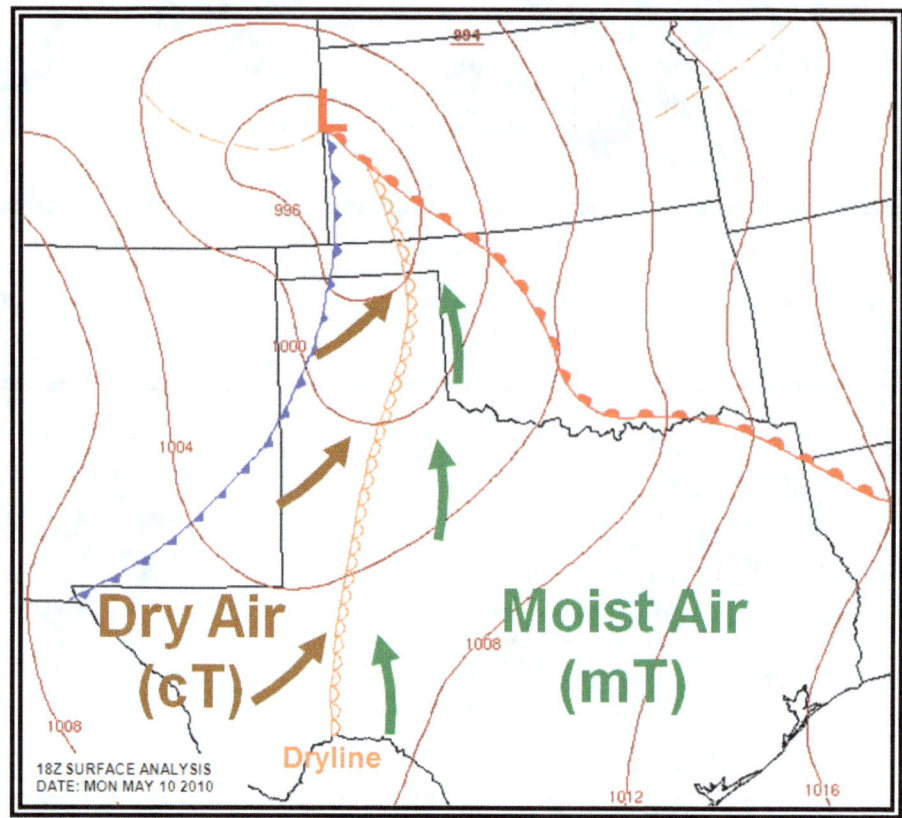

Figure 11-14. Dryline Example

The dryline typically advances eastward during the afternoon and retreats westward at night. However, a strong wave cyclone can sweep the dryline eastward into the Mississippi Valley, or even farther east, regardless of the time of day. Low-level clouds and early morning fog often prevail in the moist air, while generally clear skies mark the dry side. Severe and sometimes tornadic thunderstorms often develop along a dryline or in the moist air just to the east of it, especially when it begins moving eastward.

A typical dryline passage results in a sharp drop in humidity (hence the name), clearing skies, and a wind shift from south or southeasterly to west or southwesterly. Blowing dust and rising temperatures also may follow, especially if the dryline passes during the daytime. These changes occur in reverse order when the dryline retreats westward.

12 Vertical Motion and Clouds

12.1 Introduction

A cloud is a visible aggregate of minute water droplets and/or ice particles in the atmosphere above the Earth's surface. Fog differs from clouds only in that the base of fog is at the Earth's surface while clouds are above the surface. Clouds are like signposts in the sky that provide information on air motion, stability, and moisture. Clouds help pilots visualize weather conditions and potential weather hazards.

Clouds form in the atmosphere as a result of condensation of water vapor in rising currents of air, or by the evaporation of the lowest layer of fog. Rising currents of air are necessary for the formation of vertically deep clouds capable of producing precipitation heavier than light intensity.

12.2 Vertical Motion Effects on an Unsaturated Air Parcel

As a bubble or parcel of air ascends (rises), it moves into an area of lower pressure (pressure decreases with height). As this occurs, the parcel expands. This requires energy (or work), which takes heat away from the parcel, so the air cools as it rises (see Figure 12-1). This is called an adiabatic process. The term "adiabatic" means that no heat transfer occurs into, or out of, the parcel. Air has low thermal conductivity (see Table 5-3), so transfer of heat by conduction is negligibly small.

The rate at which the parcel cools as it is lifted is called the lapse rate. The lapse rate of a rising, unsaturated parcel (air with relative humidity less than 100 percent) is approximately 3 °C per 1,000 ft (9.8 °C per km). This is called the *dry adiabatic* lapse rate. This means that for each 1,000-ft increase in elevation, the parcel's temperature decreases by 3 °C. Concurrently, the dewpoint decreases approximately 0.5 °C per 1,000 ft (1.8 °C per km). The parcel's temperature-dewpoint spread decreases, while its relative humidity increases.

This process is reversible if the parcel remains unsaturated and, thus, does not lose any water vapor. A descending (subsiding) air parcel compresses as it moves into an area of higher pressure. The atmosphere surrounding the parcel does work on the parcel, and energy is added to the compressed parcel, which warms it. Thus, the temperature of a descending air parcel increases approximately 3 °C per 1,000 ft (9.8 °C per km). Concurrently, the dewpoint increases approximately 0.5 °C per 1,000 ft (1.8 °C per km). The parcel's temperature-dewpoint spread increases, while its relative humidity decreases.

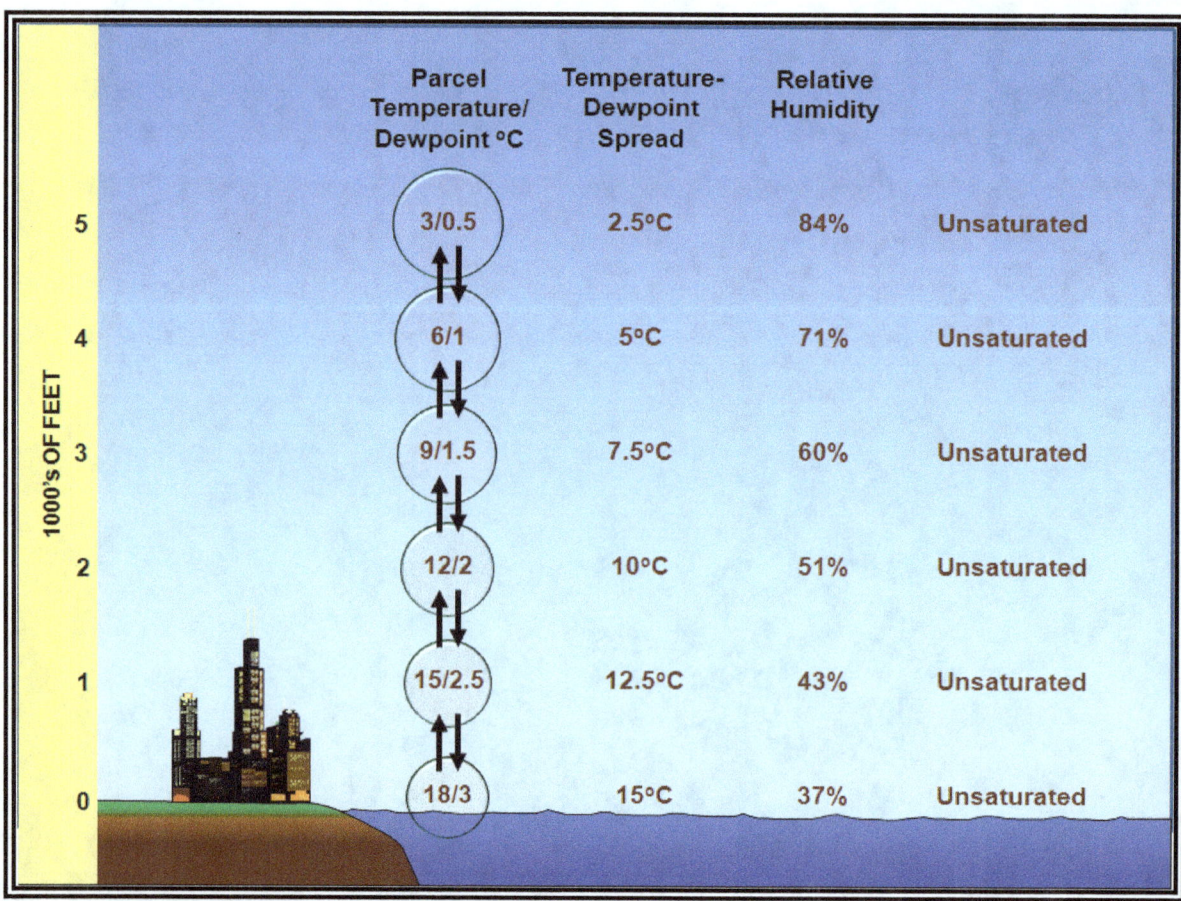

Figure 12-1. Unsaturated Ascending/Descending Air Parcel Example

12.3 Vertical Motion Effects on a Saturated Air Parcel

The Lifted Condensation Level (LCL) is the level at which a parcel of moist air lifted dry adiabatically becomes saturated. At this altitude, the temperature-dewpoint spread is zero and relative humidity is 100 percent.

Further lifting of the saturated parcel results in condensation, cloud formation, and latent heat release. Because the heat added during condensation offsets some of the cooling due to expansion, the parcel now cools at the moist adiabatic lapse rate, which varies between approximately 1.2 °C per 1,000 ft (4 °C per km) for very warm saturated parcels to 3 °C per 1,000 ft (9.8 °C per km) for very cold saturated parcels. Concurrently, the parcel's dewpoint decreases at an identical rate. For simplicity, examples shown in this handbook use a moist adiabatic lapse rate of 2 °C per 1,000 ft. Regardless of temperature, the relative humidity remains constant at about 100 percent.

As the saturated air parcel expands and cools, its water vapor content decreases (see Figure 12-2). This occurs because some of the water vapor is condensed to water droplets or deposited into ice crystals to form a cloud. This process is triggered by the presence of microscopic cloud condensation (and ice) nuclei, such as dust, clay, soot, sulfate, and sea salt particles. The cloud grows vertically deeper as the parcel continues to rise.

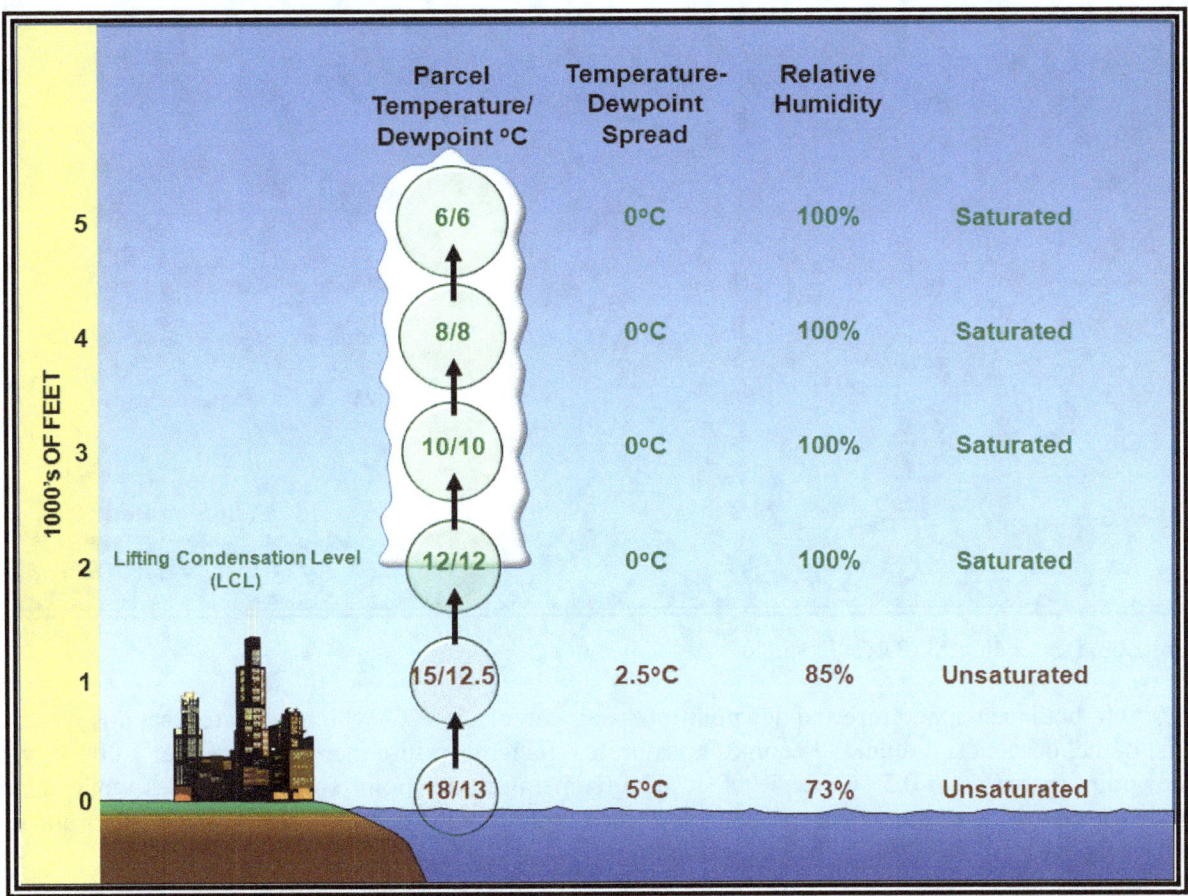

Figure 12-2. Ascending Air Parcel That Becomes Saturated Example

In Figure 12-2, at the surface, the air parcel has a temperature of 18 °C and a dewpoint of 13 °C, which indicates that it is unsaturated. As the parcel ascends, its temperature decreases at the dry adiabatic lapse rate of 3 °C per 1,000 ft, while the dewpoint decreases at 0.5 °C per 1,000 ft. The temperature-dewpoint spread decreases while relative humidity increases until the parcel achieves saturation at its LCL of 2,000 ft. As the parcel continues to ascend, condensation produces cloud formation. Because the heat added during condensation offsets some of the cooling due to expansion, the parcel now cools at the moist adiabatic lapse rate of 2 °C per 1,000 ft. The parcel's dewpoint decreases at an identical rate as the lost water vapor condenses to form the cloud. The relative humidity of the ascending saturated (i.e., cloudy) parcel remains constant at about 100 percent.

A descending saturated air parcel quickly becomes unsaturated (see Figure 12-3). Its temperature increases at 3 °C per 1,000 ft, while its dewpoint increases at 0.5 °C per 1,000 ft (see Table 12-1). The temperature-dewpoint spread increases while relative humidity decreases.

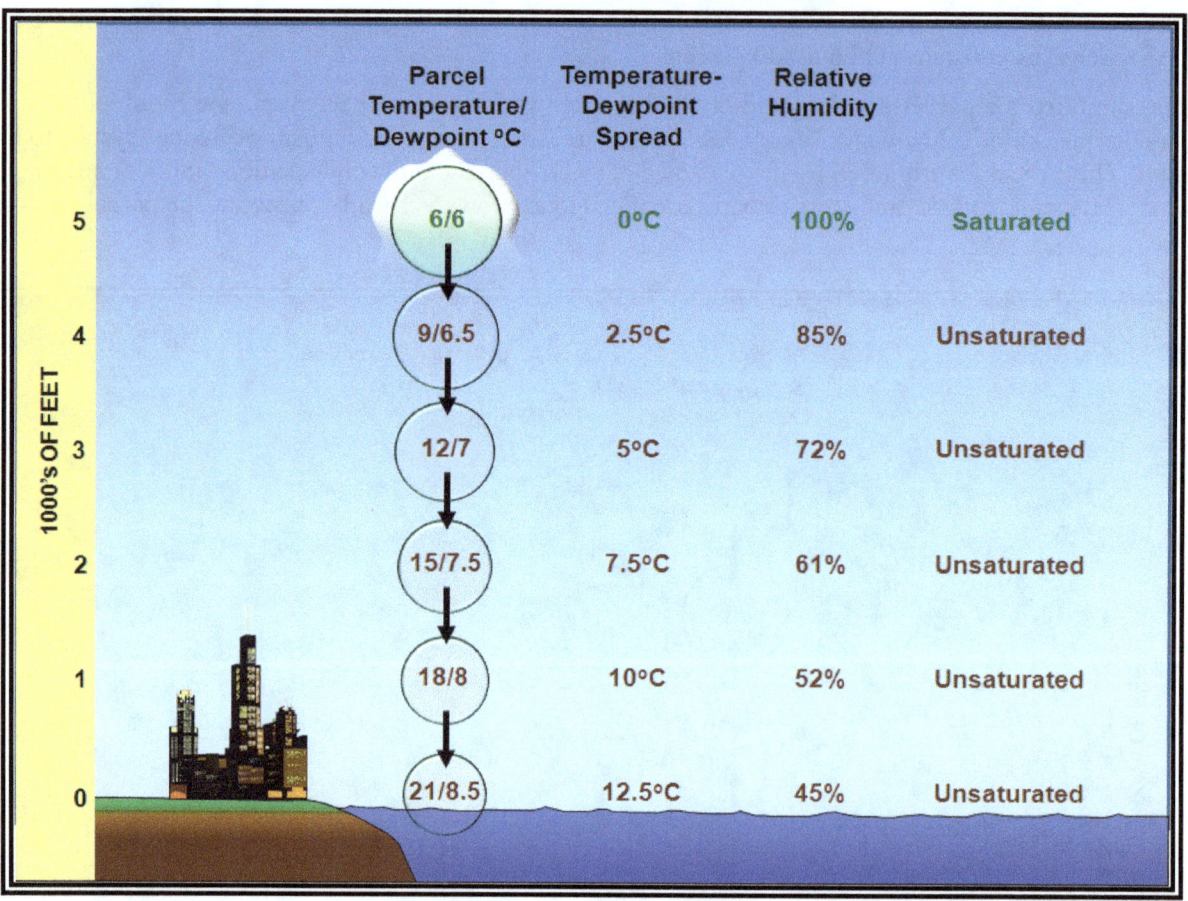

Figure 12-3. Descending Air Parcel Example

At 5,000 ft, both the temperature and dewpoint of the air parcel are 6 °C, which indicates that it is saturated. As the parcel descends, it quickly becomes unsaturated. Its temperature increases 3 °C per 1,000 ft, while its dewpoint increases at 0.5 °C per 1,000 ft. The temperature-dewpoint spread increases while relative humidity decreases until the parcel reaches the surface. Note that the parcel is now much warmer and drier at the surface than when it began the vertical motion process in Figure 12-2.

Chapter 12, Vertical Motion and Clouds

Table 12-1. Air Parcel Vertical Motion Characteristics

Parcel	Unsaturated			Saturated		
	Temperature Change	Dewpoint Change	Relative Humidity	Temperature Change	Dewpoint Change	Relative Humidity
Ascending (rising)	-3 °C/1,000 ft	-0.5 °C/1,000 ft	Increases	-1.2 °C to -3 °C/1,000 ft	Identical to temperature change	100%
Descending (subsiding)	+3 °C/1,000 ft	+0.5 °C/1,000 ft	Decreases			

12.4 Common Sources of Vertical Motion

There are many sources of vertical motion in the atmosphere. Four of the most common types of vertical motion are orographic effects, frictional effects, frontal lift, and buoyancy.

12.4.1 Orographic Effects

Winds blowing across mountains and valleys cause the moving air to alternately ascend and descend. If relief is sufficiently great, the resulting expansional cooling and compressional warming of air affects the development and dissipation of clouds and precipitation.

For example, a mountain range that is oriented perpendicular to the prevailing wind flow forms a barrier that results in a cloudier and wetter climate on one side of the range than on the other side (see Figure 12-4). As air is forced to rise along the windward slope, it expands and cools, which increases its relative humidity. With sufficient cooling, clouds and precipitation develop at and above the LCL. Conversely, on the mountain's leeward slope, air descends and warms, which reduces its relative humidity, and tends to dissipate clouds and precipitation. In this way, mountain ranges induce two contrasting climatic zones: a moist climate on the windward slope and a dry climate on the leeward slope. Dry conditions often extend hundreds of miles to the lee of a prominent mountain range in a region known as the rain shadow.

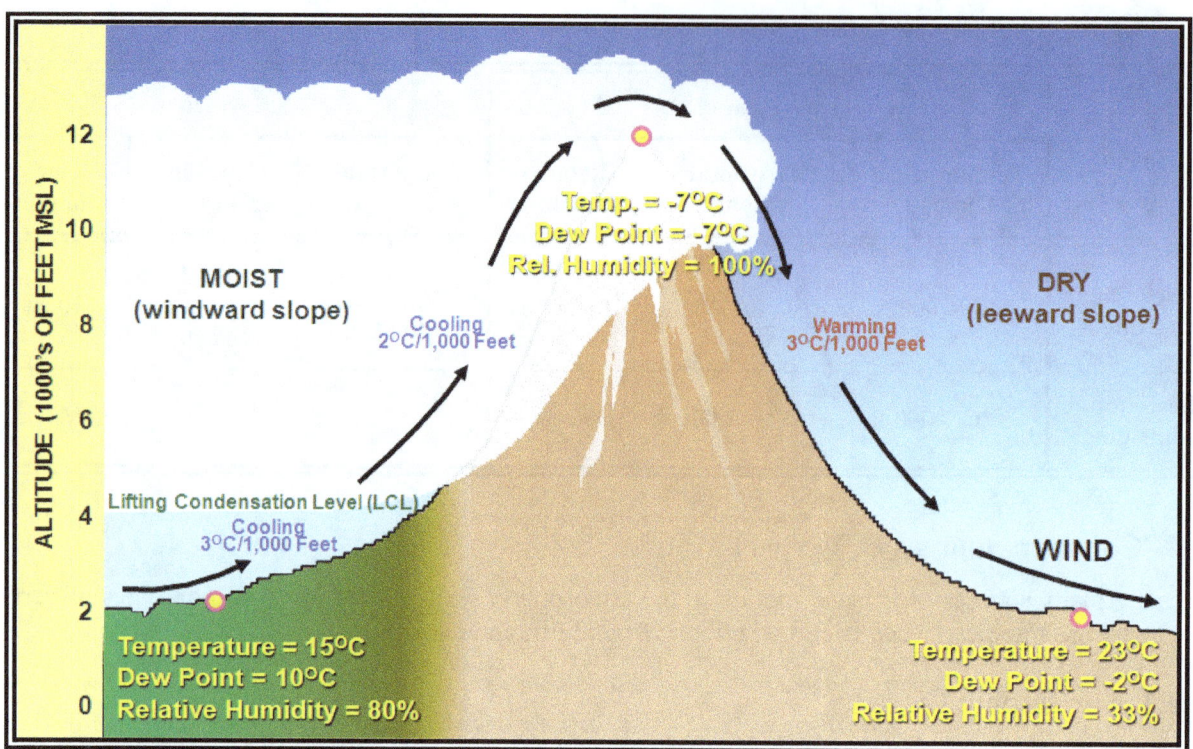

Figure 12-4. Orographic Effects Example

The air parcel begins with a temperature of 15 °C, dewpoint of 10 °C, and a relative humidity of 80 percent at 2,000 ft. As the parcel is lifted on the windward slope, the temperature cools at the dry adiabatic lapse rate of 3 °C per 1,000 ft, and the dewpoint cools at a rate of 0.5 °C per 1,000 ft until it becomes saturated at the LCL at 4,000 ft. Then, the air parcel's temperature and dewpoint both cool at the moist adiabatic lapse rate of 2 °C per 1,000 ft until the parcel reaches the summit at 12,000 ft. At that altitude, the parcel's temperature is -7 °C, the dewpoint is -7 °C, and the relative humidity is 100 percent. As the air parcel descends the leeward slope, the temperature increases at a rate of 3 °C per 1,000 ft while the dewpoint increases 0.5 °C per 1,000 ft. The air parcel ends with a temperature of 23 °C, dewpoint of -2 °C, and a relative humidity of 33 percent at 2,000 ft, much warmer and drier than at the beginning.

Orographic effects are especially apparent from west to east across the Pacific Northwest, where the north–south Cascade Range intercepts the prevailing flow of humid air from the Pacific Ocean. Exceptionally cloudy, rainy weather prevails western slopes, whereas semiarid weather characterizes the eastern slopes and areas farther east.

12.4.2 Frictional Effects

In the Northern Hemisphere, the surface wind spirals clockwise and outward from high pressure, and counterclockwise and inward into low pressure due to frictional force. The end result is that winds diverge away from surface high pressure, causing the air to sink, compress, and warm, which favors the dissipation of clouds and precipitation. Conversely, winds converge into surface low pressure, causing the air to rise, expand, and cool, which favors the formation of clouds and precipitation given sufficient moisture (see Figure 12-5).

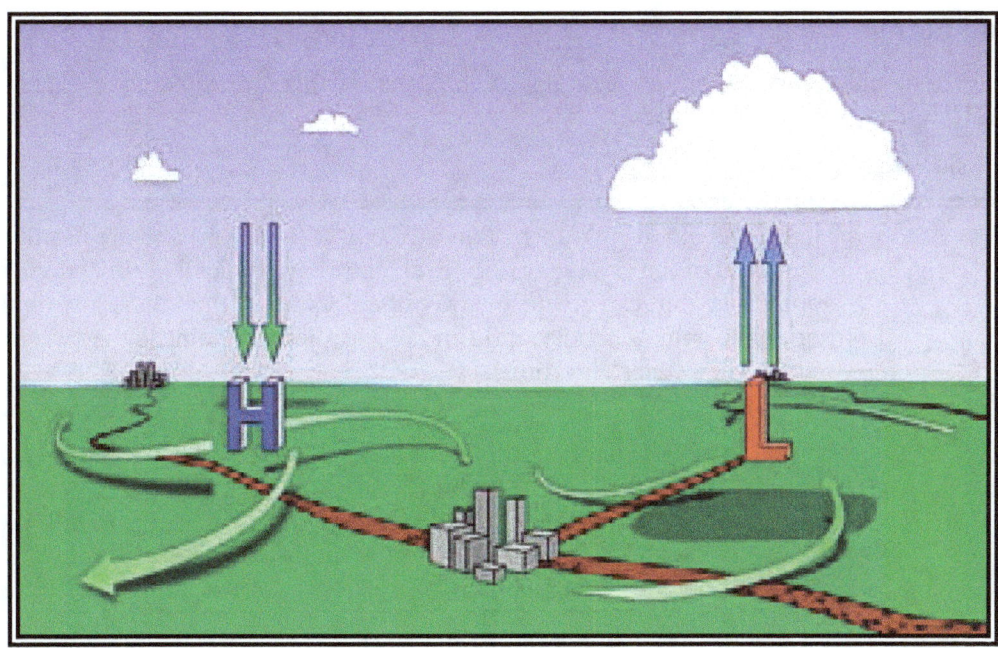

Figure 12-5. Frictional Effects

12.4.3 Frontal Lift

Frontal lift (see Figure 12-6) occurs when the cold, denser air wedges under the warm, less dense air, plowing it upward, and/or the warmer air rides up and over the colder air in a process called overrunning. Clouds and precipitation will form given sufficient lift and moisture content of the warm air.

Figure 12-6. Frontal Lift

12.4.4 Buoyancy

Air near the ground can warm at different rates depending on the insular properties of the ground with which it is in contact. A newly plowed field will warm faster than an adjacent lake. These temperature differences result in different densities, allowing the warm air to become buoyant. The denser cool air will tend to push (i.e., lift) the less dense warm air aloft. On a grand scale, the tendency of air to rise due to heating, and how high it will rise, is referred to as stability and is covered in Chapter 13, Atmospheric Stability.

12.5 Cloud Forms

There are four basic cloud forms (appearances) observed in the Earth's atmosphere (see Table 12-2). See Appendix A, Cloud Types, for cloud types.

Table 12-2. Cloud Forms

Cirri-form	High-level clouds that form above 20,000 ft (6,000 m) and are usually composed of ice crystals. High-level clouds are typically thin and white in appearance, but can create an array of colors when the Sun is low on the horizon. Cirrus generally occur in fair weather and point in the direction of air movement at their elevation.
Nimbo-form	Nimbus comes from the Latin word meaning "rain." These clouds typically form between 7,000 and 15,000 ft (2,100 to 4,600 m) and bring steady precipitation. As the clouds thicken and precipitation begins to fall, the bases of the clouds tend to lower toward the ground.
Cumuli-form	Clouds that look like white, fluffy cotton balls or heaps and show the vertical motion or thermal uplift of air taking place in the atmosphere. The level at which condensation and cloud formation begins is indicated by a flat cloud base, and its height will depend upon the humidity of the rising air. The more humid the air, the lower the cloud base. The tops of these clouds can reach over 60,000 ft (18,000 m).
Strati-form	Stratus is Latin for "layer" or "blanket." The clouds consist of a featureless low layer that can cover the entire sky like a blanket, bringing generally gray and dull weather. The cloud bases are usually only a few hundred feet above the ground. Over hills and mountains, they can reach ground level when they may be called fog. Also, as fog lifts off the ground due to daytime heating, the fog forms a layer of low stratus clouds.

Source: NWS JetStream – Online School for Weather

12.6 Cloud Levels

By convention, the part of the atmosphere in which clouds are usually present has been divided into three levels: high, middle, and low (see Table 12-3). Each level is defined by the range of heights at which the cloud of a certain type occurs most frequently. The levels overlap, and their limits vary with latitude. The approximate heights of the limits are included in Table 12-3.

Table 12-3. Approximate Height of Cloud Bases Above the Surface

Level	Polar Regions	Temperate Regions	Tropical Regions
High Clouds	10,000–25,000 ft (3–8 km)	16,500–40,000 ft (5–13 km)	20,000–60,000 ft (6–18 km)
Middle Clouds	6,500–13,000 ft (2–4 km)	6,500–23,000 ft (2–7 km)	6,500–25,000 ft (2–8 km)
Low Clouds	Surface–6,500 ft (0–2 km)	Surface–6,500 ft (0–2 km)	Surface–6,500 ft (0–2 km)

13 Atmospheric Stability

13.1 Introduction

Convective clouds and precipitation pose a distinctly different flying environment than stratiform clouds and precipitation. These sharply contrasting conditions result from the atmosphere either resisting or accelerating the vertical motion of air parcels. Atmospheric stability is the property of the ambient air that either enhances or suppresses vertical motion of air parcels and determines which type of clouds and precipitation a pilot will encounter.

13.2 Using a Parcel as a Tool to Evaluate Stability

An air parcel can be used as a tool to evaluate atmospheric stability within a specified vertical column of air in the atmosphere. A parcel is selected from a specified altitude (usually the surface) and hypothetically lifted upward to a specified test altitude. As the parcel is lifted, its temperature decreases due to the expansion and latent heat effects discussed in Chapter 12, Vertical Motion and Clouds.

The parcel and the surrounding environmental air temperatures are then compared. If the lifted parcel is colder than the surrounding air, it will be denser (heavier) and sink back to its original level. In this case, the parcel is stable because it resists upward displacement. If the lifted parcel is the same temperature as the surrounding air, it will be the same density and remain at the same level. In this case, the parcel is neutrally stable. If the lifted parcel is warmer and, therefore, less dense (lighter) than the surrounding air, it will continue to rise on its own until it reaches the same temperature as its environment. This final case is an example of an unstable parcel. Greater temperature differences result in greater rates of vertical motion.

13.3 Stability Types

The stability of a column of air in the atmosphere is classified by the distribution of parcel stabilities within the column. The Earth's surface is typically selected as the base while the top determines the column's depth. Four unique types of atmospheric stability can be identified and are discussed in the following sections.

13.3.1 Absolute Stability

Absolute stability (see Figure 13-1) is the state of a column of air in the atmosphere when its lapse rate of temperature is less than the moist adiabatic lapse rate. This includes both isothermal and inversion temperature profiles. An air parcel lifted upward would be colder (denser) than the surrounding environmental air and would tend to sink back to its level of origin.

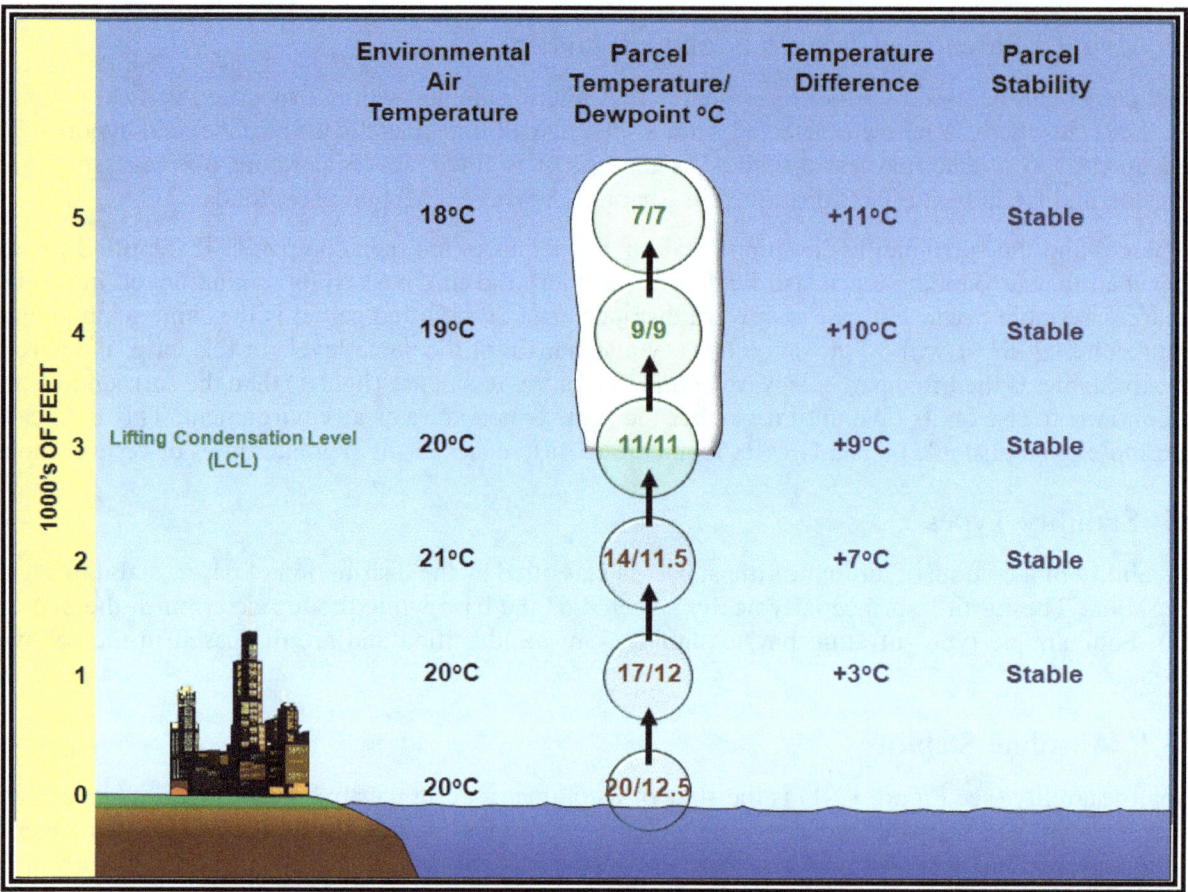

Figure 13-1. Absolute Stability Example

13.3.2 Neutral Stability

Neutral stability (see Figure 13-2) is the state of a column of air in the atmosphere in which an ascending (or descending) air parcel always has the same temperature (density) as the surrounding environmental air. If the column of air is unsaturated, then neutral stability exists when its lapse rate of temperature equals the dry adiabatic lapse rate. If the column of air is saturated, then neutral stability exists when its lapse rate of temperature equals the moist adiabatic lapse rate.

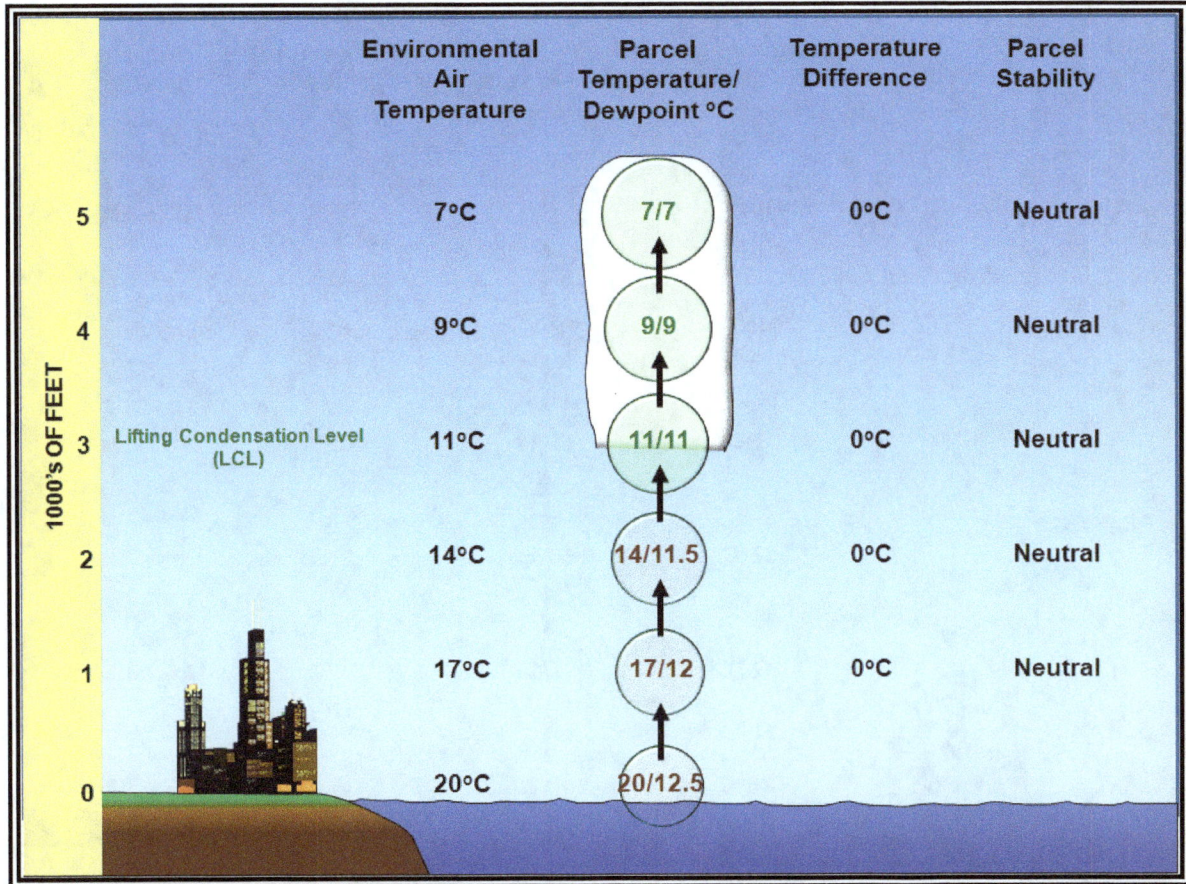

Figure 13-2. Neutral Stability Example

13.3.3 Absolute Instability

Absolute instability (see Figure 13-3) is the state of a column of air in the atmosphere when it has a superadiabatic lapse rate of temperature (i.e., greater than the dry adiabatic lapse rate). An air parcel displaced vertically would be accelerated in the direction of the displacement. The kinetic energy of the parcel would consequently increase with increasing distance from its level of origin.

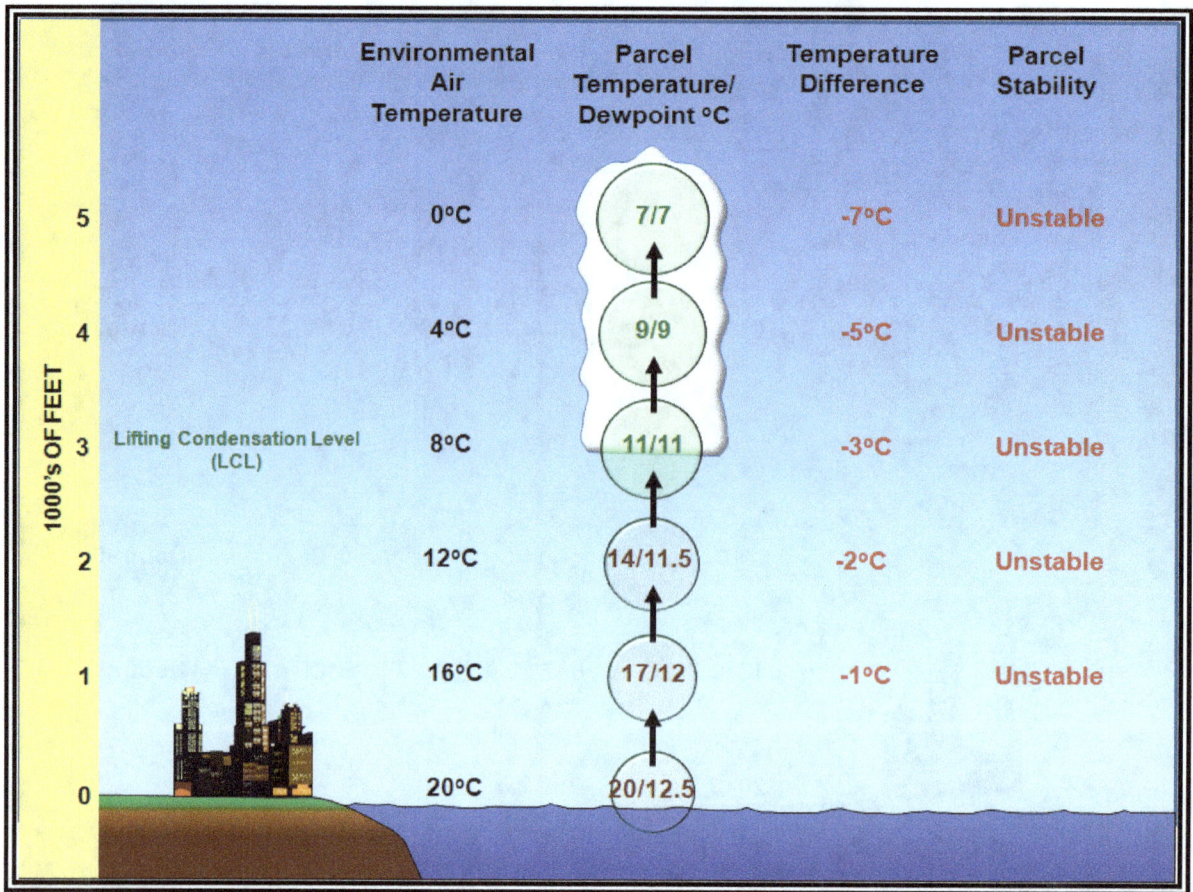

Figure 13-3. Absolute Instability Example

13.3.4 Conditional Instability

Conditional instability (see Figure 13-4) is the state of a column of unsaturated air in the atmosphere when its lapse rate of temperature is less than the dry adiabatic lapse rate, but greater than the moist adiabatic lapse rate. An air parcel lifted upward would be initially stable, but at some point above its LCL it would become unstable. The term "conditional" means the parcel must be lifted to a particular level (altitude) before it becomes unstable and rises because of its own buoyancy. The Level of Free Convection (LFC) is the level at which a parcel of air lifted dry adiabatically until saturated and moist adiabatically thereafter would first become warmer than the surrounding environmental air (i.e., unstable) in a conditionally unstable column of air in the atmosphere. The LFC is a defining feature of a conditionally unstable column of air.

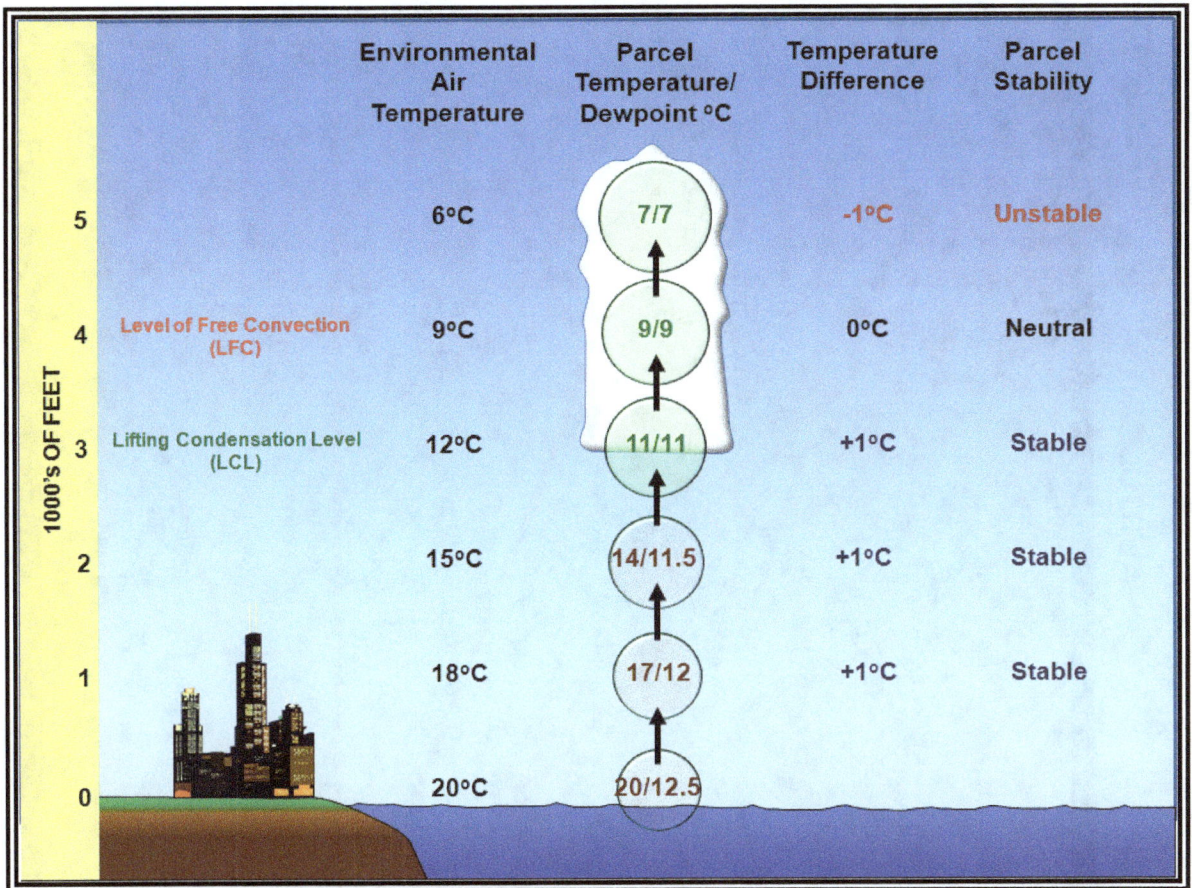

Figure 13-4. Conditional Instability Example

13.3.5 Summary of Stability Types

Figure 13-5 below summarizes the possible atmospheric stability types.

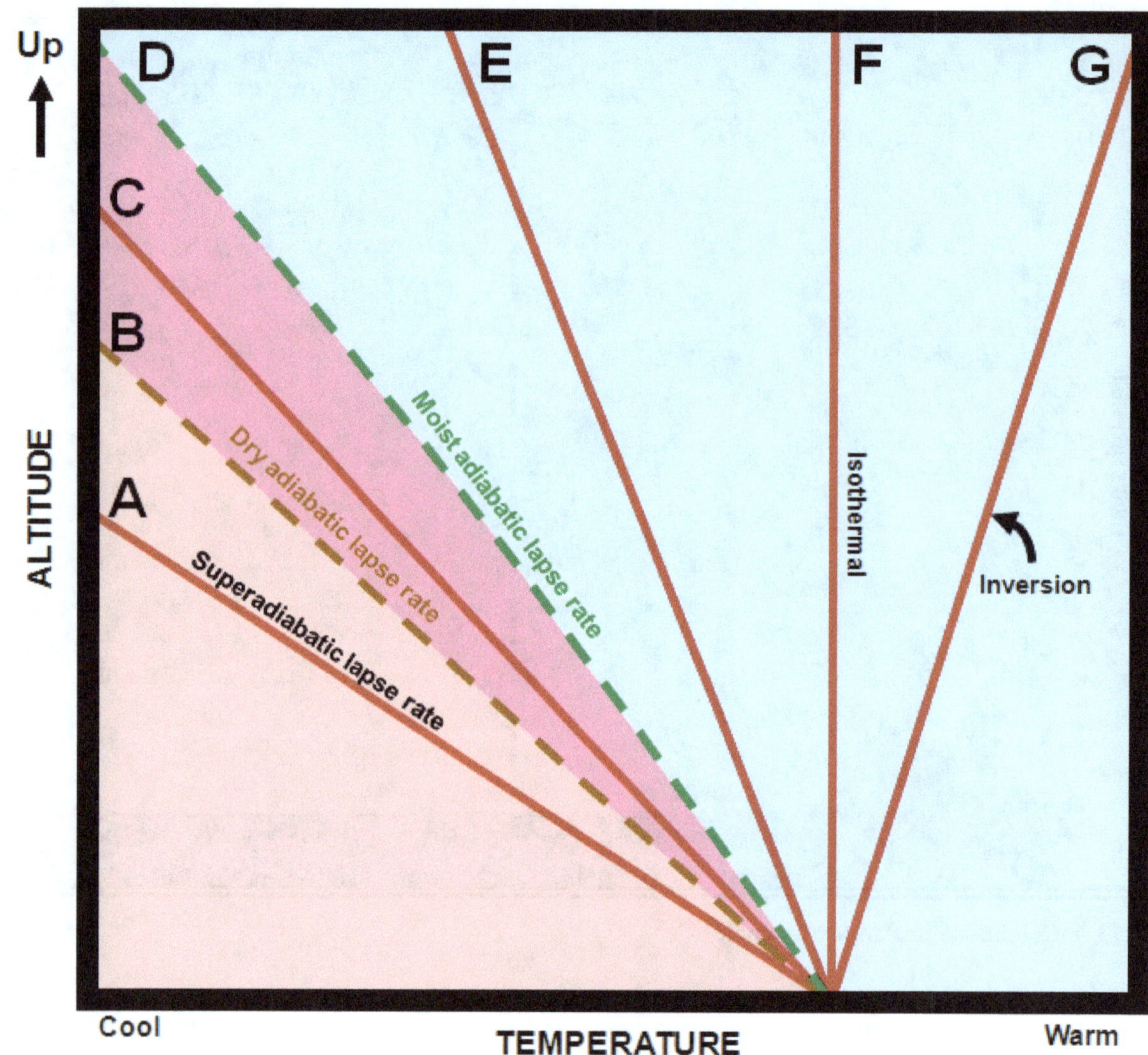

Figure 13-5. Stability Types

Chapter 13, Atmospheric Stability

13.4 Processes That Change Atmospheric Stability

Changes in atmospheric stability are inversely related to temperature (density) changes with height (see Figure 13-6). If temperature lapse rates increase, then stability decreases. Conversely, if temperature lapse rates decrease, then stability increases. Most of these changes occur as a result of the movement of air, but diurnal (day/night) temperature variations can play a significant role.

In Figure 13-6, the column of air on the right is less stable because its temperature lapse rate is higher.

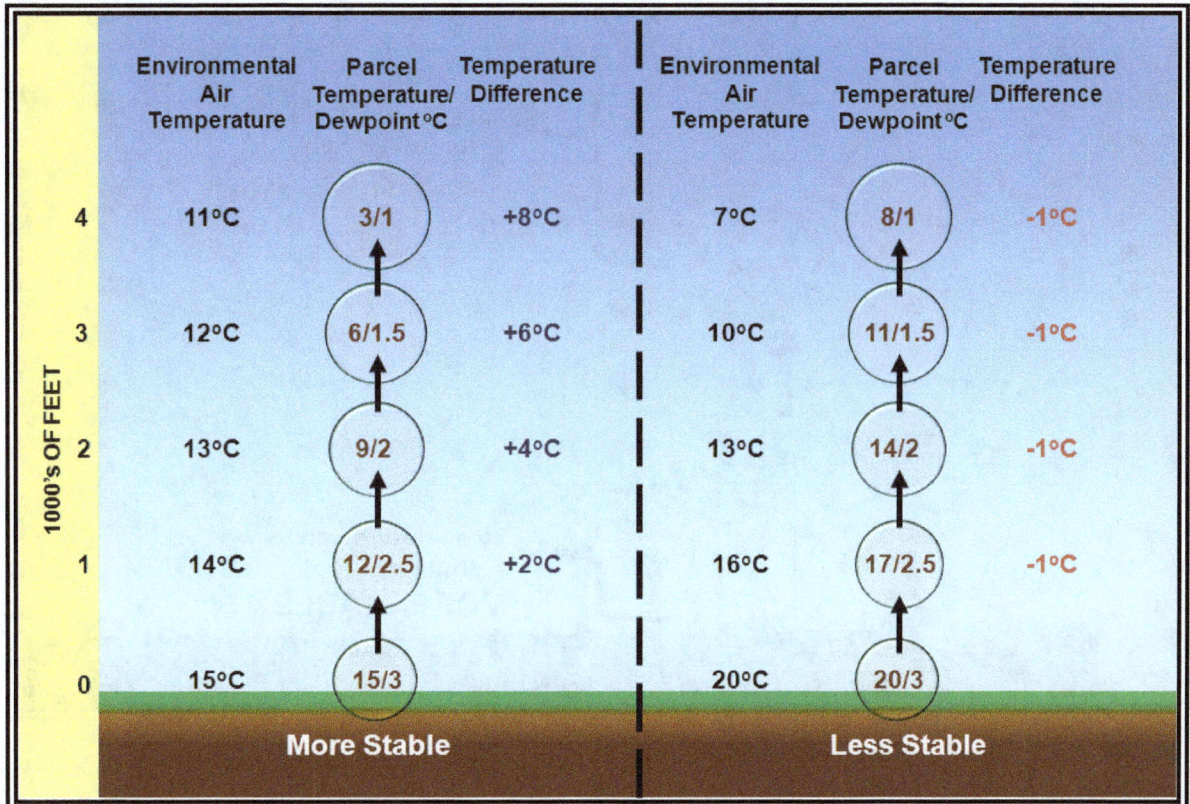

Figure 13-6. Temperature Lapse Rate Effects on Stability

13.4.1 Wind Effects on Stability

Wind can act to change the stability of a column of air in the atmosphere by changing the temperature lapse rate. Stability increases when wind blows colder air into the bottom of the air column (cold air advection) and/or warmer air at the top (warm air advection). Conversely, stability decreases when wind blows warmer air into the bottom of the air column and/or colder air at the top.

13.4.2 Vertical Air Motion Effects on Stability

A column of air in the atmosphere will become more stable when it descends (subsides) (see Figure 13-7). As it subsides, it becomes compressed by the weight of the atmosphere and shrinks vertically. The entire layer warms due to adiabatic compression. However, the upper part of the column sinks farther and, thus, warms more than the bottom part. This process acts to decrease the temperature lapse rate and increase stability.

Conversely, a column of air in the atmosphere will become less stable when it ascends (rises). As it rises, the rapid decrease in air density aloft causes the column to stretch out vertically. As long as the layer remains unsaturated, the entire layer cools at the dry adiabatic lapse rate (see Figure 13-7). However, due to the stretching effect, air at the top of the column cools more than the air at the bottom of the column. This process acts to increase the temperature lapse rate and decrease stability.

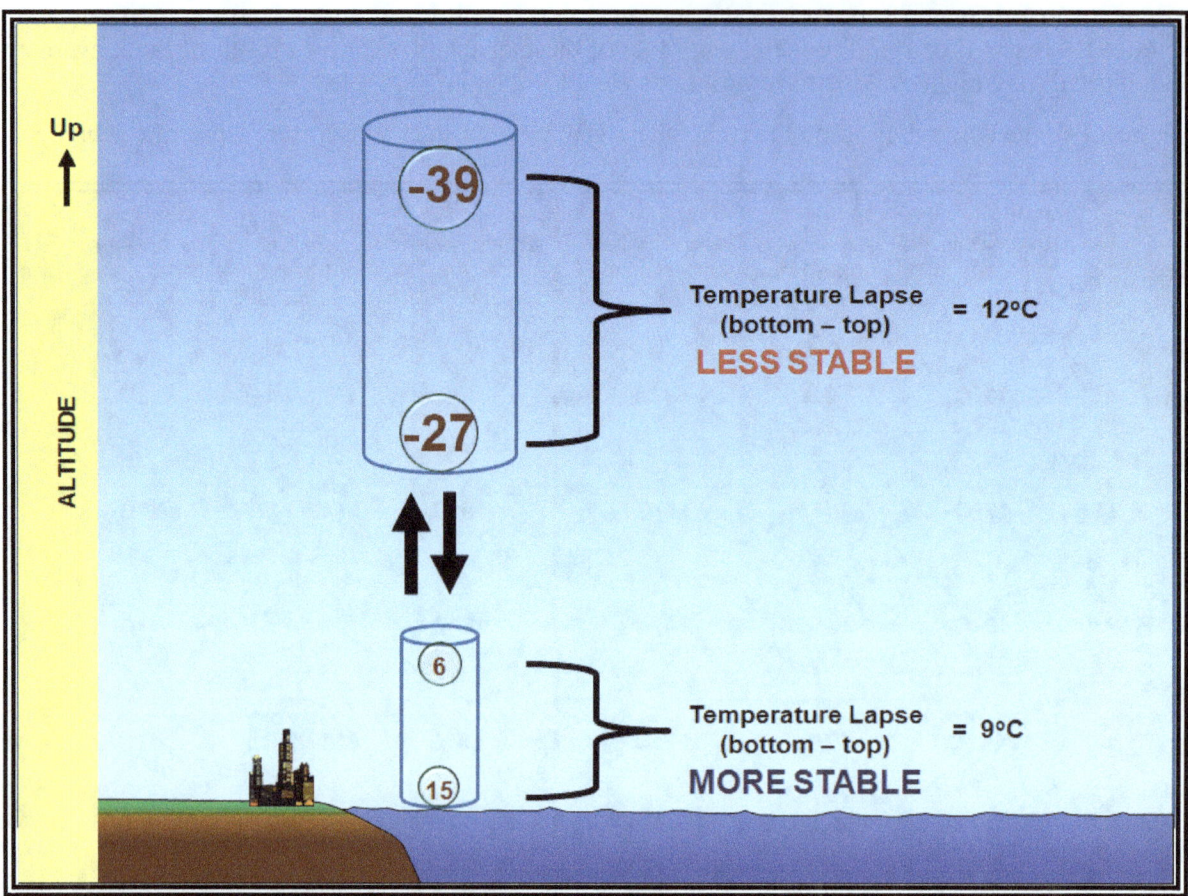

Figure 13-7. Vertical Motion Effects on Stability

A rising column of air will become less stable when air at the bottom has a higher relative humidity than air at the top. As the air moves upward, the bottom becomes saturated first and cools at the lesser moist adiabatic lapse rate. The net effect is to increase the lapse rate within the column and decrease stability. This process is called convective instability, and is associated with the development of thunderstorms.

13.4.3 Diurnal Temperature Variation Effects on Stability

Diurnal (day/night) temperature variations can have a significant impact on atmospheric stability (see Figure 13-8). Daytime heating of the surface increases temperature lapse rates and decreases stability. Conversely, nighttime cooling of the surface decreases temperature lapse rates and increases stability. Diurnal temperature variations are most pronounced in the lower troposphere because air is a poor conductor of heat (see Table 5-3).

The magnitude of diurnal temperature (and stability) variation is primarily influenced by surface type, latitude, sky cover (e.g., clouds and pollutants), water vapor content of the air, and wind speed. Temperature variation is maximized over land, at low latitudes, with a clear sky, dry air, and light wind. Conversely, temperature variation is minimized over large bodies of water, at high latitudes, with a cloudy sky, moist air, and strong wind.

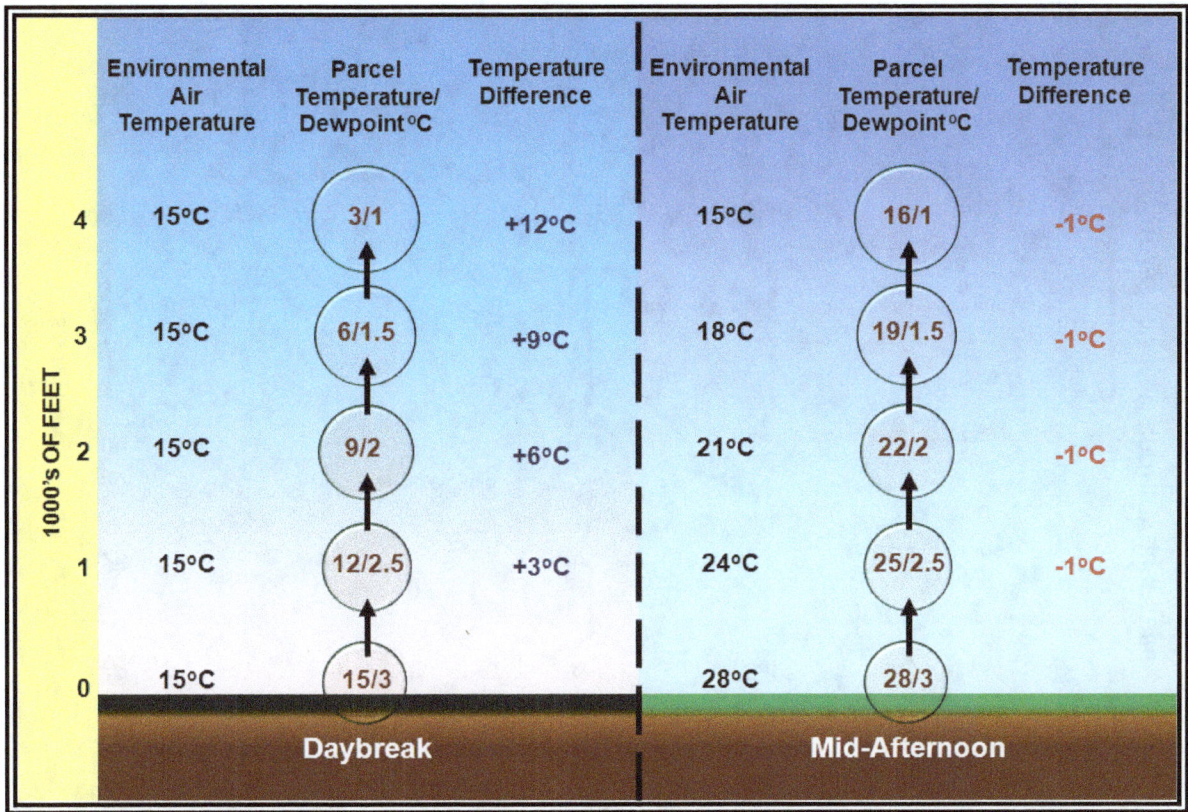

Figure 13-8. Diurnal Temperature Variation Effects on Stability

13.5 Measurements of Stability

Several stability indexes and other quantities exist that evaluate atmospheric stability and the potential for convective storms. The most common of these are Lifted Index (LI) and Convective Available Potential Energy (CAPE).

13.5.1 Lifted Index (LI)

The LI (see Figure 13-9) is the temperature difference between an air parcel (usually at the surface) lifted adiabatically (see Chapter 12) and the temperature of the environment at a given pressure (usually 500 mb) in the atmosphere. A positive value indicates a stable column of air (at the respective pressure), a negative value indicates an unstable column of air, and a value of zero indicates a neutrally stable column of air. The larger the positive (negative) LI value, the more stable (unstable) the column of air.

LI is generally used in thunderstorm forecasting; however, CAPE is generally considered a superior measurement of instability. However, LI is easier to determine without using a computer.

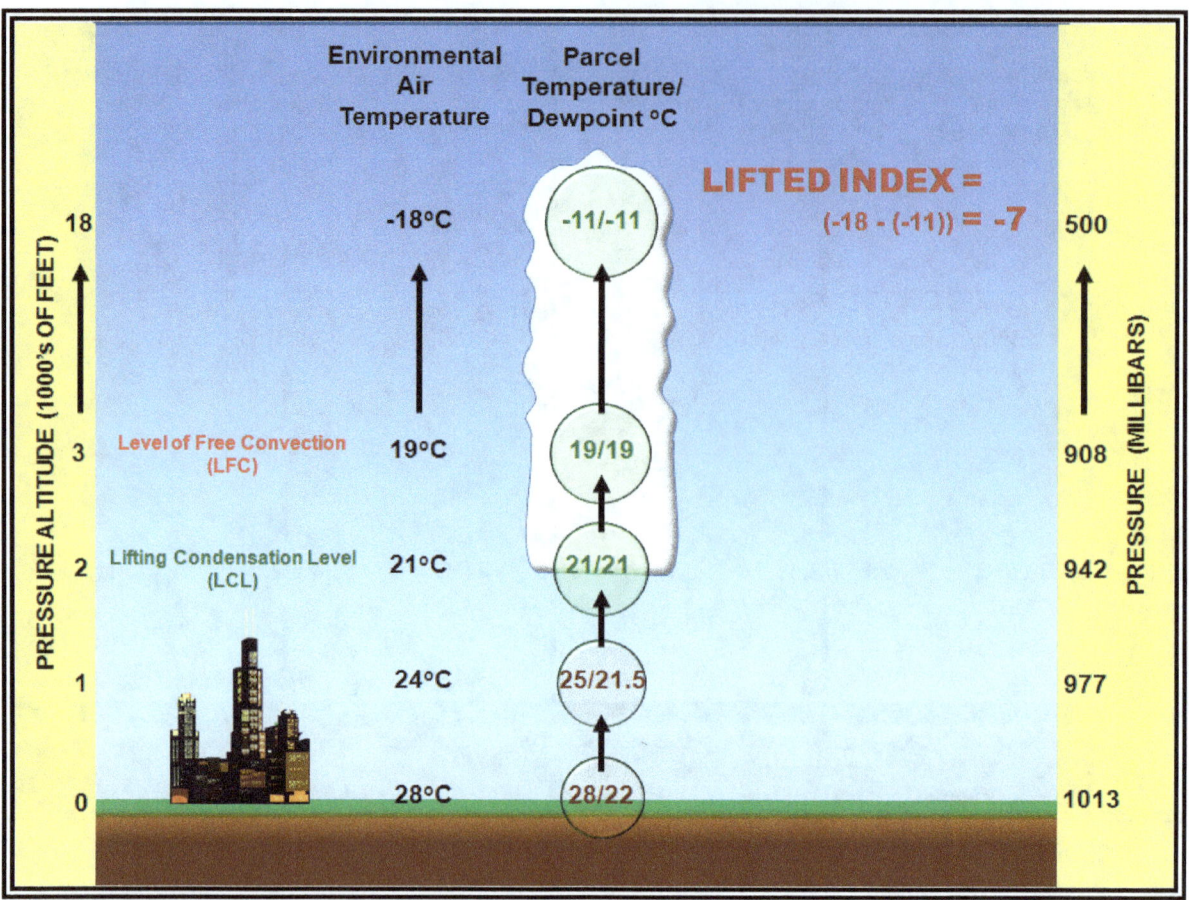

Figure 13-9. Lifted Index Example

13.5.2 Convective Available Potential Energy (CAPE)

CAPE is the maximum amount of energy available to an ascending air parcel for convection. CAPE is represented on a sounding by the area enclosed between the environmental temperature profile and the path of a rising air parcel over the layer within which the latter is warmer than the former. Units are joules per kilogram (J/kg) of air. Any value greater than 0 J/kg indicates instability and the possibility of thunderstorms.

CAPE is directly related to the maximum potential vertical speed within an updraft; thus, higher values indicate the potential for stronger updrafts. Observed values in thunderstorm environments often exceed 1,000 J/kg, and in extreme cases may exceed 5,000 J/kg.

13.6 Summary

Atmospheric stability influences weather by affecting the vertical motion of air. Stable air suppresses vertical motion while unstable air enhances it.

14 Precipitation

14.1 Introduction

Precipitation is any of the forms of water particles, whether liquid or solid, that fall from the atmosphere and reach the ground. This chapter covers the necessary ingredients for formation, the growth process, and the types of precipitation. Some precipitation types include: drizzle, rain, freezing rain, freezing drizzle, snow, snow grains, ice crystals, ice pellets, hail, and small hail and/or snow pellets.

14.2 Necessary Ingredients for Formation

Precipitation formation requires three ingredients: water vapor, sufficient lift to condense the water vapor into clouds, and a growth process that allows cloud droplets to grow large and heavy enough to fall as precipitation. Significant precipitation usually requires clouds to be at least 4,000 ft thick. The heavier the precipitation, the thicker the clouds are likely to be. When arriving or departing from an airport reporting precipitation of light or greater intensity, expect clouds to be more than 4,000 ft thick.

14.3 Growth Process

All clouds contain water, but only some produce precipitation. This is because cloud droplets and/or ice crystals are too small and light to fall to the ground as precipitation. Because of their microscopic size, the rate at which cloud droplets fall is incredibly slow. An average cloud droplet falling from a cloud base at 3,300 ft (1,000 m) would require about 48 hours to reach the ground. It would never complete this journey because it would evaporate within minutes after falling below the cloud base.

Two growth processes exist that allow cloud droplets (or ice crystals) to grow large enough to reach the ground as precipitation before they evaporate (or sublimate). One process is called the collision-coalescence, or warm rain process (see Figure 14-1). In this process, collisions occur between cloud droplets of varying sizes and different fall speeds, sticking together or coalescing to form larger drops. Finally, the drops become too large to be suspended in the air, and they fall to the ground as rain. This is thought to be the primary growth process in warm, tropical air masses where the freezing level is very high.

Most cloud droplets are too small and light to fall to the ground as precipitation. However, the larger cloud droplets fall more rapidly and are able to sweep up the smaller ones in their path and grow.

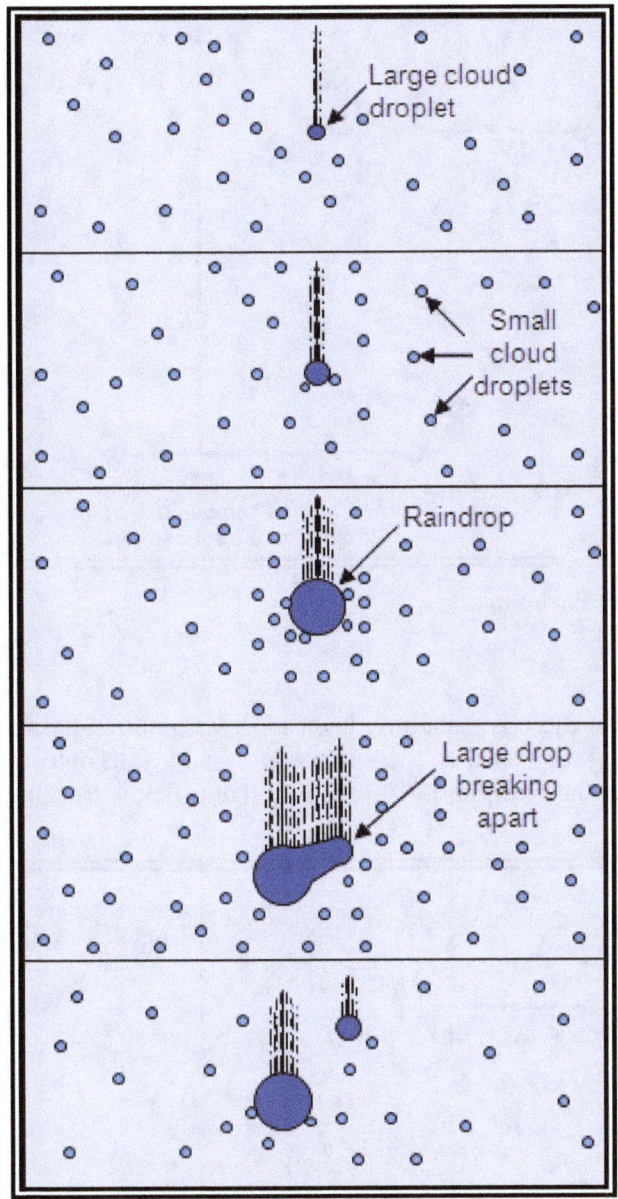

Figure 14-1. The Collision-Coalescence or Warm Rain Process

The other process is called the ice crystal process. This occurs in colder clouds when both ice crystals and water droplets are present. In this situation, it is easier for water vapor to deposit directly onto the ice crystals so the ice crystals grow at the expense of the water droplets. The crystals eventually become heavy enough to fall. If it is cold near the surface, it may snow; otherwise, the snowflakes may melt to rain. This is thought to be the primary growth process in mid-latitudes and high latitudes.

14.4 Precipitation Types

14.4.1 Snow

The vertical distribution of temperature will often determine the type of precipitation that occurs at the surface. Snow occurs when the temperature remains below freezing throughout the entire depth of the atmosphere (see Figure 14-2).

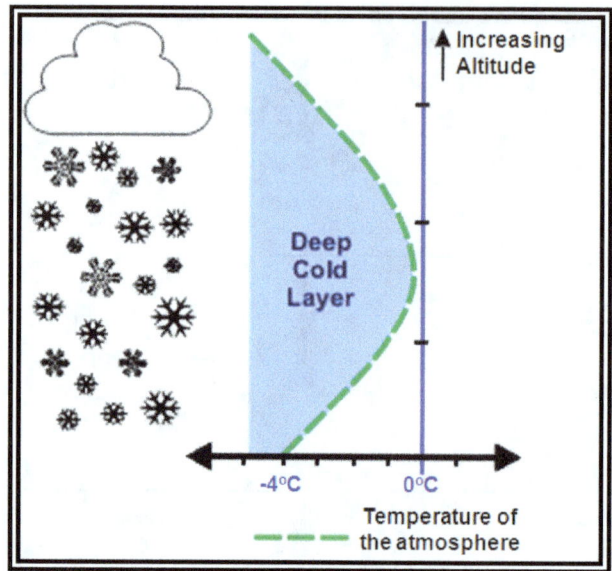

Figure 14-2. Snow Temperature Environment

14.4.2 Ice Pellets

Ice pellets (sleet) occur when there is a shallow layer aloft with above-freezing temperatures and with a deep layer of below-freezing air based at the surface. As snow falls into the shallow warm layer, the snowflakes partially melt. As the precipitation reenters air that is below freezing, it refreezes into ice pellets (see Figure 14-3).

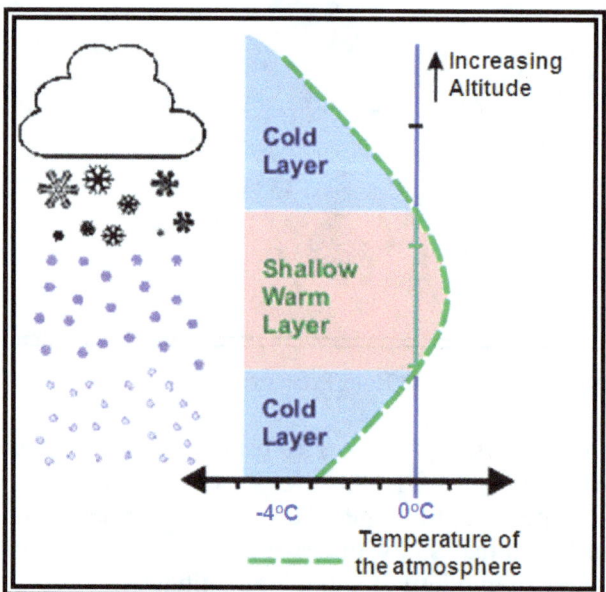

Figure 14-3. Ice Pellets Temperature Environment

14.4.3 Freezing Rain

Freezing rain occurs when there is a deep layer aloft with above-freezing temperatures and with a shallow layer of below-freezing air at the surface. Ordinarily, air temperatures decrease with increasing altitude, but freezing rain requires a temperature inversion, which can occur when a warmer air mass overlies a colder air mass. This situation can occur along a warm front, where a warm air mass overruns a cold air mass. It can begin as rain and/or snow, but becomes all rain in the warm layer. The rain falls back into below-freezing air, but since the depth is shallow, the rain does not have time to freeze into ice pellets (see Figure 14-4). The drops freeze on contact with the ground or exposed objects, such as aircraft.

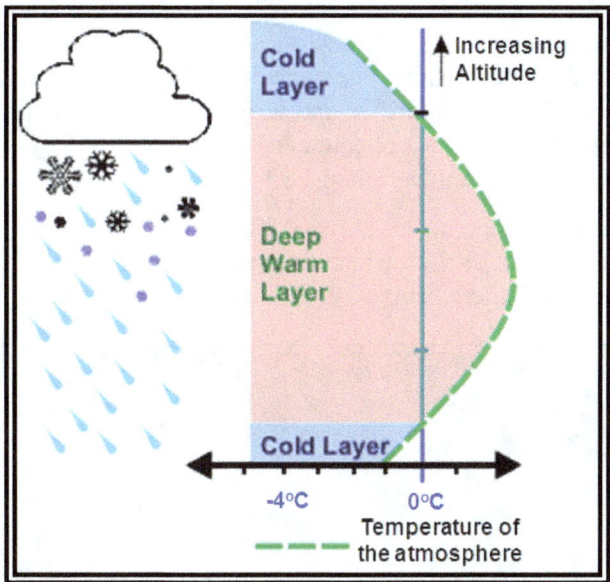

Figure 14-4. Freezing Rain Temperature Environment

14.4.4 Rain

Rain occurs when there is a deep layer of above freezing air based at the surface (see Figure 14-5).

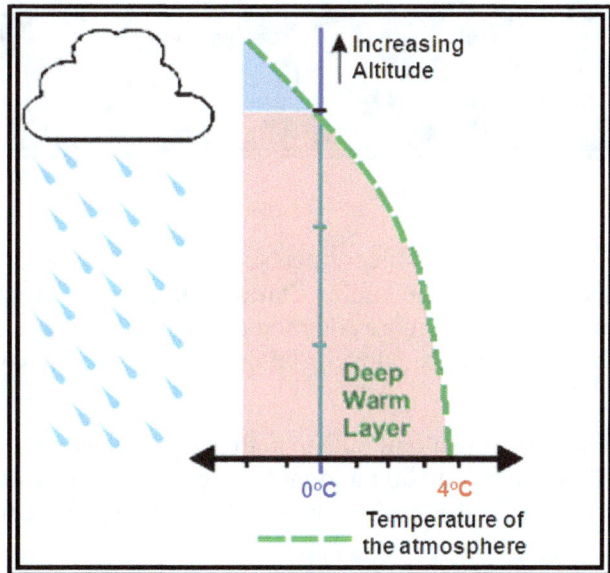

Figure 14-5. Rain Temperature Environment

14.4.5 Hail

Hail is precipitation in the form of balls or other irregular lumps of ice produced by thunderstorms. Thunderstorms that are characterized by strong updrafts, large Supercooled Liquid Water Content (SLWC), large cloud-drop sizes, and great vertical height are favorable to hail formation.

Hail forms when supercooled water droplets above the freezing level begin to freeze. Once a droplet has frozen, other droplets latch on and freeze to it, so the hailstone grows—sometimes into a huge iceball. Large hail occurs with severe thunderstorms with strong updrafts that have built to great heights. Eventually, the hailstones fall and may be encountered in clear air several miles from the thunderstorm.

An individual unit of hail is called a hailstone. Hailstones can range in size from a pea (0.25-in diameter) to larger than a softball (4.5-in diameter). Hail competes with turbulence as the greatest thunderstorm hazard to aircraft. Hailstones that are 0.75 in in diameter and larger can cause significant damage to aircraft and make it difficult to control. A hailstone was collected at Vivian, SD, on July 23, 2010, that measured 8 in in diameter, 18.62 in in circumference, and weighed 1.93 lb (see Figure 14-6).

As hailstones fall through air with temperatures above 0 °C, they begin to melt, and precipitation may reach the ground as either hail or rain. Rain at the surface does not mean the absence of hail aloft. Pilots should anticipate possible hail with any thunderstorm, especially beneath the anvil of a large cumulonimbus.

Figure 14-6. Vivian, South Dakota, Record Hailstone

Hail is most frequently found in the interior of continents within the mid-latitudes and generally confined to higher elevations within the tropics. In the United States, hail is most common across the Great Plains region east of the Rocky Mountains. Hail is more common aloft and at higher elevations because the stones begin to melt when they fall below the freezing level, and the smaller stones may melt into raindrops before they reach the surface.

When viewed from the air, it is evident that hail falls in paths known as hail swaths. They can range in size from a few acres to an area 10 mi wide and 100 mi long. Piles of hail in hail swaths have been deep enough to need a snowplow to remove them, and occasionally hail drifts have been reported.

15 Weather Radar

15.1 Introduction

The most effective tool to detect precipitation is radar. Radar, which is an acronym that stands for "radio detection and ranging," has been utilized to detect precipitation since the 1940s. Radar enhancements have enabled more precision in detecting and displaying precipitation.

15.2 Principles of Weather Radar

The radar used by the NWS is called the Weather Surveillance Radar—1988 Doppler (WSR-88D). The prototype radar was built in 1988.

It is essential to understand some principles of weather radar. This will allow correct interpretation of WSR-88D images. This chapter will also include a comparison between some WSR-88D principles and aircraft radar principles. These comparisons will help explain the strengths and limitations of the WSR-88D and aircraft radar.

15.2.1 Antenna

The antenna (see Figure 15-1) alternately emits and receives radio waves into the atmosphere. Pulses of energy from the radio waves may strike a target. If they do, part of that energy will return to the antenna.

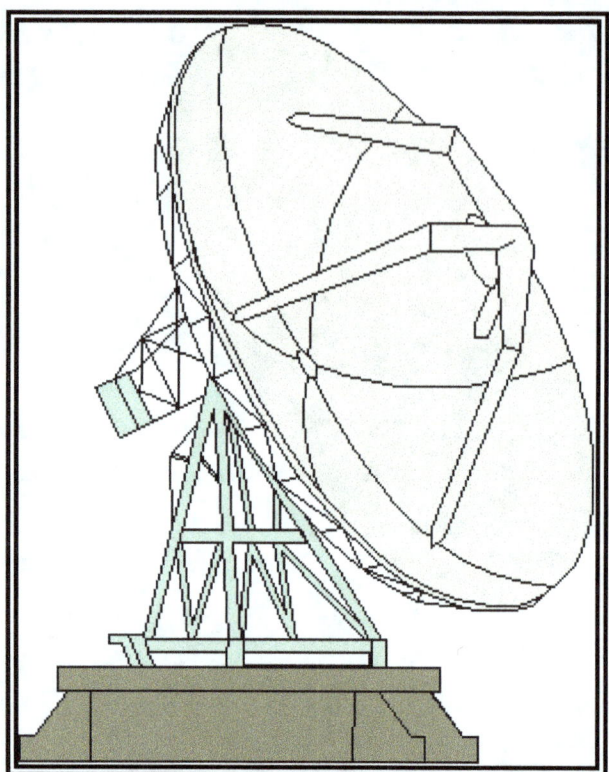

Figure 15-1. Radar Antenna

The shape of an antenna determines the shape of a beam. The WSR-88D has a parabolic-shaped antenna. This focuses the radio waves into a narrow, coned-shaped beam. The antenna can be tilted to scan many altitudes of the atmosphere.

15.2.2 Backscattered Energy

The amount of energy returned directly back to the radar after striking a target is called backscattered energy (see Figure 15-2).

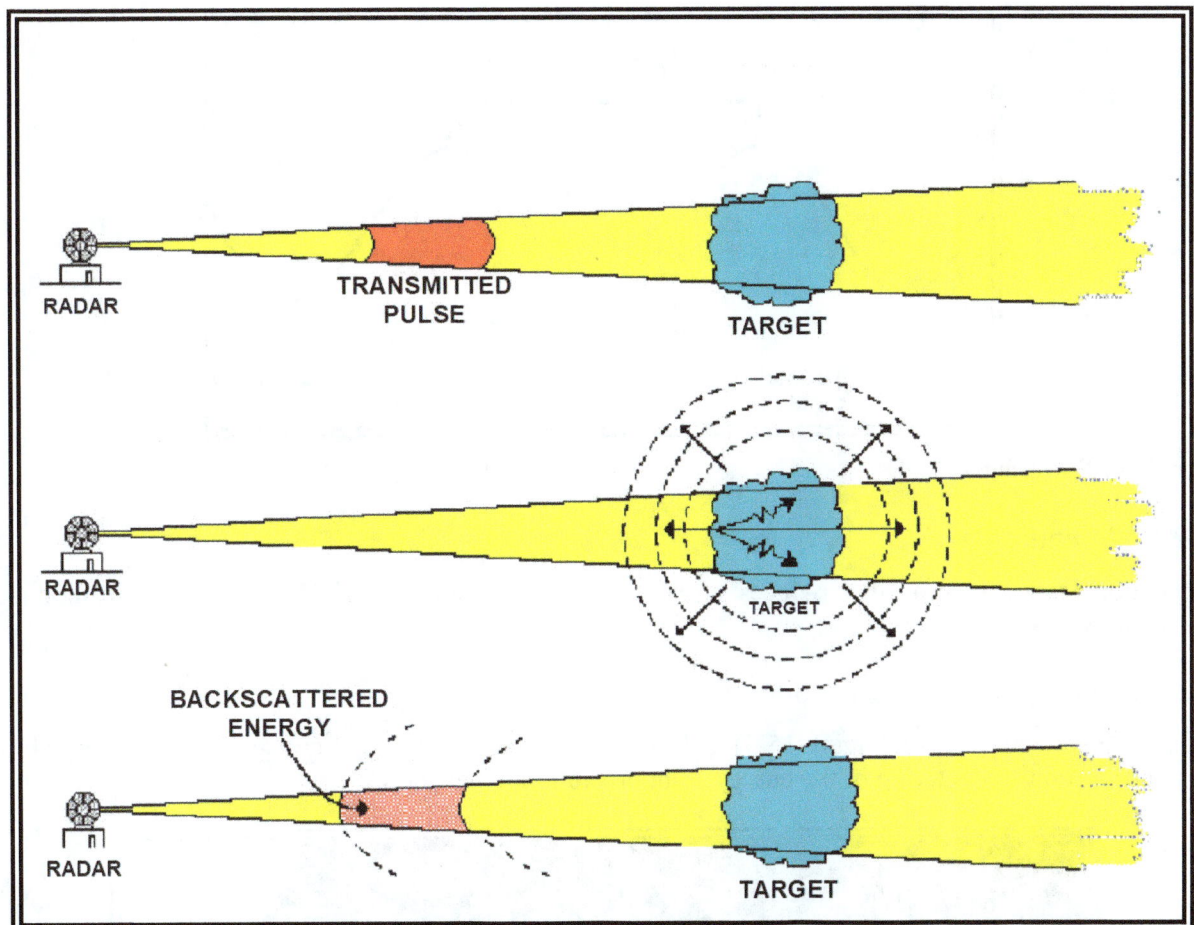

Figure 15-2. Backscattered Energy

Targets may include precipitation, clouds, dust, birds, insects, buildings, air mass boundaries, terrain features, wind farms/turbines, etc. Reflectivity is a measurement of the amount of backscattered energy. An echo is the appearance, on a radar display, of the backscattered energy (i.e., reflectivity).

15.2.3 Power Output

The WSR-88D has a peak power output of 750 kilowatts (kW). This allows for better detection of low reflectivity (small) targets in the atmosphere, such as clouds, dust, insects, etc.

Most aircraft radars have a peak power output of less than 50 kW. Therefore, smaller targets are difficult to detect with aircraft radar.

15.2.4 Wavelengths

The wavelength is the distance between two crests or two troughs within the radio wave emitted from the radar (see Figure 15-3). The WSR-88D has a wavelength of 10 cm. Most aircraft radars have a wavelength of 3 cm. Although shorter wavelengths are better at detecting smaller targets, they are significantly more attenuated than longer wavelengths.

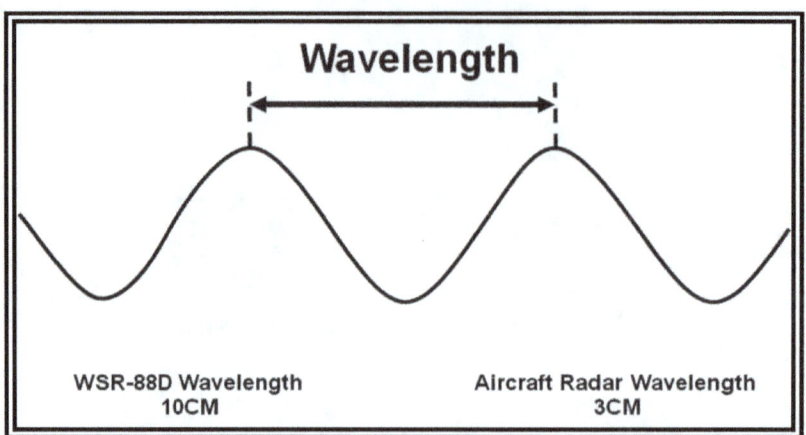

Figure 15-3. Wavelengths

15.2.5 Attenuation

Attenuation is any process that reduces energy within the radar beam. This reduces the amount of backscattered energy.

15.2.5.1 Precipitation Attenuation

Precipitation attenuation (see Figure 15-4) is the decrease of the intensity of energy within the radar beam due to absorption or scattering of the energy from precipitation particles.

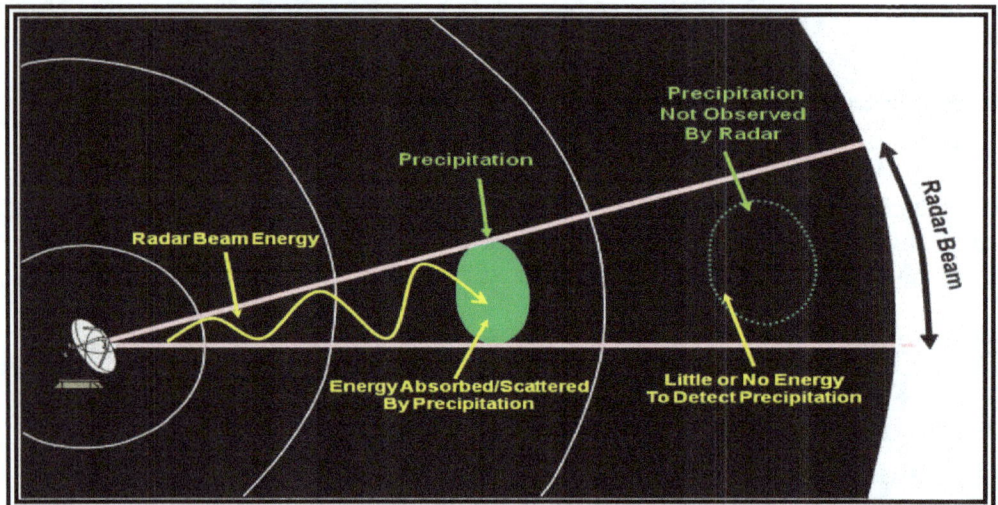

Figure 15-4. Precipitation Attenuation

Precipitation close to the radar absorbs and scatters energy within the radar beam. Therefore, very little, if any, energy will reach targets beyond the initial area of precipitation. Because of precipitation attenuation, distant targets (i.e., precipitation) may not be displayed on a radar image.

The amount of precipitation attenuation is related to the wavelength of the radar (see Figure 15-5).

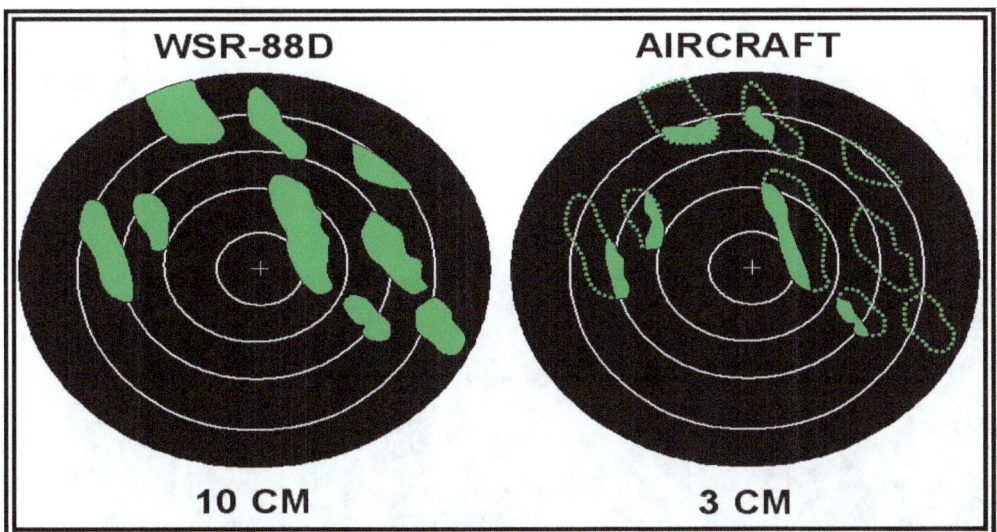

Figure 15-5. Precipitation Attenuation versus Wavelength

As the wavelength of the radar decreases, the amount of precipitation attenuation increases.

The WSR-88D's 10-cm wavelength is not significantly attenuated by precipitation. However, aircraft radars, which typically have 3-cm wavelengths, have a significant precipitation attenuation problem. As a result, aircraft weather radar typically only shows the leading edge of extreme intensity echoes.

15.2.5.2 Range Attenuation

Range attenuation is the decrease of the intensity of energy within the radar beam as the beam gets farther away from the antenna. If not compensated for, a target that is farther away from the radar will appear less intense than an identical target closer to the radar.

Range attenuation is automatically compensated for by the WSR-88D. However, most airborne radars only compensate for range attenuation out to a distance of 50 to 75 NM. Targets beyond these ranges will appear less intense than they actually are.

15.2.6 Resolution

Resolution is the ability of the radar to show targets separately.

15.2.6.1 Beam Resolution

Beam resolution is the ability of the radar to identify targets separately at the same range, but different azimuths (see Figure 15-6).

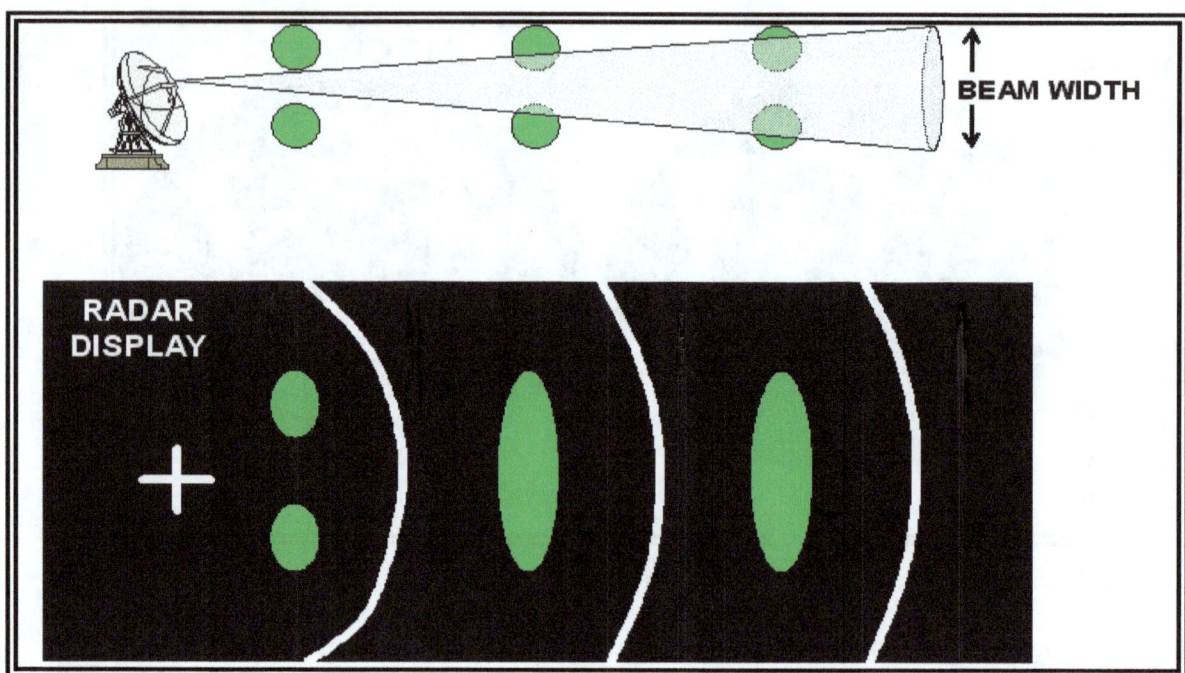

Figure 15-6. Beam Resolution

Two targets must be separated by at least one beam width (diameter) in order to be displayed as two separate echoes on a radar image.

The WSR-88D has a beam width of 0.95°. Therefore, at a range of 60 NM, targets separated by at least 1 NM will be displayed separately. At a range of 120 NM, targets separated by at least 2 NM will be displayed separately.

Aircraft radars have beam widths that vary between 3° and 10°. Assuming an average beam width of 5° at a range of 60 NM, targets separated by at least 5.5 NM will be displayed separately. At a range of 120 NM, targets separated by at least 10 NM will be displayed separately.

The beam resolution is better for the WSR-88D than aircraft radar (see Figure 15-7).

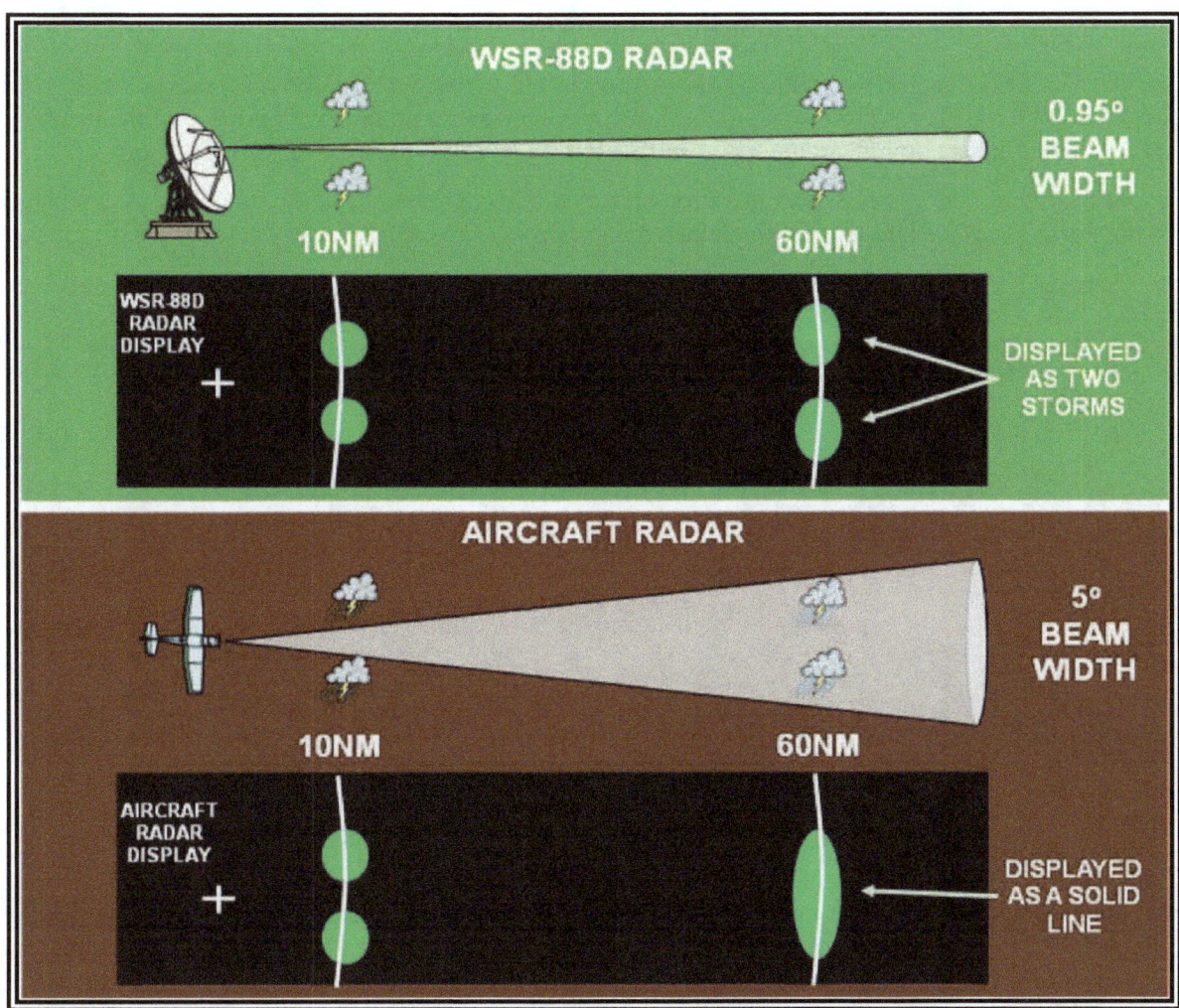

Figure 15-7. Beam Resolution Comparison between WSR-88D and Aircraft Weather Radar

In the example above, the targets (thunderstorms) are at the same range in azimuths for both the aircraft and WSR-88D radar. At 10 NM, the beam width is small enough for both the WSR-88D and aircraft radar to display the thunderstorms separately. At 60 NM, the WSR-88D beam width is still small enough to display both thunderstorms separately. However, the aircraft radar beam width is larger, which results in the two thunderstorms being displayed as one echo.

Note that the beam becomes wider at greater distances from the radar. Therefore, the beam resolution decreases with increasing range from the radar. As a result, lines of precipitation may appear to break up as they move closer to the radar. In reality, the breaks in the precipitation were most likely always there.

15.2.7 Wave Propagation

Radar beams do not travel in a straight line. The beam is bent due to differences in atmospheric density. These density differences, caused by variations in temperature, moisture, and pressure, occur in both the vertical and horizontal directions, and affect the speed and direction of the radar beam.

In a denser atmosphere, the beam travels slower. Conversely, in the less dense atmosphere, the beam travels faster. Changes in density can occur over very small distances, so it is common for the beam to be in areas of different densities at the same time as it gets larger. The beam will bend in the direction of the slower portion of the wave.

15.2.7.1 Normal (Standard) Refraction

Under normal (i.e., standard) conditions, the atmosphere's density gradually decreases with increasing height. As a result, the upper portion of a radar beam travels faster than the lower portion of the beam. This causes the beam to bend downward (see Figure 15-8).

The radar beam curvature is less than the curvature of the Earth. Therefore, the height of the radar beam above the Earth's surface increases with an increasing range.

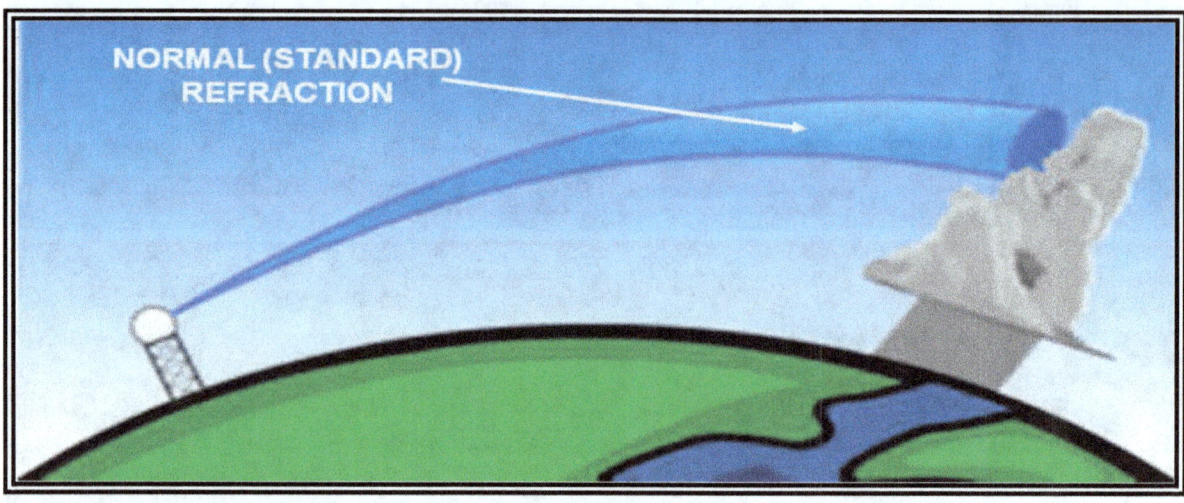

Figure 15-8. Normal Refraction

15.2.7.2 Subrefraction

Atmospheric conditions are never normal or standard. Sometimes, the density of the atmosphere decreases with height at a more-than-normal rate (actual density is less than normal). When this occurs, the radar beam bends less than normal and climbs skyward. This phenomenon is known as subrefraction (see Figure 15-9).

Subrefraction may cause the radar beam to overshoot objects that would normally be detected. For example, distant thunderstorms may not be detected with subrefraction. Subrefraction may also cause radar to underestimate the true strength of a thunderstorm. Thunderstorms may appear weaker on radar because subrefraction causes the radar beam to strike the thunderstorm near the top of the cumulonimbus cloud, where the precipitation particles tend to be smaller.

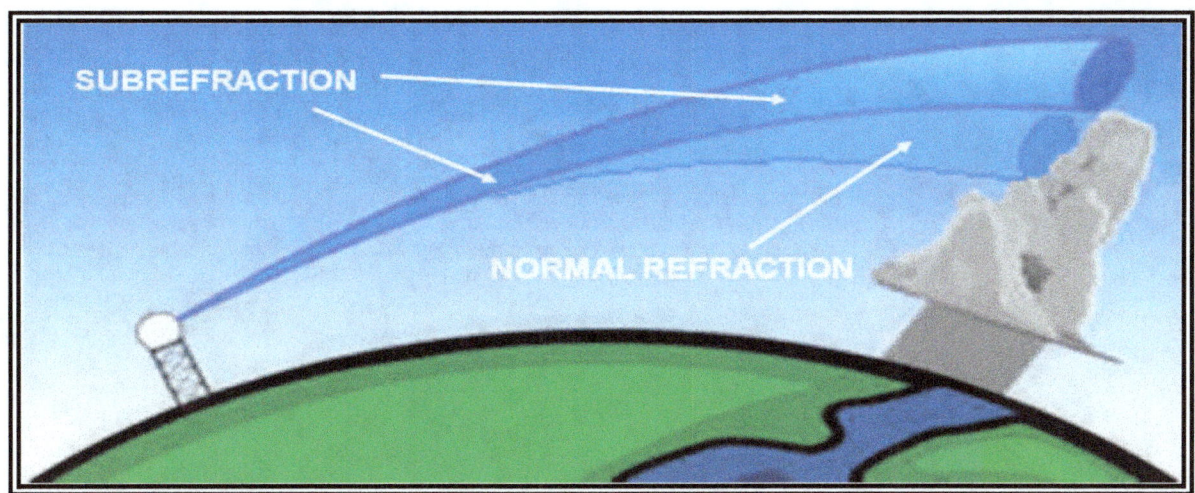

Figure 15-9. Subrefraction

15.2.7.3 Superrefraction

Conversely, sometimes the density of the atmosphere decreases with height at a less-than-normal rate (actual density is greater than normal) or even increases with height. When this occurs, the radar beam will bend more than normal toward the Earth's surface. This phenomenon is called superrefraction (see Figure 15-10).

Superrefraction causes the radar beam to travel closer to the Earth's surface than what would occur in a normal atmosphere. This can lead to overestimating the strength of a thunderstorm, as the beam would detect the stronger core of the storm, where precipitation-sized particles are larger.

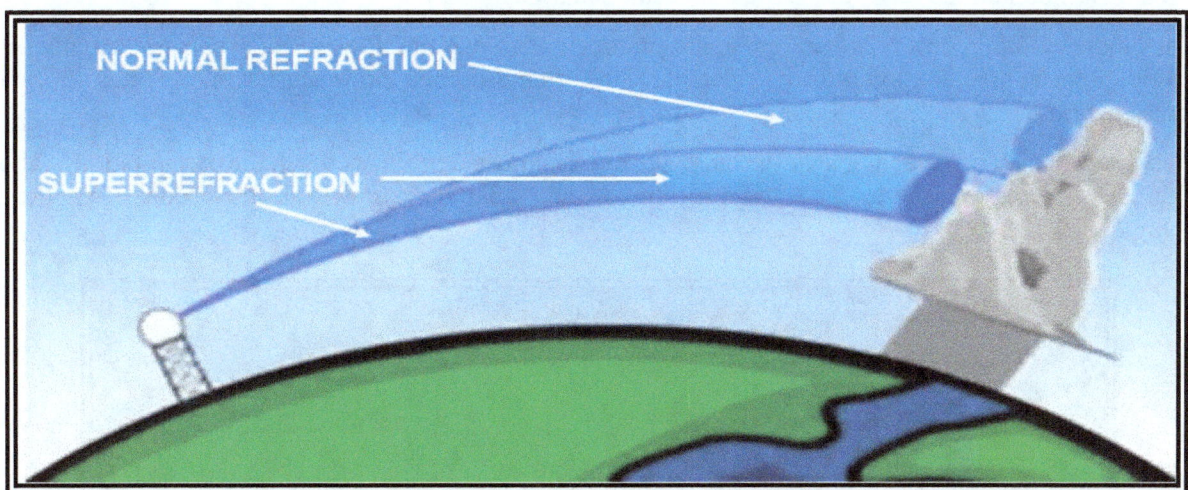

Figure 15-10. Superrefraction

15.2.7.4 Ducting

If the atmospheric condition that causes superrefraction bends the beam equal to, or greater than, the Earth's curvature, then a condition called ducting (or trapping) occurs (see Figure 15-11).

When ducting occurs, the radar beam will hit the surface of the Earth, causing some of the beam's energy to backscatter. This often leads to false echoes, also known as anomalous propagation (AP), to appear in the radar display.

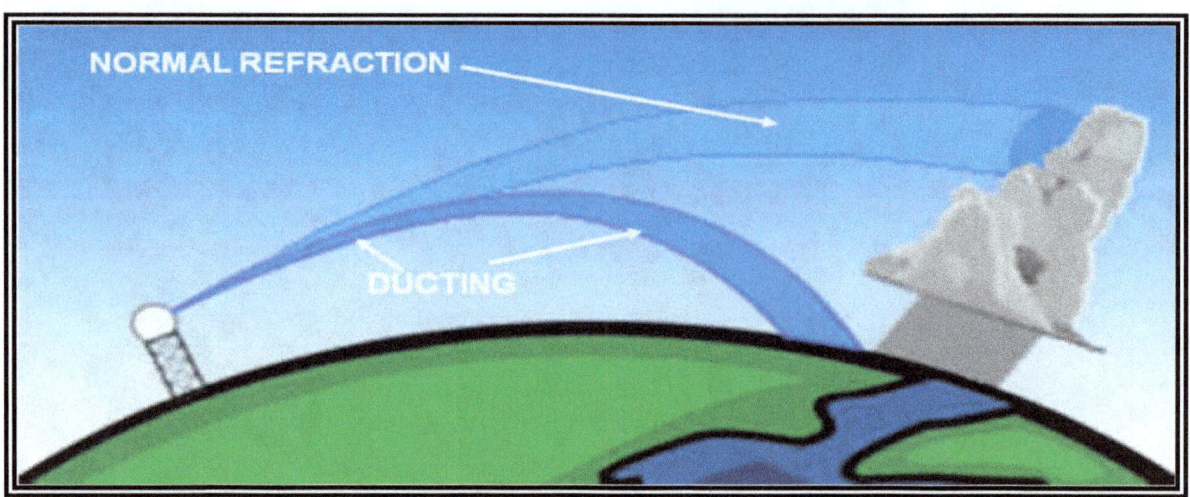

Figure 15-11. Ducting

15.2.8 Radar Beam Overshooting and Undershooting

Radar beam overshooting may occur because the lowest radar beam can be higher than the top of precipitation. This will most likely occur with stratiform precipitation and low-topped convection. For example, at a distance of 124 NM from the radar, the lowest radar beam is at an altitude of approximately 18,000 ft; at 248 NM the beam height is approximately 54,000 ft. Any precipitation with tops below these altitudes and distances will **not** be displayed on a single-site radar image. Therefore, it is quite possible that precipitation may be occurring where none appears on the single-site radar image.

Radar overshooting occurs more often in the mountainous western United States where some radars are located on a mountaintop (e.g., the WSR-88D near Cedar City, UT).

Radar undershooting occurs when precipitation occurs above the lowest radar beam, usually with high-cloud-based precipitation near the radar site. This often occurs in the western United States during the summer months.

Undershooting may occur at and near the radar site even in mosaic products when the precipitation is above the highest elevation angle. This region above the radar is known as the "cone of silence" (see Figure 15-12).

Figure 15-12. Cone of Silence

15.2.9 Beam Blockage

Beam blockage (see Figure 15-13) occurs when the radar beam is blocked by terrain and is particularly predominant in mountainous terrain. (See Section 24.6.1 for more information on the WSR-88D.)

Beam blockage is most easily seen on the lowest radar beam (also known as "Base Reflectivity," "Lowest Tilt," and "Reflectivity at Lowest Altitude") images where it appears as a pie-shaped area (or areas) perpetually void of echoes. When animating the imagery, the beam blockage area will remain clear of echoes even as precipitation and other targets pass through. In many cases, the beam blockage effect seen on a single-site radar can be minimized by viewing mosaic images.

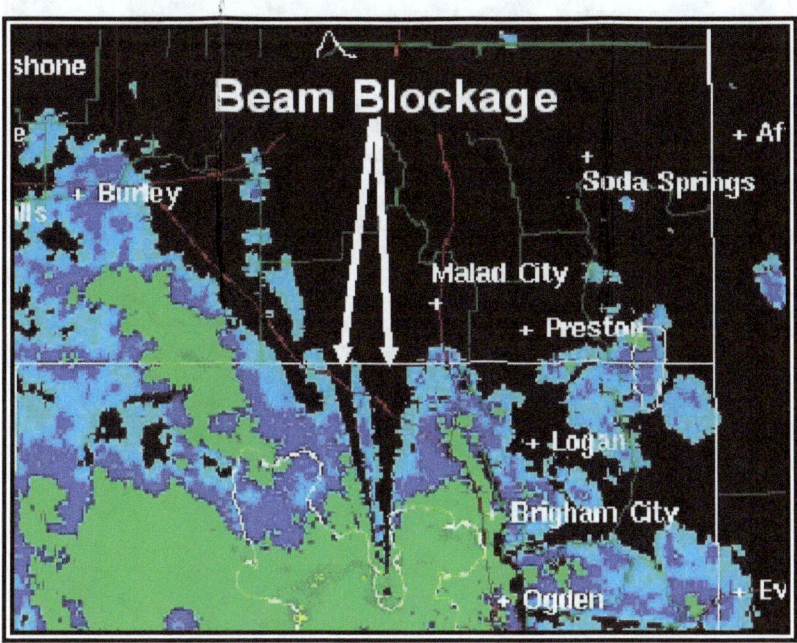

Figure 15-13. WSR-88D Weather Radar Beam Blockage on Base Reflectivity Product Example

15.2.10 Ground Clutter

Ground clutter (see Figure 15-14) is radar echoes' returns from trees, buildings, or other objects on the ground. It appears as a roughly circular region of high reflectivity at ranges close to the radar. Ground clutter appears stationary when animating images and can mask precipitation located near the radar. Most ground clutter is automatically removed from WSR-88D imagery, so typically it is does not interfere with image interpretation.

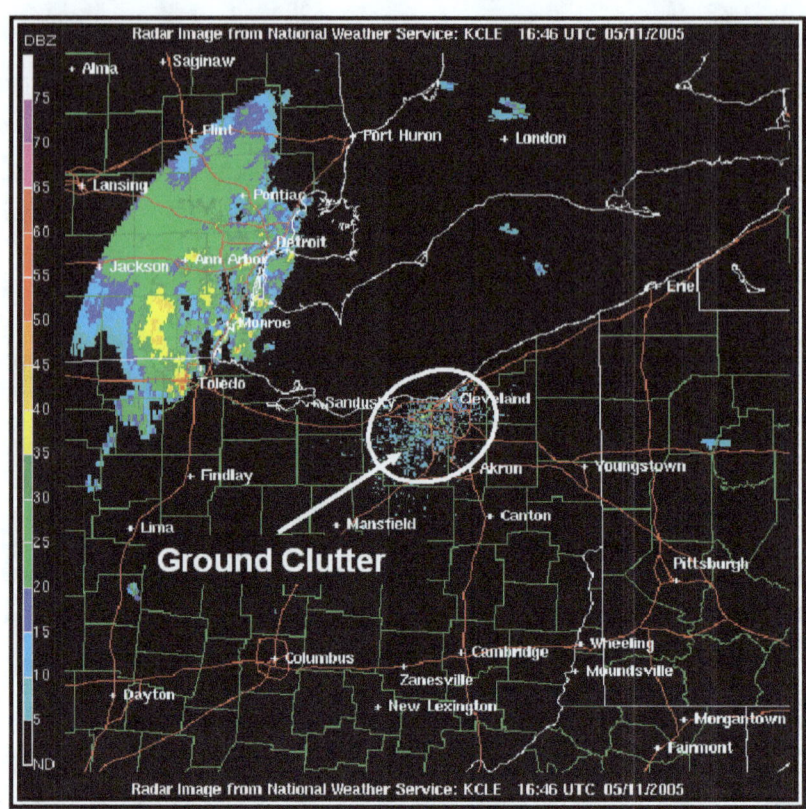

Figure 15-14. WSR-88D Weather Radar Ground Clutter Example

15.2.11 Ghost

A ghost (see Figure 15-15) is a diffused echo in apparently clear air caused by a "cloud" of point targets, such as insects, or by refraction returns of the radar beam in truly clear air.

The latter case commonly develops at sunset due to superrefraction during the warm season. The ghost develops as an area of low reflectivity echoes (typically less than 15 decibels of Z (dBZ)) near the radar site and quickly expands. When animating the imagery, the ghost echo shows little movement.

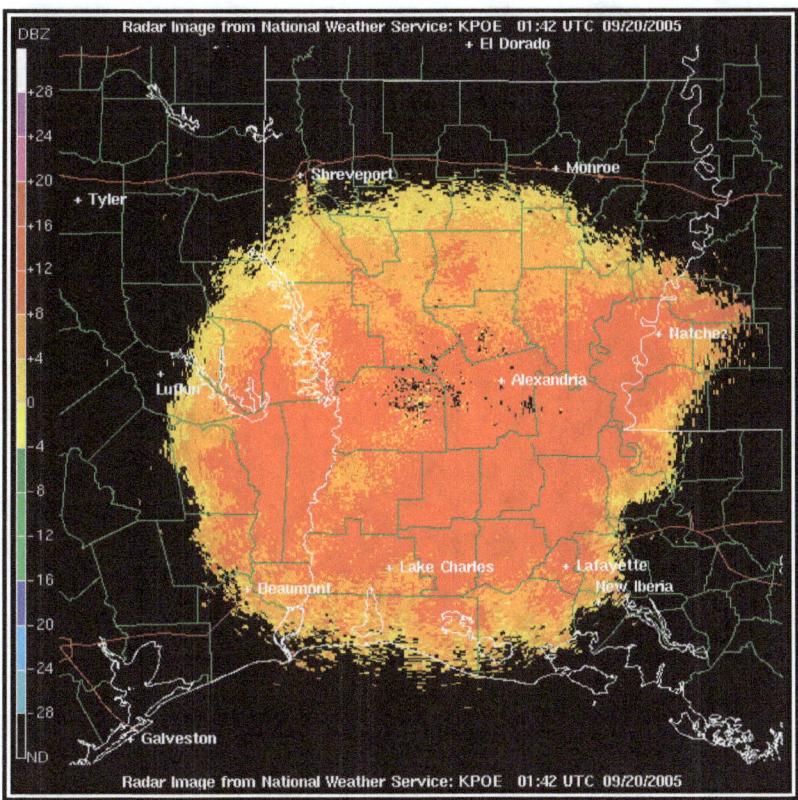

Figure 15-15. WSR-88D Weather Radar Ghost Example

15.2.12 Angels

Angels are echoes caused by a physical phenomenon not discernible by the eye at the radar site. They are usually caused by bats, birds, or insects. Angels typically appear as a donut-shaped echo with low reflectivity values (see Figure 15-16). When animated, the echo expands and becomes more diffuse with time.

Angels typically only appear when the WSR-88D radar is in Clear Air Mode because of their weak reflectivity (see Section 24.6.1.3 for additional information on modes). Echoes caused by birds are typically detected in the morning when they take flight for the day. Echoes caused by bats are typically detected in the evening when they are departing from caves.

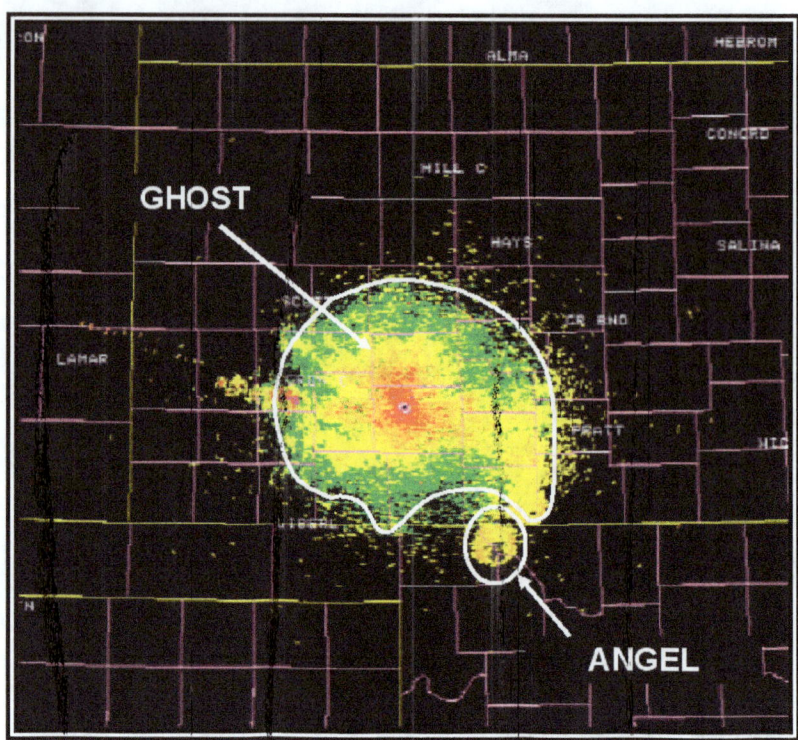

This angel was caused by bats departing Selman Bat Cave at Alabaster Caverns State Park, OK, around sunset.

Figure 15-16. WSR-88D Weather Radar Angel Example

15.2.13 Anomalous Propagation (AP)

AP (see Figure 15-17) is an extended pattern of ground echoes caused by superrefraction of the radar beam. Superrefraction causes the radar beam to bend downward and strike the ground. It differs from ground clutter because it can occur anywhere within the radar's range, not just at ranges close to the radar.

AP typically appears as speckled or blotchy high-reflectivity echoes. When animating images, AP tends to "bloom up" and dissipate, and has no continuity of motion. AP can sometimes be misinterpreted as thunderstorms; differentiating between the two is determined by animating the images. Thunderstorms move with a smooth, continuous motion while AP appears to "bloom up" and dissipate randomly.

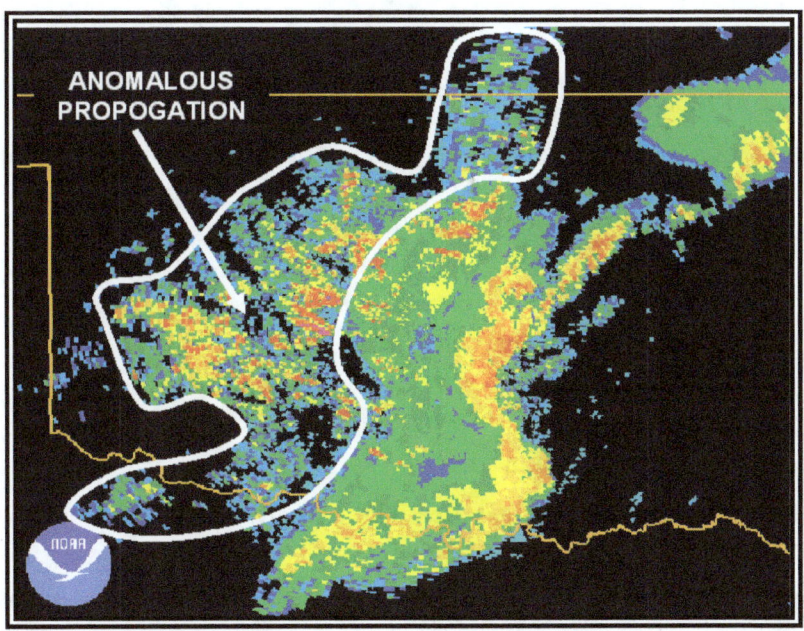

Figure 15-17. WSR-88D Weather Radar AP Example

15.2.14 Other Nonmeteorological Phenomena

15.2.14.1 Wind Farms

Wind farms can affect the return signal of the radar beam. Depending on the proximity of the wind farm to the radar site (generally within 10 NM), wind farm turbines can result in beam blockage, false echoes, or high reflectivity values (see Figure 15-18).

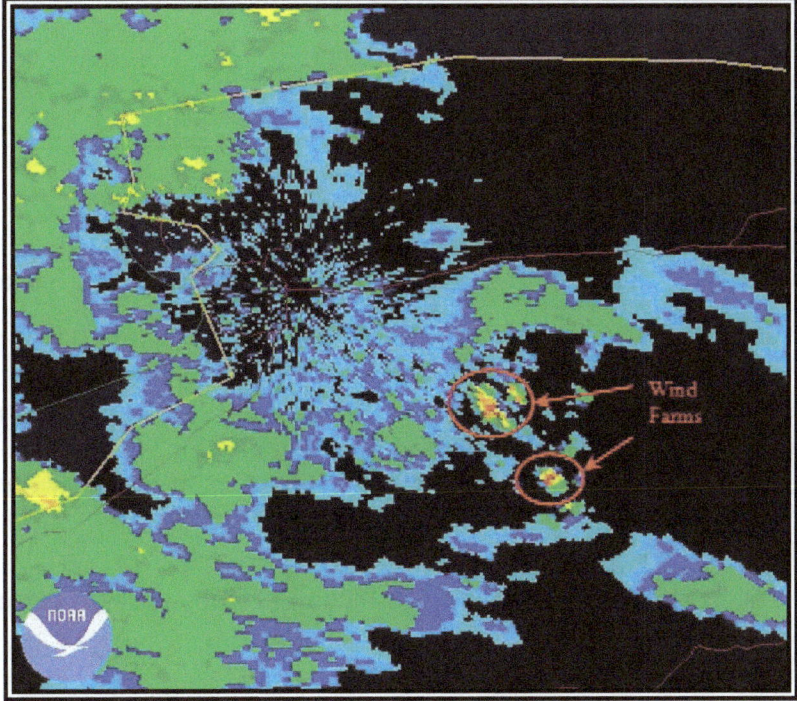

Figure 15-18. Wind Farms Can Make Benign Echoes Appear as Small Storms

15.2.15 Precipitation

15.2.15.1 Intensity of Precipitation

The intensity of precipitation is determined from the amount of energy backscattered by precipitation, also known as reflectivity. Reflectivity is determined by:

- The size of precipitation particles;
- The precipitation state (liquid or solid);
- The concentration of precipitation (particles per volume); and
- The shape of the precipitation.

15.2.15.1.1 Intensity of Liquid Precipitation

The most significant factor in determining the reflectivity of liquid particles is the size of the precipitation particle (see Figure 15-19).

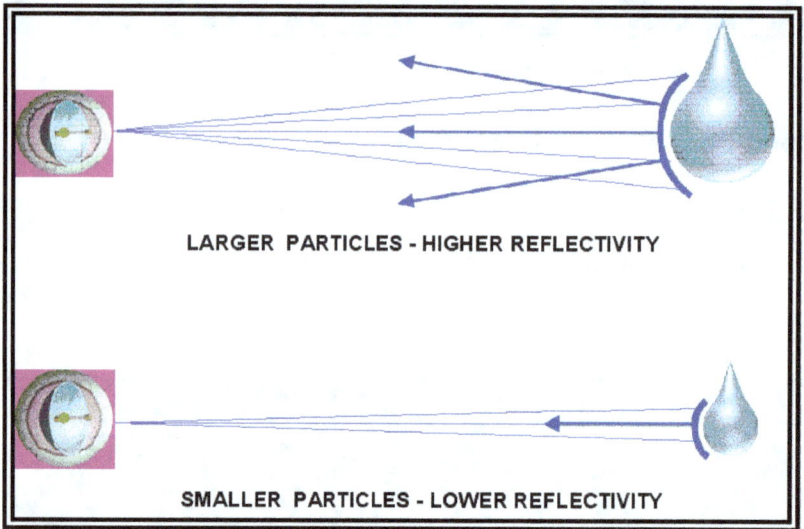

Figure 15-19. Reflectivity Associated with Liquid Targets

Larger particles have greater reflectivity than smaller particles. For example, a particle with a 1/4-in diameter backscatters the same amount of energy as 64 particles that each have a 1/8-in diameter.

Radar images/intensity scales are associated with reflectivities that are measured in dBZ. The dBZ values increase based on the strength of the return signal from targets in the atmosphere.

Typically, liquid precipitation-sized particle reflectivities are associated with values that are 15 dBZ or greater. Values less than 15 dBZ are typically associated with liquid cloud-sized particles. However, these lower values can also be associated with dust, pollen, insects, or other small particles in the atmosphere.

15.2.15.1.2 Convective Precipitation

Convective precipitation (see Figure 15-20) is distinguished by the following radar characteristics:

- Echoes tend to form as lines or cells;
- Reflectivity gradients are strong;

- Precipitation intensities generally vary from moderate to extreme;
- Occasionally, precipitation intensities can be light; and
- Echo patterns change rapidly when animating the image.

Numerous hazards are associated with convective precipitation. These hazards include: turbulence, Low-Level Wind Shear (LLWS), strong and gusty surface winds, icing above the freezing level, hail, lightning, tornadoes, and localized IFR conditions below the cloud base due to heavy precipitation.

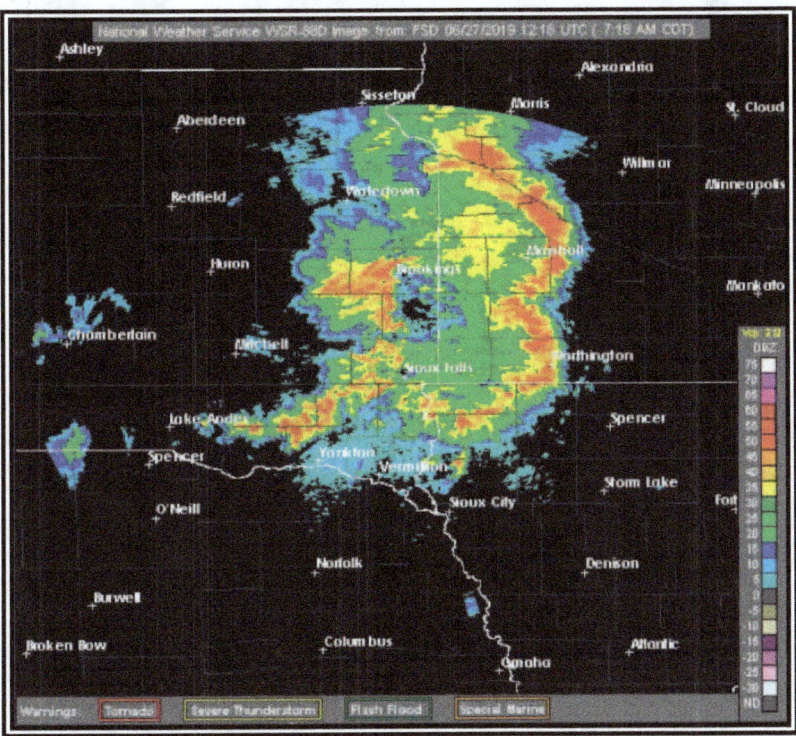

Figure 15-20. WSR-88D Weather Radar Convective Precipitation on the 0.5° Base Reflectivity Product Example

15.2.15.1.3 Stratiform Precipitation

Stratiform precipitation (see Figure 15-21) has the following radar characteristics:

- Widespread in aerial coverage;
- Weak reflectivity gradients;
- Precipitation intensities are generally light or moderate (39 dBZ or less);
- Occasionally, precipitation intensities can be stronger; and
- Echo patterns change slowly when animating the image.

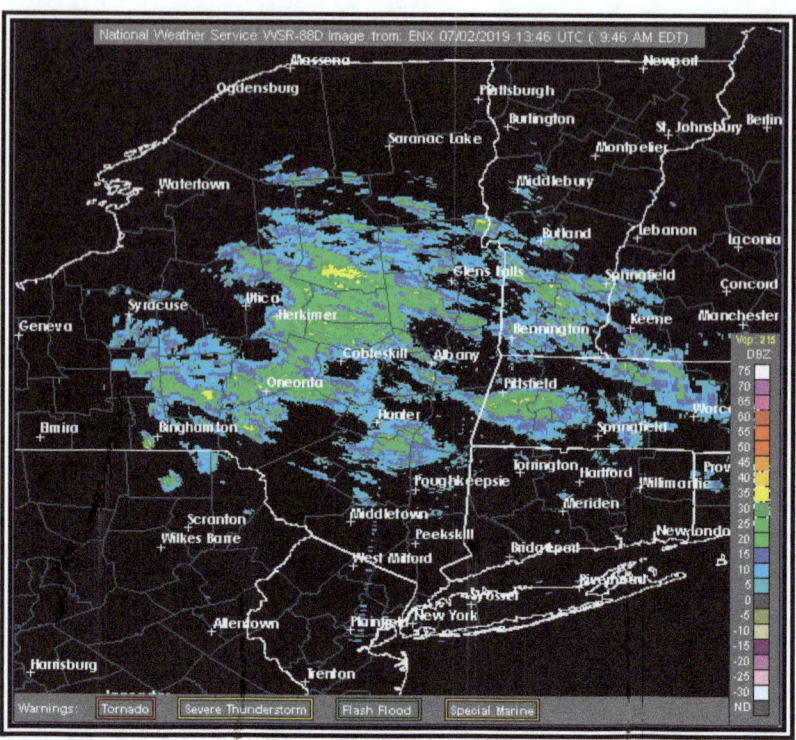

Hazards associated with stratiform precipitation include possible widespread icing above the freezing level, low ceilings, and reduced visibilities.

Figure 15-21. WSR-88D Weather Radar Stratiform Precipitation on the 0.5° Base Reflectivity Product Example

15.2.15.1.4 Intensity of Snow

A radar image cannot reliably be used to determine the intensity of snowfall. However, snowfall rates generally increase with increasing reflectivity.

15.2.15.1.5 Bright Band

Bright band is a distinct feature observed by radar that denotes the freezing (melting) level. The term originates from a band of enhanced reflectivity that can result when a radar antenna scans through precipitation. The freezing level in a cloud contains ice particles that are coated with liquid water. These particles reflect significantly more energy (appearing to the radar as large raindrops) than the portions of the cloud above and below the freezing layer.

16 Mountain Weather

16.1 Introduction

This chapter focuses on mountain waves and adverse winds. Other mountain weather phenomena are discussed in other chapters, which are noted.

16.2 Mountain Waves and Adverse Winds

The atmosphere is a fluid in motion. Just as the swiftly flowing water in a stream develops waves and eddies as it passes over and around obstructions, so does the atmosphere contain disturbances that develop as it interacts with mountainous terrain. These atmospheric eddies can range in size from a few centimeters to tens or hundreds of kilometers, and can present the pilot with relatively smooth air, or with turbulence of potentially destructive intensity, and the likelihood of loss of control.

When the atmosphere encounters a mountainous barrier, a number of responses are possible. If the wind is weak or the moving air mass exceptionally dense, the mountains may act as a dam, preventing the motion of air over the barrier. More frequently, strong winds flow over or around mountains or ridges. If the surrounding atmosphere is unstable, the vertical displacement of the air will (if sufficient moisture is present) lead to thunderstorm formation or at least the development of deep convective clouds. However, if the wind is sufficiently strong and the surrounding atmosphere is stable, a wave will develop.

The wave that results from vertical displacement of a stable air mass over a mountain or ridge can generally take one of two forms: vertically propagating mountain waves or trapped lee waves. Both types of waves can be hazardous to aviation operations. The particular type of wave or combination of waves that forms depends on the nature of the mountain range and on atmospheric properties upwind of the mountain. It is possible for both types of waves to exist at the same time. It also is possible to have hybrid or intermediate forms (i.e., waves that are only partially trapped).

The most severe mountain wind events usually occur when the large-scale (or synoptic) winds are strongest, from late autumn to early spring. During the remainder of the year, when the synoptic winds are normally much weaker, hazardous winds in the vicinity of mountains are more likely to be associated with thunderstorms and their outflow fields.

The mountain-induced flow fields to be discussed in this section are frequently accompanied by visual indicators (such as lenticular and rotor clouds or blowing dust). However, this is not always the case, and extremely severe wind events can occur with little or no visual warning of their presence.

16.2.1 Gravity Waves

In order for gravity waves to develop, the atmosphere must possess at least some degree of static stability. As stable air is deflected vertically by an obstacle (e.g., when an air mass moves over a mountain ridge), it resists the displacement because as it rises, it is heavier than the air surrounding it, and gravity is acting to return it to its equilibrium level. Because of its negative buoyancy, the deflected air begins to return to its original level once it has cleared the ridge. However, its momentum will cause it to overshoot the original altitude, warming by compression and now becoming less dense than the surrounding air. As a result, it begins to rise back to the equilibrium altitude, overshoots once more, and continues through a period of oscillations before the resulting wave motion dampens out. This process is depicted in Figure 16-1.

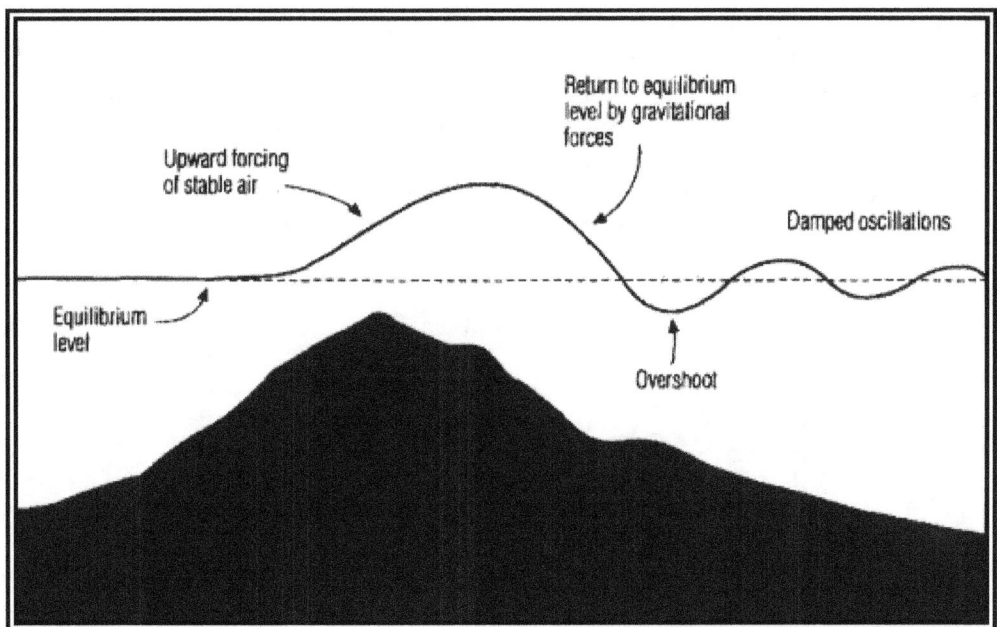

Figure 16-1. Gravity Wave Oscillations

The described gravity wave will have measurable wavelength, amplitude, phase speed, and period. The period of this type of atmospheric disturbance is related to the temperature of the air and the "spread" between the existing lapse rate and the dry adiabatic lapse rate (or, equivalently, the degree of stability present). In general, the large-scale wind (wind shear) change in altitude and temperature (lapse rate), the size and shape of the mountain or ridge over which the air is moving, and the orientation of the wind relative to the ridge line all work together in determining the character of the disturbance that develops.

16.2.2 Kelvin-Helmholtz (K-H) Waves

When wind shear is very strong, another type of wave is possible. These waves, called gravity-shear or K-H waves, can occur when the kinetic energy inherent in the shear can overcome the damping effects of a stable temperature lapse rate. This effect is illustrated in Figure 16-2.

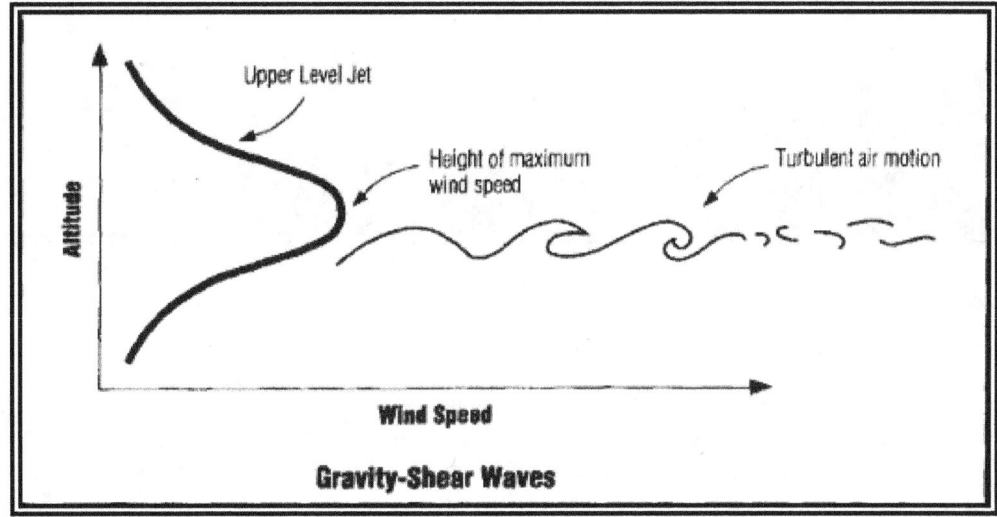

Figure 16-2. Gravity-Shear Waves

Chapter 16, Mountain Weather

If the wind shear that penetrates the layer of the atmosphere is weak (some wind shear is nearly always present), a shear-induced wave motion will not occur. However, if the magnitude of the wind shear exceeds a critical value, wave motions will begin spontaneously within the shear layer, resulting in a K-H wave. The amplitude of the resulting wave will grow with the kinetic energy in the surrounding wind field until, like an ocean wave breaking on the shore, the wave overturns and breaks down into turbulence. The resulting turbulence can have a range of effects on aircraft. The clouds associated with shear-induced gravity waves can frequently be observed in the atmosphere, as shown in Figure 16-3 and Figure 16-4.

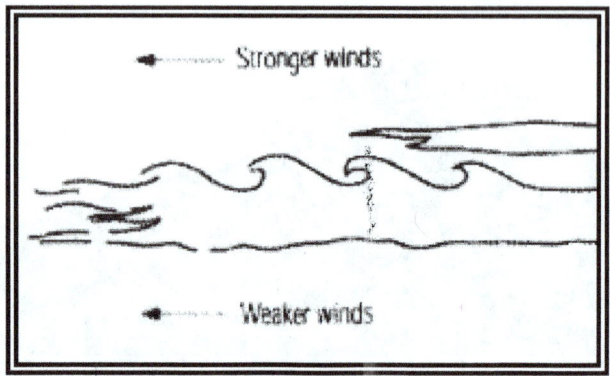

Figure 16-3. Schematic of Clouds Associated with Gravity-Shear Waves

(Photo Credit: University Corporation for Atmospheric Research (UCAR))

Figure 16-4. Clouds Associated with Gravity-Shear Waves

K-H waves are quite common in the atmosphere; they can form in the vicinity of thunderstorms, in shear layers near the jet stream, and in association with stable regions of the atmosphere that are topped by a strong wind-shear layer (such as the top of a pool of cold air on the lee side of a mountain). In fact, K-H instability induced by the wind shear associated with strong winds aloft is likely the chief source of high-level turbulence away from mountain ranges (Clear Air Turbulence (CAT)). The mechanism that causes this type of disturbance can be compared to that of a flag flapping in a breeze. The flapping is a result of instabilities created by the wind shear along the flexible surface of the flag, analogous to the wind shear through a very stable (but shallow) layer of the atmosphere.

16.2.3 Vertically Propagating Mountain Waves

Figure 16-5 shows a schematic of a vertically propagating mountain wave. This feature is essentially a standing gravity wave whose energy propagates vertically. For this class of wave, nothing is preventing vertical propagation, such as strong wind shear or neutrally stable atmospheric layers. The mere fact that a wave has developed in air moving over a mountain (or other barrier) does not in itself indicate problems for an aircraft operating in the vicinity. The potential for hazard is a function of the strength of the wave and whether or not an area of the wave "breaks" into turbulent motions that, in the extreme, can lead to structural damage or failure of an aircraft component.

With this type of wave feature, air that is moving nearly perpendicular to the barrier is deflected upward and accelerated as it passes over the crests and down the lee slopes of the terrain. Notice in Figure 16-5 that the standing wave has developed vertically above the mountain crest and that the resulting wave tilts upwind with height. This vertical propagation of the wave means that the effects of the mountain range can be felt at heights significantly above the actual altitude of the peaks (at times reaching in excess of 60,000 ft). As a result, aircraft flying at virtually any altitude may have to deal with significant turbulence and wave-induced altitude excursions. In fact, the amplitude of this type of wave actually increases with height above the mountain (in the absence of atmospheric features, such as strong inversions or shear layers that would tend to partially reflect or absorb the upward-moving wave energy). This amplification is a consequence of the normal decrease in air density with altitude.

The amplitude of the wave will be larger, for the same upstream conditions, the higher the mountain range above the surrounding terrain. Although, even very modest terrain relief can cause appreciable wave activity under the proper conditions. Wave amplitude also will tend to be larger for stronger cross-mountain wind components at mountaintop level. However, the actual amplitude depends on complex relationships between upstream atmospheric wind and temperature profiles and the height and shape of the particular mountain range. Stronger flow across the mountain leads to a deeper wave, given the same atmospheric stability. However, the greater the background stability, the shallower the resulting wave, at fixed-wind speed.

As previously noted, the primary concern for pilots with this type of feature is that the vertical motions of the air moving through the wave may become strong enough to "break" into turbulence.

What is meant by "wave breaking"? Looking again at the streamlines that show the airflow in Figure 16-5, it shows that high above the ridge there is a region of updraft. With a wave of modest amplitude (in which the vertical displacement of air moving through the wave is relatively limited), an aircraft flying through this region would likely experience appreciable "wave action," with altitude and/or airspeed fluctuations, but little turbulence. However, with sufficient amplitude, the wave breaks and localized updrafts and downdrafts occur. The consequences for a pilot flying through this region include airspeed and altitude deviations and the possible sudden onset of severe or extreme turbulence. This type of turbulence occurs typically between 20,000 ft and 39,000 ft MSL and is, therefore, primarily of importance to turboprop and jet aircraft at cruise as they approach and overfly the mountain range.

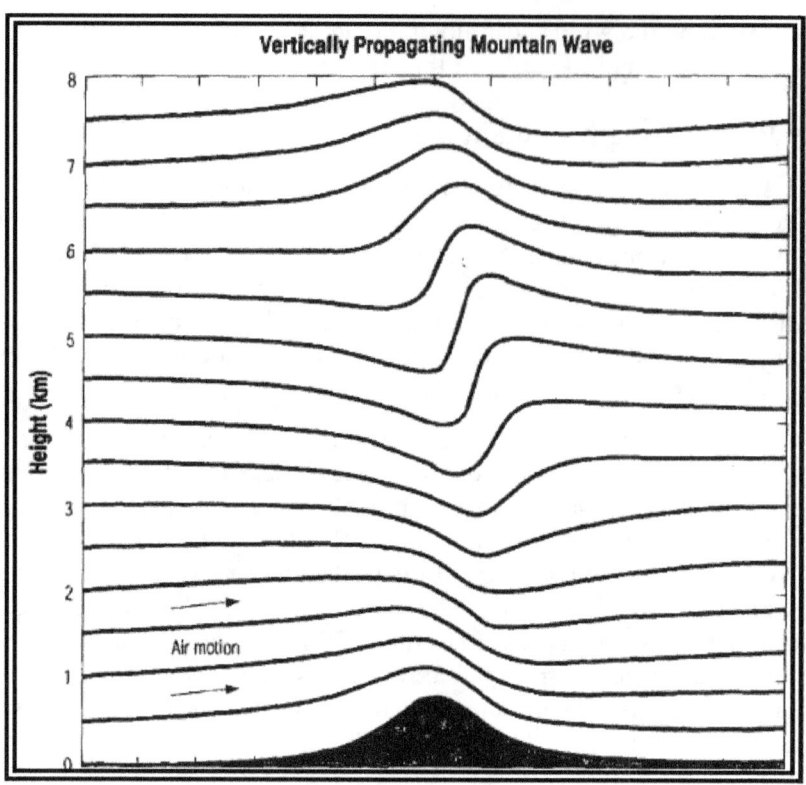

(Source: after Durran and Klemp, 1983)[4]

Figure 16-5. Schematic of a Vertically Propagating Mountain Wave

Often accompanying these high-altitude effects is the occurrence of very strong surface winds that result from the wave breaking aloft. In this case, strong downslope winds on the lee slopes can reach 100 kt in gusts, creating a low-level turbulence hazard for all aircraft. Further, these extremely strong low-level winds often abruptly terminate in a "jump" located some distance down the lee slope or well to the lee of the mountains themselves. These features are indicated schematically in Figure 16-6. The jump region is frequently an area of extreme turbulence extending to 10,000 ft or more above the surface. The area of the jump is sometimes marked by a line of ragged rotor clouds exhibiting very turbulent motion. Downwind of the jump, turbulence decreases in intensity but still may be quite strong.

---

[4] Durran, D. R., & Klemp, J. B. (1983). A compressible model for the simulation of moist mountain waves. *Monthly Weather Review, 111*, 2341-2361.

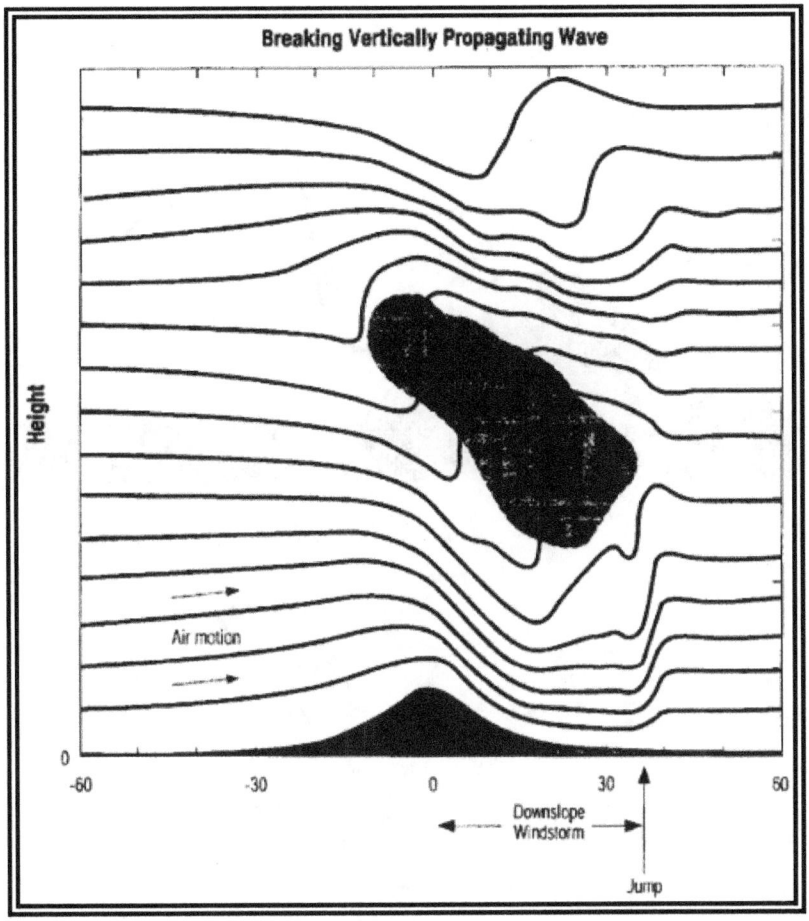

(Source: after Durran and Klemp, 1983)[5]

Figure 16-6. Schematic Showing Locations of Jump and Wave Breaking Region of a Vertically Propagating Mountain Wave

Figure 16-7 shows a schematic of the jump feature, with a pronounced wave and associated strong shear layer. The shear layer (shown in the inset) is a source of the turbulence found with the jump.

---

[5] Durran, D. R., & Klemp, J. B. (1983). A compressible model for the simulation of moist mountain waves. *Monthly Weather. Review, 111*, 2341-2361.

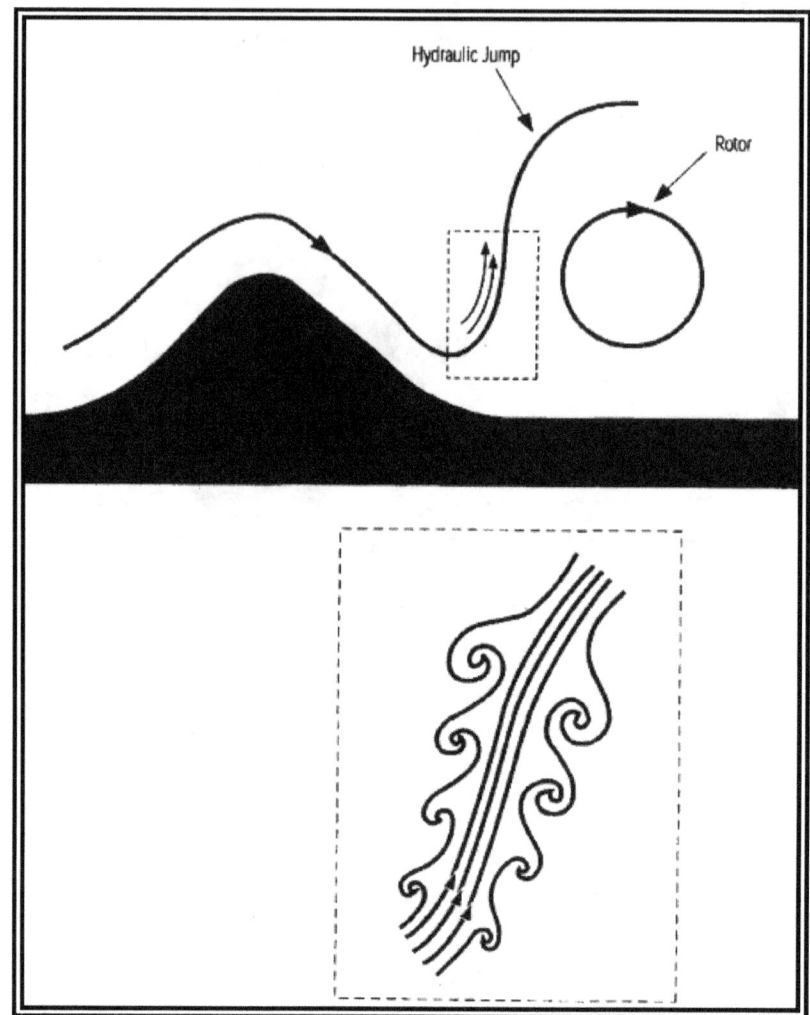

Figure 16-7. Schematic of the Strong Shear Zone Associated with a Hydraulic Jump in a Mountain Wave

16.2.4 Trapped Lee Waves

In the preceding section, an important type of mountain wave that propagates (i.e., transports its energy) vertically was discussed. Next, consider a second type of mountain wave, often manifested by a train of Altocumulus Standing Lenticular (ACSL) clouds extending far downwind of the mountain (although trapped lee waves frequently occur without clouds). These waves are of concern for takeoff and landing operations and en route flight below FL250. The associated lenticular (lens- or airfoil-shaped) clouds may appear turbulent or smooth and, depending on the moisture stratification upwind of the mountain, multilayered. They are evident as relatively straight lines or bands of clouds (with clear spaces between), parallel to the mountain range, but downstream from it.

The waves that produce these cloud features often are referred to as "trapped lee waves," because the wave energy is confined below a certain altitude. The mechanism confining this energy is strong wind shear above ridge level. Trapped lee waves are most likely to occur when the wind crosses a narrow mountain range, with a layer close to ridge level and upstream of the mountain that has strongly increasing wind speed with height and high stability, capped by a layer of strong flow and low stability.

Figure 16-8 depicts a trapped lee wave. Notice that this type of wave extends downwind from the mountain, does not develop to a high altitude, and has no upstream tilt, in contrast to the vertically propagating wave in Figure 16-5.

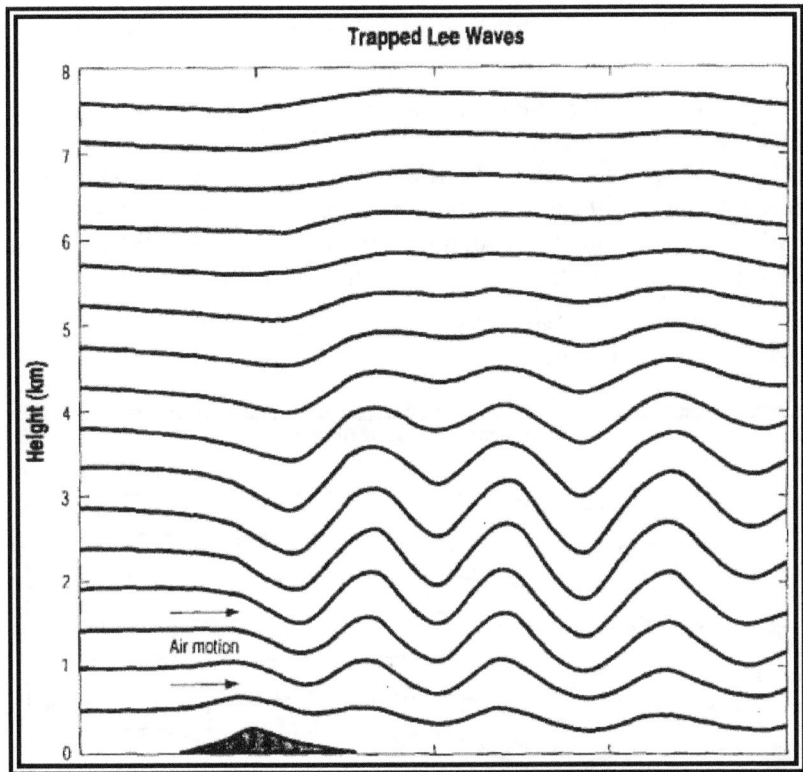

(Source: after Durran and Klemp, 1983)[6]

Figure 16-8. Computer Simulation of Trapped Lee Waves Behind a 300-m-High Mountain

This class of wave presents less turbulence hazard at high altitude than do breaking vertically propagating waves, because the wave amplitude decreases with height within the "trapping layer," typically based within a few thousand feet of the ridge crest. As a result, these waves do not extend to as great an altitude. An exception to this rule is when the atmospheric structure permits only partial trapping. This commonly occurs because the layer of wind shear that is instrumental in the trapping is weaker or shallower than necessary to do the job completely.

However, at lower altitudes, trapped lee waves can create strong turbulence encounters for aircraft. Below lenticular clouds, the wind can be quite variable and gusty, although usually not extremely strong. The gusty winds can extend from the surface up to the base of the clouds, particularly during daylight hours of spring and summer when the sky is otherwise mostly cloud-free.

Cloud bases associated with trapped lee waves are typically one to several thousand feet above ridge level, and PIREPs in the vicinity frequently indicate moderate-to-severe turbulence beneath the clouds. The turbulence associated with trapped lee waves is related to the large horizontal and vertical wind shears below cloud level.

With this type of wave, there is frequently a strong shear layer near cloud base immediately to the lee of the mountain range. This separates a turbulent wake region below mountaintop level from the faster-moving, cloud-bearing air above. In the cloud layer itself, conditions typically range from turbulent near cloud base to smooth near cloud top. The clouds themselves give some indication of the degree of turbulence within them; smooth, laminar-looking edges and tops are associated with little or no turbulence,

---

[6] Durran, D. R., & Klemp, J. B. (1983). A compressible model for the simulation of moist mountain waves. *Monthly Weather Review, 111*, 2341-2361.

while a lumpy, non-uniform appearance and a visual impression of rolling motion about an axis parallel to the cloud is indicative of turbulence.

Superimposed on the smaller-scale turbulent motions that may be present are larger-scale updraft and downdraft motions that are a part of the wave. Vertical shear of the horizontal wind is locally enhanced at the crests and troughs of the wave as a result of vertical transport (by the wave) of strong winds, leading to shear-induced turbulence. Figure 16-9 shows lenticular clouds associated with a trapped lee wave. Note the laminar appearance of the flow within the cloud that has developed from expansional cooling and condensation of water vapor in the upward-moving portion of the wave.

The rolling motions in these clouds associated with a trapped lee wave are repetitive downstream, each cloud band corresponding with a wave crest.

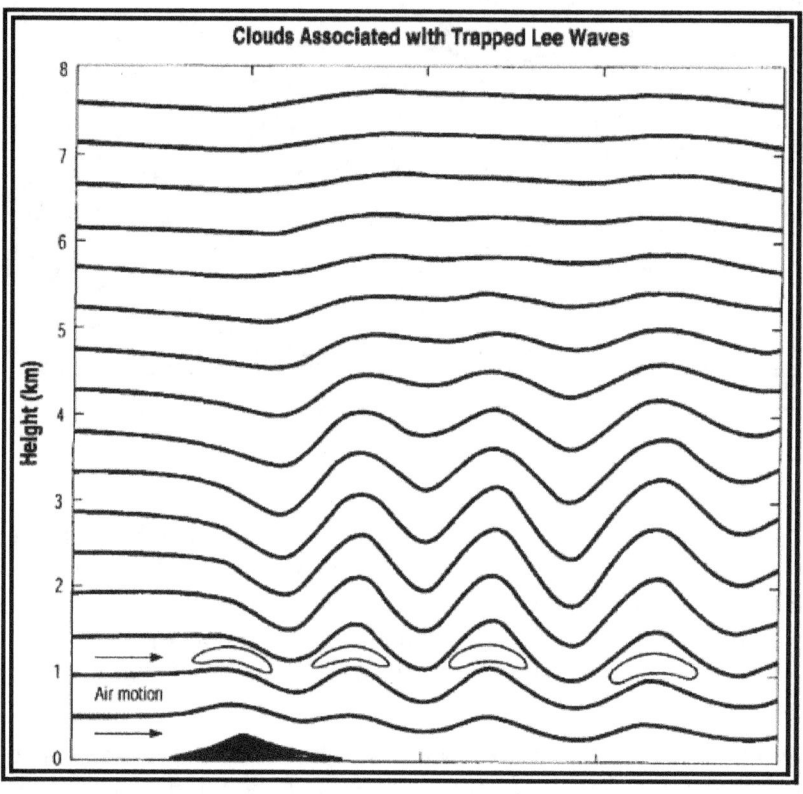

(Source: after Durran and Klemp, 1983)[7]

Figure 16-9. Lenticular Clouds Associated with a Trapped Lee Wave

16.2.5 Persistent Horizontal Roll Vortices (Rotors)

When mountain waves are present, it is quite common for a rotor zone to develop near or below ridge level on the downwind side of the mountain, under a wave crest and associated lenticular cloud (if sufficient moisture is present). This is an area of potentially severe-to-extreme wind shear and turbulence.

Figure 16-10 shows a schematic of the wind flow associated with this feature. As illustrated in this figure, rotors typically mark the downwind terminus of a downslope windstorm. When this is the case, the rotor is really part of the "jump" discussed earlier. Although strong rotation is typically present within the rotor zone and associated cloud, a pilot in a moving aircraft may not be able to detect such motion visually until

---

[7] Durran, D. R., & Klemp J. B. (1983). A compressible model for the simulation of moist mountain waves. *Monthly Weather Review*, *111*, 2341-2361.

the aircraft is quite close to the vortex. In fact, from a distance, a rotor cloud may look like a rather innocuous cumulus cloud; however, the downwind side of the rotor cloud will typically be rounded in the direction of rotation of the rotor, with cloud tags or streamers at the bottom of the cloud mass.

The latter features appear to be rapidly forming and dissipating, thereby giving some sense of rotation within the cloud.

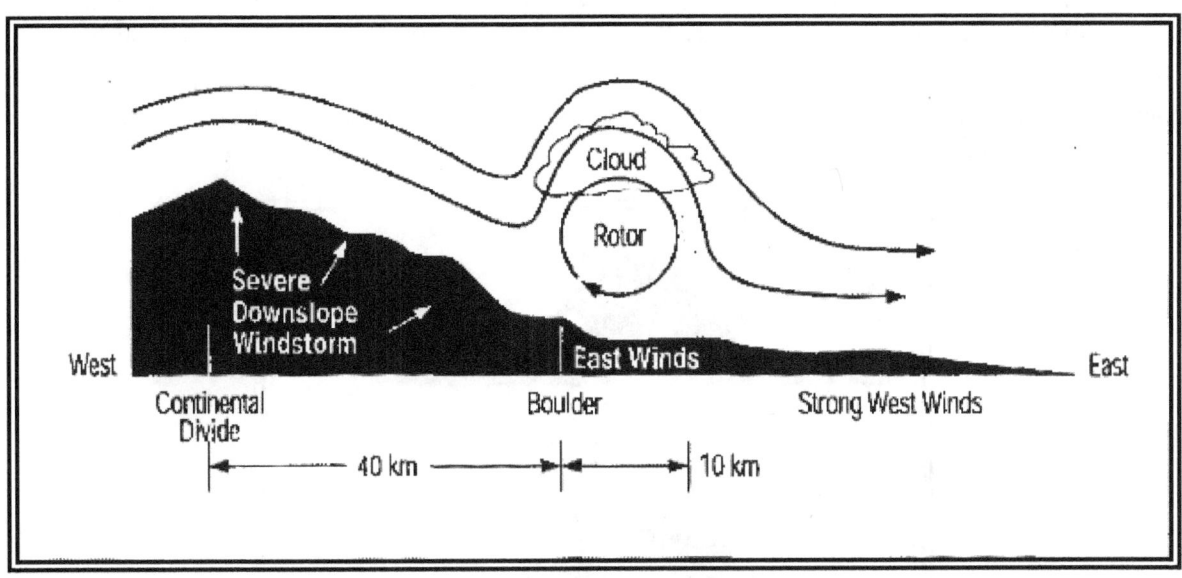

(Source: Bedard, 1993)[8]

Figure 16-10. Conceptual View of a Mountain Lee Wave Rotor Zone

Because of their potential for causing turbulence and loss of aircraft control, rotor zones should be avoided. Rotor zones are of concern not only because of the likelihood of strong turbulence in their vicinity (particularly on the upwind side of the rotor), but also because of the potential for rolling moments that could exceed the roll authority of the aircraft or otherwise lead to loss of control. Rotors are especially dangerous at low altitudes, particularly during takeoff and landing, when the aircraft is slowed and in a relatively high-drag configuration.

16.2.6 Smaller-Scale Hazards

The following smaller-scale phenomena represent specific weather hazards for aircraft operating near mountains.

16.2.6.1 Lee-Side Inversion with Shear Flow (Mountain-Induced Shear with No Wave Development)

Occasionally, an extremely strong low-level temperature inversion can occur in mountainous areas, with the inversion top below ridge level (perhaps 900 to 1,000 ft AGL) and a pool of very cold air at the surface. If this phenomenon occurs with strong wind flow above the inversion layer, there will be a concentrated shear zone near the inversion, which can lead to both significant turbulence encounters and abrupt airspeed changes for aircraft that penetrate the inversion on climbout or during descent. This situation is true particularly when significant mountain wave activity is present above the inversion in the strong flow aloft.

---

[8] Bedard, A. J., Jr. (1993). *Atmospheric turbulence aloft: A review of possible methods for detection, warning, and validation of prediction models.* Washington, DC: American Institute of Aeronautics and Astronautics.

In this case, the surface-based pool of cold air and the inversion above it shelter the surface from what might otherwise be a damaging windstorm.

16.2.6.2 Nonsteady Horizontal Roll Vortices (Moving Horizontal Vortices)

The surge of wind across a ridge can initiate a vortex downwind of the ridge (Figure 16-11). The vortex rolls up to maximum strength of rotation as it continues to move downwind away from the ridge and slowly dissipates. In its wake, with a return to steady flow, K-H waves develop at the top of the shear layer.

Extreme gustiness is a characteristic of the surface winds during severe downslope windstorms. The interaction of these gusts with strong large-scale winds moving perpendicular to a ridge may produce strong horizontal vortices of small scale.

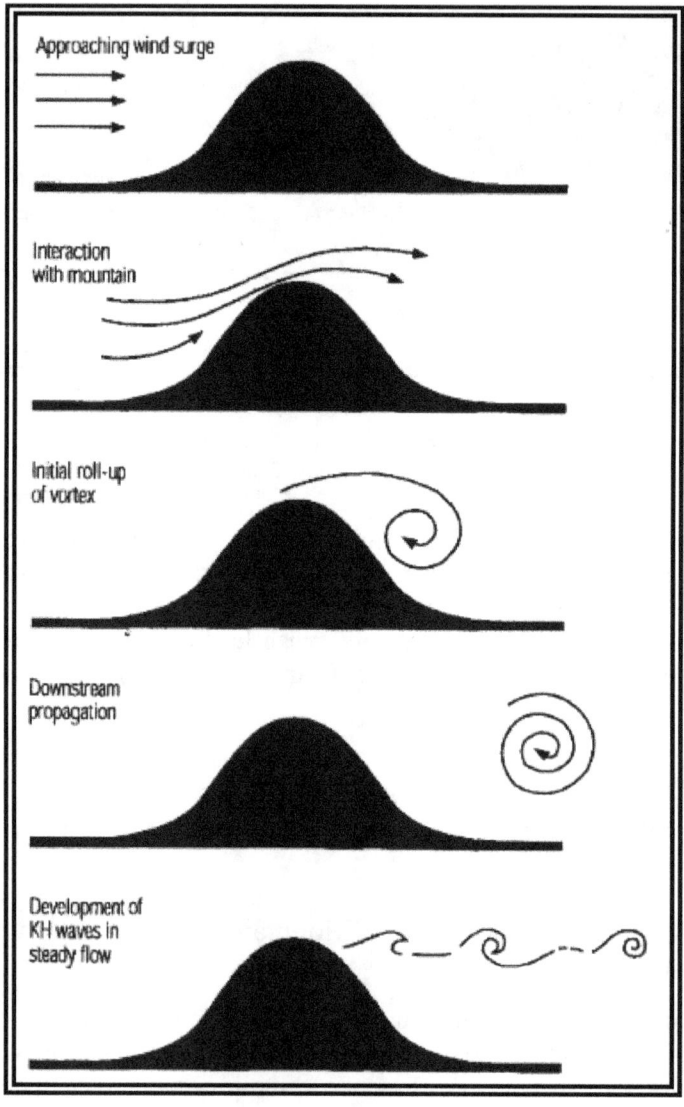

(Source: Bedard, 1993)[9]

Figure 16-11. Development of a Strong Roll Vortex Associated with a Wind Surge Down the Lee Slope of a Mountain

---

[9] Bedard, A. J., Jr. (1993). *Atmospheric turbulence aloft: A review of possible methods for detection, warning, and validation of prediction models.* Washington, DC: American Institute of Aeronautics and Astronautics.

Flight operations may be conducted in the vicinity of strong horizontal vortices without any encounters because they are highly localized, short-lived, and generally cloud-free. Conversely, one or more aircraft may encounter a strong, but invisible, vortex (that might be described as being like a "horizontal tornado," even though it is not) and undergo rolling moments and localized turbulence that make it impossible for the pilot to maintain aircraft control.

16.2.6.3 Intense Vertical-Axis Vortices

Analogous to the horizontal vortices described in the previous section are vertically oriented vortices of great intensity, similar to a short-lived, tornado-like event. They can form downwind of localized rugged terrain as well as isolated peaks (see Figure 16-12).

These vortices are not associated with thunderstorms and are therefore not tornadoes, but their wind speeds can reach 150 kt or more. As is the case with horizontal vortices, there may be no visual indications (i.e., visible cloud) of the presence of such a strong vertically oriented vortex.

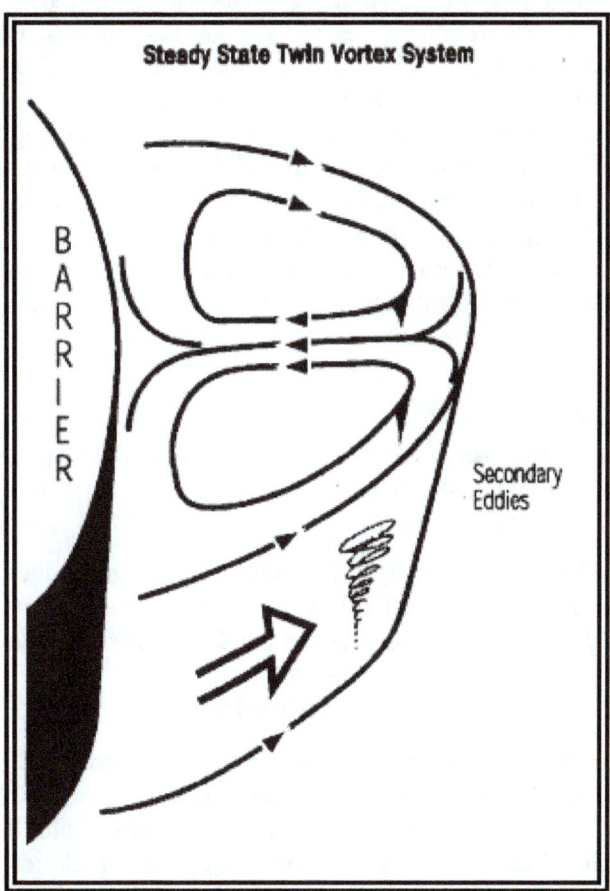

Figure 16-12. Schematic of Vertically Oriented Vortices Generated in the Lee of an Isolated Mountain Peak

16.2.6.4 Boras

The Glossary of Meteorology[10] defines a bora as a "fall wind with a source so cold that, when the air reaches the lowlands or coast, the dynamic warming is insufficient to raise the air temperature to the normal level for the region; hence it appears as a cold wind." Cold air building up on one side of a mountain range will

---

[10] American Meteorological Society, cited 2022: Bora. Glossary of Meteorology. [Available online at https://glossary.ametsoc.org/wiki/bora.]

often be blocked. However, if it deepens sufficiently, it will eventually spill over the mountain barrier and accelerate down the opposite slope, on rare occasions reaching speeds as high as 80 kt.

The resulting low-level winds and turbulence can be a significant hazard for aircraft that are flying in the vicinity of the down-rush of air caused by the bora. The danger is heightened by the fact that the exact timing and location of the air surge is difficult to forecast. There are at least two primary causes of boras: 1) cold fronts aligned parallel to the mountain range and moving perpendicular to it, with the cold air eventually spilling over; and 2) cold outflow, from thunderstorms over or near a mountain range, that builds up to sufficient depth to spill over and down the opposite slope. The latter phenomenon is short-lived and very difficult to predict; the strong thunderstorm winds typically last less than 1 hour. However, strong downslope winds accompanying and following cold front passages can persist for several hours. Only the initial stages of such winds have true bora or fall-wind characteristics; these winds appear to evolve into severe downslope windstorms associated with breaking waves aloft and, therefore, become potentially dangerous at all altitudes, not just within a few thousand feet of the surface.

In many areas along the eastern slopes of the Rocky Mountains, and in particular in Colorado, prefrontal windstorms with very warm lee-side temperatures are known as chinooks; post-cold frontal windstorms with cold lee-side winds are often called bora windstorms, or boras. Thus, the term bora in these areas can mean both the initial strong burst of a cold downslope wind and any subsequent downslope windstorm. In the case of eastern-slope boras, the best indicators during the preflight briefing are the presence of a strong cold front moving through the area (i.e., with much colder air behind the front), with associated rapid frontal movement (on the order of 30 kt or more). Surface observations (as reported in a METAR), particularly special observations of strong, rapidly changing surface winds from the west or northwest, along with decreasing temperature, may warn of bora activity. The indicators for breaking internal gravity waves should not be ignored. Western-slope boras are less common and are usually associated with a strong buildup of extremely cold arctic air on the eastern slopes.

16.2.6.5 Other Phenomena

In addition to the vortex phenomena previously discussed, vortices or strong shear zones may be generated locally by strong flow past individual mountain peaks and crags, or through gaps and passes across mountain ranges.

The point is that strong wind flow in the vicinity of irregular terrain can produce a multitude of disturbances of varying size and strength, many without reliable visual indicators. Their presence should be suspected when flying downwind of rugged terrain, whenever the wind flow at ridge level exceeds about 20 kt.

16.2.7 Visual Indicators of Orographic Wind Fields

Figure 16-13 provides a schematic of mountain waves and possible associated clouds. The most distinctive clouds are the sharp-edged, lens-shaped (or almond-shaped) lenticular clouds. When sufficient moisture is present in the upstream flow, mountain waves can produce interesting cloud formations, including: cap clouds, Cirrocumulus Standing Lenticular (CCSL), ACSL, and rotor clouds (see Figure 16-14). These clouds provide visual proof that mountain waves exist. However, these clouds may be absent if the air is too dry.

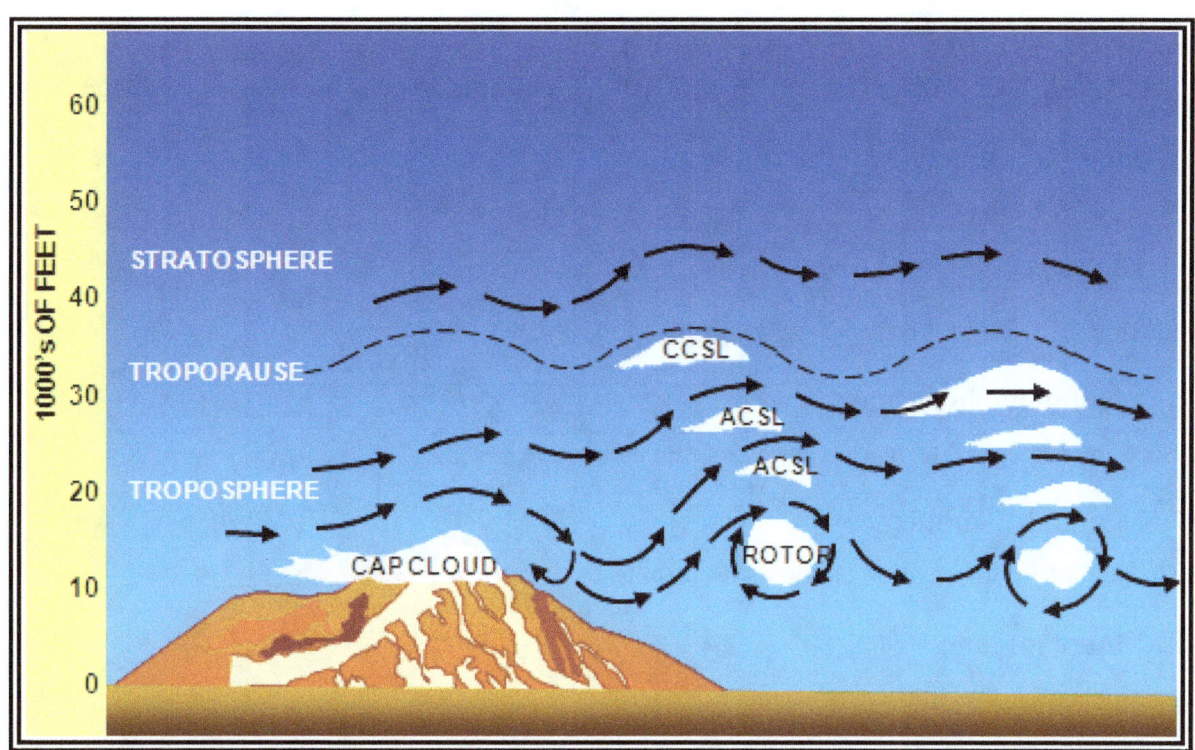

Figure 16-13. Schematic of Mountain Waves and Associated Clouds

Figure 16-14. Examples of Mountain Wave Clouds

Chapter 16, Mountain Weather

16.3 Mountain/Valley Breezes and Circulation

See Sections 10.6.4, 10.6.5, and 10.6.6 for information on mountain/valley breezes and circulation.

16.4 Mountain/Valley Fog

See Section 18.1.1.1.1.1 for information on mountain/valley fog.

16.5 Upslope Fog

See Section 18.1.1.1.3 for information on upslope fog.

16.6 Mountain Obscuration

See Section 18.2.2 for information on mountain obscuration.

16.7 Mountain Turbulence

See Chapter 19 for information on mountain turbulence.

16.8 Mountain Icing

See Section 20.3.8 for information on mountain icing.

16.9 Density Altitude

See Section 8.4.1.5 for information on density altitude.

17 Tropical Weather

17.1 Introduction

Technically, the Tropics lie between latitudes 23½° N and 23½° S. However, weather typical of this region sometimes extends as much as 45° from the Equator. One may think of the Tropics as uniformly rainy, warm, and humid. The facts are, however, that the Tropics contain both the wettest and driest regions of the world.

This chapter describes the basic circulation over the Tropics, terrain influences that determine arid and wet regions, and transitory systems that invade or disturb the basic tropical circulation.

17.2 Circulation

Chapter 7, Earth-Atmosphere Heat Imbalances, stated that wind blowing out of the subtropical high-pressure belts toward the Equator form the northeast and southeast trade winds of the two hemispheres. These trade winds converge in the vicinity of the Equator where air rises. This convergence zone is referred to as the Intertropical Convergence Zone (ITCZ). In some areas of the world, seasonal temperature differences between land and water areas generate rather large circulation patterns that overpower the trade wind circulation; these areas are monsoon regions. Tropical weather discussed here includes the subtropical high-pressure belts, the trade wind belts, the ITCZ, and monsoon regions.

17.2.1 Subtropical High-Pressure Belts

If the surfaces under the subtropical high-pressure belts were all water of uniform temperature, the high-pressure belts would be continuous highs around the globe. The belts would be areas of descending or subsiding air, and would be characterized by strong temperature inversions and very little precipitation. However, land surfaces at the latitudes of the high-pressure belts are generally warmer throughout the year than are water surfaces. Thus, the high-pressure belts are broken into semipermanent high-pressure areas over oceans with troughs or lows over continents, as shown in Figure 17-1 and Figure 17-2. The subtropical highs shift southward during the Northern Hemisphere winter and northward during summer. The seasonal shift, the height and strength of the inversion, and terrain features determine the weather in the subtropical high-pressure belts.

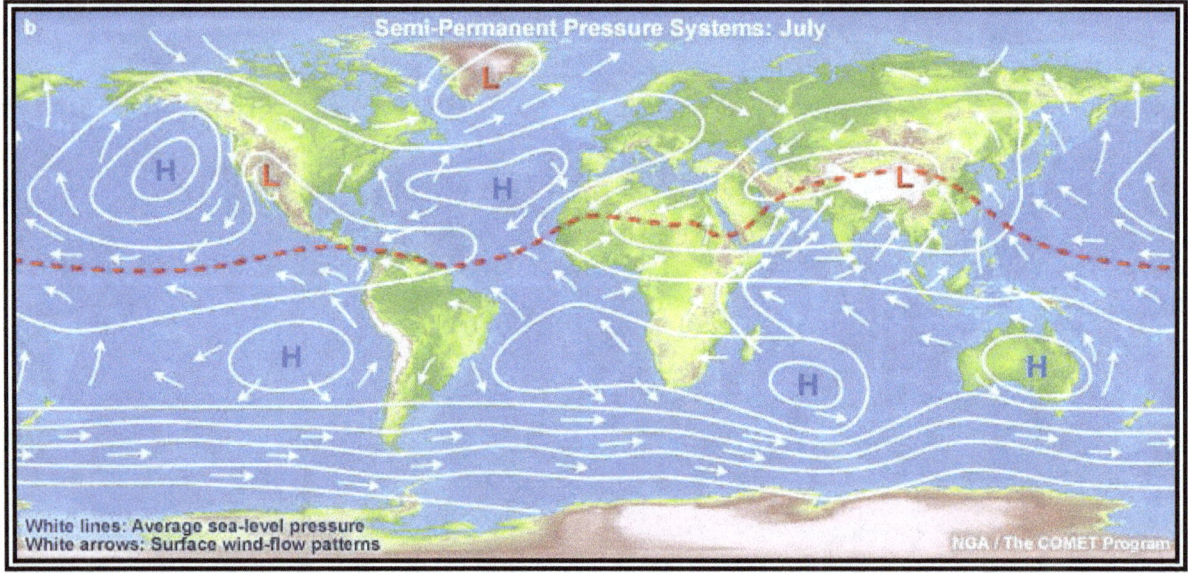

In the warm Northern Hemisphere, warm land areas tend to have low pressure and cool oceanic areas tend to have high pressure. In the cool Southern Hemisphere, the pattern is reversed: cool land areas tend to have high pressure and water surfaces have low pressure. However, the relationship is not so evident in the Southern Hemisphere because of relatively small amounts of land. The subtropical high-pressure belts are clearly evident at about 30° latitude in both hemispheres. The red dashed line shows the ITCZ.

Figure 17-1. Mean Worldwide Surface Pressure Distribution and Prevailing Winds Throughout the World in July

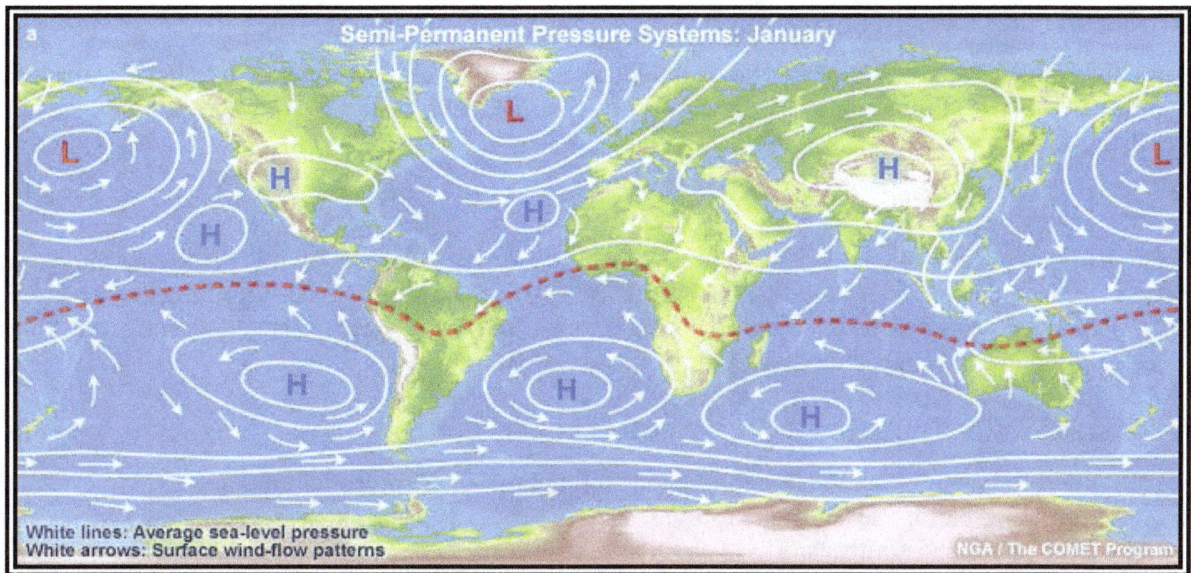

In this season, the pattern from Figure 17-1 is reversed. In the cool Northern Hemisphere, cold continental areas are predominantly areas of high pressure, while warm oceans tend to be low-pressure areas. In the warm Southern Hemisphere, land areas tend to have low pressure and oceans have high pressure. The subtropical high-pressure belts are evident in both hemispheres. Note that the pressure belts shift southward in January and northward in July, with the shift in the zone of maximum heating. The red dashed line shows the ITCZ.

Figure 17-2. Mean Worldwide Surface Pressure Distribution and Prevailing Winds Throughout the World in January

17.2.1.1 Continental Weather

Along the west coasts of continents under a subtropical high, the air is stable. The inversion is strongest and lowest where the east side of the subtropical high-pressure area overlies the west side of a continent. Moisture is trapped under the inversion; fog and low stratus occur frequently. However, precipitation is rare, since the moist layer is shallow and the air is stable. Heavily populated areas also add contaminants to the air which, when trapped under the inversion, add to the visibility problem.

The extreme southwestern United States, for example, is dominated in summer by a subtropical high. Most people are familiar with the semiarid summer climate of southern California. Rainfall is infrequent, but fog is common along the coast.

In winter, the subtropical high-pressure belts shift southward. Consider southern California as an example. In winter, the area comes under the influence of mid-latitude circulation, which increases the frequency of rain. Also, an occasional wintertime outbreak of polar air brings clear skies with excellent visibility.

The situation on eastern continental coasts is just the opposite. The inversion is weakest and highest where the west side of the subtropical high-pressure area overlies the eastern coast of a continent. Convection can penetrate the inversion, and showers and thunderstorms often develop. Precipitation is generally sufficient to support considerable vegetation. For example, in the United States, Atlantic coastal areas at the same latitude as southern California are far from arid in summer.

Low ceiling and fog often prevent landing at a west coast destination, but a suitable alternate generally is available a few miles inland. Alternate selection may be more critical for an east coast destination because of widespread instability and associated hazards.

17.2.1.2 Weather Over Open Sea

Under a subtropical high over the open sea, cloudiness is scant. The few clouds that do develop have tops from 3,000 to 6,000 ft, depending on height of the inversion. Ceiling and visibility are generally sufficient for VFR flight.

17.2.1.3 Island Weather

An island under a subtropical high receives very little rainfall because of the persistent temperature inversion. Surface heating over some larger islands causes light convective showers. Cloud tops are only slightly higher than those over open water. Temperatures are mild, showing small seasonal and diurnal changes. A good example is the pleasant, balmy climate of Bermuda.

17.2.2 Trade Wind Belts

Figure 17-1 and Figure 17-2 show prevailing winds throughout the Tropics for July and January. Note that trade winds blowing out of the subtropical highs over ocean areas are predominantly northeasterly in the Northern Hemisphere and southeasterly in the Southern Hemisphere. The inversion from the subtropical highs is carried into the trade winds and is known as the trade wind inversion. As in a subtropical high, the inversion is strongest where the trade winds blow away from the west coast of a continent and weakest where they blow onto an eastern continental shore. Daily variations from these prevailing directions are small, except during tropical storms. As a result, weather at any specific location in a trade wind belt varies little from day to day.

17.2.2.1 Weather Over Open Sea

In the trade wind belt, on average, about one-half of skies over open water are covered by clouds. Tops range from 3,000 to 8,000 ft, depending on the height of the inversion. Showers, although more common than under a subtropical high, are still light with comparatively little rainfall. Flying weather generally is quite good.

17.2.3 Continental Weather

Where trade winds blow offshore along the west coasts of continents, skies are generally clear and the area is quite arid. The Baja Peninsula of Baja California is a well-known example. Where trade winds blow onshore on the east sides of continents, rainfall is generally abundant in showers and occasional thunderstorms. The east coast of Mexico is a good example. Rainfall may be carried a considerable distance inland where the winds are not blocked by a mountain barrier. Inland areas blocked by a mountain barrier are deserts; examples include the Sahara Desert and the arid regions of the southwestern United States. Afternoon convective currents are common over arid regions due to strong surface heating. Cumulus and cumulonimbus clouds can develop, but cloud bases are high, and rainfall is scant because of the low moisture content.

Flying weather along eastern coasts and mountains is subject to the usual hazards of showers and thunderstorms. Flying over arid regions is good most of the time, but can be turbulent in afternoon convective currents; be especially aware of dust devils. Blowing sand or dust sometimes restricts visibility.

17.2.3.1 Island Weather

Mountainous islands have the most dramatic effect on trade wind weather. Since trade winds are consistently from approximately the same direction, they always strike the same side of the island; this side is the windward side. The opposite side is the leeward side. Winds blowing up the windward side produce copious and frequent rainfall, although cloud tops rarely exceed 10,000 ft. Thunderstorms are rare. Downslope winds on the leeward slopes dry the air, leaving relatively clear skies and much less rainfall.

Many islands in the trade wind belt have lush vegetation and even rain forests on the windward side, while the leeward is semiarid. For example, the island of Oahu, HI, is about 24 mi wide in the direction of the trade winds. Annual rainfall averages from about 60 in on the windward coast to 200 in at the mountain tops, decreasing to 10 in on the leeward shore.

The greatest flying hazard near these islands is obscured mountain tops. Ceiling and visibility occasionally restrict VFR flight on the windward side in showers. IFR weather is virtually nonexistent on leeward slopes.

Islands without mountains have little effect on cloudiness and rainfall. Afternoon surface heating increases convective cloudiness slightly, but shower activity is light. However, any island in either the subtropical high-pressure belt or trade wind belt enhances cumulus development, even though tops do not reach great heights. Therefore, a cumulus top higher than the average tops of surrounding cumulus usually marks the approximate location of an island. If it becomes necessary for a pilot to ditch in the ocean, the pilot should look for, and head toward, a tall cumulus. It probably marks a land surface, increasing chances of survival.

17.2.4 Intertropical Convergence Zone (ITCZ)

Converging winds in the ITCZ force air upward. The ITCZ appears as a band of clouds consisting of showers (with occasional thunderstorms) that encircles the globe near the Equator. The solid band of clouds may extend for many hundreds of miles and is sometimes broken into smaller line segments. It exists because of the convergence of the trade winds. In the Northern Hemisphere, the trade winds move in a southwesterly direction, while in the Southern Hemisphere, they move northwesterly. The tendency for convective storms in the tropics is to be short in duration, usually on a small scale, but they can produce intense rainfall. It is estimated that 40 percent of all tropical rainfall rates exceed 1 in per hour. Greatest rainfall typically occurs during midday. On the Equator, this occurs twice a year in March and September, and consequently there are two wet and two dry seasons.

Figure 17-1 and Figure 17-2 show the ITCZ and its seasonal shift. The ITCZ is well-marked over tropical oceans, but is weak and ill-defined over large continental areas.

Convection in the ITCZ carries huge quantities of moisture to great heights. Showers and thunderstorms frequent the ITCZ, and tops to 40,000 ft or higher are common, as shown in Figure 17-1. Precipitation is copious. Since convection dominates the ITCZ, there is little difference in weather over islands and open sea under the ITCZ.

Flying through the ITCZ usually presents no great problem if one follows the usual practice of avoiding cumulonimbus clouds and any thunderstorms.

Since the ITCZ is ill-defined over continents, this handbook will not attempt to describe ITCZ continental weather as such. Continental weather ranges from arid to rain forests and is more closely related to the monsoon than to the ITCZ.

17.2.5 Monsoon

As shown in Figure 17-1 and Figure 17-2, over the large land mass of Asia, the subtropical high pressure breaks down completely. Asia is covered by an intense high during the winter and a well-developed low during the summer. The same occurs over Australia and central Africa, although the seasons are reversed in the Southern Hemisphere.

The cold, high pressures in winter cause wind to blow from the deep interior outward and offshore. In summer, wind direction reverses, and warm moist air is carried far inland into the low-pressure area. This large-scale seasonal wind shift is the monsoon. The most notable monsoon is that of southern and southeastern Asia.

17.2.5.1 Summer or Wet Monsoon Weather

During the summer, the low over central Asia draws warm, moist, and unstable maritime air from the southwest over the continent. Strong surface heating, coupled with rising of air flowing up the higher terrain, produces extensive cloudiness, copious rain, and numerous thunderstorms. Rainfall at some stations in India exceeds 400 in per year with highest amounts between June and October.

The monsoon is so pronounced that it influences circulation many miles out over the ocean. Note in Figure 17-1 that in summer, prevailing winds from the Equator to the South Asian coast are southerly and southeasterly; without the monsoon influence, these areas would be dominated by northeasterly trade winds. Islands within the monsoon influence receive frequent showers.

17.2.5.2 Winter Monsoon Weather

Note in Figure 17-2 how the winter flow has reversed from that shown in Figure 17-1. Cold, dry air from the high plateau deep in the interior warms adiabatically as it flows down the southern slopes of the Himalayan Mountains. Virtually no rain falls in the interior in the dry winter monsoon. As the dry air moves offshore over warmer water, it rapidly takes in more moisture, becomes warmer in low levels, and is, therefore, unstable. Rain is frequent over offshore islands and even along coastal areas after the air has had a significant overwater trajectory.

The Philippine Islands are in an area of special interest. During the summer, they are definitely in southerly monsoon flow and are subjected to abundant rainfall. In the winter, wind over the Philippines is northeasterly—in the transition zone between the northeasterly trade winds and the monsoon flow. It is academic whether the phenomenon is called the trade winds or monsoon; in either case, it produces abundant rainfall. The Philippines have a year-round humid, tropical climate.

17.2.5.3 Other Monsoon Areas

Australia in July (Southern Hemisphere winter) is an area of high pressure with predominantly offshore winds, as shown in Figure 17-1. Most of the continent is dry during the winter. In January, winds are onshore into the continental low pressure, as shown in Figure 17-2. However, most of Australia is rimmed by mountains, and coastal regions are wet where the onshore winds blow up the mountain slopes. The interior is arid where downslope winds are warmed and dried.

Central Africa is known for its humid climate and jungles. Note in Figure 17-1 and Figure 17-2 that prevailing wind is onshore much of the year over these regions. Some regions are wet year-round; others have the seasonal monsoon shift and have a summer wet season and a winter dry season. The climate of Africa is so varied that only a detailed area-by-area study can explain the climate typical of each area.

In the Amazon Valley of South America during the Southern Hemisphere winter (July), southeast trade winds, as shown in Figure 17-1, penetrate deep into the valley, bringing abundant rainfall, which contributes to the jungle climate. In January, the ITCZ moves south of the valley, as shown in Figure 17-2. The northeast trade winds are caught up in the monsoon, cross the Equator, and penetrate the Amazon Valley. The jungles of the Amazon result largely from monsoon winds.

17.2.5.4 Flying Weather in Monsoons

During the winter monsoon, excellent flying weather prevails over dry interior regions. Over water, pilots should pick their way around showers and thunderstorms. In the summer monsoon, low ceilings and heavy rain often restrict VFR flight. IFR flight copes with the hazards of thunderstorms. The freezing level in the Tropics is quite high (14,000 ft or higher), so icing is restricted to high levels.

17.3 Transitory Systems

Prevailing circulations are not the only consideration in analyzing weather. Just as important, are migrating tropical weather producers—the shear line, the Tropical Upper Tropospheric Trough (TUTT), tropical waves, areas of converging northeast and southeast trade winds along the ITCZ, and tropical cyclones.

17.3.1 Remnants of Polar Fronts and Shear Lines

Remnants of a polar front can become lines of convection and occasionally generate a tropical cyclone. By the time a cold air mass originating in high latitudes reaches the Tropics, temperature and moisture are the same on both sides of the front. A shear line, or wind shift, is all that remains (see Figure 17-3). These influence storms in the Atlantic Ocean, Gulf of Mexico, or Caribbean Sea early or late in the hurricane season.

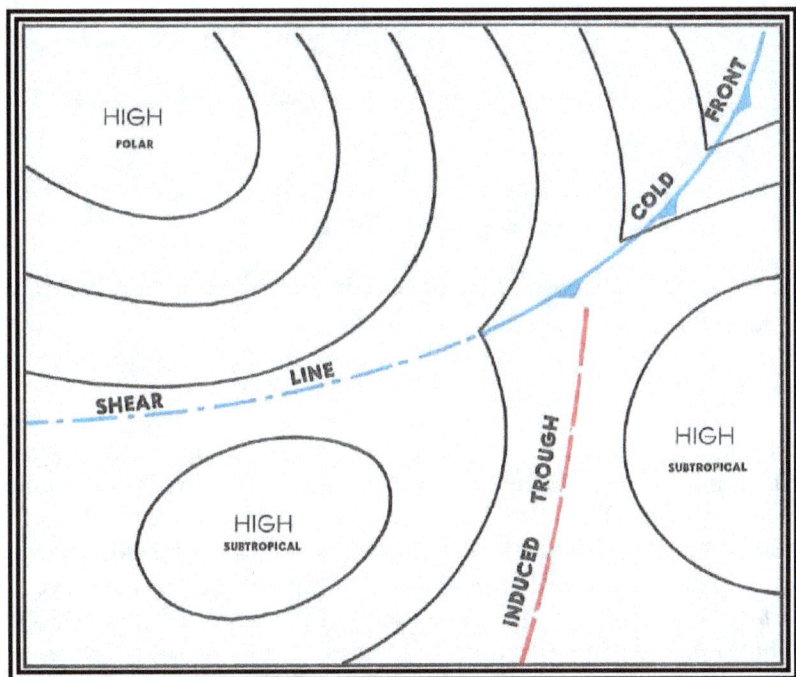

Figure 17-3. A Shear Line and an Induced Trough Caused by a Polar High Pushing into the Subtropics

A shear line, also shown in Figure 17-3, results when a semipermanent high splits into two cells, inducing a trough. These shear lines are zones of convergence creating forced upward motion. Consequently, considerable thunderstorm and rain shower activity occurs along a shear line.

17.3.2 Tropical Upper Tropospheric Trough (TUTT)

Troughs above the surface, generally at or above 10,000 ft, move through the Tropics, especially along the poleward fringes. These are known as TUTTs. Figure 17-4 shows such a trough. As a TUTT moves to the southeast or east, it spreads middle and high cloudiness over extensive areas to the east of the trough line. Occasionally, a well-developed trough will extend deep into the Tropics, and a closed low forms at the equatorial end of the trough. The low then may separate from the trough and move westward, producing a large amount of cloudiness and precipitation. If this occurs in the vicinity of a strong subtropical jet stream, extensive, and sometimes dense cirrus and some convective turbulence and CAT, often develop.

TUTTs and lows aloft produce considerable amounts of rainfall in the Tropics, especially over land areas where mountains and surface heating lift air to saturation. Low-pressure systems aloft contribute significantly to the 300+ in of annual rainfall over the higher terrain of Maui and the big island of Hawaii. Other mountainous areas of the Tropics are also among the wettest spots on Earth.

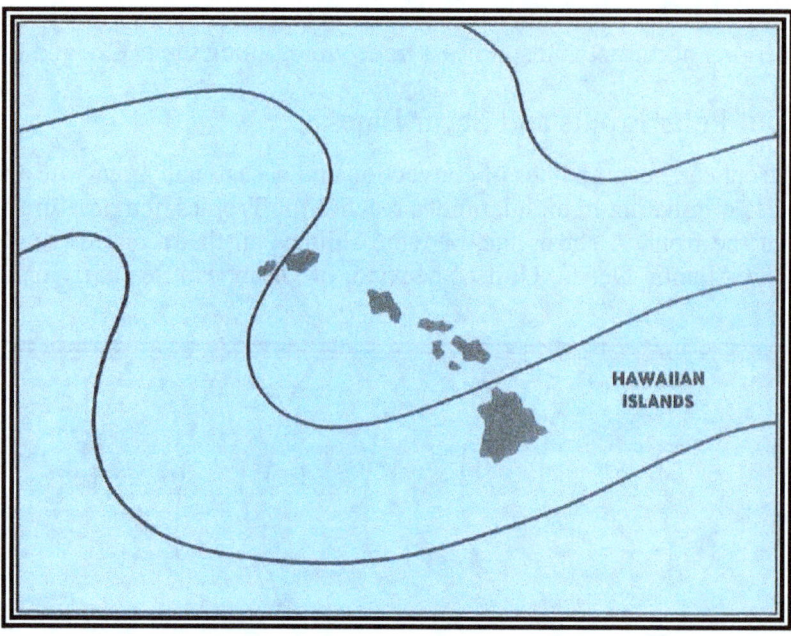

Figure 17-4. A TUTT Moves Eastward Across the Hawaiian Islands

17.3.3 Tropical Wave

Tropical waves (also called easterly waves) are common tropical weather disturbances, normally occurring in the trade wind belt. In the Northern Hemisphere, they usually develop in the southeastern perimeter of subtropical high-pressure systems. They travel from east to west around the southern fringes of these highs in the prevailing easterly circulation of the Tropics. Surface winds in advance of a wave are somewhat more northerly than the usual trade wind direction. As shown in Figure 17-5, as the wave approaches, pressure falls; as it passes, surface wind shifts to the east-southeast or southeast. The typical wave is preceded by very good weather but followed by extensive cloudiness (see Figure 17-6), and often by rain and thunderstorms. The weather activity is roughly in a north-south line.

Tropical waves occur in all seasons but are more frequent and stronger during summer and early autumn. Pacific waves frequently affect Hawaii; Atlantic waves occasionally move into the Gulf of Mexico, reaching the coast of the United States.

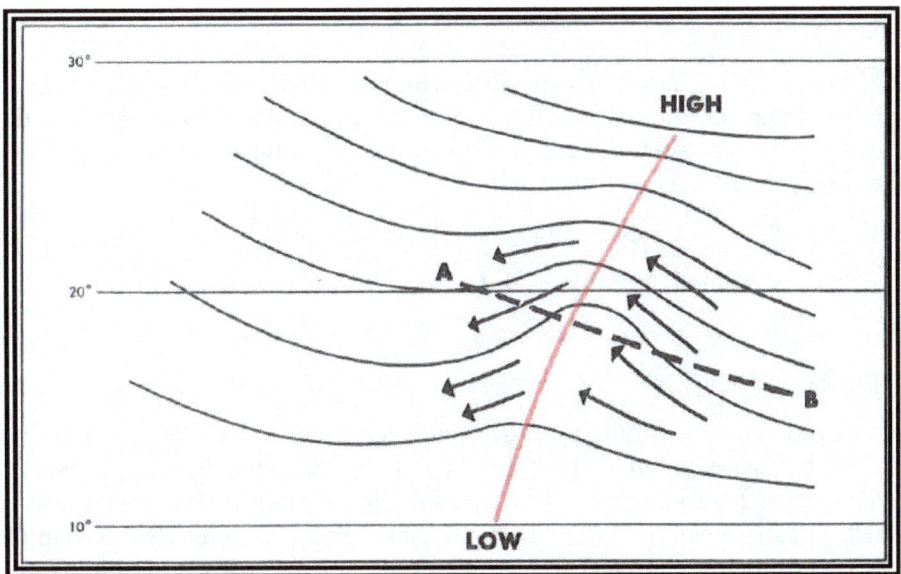

Note that winds shift generally from northeasterly to southeasterly. The wave moves toward the west, and is often preceded by good weather and followed by extensive cloudiness and precipitation.

Figure 17-5. A Northern Hemisphere Easterly Wave Progressing from A–B

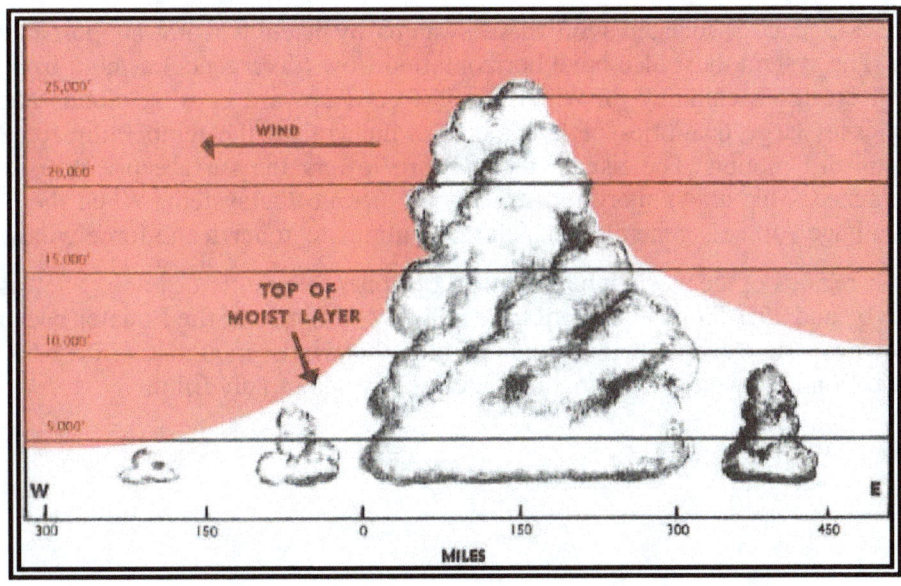

Figure 17-6. Vertical Cross-Section Along Line A–B in Figure 17-5

17.3.4 West African Disturbance Line (WADL)

On occasion, a line of convection similar to a squall line moves westward off the continent at tropical latitudes into the oceanic trade winds. In the North Atlantic, this is known as the West African Disturbance Line (WADL). A WADL can move faster than easterly waves at 20 to 40 mph. Some WADLs eventually develop into tropical storms or hurricanes.

17.3.5 Tropical Cyclones

"Tropical cyclone" is a general term for any low that originates over tropical oceans. Tropical cyclones are classified according to their intensity based on the average wind speeds. Wind gusts in these storms may be as much as 50 percent higher than the average wind speeds. Over the north Atlantic and northeast Pacific Oceans, tropical cyclone classifications are:

1. Tropical depression—sustained winds up to 34 kt (64 km/h).
2. Tropical storm—sustained winds of 35 to 64 kt (65 to 119 km/h).
3. Hurricane—sustained winds of at least 65 kt (120 km/h) or more.

The NWS NHC and the CPHC use a 1-minute average wind speed for the above classifications.

In other regions of the world, a 10-minute average wind speed is used along with different terms. Tropical cyclones meeting hurricane strength in the northwest Pacific Ocean are called "typhoons," "severe tropical cyclones" in the southwest Pacific and southeast Indian Oceans (e.g., near Australia), "severe cyclonic storms" in the north Indian Ocean, and just "tropical cyclone" in the southwest Indian Ocean. The term "super typhoon" is used if the maximum sustained winds are at least 130 kt (241 km/h).

17.3.5.1 Development

The prerequisites for tropical cyclone development are optimum sea surface temperature under low-level convergence and cyclonic wind shear. Favored breeding grounds are shear lines, TUTTs, tropical waves, and lines of convection in low latitudes moving from the continent to the tropical ocean (e.g., WADL).

The low-level convergence associated with these systems by itself will not support development of a tropical cyclone. The system must also have horizontal outflow (divergence) at high tropospheric levels. This combination creates a chimney, in which air is forced upward causing clouds and precipitation. Condensation releases large quantities of latent heat, which raises the temperature of the system and accelerates the upward motion. The rise in temperature lowers the surface pressure, which increases low-level convergence. This draws more moisture-laden air into the system. When these chain-reaction events continue, a huge vortex is generated, which may culminate in hurricane-force winds.

Figure 17-7 shows regions of the world where tropical cyclones frequently develop. They usually originate between latitudes 5° and 20°. Tropical cyclones are unlikely within 5° of the Equator because the Coriolis force is so small near the Equator that it will not turn the winds enough for them to flow around a low-pressure area. Winds flow directly into an equatorial low and rapidly fill it.

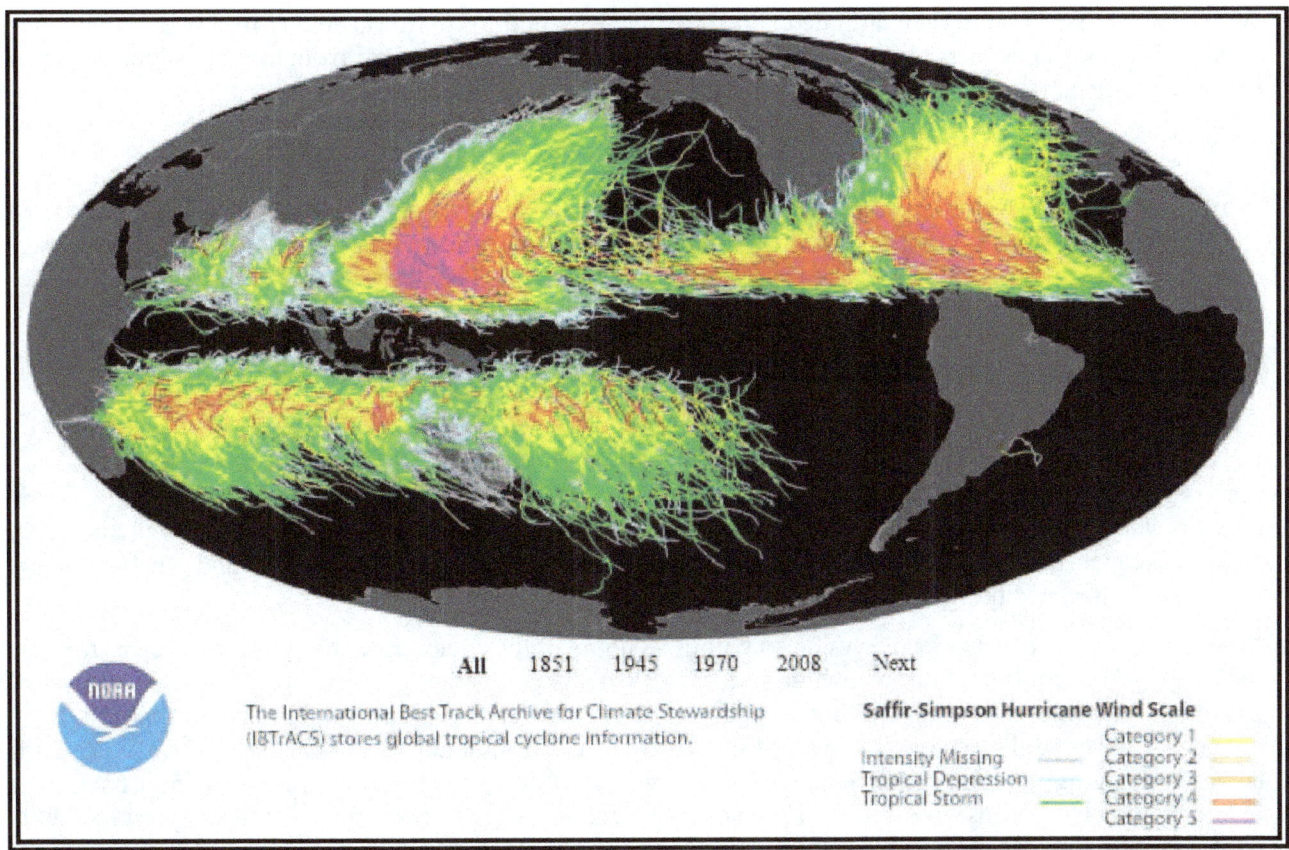

This map is based on all storm tracks available from the International Best Track Archive for Climate Stewardship (IBTrACS), a global inventory of tropical cyclones, through 2008. The accumulation of tracks reveals several details of hurricane climatology, such as where the most severe storms form and the large-scale atmospheric patterns that influence the track of hurricanes. (Note: See Table 17-1 for wind strength associated with each scale on the Saffir-Simpson Hurricane Wind Scale.)

Figure 17-7. The Tracks of Nearly 150 Years of Tropical Cyclones and Their Strength Weave Across the Globe

17.3.5.2 Movement

Tropical cyclones in the Northern Hemisphere usually move in a direction between west and northwest while in low latitudes. As these storms move toward the mid-latitudes, they come under the influence of the prevailing westerlies. At this time, the storms are under the influence of two wind systems: the trade winds at low levels and prevailing westerlies aloft. Thus, a storm may move very erratically, and may even reverse course or circle. Finally, the prevailing westerlies gain control, and the storm recurves toward the north, then to the northeast, and finally to the east-northeast. By this time, the storm is well into mid-latitudes.

17.3.5.3 Decay

As the storm curves toward the north or east (Northern Hemisphere), it usually begins to lose its tropical characteristics and acquires characteristics of lows in middle latitudes. Cooler air flowing into the storm gradually weakens it. If the storm tracks along a coastline or over the open sea, it gives up slowly, carrying its fury to areas far removed from the Tropics. However, if the storm moves well inland, it loses its moisture source and weakens from starvation and increased surface friction, usually after leaving a trail of destruction and flooding.

When a storm takes on middle latitude characteristics, it is said to be extratropical, meaning "outside the Tropics." Tropical cyclones produce weather conditions that differ somewhat from those produced by their higher latitude cousins and invite investigation.

17.3.5.4 Weather in a Tropical Depression

While in its initial developing stage, the cyclone is characterized by a circular area of broken-to-overcast clouds in multiple layers. Embedded in these clouds are numerous showers and thunderstorms. Rain shower and thunderstorm coverage varies from scattered to almost solid. The diameter of the cloud pattern varies from less than 100 mi in small systems to well over 200 mi in large ones.

17.3.5.5 Weather in Tropical Storms and Hurricanes

As cyclonic flow increases, the thunderstorms and rain showers form into broken or solid lines, paralleling the wind flow that is spiraling into the center of the storm. These lines are the spiral rain bands frequently seen on radar. These rain bands continually change as they rotate around the storm. Rainfall in the rain bands is very heavy, reducing ceiling and visibility to near zero. Winds are usually very strong and gusty and, consequently, generate violent turbulence. Between the rain bands, ceilings and visibilities are somewhat better, and turbulence generally is less intense.

Most tropical cyclones that form eyes do so within 48 hours of the cyclone reaching tropical storm strength. In the eye, skies are free of turbulent cloudiness, and wind is comparatively light. The average diameter of the eye is between 15 and 20 mi, but sometimes is as small as 7 mi, and rarely is more than 30 mi in diameter. Surrounding the eye is a wall of clouds that may extend above 50,000 ft. This wall of clouds contains deluging rain and the strongest winds of the storm. Maximum wind speeds of 175 kt have been recorded in some storms. See Figure 17-8 and Figure 17-9, which contain a radar display and satellite photograph of a mature hurricane, respectively. Note the spiral rain bands and the circular eye. Notice the similarity between these two figures.

Table 17-1 identifies the wind speed and characteristic house damage for each level on the Saffir-Simpson Hurricane Wind Scale.

Figure 17-8. Radar Image of Hurricane Katrina Observed at New Orleans, Louisiana, on August 29, 2005

Figure 17-9. Hurricane Andrew Observed by Satellite in 1992

Table 17-1. Wind Speed and Characteristic House Damage for the Saffir-Simpson Hurricane Wind Scale

Saffir-Simpson Hurricane Wind Scale	Wind Speed	Characteristic House Damage
5	≥157 mph ≥137 kt ≥252 km/h	Almost complete destruction of all mobile homes will occur, regardless of age or construction. A high percentage of frame homes will be destroyed, with total roof failure and wall collapse. Extensive damage to roof covers, windows, and doors will occur. Large amounts of windborne debris will be lofted into the air. Windborne debris damage will occur to nearly all unprotected windows and many protected windows.
4	130–156 mph 113–136 kt 209–251 km/h	Nearly all older (pre-1994 construction) mobile homes will be destroyed. A high percentage of newer mobile homes also will be destroyed. Poorly constructed frame homes can sustain complete collapse of all walls as well as the loss of the roof structure. Well-built homes also can sustain severe damage with loss of most of the roof structure and/or some exterior walls. Extensive damage to roof coverings, windows, and doors will occur. Large amounts of windborne debris will be lofted into the air. Windborne debris damage will break most unprotected windows and penetrate some protected windows.
3	111–129 mph 96–112 kt 178–208 km/h	Nearly all older (pre-1994 construction) mobile homes will be destroyed. Most newer mobile homes will sustain severe damage with potential for complete roof failure and wall collapse. Poorly constructed frame homes can be destroyed by the removal of the roof and exterior walls. Unprotected windows will be broken by flying debris. Well-built frame homes can experience major damage involving the removal of roof decking and gable ends.
2	96–110 mph 83–95 kt 154–177 km/h	Older (mainly pre-1994 construction) mobile homes have a very high chance of being destroyed, and the flying debris generated can shred nearby mobile homes. Newer mobile homes can also be destroyed. Poorly constructed frame homes have a high chance of having their roof structures removed, especially if they are not anchored properly. Unprotected windows will have a high probability of being broken by flying debris. Well-constructed frame homes could sustain major roof and siding damage. Failure of aluminum, screened-in, swimming pool enclosures will be common.
1	74–95 mph 64–82 kt 119–153 km/h	Older (mainly pre-1994 construction) mobile homes could be destroyed, especially if they are not anchored properly as they tend to shift or roll off their foundations. Newer mobile homes that are anchored properly can sustain damage involving the removal of shingle or metal roof coverings, and loss of vinyl siding, as well as damage to carports, sunrooms, or lanais. Some poorly constructed frame homes can experience major damage, involving loss of the roof covering and damage to gable ends, as well as the removal of porch coverings and awnings. Unprotected windows may break if struck by flying debris. Masonry chimneys can be toppled. Well-constructed frame homes could have damage to roof shingles, vinyl siding, soffit panels, and gutters. Failure of aluminum, screened-in, swimming pool enclosures can occur.

18 Weather and Obstructions to Visibility

18.1 Introduction

Weather and obstructions to visibility include: fog, mist, haze, smoke, precipitation, blowing snow, dust storm, sandstorm, and volcanic ash. This chapter will discuss each in detail.

18.1.1 Fog

Fog is a visible aggregate of minute water droplets that are based at the Earth's surface, and it reduces horizontal visibility to less than 5/8 sm (1 km); unlike drizzle, it does not fall to the ground. Fog differs from a cloud only in that its base must be at the Earth's surface, while clouds are above the surface.

Cloud droplets can remain liquid even when the air temperature is below freezing. Fog composed of water droplets and occurring with temperatures at or below freezing is termed freezing fog. When fog is composed of ice crystals, it is termed ice fog. If fog is so shallow that it is not an obstruction to vision at a height of 6 ft (2 m) above the surface, it is called shallow (ground) fog.

Fog forms when the temperature and dewpoint of the air become identical (or nearly so). This may occur through cooling of the air to its dewpoint (producing radiation fog, advection fog, or upslope fog), or by adding moisture and thereby elevating the dewpoint (producing frontal fog or steam fog). Fog seldom forms when the temperature-dewpoint spread is greater than 2 °C (4 °F).

18.1.1.1 Fog Types

Fog types are named according to their formation mechanism.

18.1.1.1.1 Radiation Fog

Radiation fog (see Figure 18-1 and Figure 18-2) is a common type of fog, produced over a land area when radiational cooling reduces the air temperature to or below its dewpoint. Thus, radiation fog is generally a nighttime occurrence and often does not dissipate until after sunrise.

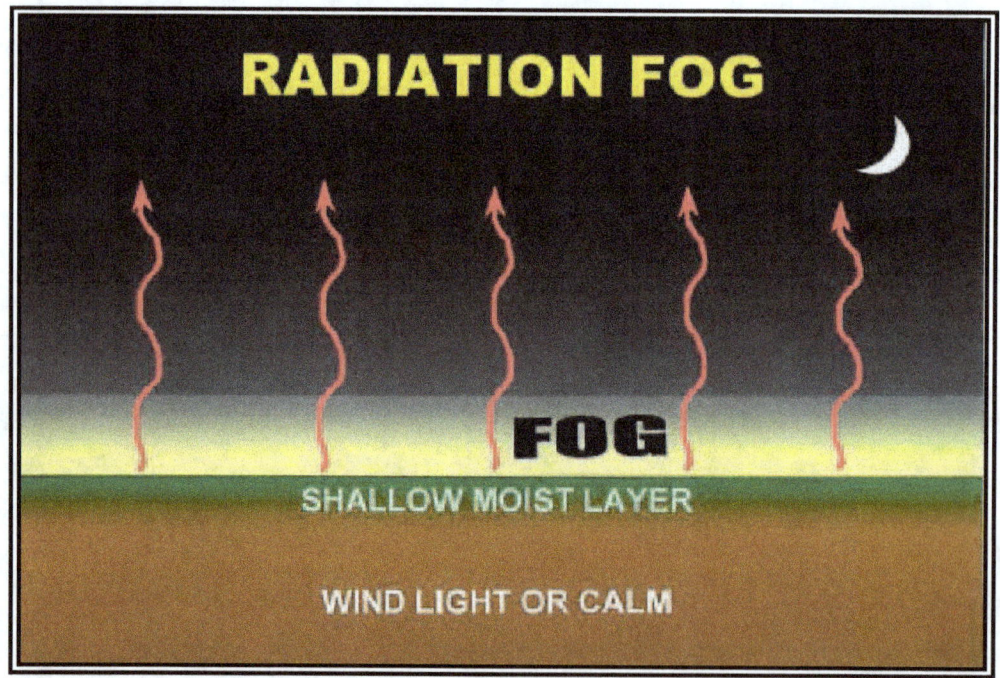

Figure 18-1. Radiation Fog Formation

Figure 18-2. Radiation Fog

Radiation fog is relatively shallow fog. It may be dense enough to hide the entire sky, or it may conceal only part of the sky. Ground fog is a form of radiation fog that is confined to near ground level.

Factors favoring the formation of radiation fog are: 1) a shallow surface layer of relatively moist air beneath a dry layer, 2) clear skies, and 3) light surface winds. Terrestrial radiation cools the ground; in turn, the ground cools the air in contact with it. When the air is cooled to its dewpoint, fog forms. When rain soaks the ground, followed by clearing skies, radiation fog is not uncommon the following morning.

Radiation fog is restricted to land because water surfaces cool little from nighttime radiation. It is shallow when wind is calm. Winds up to about 5 kt mix the air slightly and tend to deepen the fog by spreading the cooling through a deeper layer. Stronger winds disperse the fog or mix the air through a still deeper layer, with stratus clouds forming at the top of the mixing layer.

Ground fog usually burns off rather rapidly after sunrise. Other radiation fog generally clears before noon unless clouds move in over the fog. It can be difficult at times to differentiate between this and other types of fog, especially since nighttime cooling intensifies all fogs.

18.1.1.1.1.1 Mountain/Valley Fog

Mountain tops jutting skyward out of the fog can be a beautiful sight, but it can also be dangerous. There are two ingredients that add to the formation of mountain/valley fog in areas of variable terrain.

First, overnight, the ground cools as the heat that was gathered from the Sun's rays during the day is released back into the air near the ground level. The denser, cooler air on mountaintops sinks into valleys and collects there.

Second, over the course of the night, the valley begins to fill from the bottom with cold layers of air. This phenomenon is known as "cold air drainage." This cooler air lowers the surrounding air temperatures closer to the dewpoint and subsequently saturation. If there is sufficient moisture in the air, fog will begin to form in these valleys as the night progresses. This type of fog is most commonly observed in the autumn and spring months, and is densest around sunrise when surface temperatures are often lowest. See Figure 18-3 and Figure 18-4.

Figure 18-3. Mountain/Valley Fog Formation

Figure 18-4. Mountain/Valley Fog

18.1.1.1.2   Advection Fog

Advection fog (see Figure 18-5 and Figure 18-6) forms when moist air moves over a colder surface and the subsequent cooling of that air to below its dewpoint. It is most common along coastal areas, but often moves deep into continental areas. At sea, it is called sea fog. Advection fog deepens as wind speed increases up to about 15 kt. Wind much stronger than 15 kt lifts the fog into a layer of low stratus or stratocumulus clouds.

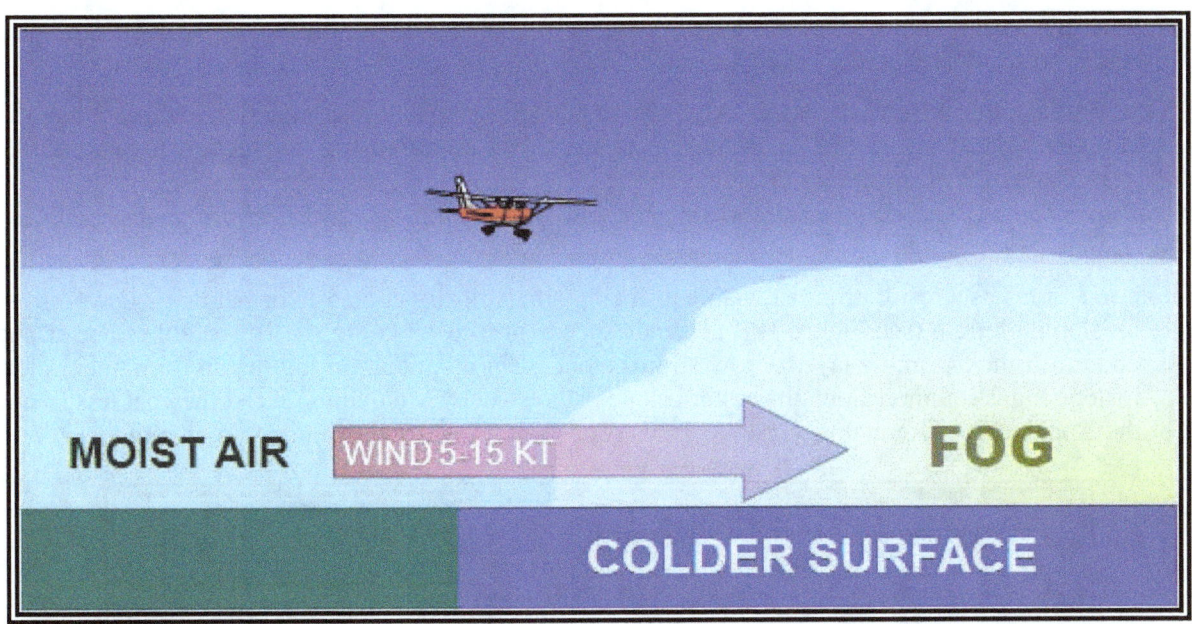

Figure 18-5. Advection Fog Formation

Figure 18-6. Advection Fog

The west coast of the United States is quite vulnerable to advection fog. This fog frequently forms offshore as a result of cold water and then is carried inland by the wind. It can remain over the water for weeks, advancing over the land during night and retreating back over the water the next morning.

During the winter, advection fog over the central and eastern United States results when moist air from the Gulf of Mexico spreads northward over cold ground. The fog may extend as far north as the Great Lakes.

Water areas in northern latitudes have frequent dense sea fog in summer as a result of warm, moist, tropical air flowing northward over colder Arctic waters.

A pilot will notice little difference between flying over advection fog and over radiation fog. Also, advection fog is usually more extensive and much more persistent than radiation fog. Advection fog can move in rapidly regardless of the time of day or night.

18.1.1.1.3 Upslope Fog

Upslope fog forms as a result of moist, stable air being adiabatically cooled to or below its dewpoint as it moves up sloping terrain (see Figure 18-7). Winds speeds of 5 to 15 kt are most favorable since stronger winds tend to lift the fog into a layer of low stratus clouds. Unlike radiation fog, it can form under cloudy skies. Upslope fog is common along the eastern slopes of the Rocky Mountains, and somewhat less frequent east of the Appalachian Mountains. Upslope fog is often quite dense and extends to high altitudes.

Figure 18-7. Upslope Fog Formation

18.1.1.1.4 Frontal Fog

When warm, moist air is lifted over a front, clouds and precipitation may form. If the cold air below is near its dewpoint, evaporation (or sublimation) from the precipitation may saturate the cold air and form fog (see Figure 18-8). A fog formed in this manner is called frontal (or precipitation-induced) fog. The result is a more or less continuous zone of condensed water droplets reaching from the ground up through the clouds. Frontal fog can become quite dense and continue for an extended period of time. This fog may extend over large areas, completely suspending air operations. It is most commonly associated with warm fronts, but can occur with other fronts as well.

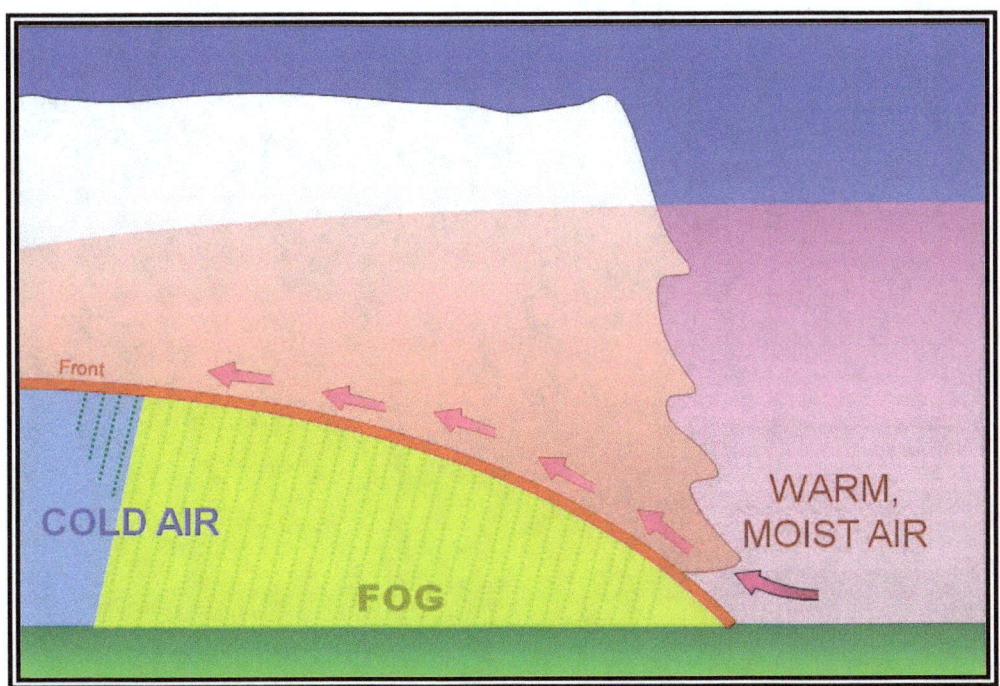

Figure 18-8. Frontal Fog Formation

18.1.1.1.5 Steam Fog

When very cold air moves across relatively warm water, enough moisture may evaporate from the water surface to produce saturation. As the rising water vapor meets the cold air, it immediately recondenses and rises with the air that is being warmed from below. Because the air is destabilized, fog appears as rising filaments or streamers that resemble steam. This phenomenon is called steam fog (see Figure 18-9). It is commonly observed over lakes and streams on cold autumn mornings, and over the ocean during the winter when cold air masses move off the continents and ice shelves. Steam fog is often very shallow, for as the steam rises, it reevaporates in the unsaturated air above. However, it can be dense and extend over large areas.

Steam fog is associated with a shallow layer of unstable air; thus, pilots can expect convective turbulence flying through it. On occasion, columns of condensed vapor rise from the fog layer, forming whirling steam devils, which appear similar to the dust devils on land.

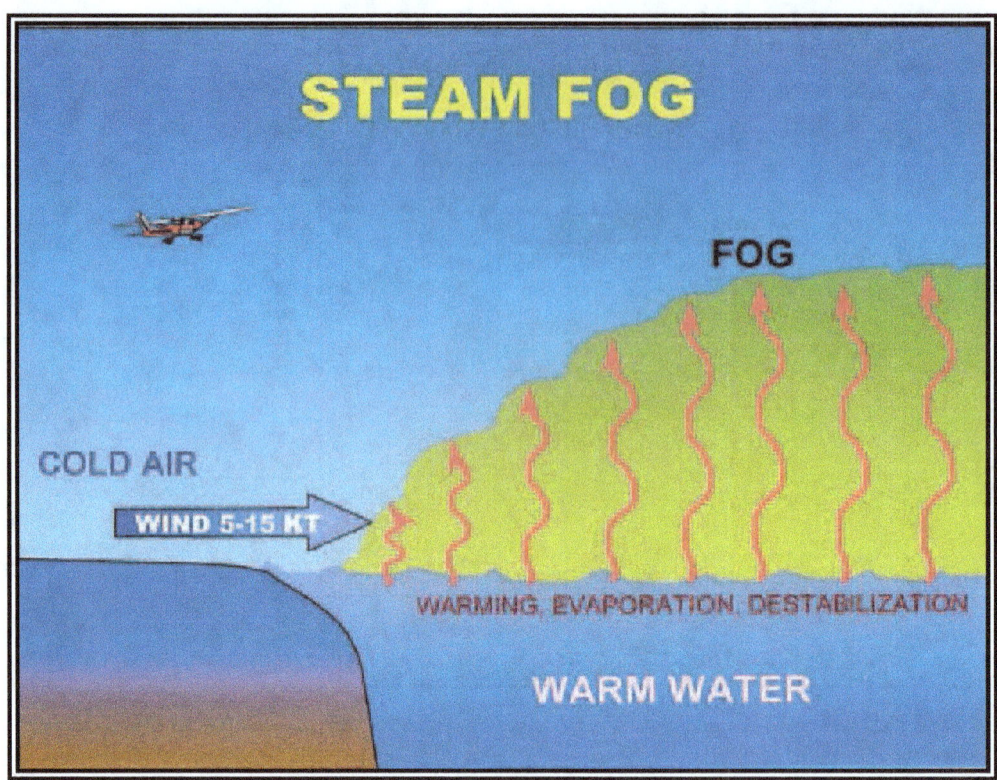

Figure 18-9. Steam Fog Formation

18.1.1.1.6 Freezing Fog

Freezing fog occurs when the temperature falls to 32 °F (0 °C) or below. Tiny, supercooled liquid water droplets in fog can freeze instantly on exposed surfaces when surface temperatures are at or below freezing. Some surfaces that these droplets may freeze on include tree branches, stairs and rails, sidewalks, roads, and vehicles (see Figure 18-10). For those flying, or even taxiing, a layer of ice can form on the aircraft, making flight very dangerous unless the aircraft is treated or has effective deicing equipment.

Figure 18-10. Freezing Fog

18.1.2 Mist

Mist is a visible aggregate of minute water droplets or ice crystals suspended in the atmosphere that reduces visibility to less than 7 sm (11 km), but greater than, or equal to, 5/8 sm (1 km). Mist forms a thin grayish veil that covers the landscape. It is similar to fog, but does not obstruct visibility to the same extent.

Mist may be considered an intermediate between fog and haze. It has lower relative humidity (95 to 99 percent) than fog and does not obstruct visibility to the same extent. However, there is no distinct line between any of these categories.

18.1.3 Haze

Haze is a suspension in the air of extremely small particles invisible to the naked eye and sufficiently numerous to give the air an opalescent appearance. It reduces visibility by scattering the shorter wavelengths of light. Haze produces a bluish color when viewed against a dark background and a yellowish veil when viewed against a light background. Haze may be distinguished by this same effect from mist, which yields only a gray obscuration. Certain haze particles increase in size with increasing relative humidity, drastically decreasing visibility. While visibility is a measure of how far one can see, including the ability to see the textures and colors therein, haze is the inability to view a similar scene with equal clarity.

Haze occurs in stable air and is usually only a few thousand feet thick, but may extend upwards to 15,000 ft (4,600 m). A haze layer has a definite ceiling above which in-flight (air-to-air) visibility is unrestricted. At or below this level, the slant range (air-to-ground) visibility is poor. Visibility in haze varies greatly, depending on whether the pilot is facing into or away from the Sun.

18.1.4 Smoke

Smoke is a suspension in the air of small particles produced by combustion due to fires, industrial burning, or other sources. It may transition to haze when the particles travel 25 to 100 mi (40 to 160 km) or more, and the larger particles have settled and others become widely scattered through the atmosphere.

Not only can smoke reduce visibility to zero, but many of its compounds are highly toxic and/or irritating. The most dangerous is carbon monoxide, which can lead to carbon monoxide poisoning, sometimes with supporting effects of hydrogen cyanide and phosgene.

When skies are clear above a surface-based layer of haze or smoke, visibility generally improves during the day. Heating during the day may cause convective mixing, spreading the smoke or haze to a higher altitude, and decreasing the concentration near the surface. However, the improvement is slower than the clearing of fog. Fog evaporates, but haze and smoke must be dispersed by the movement of air. A thick layer of clouds above haze or smoke may block sunlight, preventing dissipation. Visibility will improve little, if any, during the day.

18.1.5 Precipitation

Precipitation is any of the forms of water particles, whether liquid or solid, that fall from the atmosphere and reach the ground. Snow, rain, and drizzle are types of precipitation. Heavy snow may reduce visibility to zero. Rain seldom reduces surface visibility below 1 mi except in brief, heavy showers.

Drizzle usually restricts visibility to a greater degree than rain. It forms in stable air, falls from stratiform clouds, and is typically accompanied by fog. When drizzle changes to light rain, visibility usually improves because the droplet size increases, meaning there are fewer droplets per unit area.

For more information on precipitation, see Chapter 14, Precipitation.

18.1.6 Blowing Snow

Blowing snow is snow lifted from the surface of the Earth by the wind to a height of 6 ft (2 m) or more above the ground, and blown about in such quantities that the reported horizontal visibility is reduced to less than 7 sm (11 km). Light, dry powder snow is most prone to being blown by the wind. When strong winds keep the snow suspended up to 50 ft (15 m) or so, obscuring the sky, and reducing surface visibility to near zero, it is called a whiteout. Visibility improves rapidly when the wind subsides.

18.1.7 Dust Storm

A dust storm is a severe weather condition characterized by strong winds and dust-filled air over an extensive area. Dust storms originate over regions when fine-grained soils, rich in clay and silt, are exposed to strong winds and lofted airborne. Fine-grained soils are commonly found in dry lake beds (called playas), river flood plains, ocean sediments, and glacial deposits.

Most of the dust originates from a number of discrete point sources. Intense dust storms reduce visibility to near zero in and near source regions, with visibility improving away from the source.

A dust storm is favored with extreme daytime heating of barren ground and a turbulent, unstable air mass that allows the dust to be lofted. Surface winds need to be 15 kt or greater to mobilize dust. A speed of 35 kt may be needed over a desert surface that is covered with closely packed rock fragments called desert pavement. The average height of a dust storm is 3,000 to 6,000 ft (about 1 km); however, they can frequently extend up to 15,000 ft (4,600 m).

Strong cooling after sunset quickly stabilizes the lowest atmosphere, forming a temperature inversion and settling the dust. Without turbulence, dust generally settles at a rate of 1,000 ft (300 m) per hour. It can take many hours (or days) for the dust to completely settle. However, precipitation will very effectively remove dust from the atmosphere.

Aircraft operation in a dust storm can be very hazardous. Visibility can drop to zero in a matter of seconds. Dust can also clog the air intake of engines, damage electro-optical systems, and cause problems with human health.

From a pilot's point of view, it is important to recognize that slant range (air-to-ground) visibility (see Figure 24-3) in dust is generally reduced compared to reported surface (horizontal) visibility. Therefore, it may not be possible to pick out an airfield from above, even when reported surface visibility is 3 mi or more.

18.1.8 Sandstorm

A sandstorm is particles of sand carried aloft by a strong wind. The sand particles are mostly confined to the lowest 10 ft (3.5 m), and rarely rise more than 50 ft (15 m) above the ground. Sandstorms are similar to dust storms, but occur on a localized scale. This is because sand particles are larger and heavier than dust particles. Sandstorms are best developed in desert regions where there is loose sand, often in dunes, without much admixture of dust.

18.1.8.1 Haboob

A haboob (see Figure 18-11) is a dust storm or sandstorm that forms as cold downdrafts from a thunderstorm turbulently lift dust and sand into the air. While haboobs are often short-lived, they can be quite intense. The dust wall may extend horizontally for more than 60 mi (100 km) and rise vertically to the base of the thunderstorm. Spinning whirlwinds of dust frequently form along the turbulent cold air outflow, giving rise to huge dust/sand whirls.

Figure 18-11. Haboob

18.1.9 Volcanic Ash

Volcanic ash is made up of fine particles of rock powder that originate from a volcano and that may remain suspended in the atmosphere for long periods. Severe volcanic eruptions that send ash into the upper atmosphere occur somewhere around the world several times per year. The explosive eruption from the volcano in Tonga, South Pacific Ocean in January 2022 sent an ash cloud into the mesosphere. Weather satellites estimated the ash cloud reached an altitude of 190,000 ft, which was the highest ash cloud ever observed.

Visible ash is what an observer or aircrew member sees with their eyes. The lower limit of visible ash ranges from an ash concentration of approximately 0.01 milligrams per cubic meter (mg/m^3) to 10 mg/m^3, depending on many factors such as time of day, sky background, and position of the sun to the observer (pilot), as well as the angle from which the ash cloud is viewed (e.g., viewed from the side).

Discernible ash is what a satellite or other remote sensing instrument detects. Forecasters at the world's nine VAACs (see Section 26.5.1) use discernible ash from satellites to define the observed area of the ash cloud in the VAA product (see Section 26.5). The lower limit of discernible ash from satellites is approximately 0.1 to 0.2 mg/m3, depending on the satellite and other factors.

The ash cloud may not be visible, especially at night or in instrument meteorological conditions (IMC). Even if visible, it is difficult to distinguish visually between an ash cloud and an ordinary cloud. Radar may be able to detect heavy concentrations of airborne ash near the volcano, but is not able to detect fine airborne ash, and is not likely to detect the ash cloud as it spreads downwind of the volcano.

Flying into a volcanic ash cloud can be hazardous. Volcanic ash is composed of silica (glass). When ash is ingested into a jet engine, it melts to produce a soft, sticky molten product that adheres to the compressor turbine blades and fuel injectors/igniters. With no air going into the engine, the fuel cannot ignite, the engine comes to a slow spinning stop by spooling down, and a flameout occurs. As the aircraft exits the ash cloud and enters colder temperatures, the cooled, hardened silicas on the turbine blades become dislodged, allowing the fan blades to rotate, and allowing for an engine relight as the air moves through the engine again. Piston-powered aircraft are less likely to lose power, but severe engine damage is likely after an encounter with a volcanic ash cloud that is only a few hours old.

Volcanic ash also causes abrasive damage to aircraft flying through it at hundreds of miles per hour. Particles impacting the windshield can sandblast the surface into a frosted finish that obscures the pilot's view. The sandblasting can also remove paint and pit metal on the nose and leading edges of wings and navigation equipment. Ash contaminates aircraft ventilation, hydraulic, instrument, electronic, and air data systems. Ash covering a runway can cover its markings and cause aircraft to lose traction during takeoffs and landings.

18.2 Low Ceiling and Mountain Obscuration

18.2.1 Low Ceiling

Stratus is the most frequent cloud associated with low ceilings. Stratus clouds, like fog, are composed of extremely small water droplets or ice crystals suspended in air. An observer on a mountain in a stratus layer would call it fog. Stratus and fog frequently exist together. In many cases, there is no real line of distinction between the fog and stratus; rather, one gradually merges into the other. Flight visibility may approach zero in stratus clouds. Stratus over land tends to be lowest during night and early morning, lifting or dissipating due to solar heating by late morning or early afternoon. Low stratus clouds often occur when moist air mixes with a colder air mass, or in any situation where the temperature-dewpoint spread is small.

18.2.2 Mountain Obscuration

A mountain obscuration is a condition in which mountains or mountain ridges are obscured due to clouds, precipitation, smoke, or other obscurations.

Flight can be especially hazardous over mountain routes when the mountains are obscured. The large elevation variations around mountains can cause surface weather observations to mislead. For example, a weather station located in a valley could report a VFR cloud ceiling, while a hiker in the mountains sees fog.

19 Turbulence

19.1 Introduction

Aircraft turbulence is irregular motion of an aircraft in flight, especially when characterized by rapid up-and-down motion caused by a rapid variation of atmospheric wind velocities. Turbulence varies from annoying bumpiness to severe jolts, which cause structural damage to an aircraft and/or injury to its passengers. It is important to note that the effect of turbulence varies based on the size of the aircraft. Turbulence intensities and their associated aircraft reactions are described below:

- Light – Causes slight, erratic changes in altitude and/or attitude (pitch, roll, or yaw). Report as Light Turbulence. Or causes slight, rapid, and somewhat rhythmic bumpiness without appreciable changes in altitude or attitude. Report as Light Chop.

- Moderate – Similar to Light but of greater intensity. Changes in altitude and/or attitude occur but the aircraft remains in positive control at all times. It usually causes variations in indicated airspeed. Report as Moderate Turbulence. Or turbulence that is similar to Light Chop but of greater intensity. It causes rapid bumps or jolts without appreciable changes in aircraft altitude or attitude. Report as Moderate Chop.

- Severe – Causes large, abrupt changes in altitude and/or attitude. It usually causes large variations in indicated airspeed. Aircraft may be momentarily out of control.

- Extreme – The aircraft is violently tossed about and is practically impossible to control. It may cause structural damage.

19.2 Causes of Turbulence

Turbulence is caused by convective currents (called convective turbulence), obstructions in the wind flow (called mechanical turbulence), and wind shear.

19.2.1 Convective Turbulence

Convective turbulence is turbulent vertical motions that result from convective currents and the subsequent rising and sinking of air. For every rising current, there is a compensating downward current. The downward currents frequently occur over broader areas than do the upward currents; therefore, they have a slower vertical speed than do the rising currents.

Convective currents are most active on warm summer afternoons when winds are light. Heated air at the surface creates a shallow, absolutely unstable layer within which bubbles of warm air rise upward. Convection increases in strength and to greater heights as surface heating increases. Barren surfaces such as sandy or rocky wastelands and plowed fields become hotter than open water or ground covered by vegetation. Thus, air at and near the surface heats unevenly. Because of uneven heating, the strength of convective currents can vary considerably within short distances.

As air moves upward, it cools by expansion. A convective current continues upward until it reaches a level where its temperature cools to the same as that of the surrounding air. If it cools to saturation, a cumuliform cloud forms.

Billowy cumuliform clouds, usually seen over land during sunny afternoons, are signposts in the sky indicating convective turbulence. The cloud top usually marks the approximate upper limit of the convective current. A pilot can expect to encounter turbulence beneath or in the clouds, while above the clouds, air generally is smooth (see Figure 19-1). When convection extends to great heights, it develops larger towering cumulus clouds and cumulonimbus with anvil-like tops. The cumulonimbus gives visual warning of violent convective turbulence.

Figure 19-1. Convective Turbulence

When the air is too dry for cumuliform clouds to form, convective currents can still be active. This is called dry convection, or thermals (see Figure 19-2). A pilot has little or no indication of their presence until encountering the turbulence.

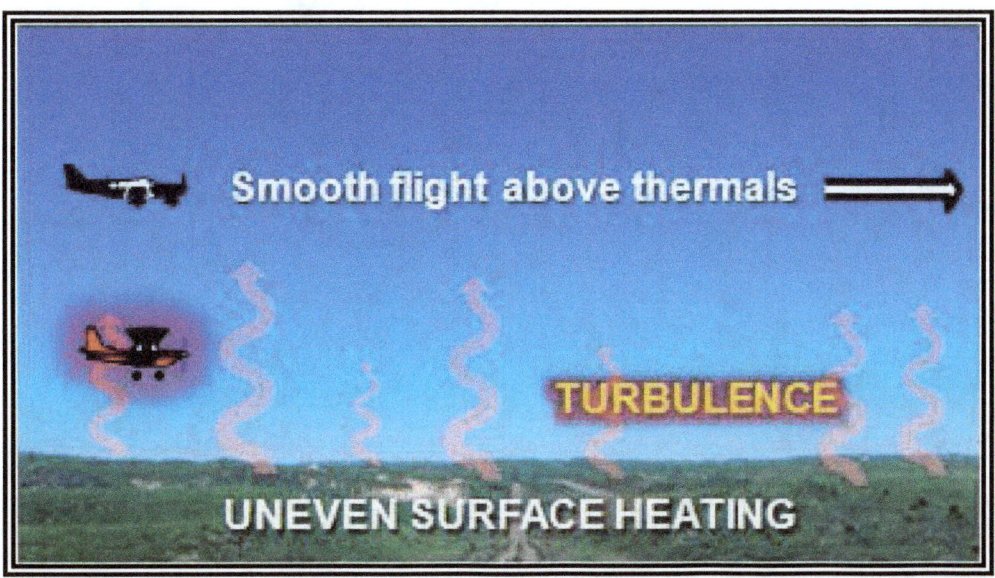

Figure 19-2. Thermals

19.2.1.1 Thunderstorms

Turbulence is present in all thunderstorms, and severe or extreme turbulence is common. A severe thunderstorm can destroy an aircraft. Gust loads can be severe enough to stall an aircraft at maneuvering speed or to cause structural damage at cruising speed. The strongest turbulence within the cloud occurs between updrafts and downdrafts.

Outside the cloud, shear turbulence has been encountered several thousand feet above and up to 20 mi laterally from a severe storm. Additionally, CAT may be encountered 20 or more miles from the anvil cloud edge. These kinds of turbulence are sometimes referred to as Convectively Induced Turbulence (CIT).

It is almost impossible to hold a constant altitude in a thunderstorm, and maneuvering to do so greatly increases stress on the aircraft. Stresses are least if the aircraft is held in a constant attitude.

The low-level, wind-shear zone between the gust front and surrounding air is very turbulent airspace. Oftentimes, the surface position of the gust front is denoted by a line of dust or debris along the ground or a line of spray along bodies of water. Gust fronts often move far ahead (up to 15 mi) of associated precipitation. The gust front causes a rapid and sometimes drastic change in surface wind ahead of an approaching storm. Often, a "roll cloud" or "shelf cloud" on the leading edge of the storm (see Figure 19-3) marks the top of the extreme turbulence zone, which forms as warm, moist air is lifted by the gust front. Shelf clouds are most common with multicell line thunderstorms.

Figure 19-3. Thunderstorm with Shelf Cloud

19.2.2 Mechanical Turbulence

Mechanical turbulence is turbulence caused by obstructions to the wind flow, such as trees, buildings, mountains, and so on. Obstructions to the wind flow disrupt smooth wind flow into a complex snarl of eddies (see Figure 19-4). An aircraft flying through these eddies experiences mechanical turbulence.

Figure 19-4. Mechanical Turbulence

The intensity of mechanical turbulence depends on wind speed and roughness of the obstructions. The higher the speed and/or the rougher the surface, the greater the turbulence.

The wind carries the turbulent eddies downstream; how far depends on wind speed and stability of the air. Unstable air allows larger eddies to form than those that form in stable air; but the instability breaks up the eddies quickly, while in stable air they dissipate slowly.

19.2.2.1 Mountain Waves

Mountain waves are a form of mechanical turbulence that develop above and downwind of mountains. See Chapter 16, Mountain Weather, for information on mountain waves.

19.2.3 Wind Shear Turbulence

Wind shear is defined in Section 19.2.4. Wind shear generates turbulence between two wind currents of different directions and/or speeds (see Figure 19-5). Wind shear may be associated with either a wind shift or a wind speed gradient at any level in the atmosphere.

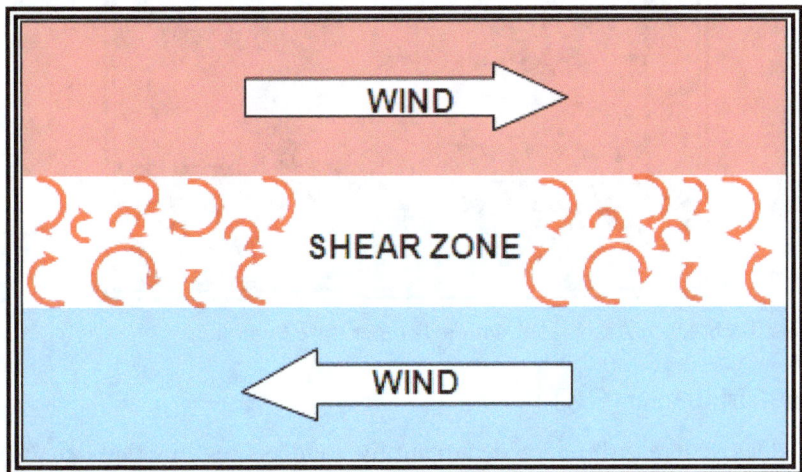

Figure 19-5. Wind Shear Turbulence

19.2.3.1 Temperature Inversion

A temperature inversion is a layer of the atmosphere in which temperature increases with altitude. Inversions commonly occur within the lowest few thousand feet above ground due to nighttime radiational cooling, along frontal zones, and when cold air is trapped in a valley. Strong wind shears often occur across temperature inversion layers, which can generate turbulence (see Figure 19-6).

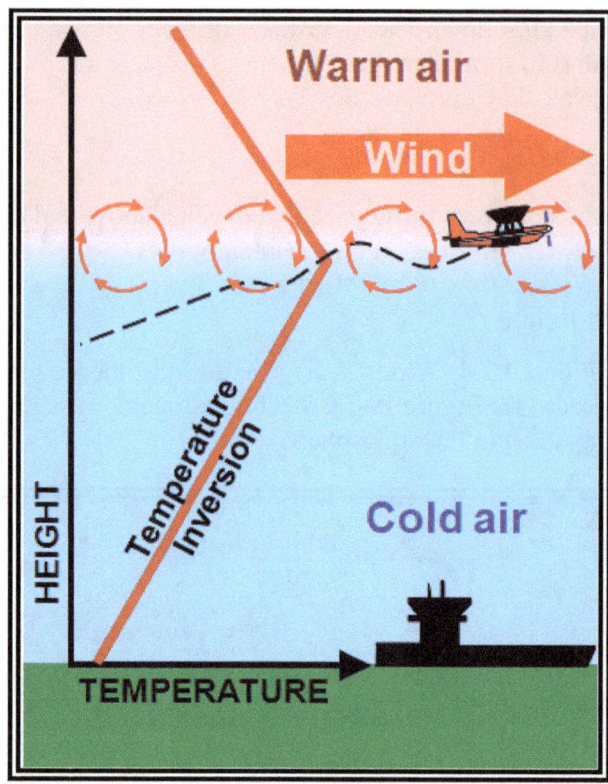

Figure 19-6. Wind Shear Turbulence Associated with a Temperature Inversion

19.2.3.2 Clear Air Turbulence (CAT)

CAT is defined as sudden severe turbulence occurring in cloudless regions that causes violent buffeting of aircraft. CAT is a higher altitude turbulence (normally above 15,000 ft) particularly between the core of a jet stream and the surrounding air. This includes turbulence in cirrus clouds, within and in the vicinity of standing lenticular clouds and, in some cases, in clear air in the vicinity of thunderstorms. Generally, though, CAT definitions exclude turbulence caused by thunderstorms, low-altitude temperature inversions, thermals, strong surface winds, or local terrain features.

CAT is a recognized problem that affects all aircraft operations. CAT is especially troublesome because it is often encountered unexpectedly and frequently without visual clues to warn pilots of the hazard.

19.2.3.2.1 CAT Discussion

One of the principal areas where CAT is found is in the vicinity of the jet streams. There are three jet streams: the polar front jet stream, the subtropical jet stream, and the polar night jet stream. (This handbook does not address the polar night jet stream, as it is a phenomenon in the stratosphere.) See Chapter 9, Global Circulations and Jet Streams, and Figure 9-4 and Figure 9-5 for more information and the polar front jet stream and the subtropical jet stream locations.

CAT associated with a jet stream is most commonly found in the vicinity of the tropopause. CAT is most frequently found on the poleward side of the jet stream (over the United States, this is to the left side when facing downwind). CAT is also common in the vicinity of a jet stream maxima, a region of stronger winds within the jet stream that translates along the jet stream core.

There are several patterns of upper-level winds that are associated with CAT. One of these is a deep, upper trough. CAT is found most frequently at, and just upwind of, the base of the trough, especially just downwind of an area of strong temperature advection. Another area of the trough in which to suspect CAT is along the centerline of a trough area, where there is a strong horizontal wind shear between the jet core and winds to the poleward side of the jet core. CAT is also found in the west side of a trough in the vicinity of a wind maxima as the maxima passes along the trough.

One noteworthy generator of CAT is the confluence of two jet streams. On occasion, the polar front jet stream will dip south and pass under the subtropical jet stream. The wind shear effect between the two jet streams in the region of confluence and immediately downstream is often highly turbulent.

CAT intensity can vary significantly along any flightpath. Common dimensions of a turbulent area associated with a jet stream are on the order of 100 to 300 mi long, elongated in the direction of the wind, 50 to 100 mi wide, and 5,000 ft deep. These areas may persist from 30 minutes to 1 day.

The threshold wind speed in the jet stream for CAT is generally considered to be 110 kt. The probability of encountering CAT increases proportionally with the rapidity of the decrease in wind speed away from the jet core. This is known as wind shear. It is not the wind speed itself that causes CAT; it is the wind shear that is turbulent to an aircraft as the atmosphere bounces in waves or actually overturns. Moderate CAT is considered likely when the vertical wind shear is 5 kt per 1,000 ft or greater, and/or the horizontal wind shear is 40 kt per 150 mi or greater.

Jet streams stronger than 110 kt (at the core) have potential for generating significant turbulence near the sloping tropopause above the core, in the jet stream front below the core, and on the low pressure side of the core.

Wind shear and its accompanying CAT in jet streams are more intense above, and to the lee of, mountain wave ranges. CAT should be anticipated whenever the flightpath traverses a strong jet stream in the vicinity of mountainous terrain.

Both vertical and horizontal wind shear are, of course, greatly intensified in mountain wave conditions. Therefore, when the flightpath traverses a mountain-wave-type of flow, it is desirable to fly at turbulence penetration speed and avoid flight over areas where the terrain drops abruptly, even though there may be no lenticular clouds to identify the condition.

CAT is also related to vertical shear. If vertical shear is greater than 5 kt per 1,000 ft, turbulence is likely.

Curving jet streams are more apt to have turbulent edges than straight ones, especially jet streams that curve around a deep pressure trough.

Wind shift areas associated with pressure troughs and ridges are frequently turbulent. The magnitude of the wind shear is the important factor.

19.2.4 Wind Shear

Wind shear is the sudden, drastic change in wind speed and/or direction over a small area, from one level or point to another, usually in the vertical (see Figure 19-7). Wind shear occurs in all directions, but for convenience, it is measured along vertical and horizontal axes, thus becoming horizontal and vertical wind shear.

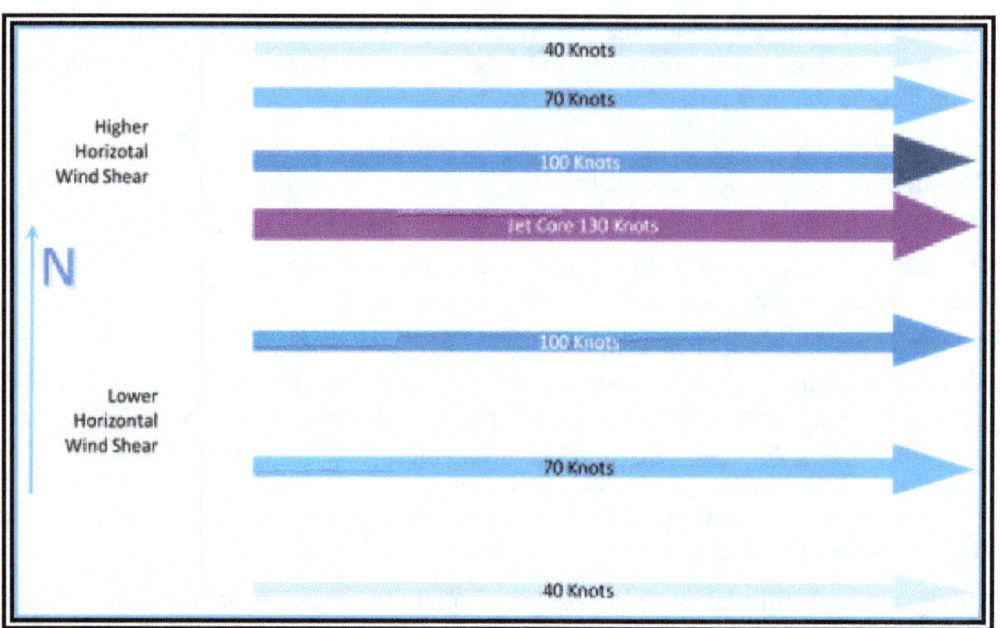

Figure 19-7. Wind Shear Example

It is important to remember that wind shear can affect any flight at any altitude (e.g., at upper levels near jet steams or near the ground due to convection). Wind shear can subject an aircraft to violent updrafts and downdrafts, as well as abrupt changes to the horizontal movement of the aircraft. While wind shear may be reported, it often remains undetected and is a silent aviation weather hazard. Always be alert to the possibility of wind shear, especially when flying in and around thunderstorms and frontal systems.

Some references or publications may use the term "severe wind shear." They may define the term as a wind shear that exceeds the performance capability of the aircraft or a wind shear producing airspeed changes greater than 15 kt or vertical speed changes greater than 500 feet per minute (fpm).

19.2.4.1 Nonconvective Low-Level Wind Shear (LLWS)

Wind variations at low altitude have long been recognized as a serious hazard to airplanes during takeoff and approach. These wind variations can result from a large variety of meteorological conditions such as topographical conditions, temperature inversions, sea breezes, frontal systems, and strong surface winds.

While wind shear can occur at any altitude, nonconvective LLWS is especially hazardous due to the proximity of an aircraft to the ground. Nonconvective LLWS is defined as a wind shear of 10 kt or more per 100 ft in a layer more than 200 ft thick that occurs within 2,000 ft of the surface. So what does this mean? It means that within the lowest 2,000 ft, the wind speed and/or direction is changing rapidly in a 200-ft layer (see Figure 19-8). Nonconvective LLWS is commonly associated with passing frontal systems, temperature inversions, and strong upper-level winds (greater than 25 kt).

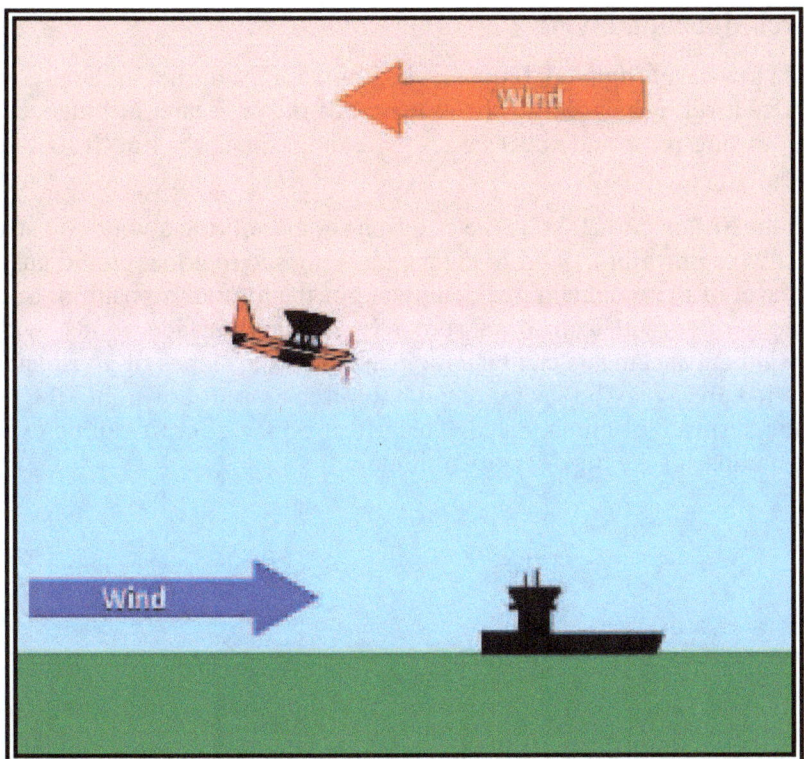

Figure 19-8. LLWS Example

19.2.4.2 Convective Wind Shear

See Section 22.7.3 for information on convective wind shear.

19.2.5 The Effects of Orographic Winds and Turbulence on Aviation Operations

19.2.5.1 High-Altitude Operations

Turbine-powered aircraft operating at cruise altitudes above FL180 in the vicinity of mountainous terrain may encounter moderate or greater turbulence associated with orographic winds. This type of turbulence may be characterized by relatively rapid onset and can lead to structural damage or airframe failure.

Structural damage is not the only danger associated with high-altitude turbulence encounters. It is possible to operate some turbine-powered aircraft at such weights and altitudes so that their cruise airspeed is only a few knots below the onset of Mach buffet and a like speed above stall buffet. In this situation (the so-called "coffin corner"), turbulent airspeed excursions of moderate or greater intensity (15 kt or more) can quickly lead to high-speed upset, Mach tuck, and loss of control.

19.2.5.2 Takeoff and Landing

Takeoff and landing concerns include experiencing turbulent air with inadequate stall margins, loss of directional control on or near the runway, rolling moments that surpass aircraft roll authority, and downdraft velocities that exceed the climb capability of the aircraft, particularly for airplanes with high wing and power-loading. Localized gusts in excess of 50 kt, with downdrafts greater than 1,500 fpm, are also possible.

Vortices spawned by the interaction of strong winds and high terrain can lead to severe turbulence and aircraft rolling moments that may exceed the pilot's ability to maintain aircraft control.

19.2.5.3 Low-Level Mountain Flying

Aircraft that engage in low-level flight operations over mountainous terrain in the presence of strong winds (20 kt or greater at ridge level) can expect to encounter moderate or greater turbulence, strong updrafts and downdrafts, and very strong rotor and shear zones. This is particularly true for General Aviation (GA) aircraft.

Mountain flying literature often cite 20 kt as the criterion for classifying a wind as "strong." This criterion refers to the large-scale (or prevailing wind in the area as opposed to a local wind gust) wind speed at the crest of the ridge or level of the mountain peaks, upwind of the aircraft's position. Such an ambient wind flow perpendicular to a ridge will lead to substantially stronger surface winds, with the likelihood of turbulence. Similar wind enhancements can be anticipated near the slopes of an isolated peak. In contrast, downdrafts over forested areas may be strong enough to force aircraft down into the trees, even when the aircraft is flown at the best rate-of-climb speed. This effect on the aircraft can be exacerbated by loss of aircraft performance because of the high density altitude.

20 Icing

20.1 Introduction

In general, icing is any deposit of ice forming on an object. It is one of the major weather hazards to aviation. Icing is a cumulative hazard. The longer an aircraft collects icing, the worse the hazard becomes.

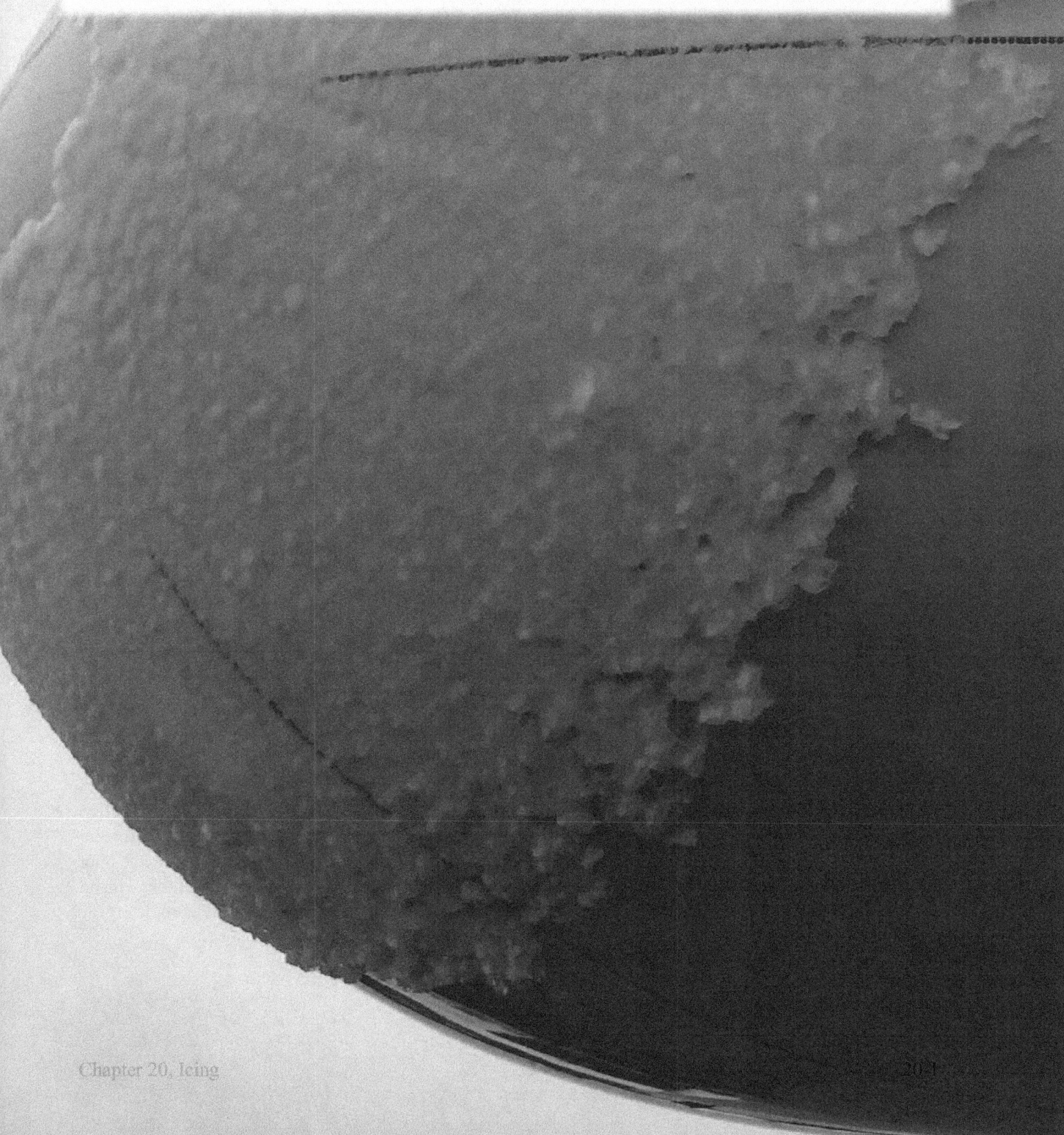

20.2 Supercooled Water

Freezing is a complex process. Pure water suspended in the air does not freeze until it reaches a temperature of -40 °C. This occurs because surface tension of the droplets inhibits freezing. The smaller and purer the water droplet, the more likely it is supercooled. Also, supercooled water can exist as large drops known as Supercooled Large Drops (SLD). SLDs are common in freezing rain and freezing drizzle situations (see Section 14.4.3).

Supercooled water content of clouds varies with temperature. Between 0 and -10 °C, clouds consist mainly of supercooled water droplets. Between -10 and -20 °C, liquid droplets coexist with ice crystals. Below -20 °C, clouds are generally composed entirely of ice crystals. However, strong vertical currents (e.g., cumulonimbus) may carry supercooled water to great heights where temperatures are as low as -40 °C.

Supercooled water will readily freeze if sufficiently agitated. This explains why airplanes collect ice when they pass through a liquid cloud or precipitation composed of supercooled droplets.

20.3 Structural Icing

Structural icing is the ice that sticks to the outside of the airplane. It occurs when supercooled water droplets strike the airframe and freeze. Structural icing can be categorized into three types: rime, clear (or glaze), and mixed.

Icing intensities are described in the AIM, Chapter 7, Section 1, Meteorology.

20.3.1 Rime Icing

Rime ice is rough, milky, and opaque ice formed by the instantaneous freezing of small, supercooled water droplets after they strike the aircraft. It is the most frequently reported icing type. Rime ice can pose a hazard because its jagged texture can disrupt an aircraft's aerodynamic integrity.

Rime icing formation favors colder temperatures, lower liquid water content, and small droplets. It grows when droplets rapidly freeze upon striking an aircraft. The rapid freezing traps air and forms a porous, brittle, opaque, and milky-colored ice. Rime ice grows into the air stream from the forward edges of wings and other exposed parts of the airframe.

20.3.2 Clear Icing

Clear ice (or glaze ice) is a glossy, clear, or translucent ice formed by the relatively slow freezing of large, supercooled water droplets. Clear icing conditions exist more often in an environment with warmer temperatures, higher liquid water contents, and larger droplets.

Clear ice forms when only a small portion of the drop freezes immediately while the remaining unfrozen portion flows or smears over the aircraft surface and gradually freezes. Few air bubbles are trapped during this gradual process. Thus, clear ice is less opaque and denser than rime ice. It can appear either as a thin smooth surface or as rivulets, streaks, or bumps on the aircraft.

Clear icing is a more hazardous ice type for many reasons. It tends to form horns near the top and bottom of the airfoils' leading edge, which greatly affects airflow. This results in an area of disrupted and turbulent airflow that is considerably larger than that caused by rime ice. Since it is clear and difficult to see, the pilot may not be able to quickly recognize that it is occurring. It can be difficult to remove since it can spread beyond the deicing or anti-icing equipment, although in most cases, it is removed nearly completely by deicing devices.

20.3.2.1 Supercooled Large Drops (SLD)

A type of clear icing that is especially dangerous to flight operations is ice formed from SLDs. These are water droplets in a subfreezing environment with diameters larger than 40 microns, such as freezing drizzle (40 to 200 microns) and freezing rain (>200 microns). These larger droplets can flow along the airfoil for some distance prior to freezing. SLDs tend to form a very lumpy, uneven, and textured ice similar to glass in a bathroom window.

SLD ice tends to form aft, beyond the reach of deicing equipment. Thus, ice remaining on the airfoil continues to disrupt the airflow and reduce the aircraft's aerodynamic integrity. Even a small amount of ice on the lower and upper surfaces of the airfoil can seriously disrupt its aerodynamic properties. The residual ice generates turbulence along a significant portion of the airfoil. This residual ice can act as a spoiler, a device actually used to slow an aircraft in flight. In extreme cases, turbulence and flow separation bubbles can travel along the airfoil and inadvertently activate the ailerons, creating dangerously unstable flying conditions.

20.3.3 Mixed Icing

Mixed ice is a mixture of clear ice and rime ice. It forms as an airplane collects both rime and clear ice due to small-scale (tens of kilometers or less) variations in liquid water content, temperature, and droplet sizes. Mixed ice appears as layers of relatively clear and opaque ice when examined from the side.

Mixed icing poses a similar hazard to an aircraft as clear ice. It may form horns or other shapes that disrupt airflow and cause handling and performance problems. It can spread over more of the airframe's surface and is more difficult to remove than rime ice. It can also spread over a portion of airfoil not protected by anti-icing or deicing equipment. Ice forming farther aft causes flow separation and turbulence over a large area of the airfoil, which decreases the ability of the airfoil to keep the aircraft in flight.

20.3.4 Icing Factors

Structural icing is determined by many factors. The meteorological quantities most closely related to icing type and severity are, in order of importance: SLWC, temperature (altitude), and droplet size. However, aircraft type/design and airspeed are also important factors.

SLWC is important in determining how much water is available for icing. The highest quantities can be found in cumuliform clouds, with the lowest quantities found in stratiform clouds. However, in most icing cases, SLWC is low.

Icing potential is very temperature dependent. For icing to occur, the OAT must be below 0 °C. As clouds get colder, SLWC decreases until only ice crystals remain. Thus, almost all icing tends to occur in the temperature interval between 0 °C and -20 °C, with about half of all reports occurring between -8 °C and -12 °C. In altitude terms, the peak of occurrence is near 10,000 ft, with approximately half of incidents occurring between 5,000 and 13,000 ft. The only physical cold limit to icing is at -40 °C because liquid droplets freeze without nuclei present.

In general, rime icing tends to occur at temperatures colder than -15 °C, clear ice when the temperature is warmer than -10 °C, and mixed ice at temperatures in between. This is only general guidance. The type of icing will vary depending on the liquid water content, droplet size, and aircraft-specific variables.

An airframe can remain cold (temperature below 0 °C) in a warm (temperature above 0 °C) atmosphere if it is cold-soaked. For example, if an aircraft has been flying in a cold environment, but then descends into warmer temperatures, the airframe does not heat up immediately to the air temperature. For some aircraft, the airframe can remain colder than 0 °C for some time, even after landing. Aircraft with fuel tanks mounted flush to the airframe are particularly susceptible to icing, even in an environment where the air temperature is slightly above 0 °C. Because these characteristics vary from airframe to airframe, it is important for pilots to be aware of the limitations of their aircraft.

Droplet size can influence icing, but is not as important as SLWC and temperature, unless the droplets are larger than cloud droplets in size (e.g., freezing drizzle and freezing rain). Droplet size affects the collection of drops by the airframe. Small droplets tend to impact the airfoil near the plane's leading edge. Larger drops, including freezing rain and freezing drizzle, can cross the streamlines and impact farther back.

Aircraft airspeed is an important nonmeteorological factor that determines icing type and severity. The rate of supercooled water droplet impact increases with airspeed, which acts to increase ice accumulation, but this is counteracted by the increase of airframe skin surface heating due to friction. Typically, airframe icing is negligible at speeds above 575 kt.

Aircraft type and design are also important factors. Because these characteristics vary, it is important for pilots to be aware of the limitations of their aircraft.

Commercial jet aircraft are generally less vulnerable to structural icing than light turboprop aircraft. This is due to their rapid airspeed, powerful deicing equipment, and tendency to cruise at higher altitudes where temperatures are typically too cold for icing. Conversely, light turboprop aircraft are more susceptible to icing because they typically fly at lower altitudes where icing is more common and at slower speeds.

20.3.5 Icing in Stratiform Clouds

Icing in middle and low-level stratiform clouds is confined, on the average, to a layer between 3,000 and 4,000 ft thick. Thus, a change in altitude of only a few thousand feet may take the aircraft out of icing conditions, even if it remains in clouds. Icing intensity generally ranges from a trace to light, with the maximum values occurring in the cloud's upper portions. Both rime and mixed icing are found in stratiform clouds. The main hazard lies in the great horizontal extent of stratiform clouds layers. High-level stratiform clouds (i.e., at temperatures colder than -20 °C) are composed mostly of ice crystals and produce little icing.

20.3.6 Icing in Cumuliform Clouds

The icing layer in cumuliform clouds is smaller horizontally, but greater vertically than in stratiform clouds. Icing is more variable in cumuliform clouds because many of the factors conducive to icing depend on the particular cloud's stage of development. Icing intensities may range from a trace in small cumulus to severe in a large towering cumulus or cumulonimbus. Although icing occurs at all levels above the freezing level in a building cumuliform cloud, it is most intense in the upper portion of the cloud where the updraft is concentrated and SLDs are plentiful. Icing can extend to great heights in towering cumulus and cumulonimbus where strong updrafts allow SLDs to exist at temperatures as cold as -40 °C. Icing in a cumuliform cloud is usually clear or mixed with rime in the upper levels.

20.3.7 Icing with Fronts

Most icing reports occur in the vicinity of fronts. This icing can occur both above and below the front (see Figure 20-1).

For significant icing to occur above the front, the warm air must be lifted and cooled to saturation at temperatures below zero, making it contain supercooled water droplets. The supercooled water droplets freeze on impact with an aircraft. If the warm air is unstable, icing may be sporadic; if it is stable, icing may be continuous over an extended area. A line of showers or thunderstorms along a cold front may produce icing, but only in a comparatively narrow band along the front.

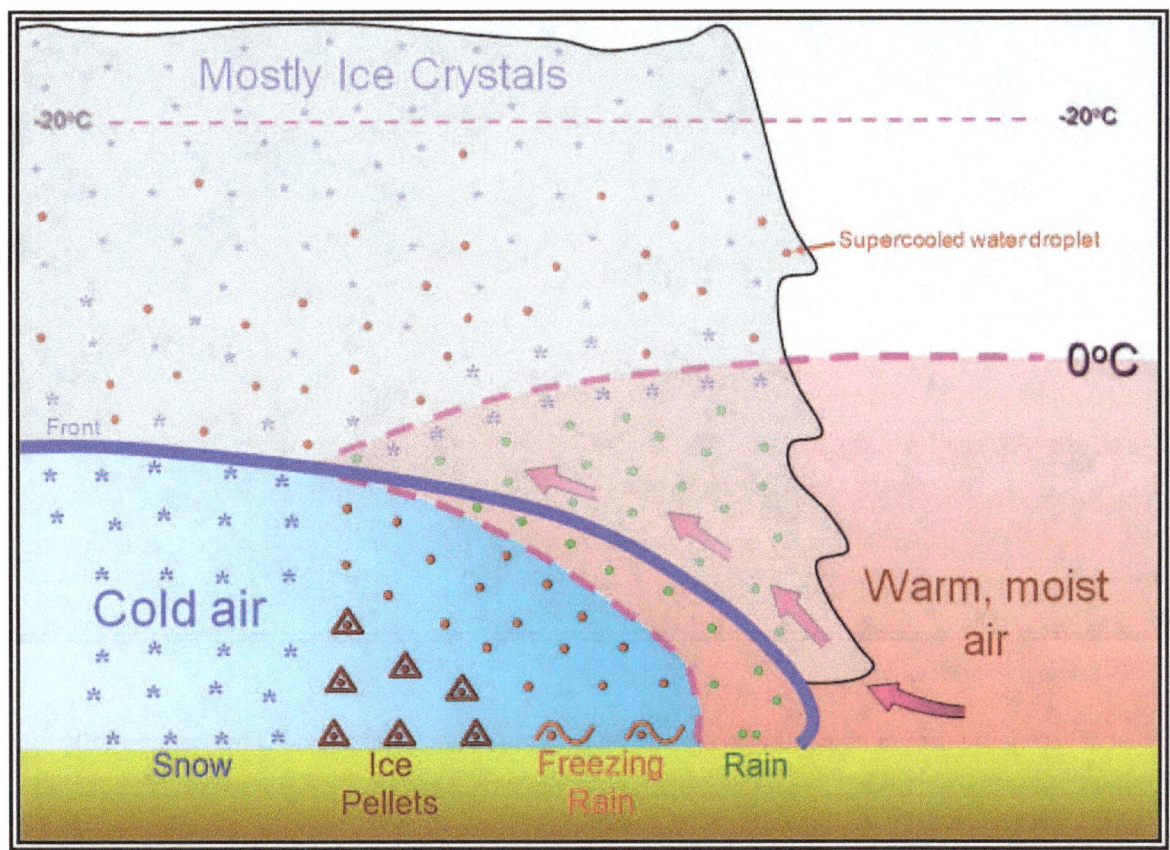

Figure 20-1. Icing with Fronts

A favored location for severe clear icing is freezing rain and/or freezing drizzle below a front. Rain forms above the frontal surface at temperatures warmer than freezing. Subsequently, it falls through air at temperatures below freezing and becomes supercooled. The SLDs freeze on impact with an aircraft. If the below-freezing layer is shallow, freezing rain will occur at the surface. If the below-freezing layer is deep, the supercooled droplets may freeze into ice pellets. Ice pellets indicate icing above. The icing can be severe because of the large amount of supercooled water. Icing in freezing precipitation is especially dangerous because it often extends horizontally over a broad area and a pilot may be unable to escape it by descending to a lower altitude.

20.3.8 Icing with Mountains

Icing is more likely and more severe in mountainous regions. Mountain ranges cause upward air motions on their windward side. These vertical currents support large supercooled water droplets above the freezing level. The movement of a front across a mountain range combines frontal lift with the mountain's upslope flow effect to create extremely hazardous icing zones (see Figure 20-2).

Figure 20-2. Icing with Mountains

The most severe icing occurs above the crests and on the ridges' windward side. This zone usually extends to about 5,000 ft above the mountaintops, but can extend much higher if cumuliform clouds develop.

Icing with mountains can be especially hazardous because a pilot may be unable to descend to above-freezing temperatures due to terrain elevation. If a pilot approaches a mountain ridge from the windward side, the pilot's aircraft may be unable to climb above the mountaintops, or even maintain altitude due to severe ice accumulation. The end result may be a crash.

20.3.9 Convective Icing

Thunderstorms produce abundant supercooled liquid water above the freezing level. When temperature in the upward current cools to about -15 °C, much of the remaining water vapor deposits as ice crystals; above this level, the amount of supercooled water decreases. However, supercooled water can exist at temperatures as cold as -40 °C in the presence of vigorous upward vertical motion, such as in a thunderstorm updraft.

Supercooled water freezes on impact with an aircraft. Clear icing can occur at any altitude above the freezing level, but at high levels, icing may be rime or mixed rime and clear. The abundance of supercooled water makes clear icing very rapid between 0 °C and -15 °C, and encounters can be frequent in a cluster of cells. Thunderstorm icing can be extremely hazardous.

20.3.10 Icing Hazards

Structural icing degrades an aircraft's performance. It destroys the smooth flow of air, increasing drag while decreasing the ability of the airfoil to create lift. The actual weight of ice on an airplane is insignificant when compared to the airflow disruption it causes. As power is added to compensate for the additional drag and the nose is lifted to maintain altitude, the angle of attack is increased. This allows the underside of the wings and fuselage to accumulate additional ice.

Wind tunnel and flight tests have shown that frost, snow, and ice accumulations (on the leading edge or upper surface of the wing) no thicker or rougher than a piece of coarse sandpaper can reduce lift by 30 percent and increase drag up to 40 percent. Larger accretions can reduce lift even more and can increase drag by 80 percent or more.

Ice accumulates on every exposed frontal surface of the airplane: wings, propeller, windshield, antennas, vents, intakes, and cowlings. It can build in flight where no heat or boots can reach it. It can cause antennas to vibrate so severely that they break. In moderate-to-severe icing, a light aircraft could be subject to enough ice accumulation or accretion that continued flight is impossible. The airplane may stall at much higher speeds and lower angles of attack than normal. It can roll or pitch uncontrollably, and recovery might be impossible.

Regardless of anti-ice or deice protection offered by the aircraft, a pilot's first course of action should be to leave the area of visible moisture. This might mean descending to an altitude below the cloud bases, climbing to an altitude that is above the cloud tops, or turning to a different course. If this is not possible, then the pilot should move to an altitude where the temperature is above freezing.

20.4 Engine Icing

20.4.1 Carburetor Icing

In an aspirated engine, the carburetion process can lower the temperature of the incoming air by as much as 38 °C (100 °F). If the moisture content is high enough, ice will form on the throttle plate and venturi, shutting off the supply of air to the engine. Even a small amount of carburetor ice will cause power loss and may make the engine run rough. It is possible for carburetor ice to form even when the skies are clear and the outside temperature is as high as 38 °C (100 °F) if the relative humidity is 50 percent or more. The icing risk does not go away until the humidity falls below roughly 25 percent and/or the OAT drops well below freezing.

20.4.2 High Ice Water Content (HIWC)

High Ice Water Content (HIWC) is a relatively new icing hazard, at least from the standpoint of research and understanding. HIWC refers to high-altitude ice crystals, which may exist in the tops and anvils of cumulonimbus clouds and thunderstorms. Under certain HIWC conditions, turbine engine performance can be affected, including flameouts. Research into HIWC conditions continues as of the writing of this section.

20.5 Additional Information

Refer to AC 91-74, Pilot Guide: Flight in Icing Conditions, for additional information on icing.

21 Arctic Weather

21.1 Introduction

Strictly speaking, the Arctic is the region shown in Figure 21-1, which lies within the Arctic Circle (66.5° N latitude). However, it is loosely defined as the northern regions, in general. This chapter includes Alaska weather, even though much of Alaska lies south of the Arctic Circle.

As an introduction to Arctic weather, this chapter surveys climate, air masses, and fronts of the Arctic, and introduces some Arctic weather peculiarities.

This chapter also covers Arctic aviation weather hazards.

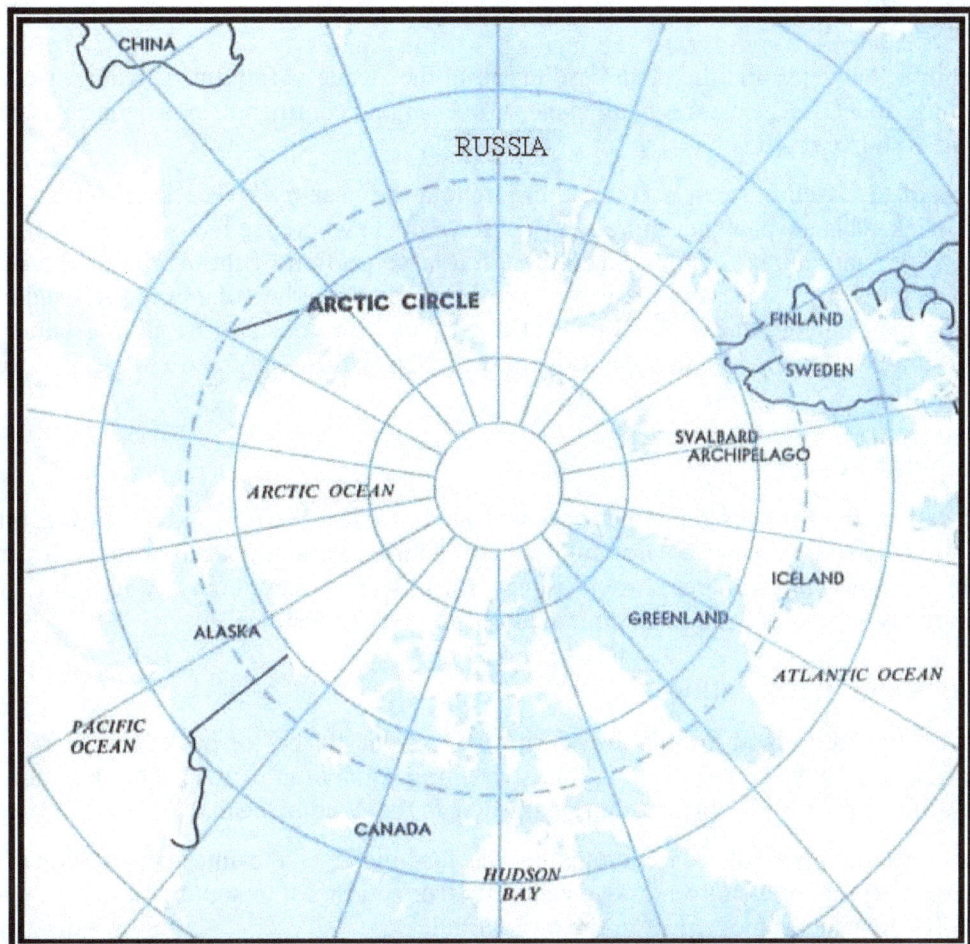

Figure 21-1. The Arctic Circle

21.2 Climate, Air Masses, and Fronts

There are a number of factors that influence Arctic climate. Climate of any region, especially in the Arctic, is largely determined by the amount of energy received from the Sun; however, local characteristics of the area, such as mountains and ice cover, also influence climate.

21.2.1 Long Days and Nights

A profound seasonal change in length of day and night occurs in the Arctic because of the Earth's tilt and its revolution around the Sun. Any point north of the Arctic Circle has autumn and winter days when the Sun stays below the horizon all day and has spring and summer days with 24 hours of sunshine when the Sun stays above the horizon. The number of these days increases toward the North Pole; there the Sun stays below the horizon for 6 months and shines continuously during the other 6 months.

Twilight in the Arctic is prolonged because of the shallow angle of the Sun below the horizon. In more northern latitudes, it persists for days when the Sun remains just below the horizon. This abundance of twilight often makes visual reference possible at night.

21.2.2 Land and Water

Figure 21-1 shows the water and the land distribution in the Arctic. Arctic mountain ranges are effective barriers to air movement. Large masses of air stagnate over inland continental areas; thus, Arctic continental areas are air mass source regions.

A large portion of the Arctic Ocean is covered throughout the year by a deep layer of ice, known as the permanent ice pack. This ice pack goes through a seasonal cycle where ice melts in the spring and summer and increases in the autumn and winter. Even though a large portion of the Arctic Ocean is ice-covered through much of the year, the ice and the water below contain more heat than the surrounding cold land, thus moderating the climate. Oceanic and coastal areas have a milder climate during winter and a cool climate during summer. As opposed to large water bodies, large land areas show a more significant seasonal temperature variation.

21.2.3 Temperature

As one would expect, the Arctic is very cold in winter, but due to local terrain and the movement of pressure systems, occasionally some areas are surprisingly warm. During winter, coastal areas are warmer than the interior. During summer, interior areas are pleasantly warm due to many hours of sunshine, while coastal areas have relatively short, cool summers due to their proximity to water.

21.2.4 Clouds and Precipitation

Cloudiness over the Arctic is at a minimum during winter due to the ice pack being more widespread. Spring brings many cloudy days, with cloudiness reaching a maximum in summer and autumn when a portion of the sea ice melts and exposes additional water in the Arctic Ocean.

During summer afternoons, scattered cumulus clouds forming over the interior occasionally grow into thundershowers. These thundershowers move generally from northeast to southwest in the polar easterlies, which is opposite the general movement in the mid-latitudes.

During the winter, polar lows can form over the open ocean. Polar lows are small, intense low-pressure systems that can develop rapidly when cold air flows over warm water. They produce severe weather, strong surface winds, and heavy precipitation. Polar lows dissipate quickly once they move over land.

Precipitation in the Arctic can vary drastically per region, but is generally light. Some areas are known as polar deserts. In winter, the only precipitation received in the Arctic is snow, while in summer precipitation falls mostly in the form of snow over ice caps and oceanic areas and rain over interior areas. Annual amounts of precipitation over the ice pack and along the coastal areas tend to be less than the interior areas.

21.2.5 Wind

Typically, Arctic winds are light. Strong winds occur more often along the coastal areas in autumn and winter. Wind speeds are generally light in the continental interior throughout the entire year, but are normally at their strongest during summer and autumn.

21.2.6 Air Masses—Winter

In winter, air masses form over the expanded ice pack and adjoining snow-covered land areas. These air masses are characterized by very cold surface air, very low humidity, and strong low-level temperature inversions. Occasionally, air from unfrozen ocean areas flows northward over the Arctic. These intrusions of moist, cold air account for most of the infrequent wintertime cloudiness and precipitation in the Arctic.

21.2.7 Air Masses—Summer

During the summer, the top layer of the Arctic permafrost layer melts, leaving very moist ground, and the open water areas of the Polar Basin increase markedly. Thus, the entire area becomes more humid, relatively mild, and semi-maritime in character. The largest amounts of cloudiness and precipitation occur inland during the summer months.

21.2.8 Fronts

Occluded fronts are the rule. Weather conditions with occluded fronts are much the same in the Arctic as elsewhere: low clouds, precipitation, poor visibility, and sudden fog formation. Fronts are much more frequent over coastal areas than over the interior.

21.3 Arctic Peculiarities

Several Arctic phenomena are peculiar to the region.

21.3.1 Effects of Temperature Inversion

The Arctic experiences frequent low-level temperature inversions, which occur when temperature increases with height (i.e., cold air settled near the ground with warm air directly above). Inversions can slow down surface winds and trap pollutants, creating smoggy and hazy conditions that persist until the inversion ends. In addition, light rays are bent as they pass at low angles through the inversion, creating an effect known as looming, which is a form of mirage that causes objects beyond the horizon to appear above the horizon. These low-level inversion mirages distort the shape of the Sun, Moon, and other objects.

21.3.2 Light Reflection by Snow-Covered Surfaces

Much more light is reflected by snow-covered surfaces than by darker surfaces. Snow often reflects Arctic sunlight sufficiently to blot out shadows, thus decreasing the contrast between objects. Dark, distant mountains may be easily recognized, but a crevasse normally directly in view may be undetected due to lack of contrasts.

21.3.3 Light from Celestial Bodies

Illumination from the Moon and stars is much more intense in the Arctic than in lower latitudes. Even illumination from the stars creates visibility far beyond that found elsewhere. Only under heavy overcast skies does the night darkness in the Arctic begin to approach the degree of darkness in lower latitudes.

21.4 Arctic Weather Hazards

Weather hazards include visibility-restricting phenomena such as blowing snow, icing, frost, and lack of contrast (e.g., whiteout).

21.4.1 Fog and Ice Fog

Fog occurs when water droplets or ice particles are suspended in the air at the Earth's surface. Water-droplet fog occurs in coastal areas during the summer. Ice fog (also called ice-crystal fog, frozen fog, frost fog, frost flakes, air hoar, rime fog, and pogonip) is a type of fog formed by direct freezing of supercooled water droplets. Ice fog is composed of suspended particles of ice, partly ice crystals 20 to 100 microns in diameter, but chiefly (especially when dense) ice particles about 12 to 20 microns in diameter. It occurs at very low temperatures and usually in clear, calm weather in high latitudes. The Sun may cause the appearance of a halo. Effective visibility is reduced considerably more when looking toward the Sun. Ice fog is rare at temperatures warmer than -30 °C, and increases in frequency with decreasing temperature until it is almost always present at air temperatures of -45 °C in the vicinity of a source of water vapor. Such sources of water vapor are the open water of fast-flowing streams or of the sea, herds of animals, volcanoes, and especially products of combustion for heating, automobiles, and aircraft. At temperatures warmer than -30 °C, these sources can cause steam fog of liquid water droplets, which may turn into ice fog when cooled.

21.4.2 Blowing and Drifting Snow

Over the frozen Arctic Ocean and along the coastal areas, blowing snow, drifting snow, and strong winds are common hazards during autumn and winter. Blowing snow is wind-driven falling or accumulated snow that reduces surface visibility. Drifting snow is an uneven distribution of snowfall or snow depth caused by strong surface winds, which may occur during or after a snowfall. Drifting snow is usually associated with blowing snow. In the Arctic, because the snow is dry and fine, it can be picked up easily by light winds that raise the snow several feet off the ground, obliterating objects. A sudden increase in surface wind may cause an unlimited visibility to drop to near zero in a few minutes. This sudden loss of visibility occurs frequently without warning in the Arctic.

21.4.3 Frost

Frost is the formation of thin ice crystals on the ground or other surfaces on solid objects below the freezing point of water. It develops in Arctic coastal areas during spring, autumn, and winter.

21.4.4 Whiteout

Whiteout is a visibility-restricting phenomenon that occurs most often in the spring and autumn in the Arctic when the Sun is near the horizon. It occurs when a layer of overcast clouds overlies a snow- or ice-covered surface. Parallel rays of the Sun are broken up and diffused when passing through the cloud layer so that they strike the snow surface from many angles. The diffused light then reflects back and forth countless times between the snow and the cloud, eliminating all shadows. The result is a loss of depth perception. Buildings, people, and dark-colored objects appear to float in the air, and the horizon disappears.

22 Thunderstorms

22.1 Introduction

A thunderstorm is a local storm, invariably produced by a cumulonimbus (CB) cloud, and always accompanied by lightning and thunder, usually with strong gusts of wind, heavy rain, and sometimes hail. There are as many as 40,000 thunderstorm occurrences each day worldwide, and the United States certainly experiences its share.

Thunderstorms are barriers to air traffic because they are usually too tall to fly over, too dangerous to fly through or under, and can be difficult to circumnavigate. Weather recognizable as a thunderstorm should be considered hazardous, as penetration of any thunderstorm can lead to an aircraft accident and fatalities to those on board.

22.2 Necessary Ingredients for Thunderstorm Cell Formation

Thunderstorm cell formation needs three ingredients: sufficient water vapor, unstable air, and a lifting mechanism (see Figure 22-1). Sufficient water vapor (commonly measured using dewpoint) must be present to produce unstable air. Virtually all showers and thunderstorms form in an air mass that is classified as conditionally unstable. A conditionally unstable air mass needs a lifting mechanism strong enough to release the instability. Lifting mechanisms include: converging winds around surface lows and troughs, fronts, upslope flow, drylines, outflow boundaries generated by prior storms, and local winds, such as sea breeze, lake breeze, land breeze, and valley breeze circulations.

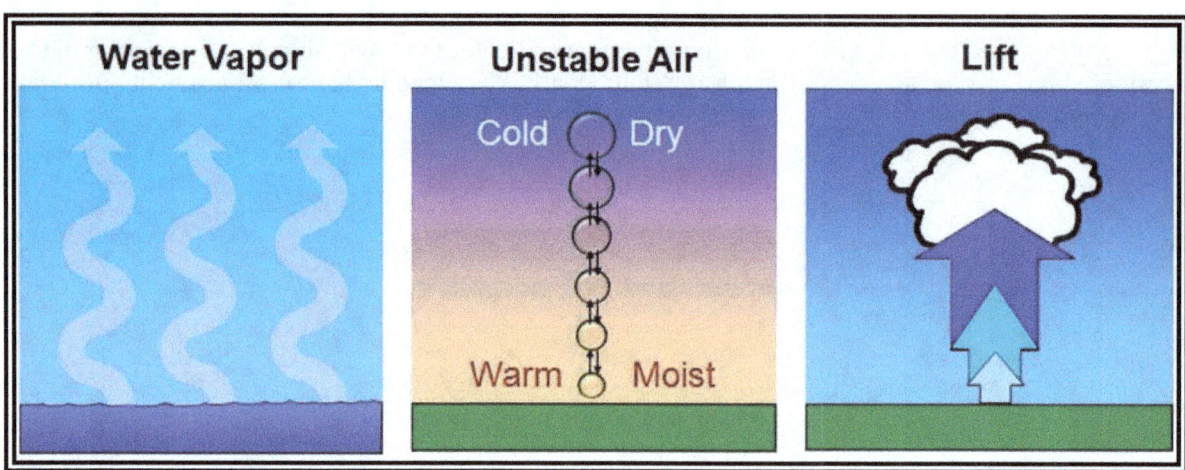

Figure 22-1. Necessary Ingredients for Thunderstorm Cell Formation

22.3 Thunderstorm Cell Life Cycle

A thunderstorm cell is the convective cell of a cumulonimbus cloud having lightning and thunder. It undergoes three distinct stages during its life cycle (see Figure 22-2): towering cumulus, mature, and dissipating. The total life cycle is typically about 30 minutes.

The distinguishing feature of the towering cumulus stage is a strong convective updraft. The updraft is a bubble of warm, rising air concentrated near the top of the cloud, which leaves a cloudy trail in its wake. Updraft speeds can exceed 3,000 fpm.

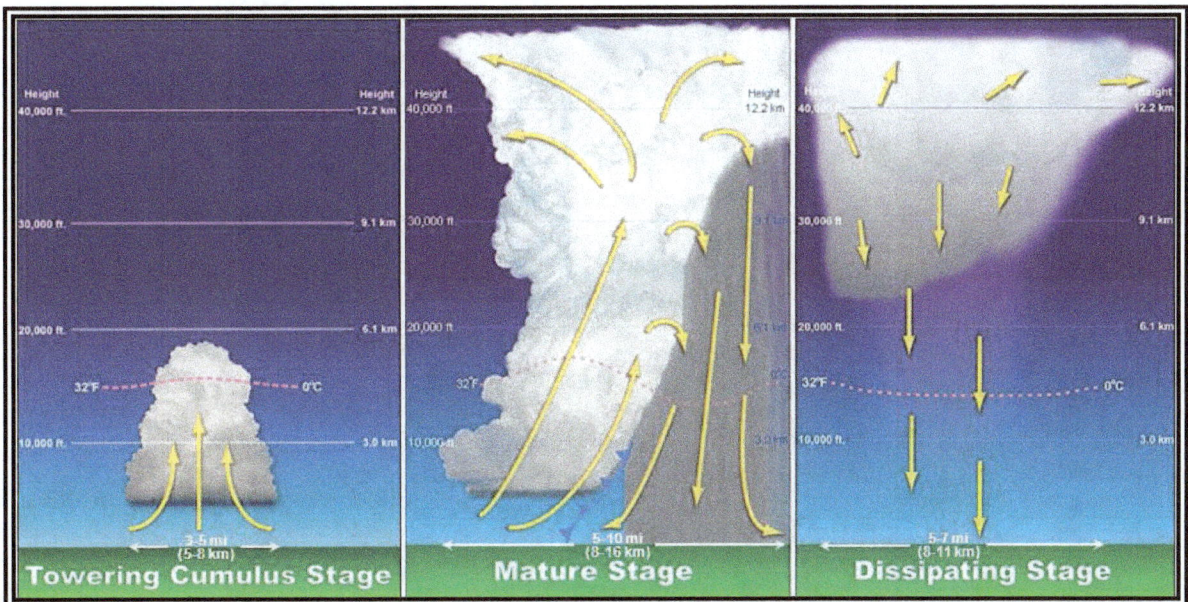

Figure 22-2. Thunderstorm Cell Life Cycle

The cell transitions to the mature stage when precipitation reaches the surface. Precipitation descends through the cloud and drags the adjacent air downward, creating a strong downdraft alongside the updraft. The downdraft spreads out along the surface, well in advance of the parent thunderstorm cell, as a mass of cool, gusty air. The arc-shaped leading edge of downdraft air resembles a miniature cold front and is called a gust front. Uplift along the gust front may trigger the formation of new cells, sometimes well ahead of the parent cell. Cumulonimbus tops frequently penetrate into the lower stratosphere as an overshooting top, where strong winds aloft distort the cloud top into an anvil shape. Weather hazards reach peak intensity toward the end of the mature stage.

The dissipating stage is marked by a strong downdraft embedded within the area of precipitation. Subsiding air replaces the updraft throughout the cloud, effectively cutting off the supply of moisture provided by the updraft. Precipitation tapers off and ends. Compression warms the subsiding air and the relative humidity drops. The convective cloud gradually vaporizes from below, leaving only a remnant anvil cloud.

22.4 Thunderstorm Types

There are three principal thunderstorm types: single-cell, multicell (cluster and line), and supercell. All thunderstorms are hazardous to aircraft.

A single-cell or common (also called ordinary-cell) thunderstorm consists of only one cell. This type of thunderstorm often develops on warm and humid summer days. These cells may be severe and produce hail and microburst winds. Its life cycle was covered in the previous section. It is easily circumnavigated by pilots, except at night or when embedded in other clouds. Single-cell thunderstorms are rare; almost all thunderstorms are multicell.

A multicell cluster thunderstorm (see Figure 22-3 and Figure 22-4) consists of a cluster of cells at various stages of their life cycle. With an organized multicell cluster, as the first cell matures, it is carried downwind, and a new cell forms upwind to take its place. A multicell cluster may have a lifetime of several hours (or more). New cells will continue to form as long as the three necessary ingredients exist (see Section 22.2). Individual cells within the cluster may move in one direction while the whole system moves in another. It can cover large areas and its persistence makes it a bit tougher to circumnavigate than a single-cell thunderstorm. An area of multicell cluster thunderstorms can be like a minefield for air traffic.

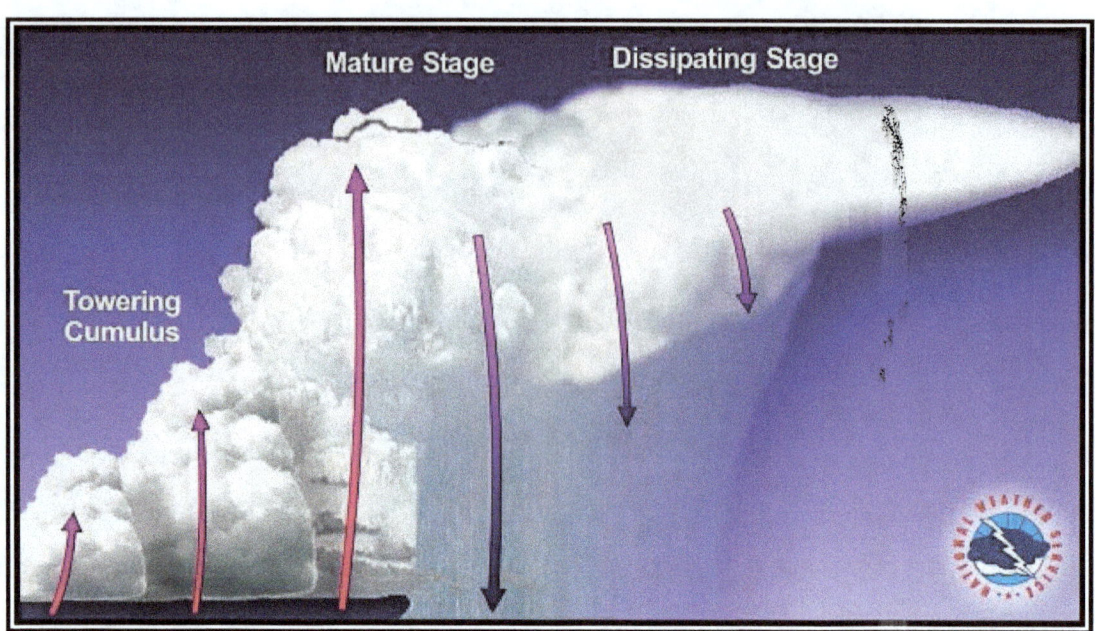

Figure 22-3. Multicell Cluster Thunderstorm

Sometimes thunderstorms will form in a narrow band or squall line that can extend laterally for hundreds of miles. Often it develops on or ahead of a cold front in moist, unstable air, but it may develop in unstable air far removed from any front. New cells continually re-form at the leading edge of the system with rain, and sometimes hail, following behind. Sometimes storms, which comprise the line, can be supercells. The line can persist for many hours (or more) as long as the three necessary ingredients continue to exist (see Section 22.2). These squall lines are the thunderstorm type which presents the most effective barrier to air traffic because the line is usually too tall to fly over, too dangerous to fly through or under, and difficult to circumnavigate. About 25 percent of all tornadoes in the United States are spawned by squall lines.

Figure 22-4. Multicell Line Thunderstorm

Chapter 22, Thunderstorms

22-4

A supercell thunderstorm (see Figure 22-5) is an often dangerous, long-lived convective storm that consists primarily of a single, quasi-steady rotating updraft that persists for an extended period of time. It has a very organized internal structure that enables it to produce especially dangerous weather for pilots who encounter them. Updraft speeds may reach 9,000 fpm (100 kt). This allows hazards to be magnified to an even greater degree. Nearly all supercells produce severe weather (e.g., large hail or damaging wind) and about 25 percent produce a tornado. A supercell may persist for many hours (or longer). New cells will continue to form as long as the three necessary ingredients exist (see Section 22.2).

A supercell's size and persistence make it a bit tougher to circumnavigate than a single-cell thunderstorm. Also, multicell clusters and lines may have supercells incorporated as part of the system as well.

Figure 22-5. Supercell Thunderstorm

22.5 Factors that Influence Thunderstorm Motion

A thunderstorm is a process, not a solid object or block of wood. Storm motion equals the combined effects of both advection and propagation (see Figure 22-6). Advection is the component of storm motion due to individual cells moving with the average wind throughout the vertical depth of the cumulonimbus cloud. The wind at FL180 (500 mb) usually provides a good approximation. Propagation is the component of storm motion due to old cell dissipation and the new cell development. Storm motion may deviate substantially from the motion of the individual cells, which comprise the storm.

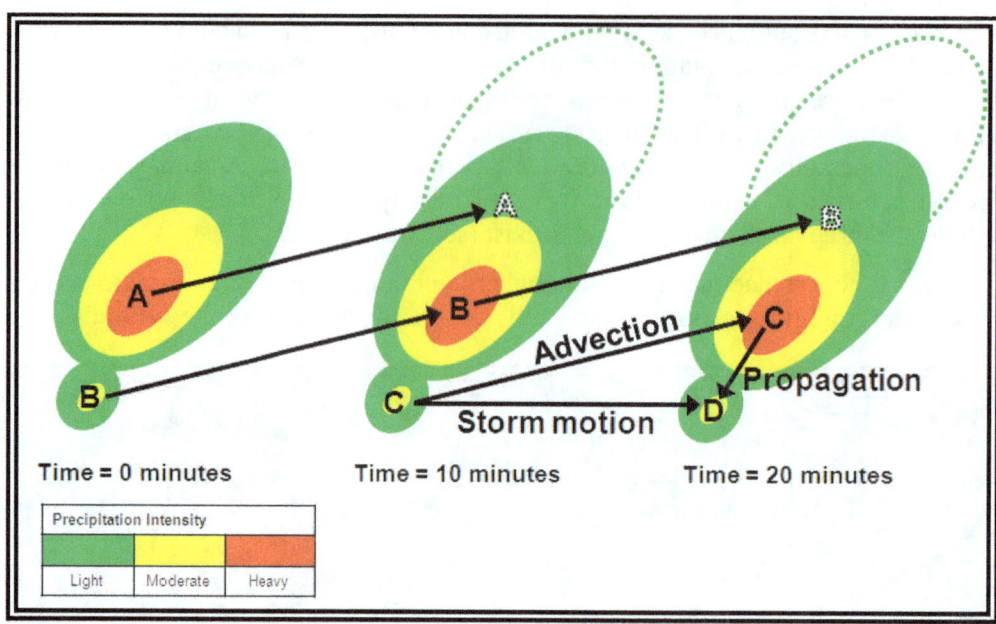

Figure 22-6. Factors that Influence Thunderstorm Motion

Individual cells that comprise the storm move northeast (advection), but dissipate and are replaced by new cells (propagation). Storm motion equals the combined effects of both advection and propagation.

22.6 Thunderstorm Terminology

Anvil. The flat, spreading top of a cumulonimbus cloud, often shaped like an anvil. Thunderstorm anvils may spread hundreds of miles downwind from the thunderstorm itself, and sometimes may spread upwind.

Bow Echo. A radar echo that is linear but bent outward in a bow shape.

Derecho. A widespread, long-lived, straight-line windstorm that is associated with a fast-moving band of severe thunderstorms.

Downdraft. A small-scale column of air that rapidly sinks toward the ground, usually accompanied by precipitation as in a shower or thunderstorm. A microburst is the result of a strong downdraft.

Gust Front. The leading edge of gusty surface winds from thunderstorm downdrafts; sometimes associated with a shelf cloud or roll cloud. May also be referred to as a gustnado or outflow boundary.

Mesoscale Convective System (MCS). A complex of multiple thunderstorms that becomes organized on a scale larger than the individual thunderstorms but smaller than extratropical cyclones, and normally persists for several hours or more.

Roll Cloud. A low, horizontal tube-shaped arcus cloud associated with a thunderstorm gust front. Roll clouds are relatively rare; they are completely detached from the thunderstorm base or other cloud features, thus differentiating them from the more familiar shelf clouds.

Severe Thunderstorm. A thunderstorm that produces hail with a diameter of 1 in (U.S. quarter size) or larger, convective winds of 50 kt (58 mph) or greater, and/or tornadoes.

Shelf Cloud. A low, horizontal wedge-shaped cloud associated with a thunderstorm gust front. Unlike a roll cloud, a shelf cloud is attached to the base of the parent cloud above it, which is usually a thunderstorm.

Updraft. A small-scale current of rising air. If the air is sufficiently moist, then the moisture condenses to become a cumulus cloud or an individual tower of a towering cumulus or cumulonimbus.

22.7 Hazards

All thunderstorms have conditions that are hazards to aviation. These hazards occur in numerous combinations. While not every thunderstorm contains all hazards, it is not possible to visually determine which hazards a thunderstorm contains. Hazards include: low ceiling and visibility, lightning, adverse winds, downbursts, turbulence, icing, hail, rapid altimeter changes, static electricity, tornadoes, and engine water ingestion.

22.7.1 Low Ceiling and Visibility

Generally, visibility is near zero within a thunderstorm cloud. Ceiling and visibility also may be restricted in precipitation and dust between the cloud base and the ground. The restrictions create the same problem as all ceiling and visibility restrictions, but the hazards are increased when associated with the other thunderstorm hazards of turbulence, hail, and lightning that make precision instrument flying virtually impossible.

22.7.2 Lightning

Every thunderstorm produces lightning and thunder by definition. Lightning is a visible electrical discharge produced by a thunderstorm. The discharge may occur within or between clouds, between a cloud and air, between a cloud and the ground, or between the ground and a cloud.

Lightning can damage or disable an aircraft. It can puncture the skin of an aircraft, and it can damage communications and electronic navigational equipment. Lightning has been suspected of igniting fuel vapors causing an explosion; however, serious accidents due to lightning strikes are extremely rare. Nearby lightning can blind the pilot, rendering the pilot momentarily unable to navigate either by instrument or by visual reference. Nearby lightning can also induce permanent errors in the magnetic compass. Lightning discharges, even distant ones, can disrupt radio communications on low and medium frequencies. Though lightning intensity and frequency have no simple relationship to other storm parameters, severe storms, as a rule, have a high frequency of lightning.

22.7.3 Downburst and Microburst

The downward moving column of air in a typical thunderstorm is large. Convective clouds, shower cells, and thunderstorm cells sometimes produce intense downdrafts called downbursts that create strong, often damaging winds and wind shear. Downbursts (see Figure 22-7) can create hazardous conditions for pilots and have been responsible for many LLWS accidents. Smaller, shorter-lived downbursts are called microbursts.

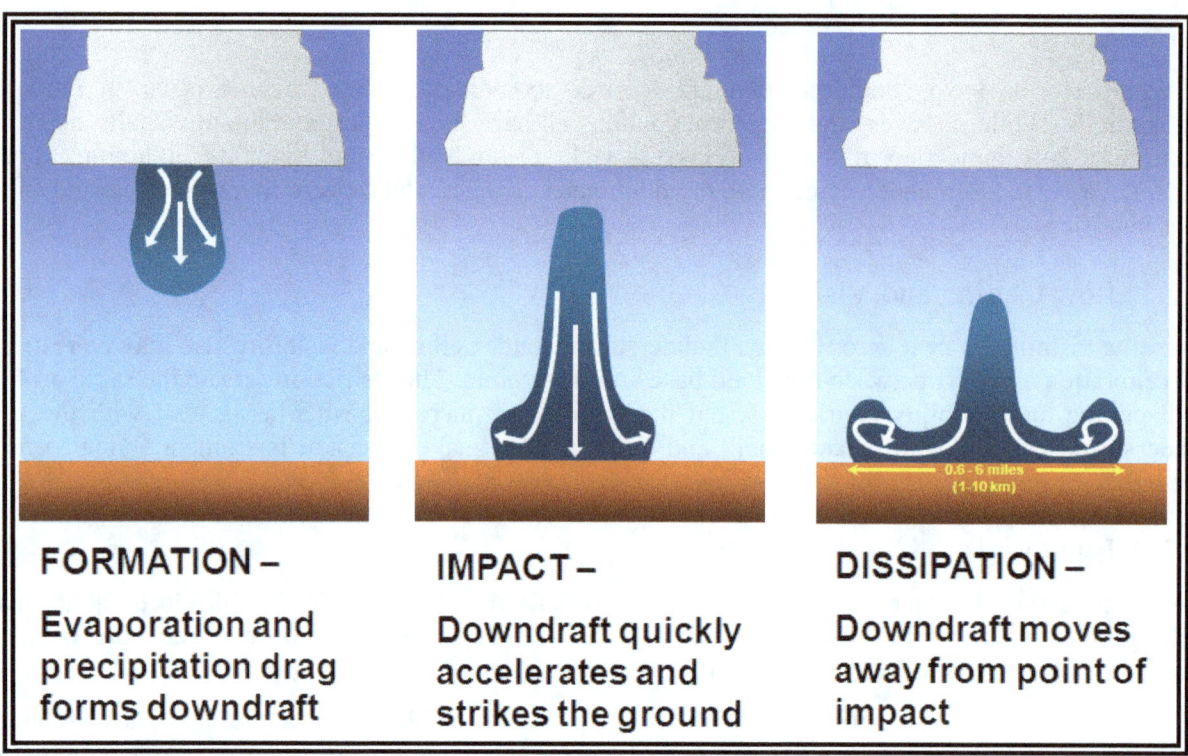

Figure 22-7. Downburst Life Cycle

A microburst (see Figure 22-8) is a small-scale, intense downdraft that, when reaching the surface, spreads outward, symmetrically (see Figure 22-9) or asymmetrically (see Figure 22-10), in all directions from the downdraft center. It is the most severe type of wind shear. Microburst activity may be indicated by an intense rain shaft at the surface, but virga (i.e., streaks of precipitation falling from a thunderstorm cloud but not reaching the ground) at the cloud base and/or a ring of blowing dust is sometimes the only visible clue (see Figure 22-11).

A typical microburst has a horizontal diameter of less than 2.5 mi and a nominal depth of 1,000 ft. The lifespan of a microburst is about 5 to 15 minutes, during which time it can produce downdrafts of up to 6,000 fpm; increasing headwind and headwind losses of 30 to 90 kt, seriously degrading performance. It can also produce strong turbulence and hazardous wind direction changes.

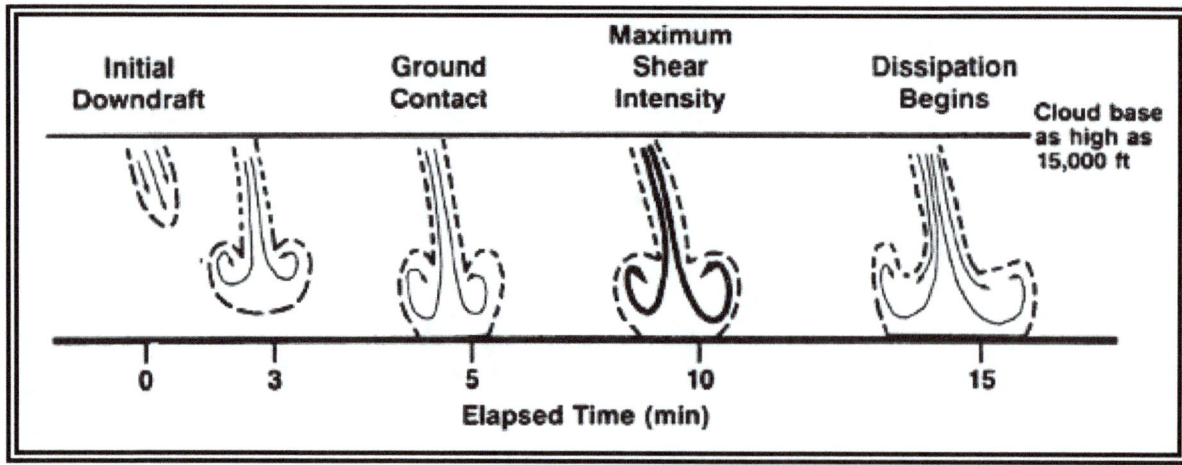

Figure 22-8. Illustration of the Evolution of a Microburst

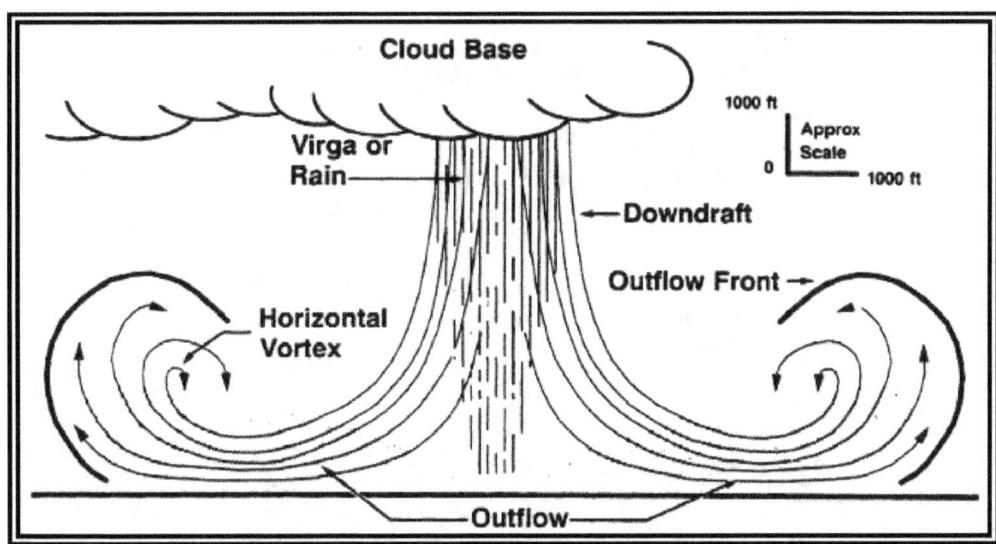

Figure 22-9. Illustration of a Symmetric Microburst

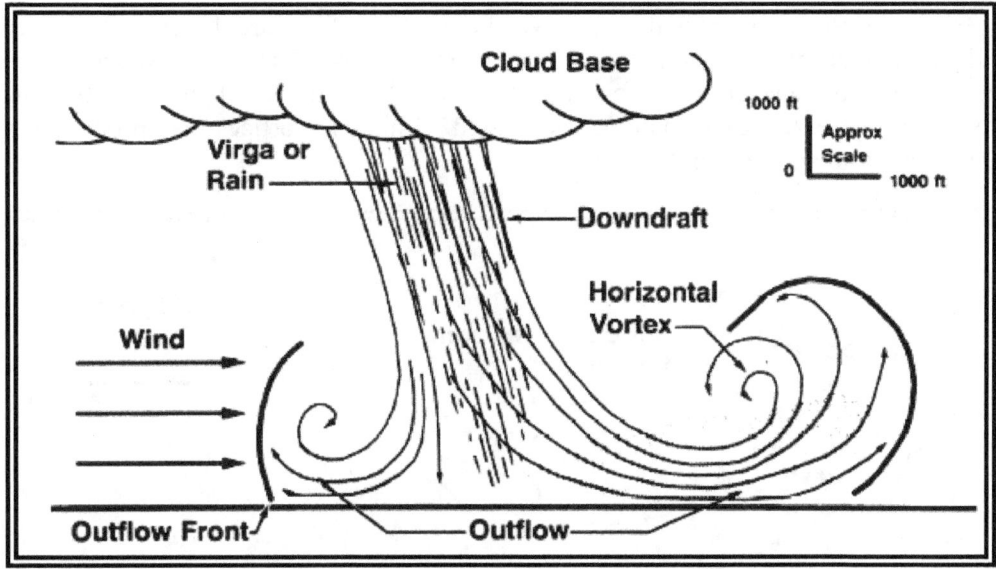

Figure 22-10. Illustration of an Asymmetric Microburst

Chapter 22, Thunderstorms

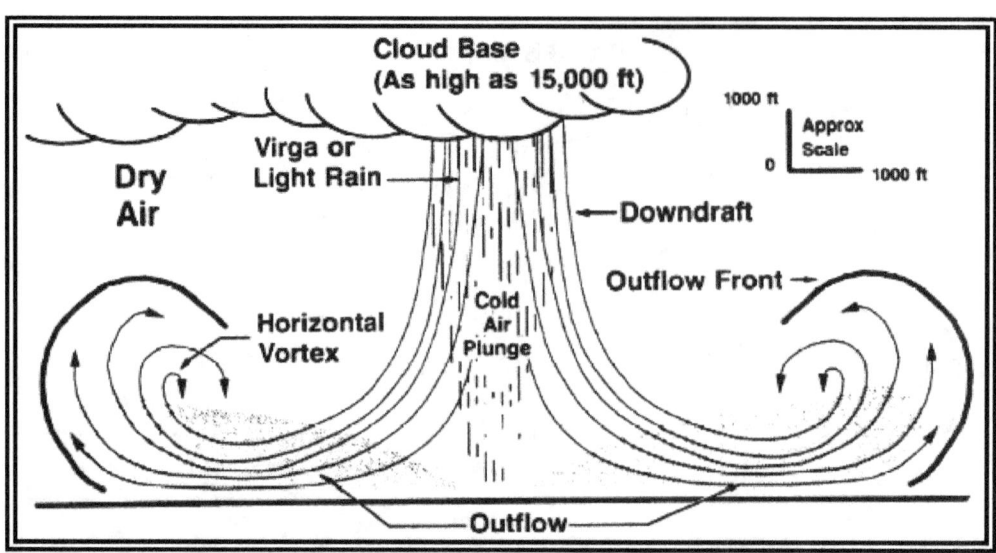

Figure 22-11. Illustration of a Dry Microburst

More than one microburst can occur in the same weather system. Pilots are therefore cautioned to be alert for additional microbursts if one has already been encountered or observed. If several microbursts are present, a series of horizontal vortices can form near the ground due to several microbursts being embedded in one another (see Figure 22-12). Conditions associated with these vortices may produce very powerful updrafts and roll forces in addition to downdrafts.

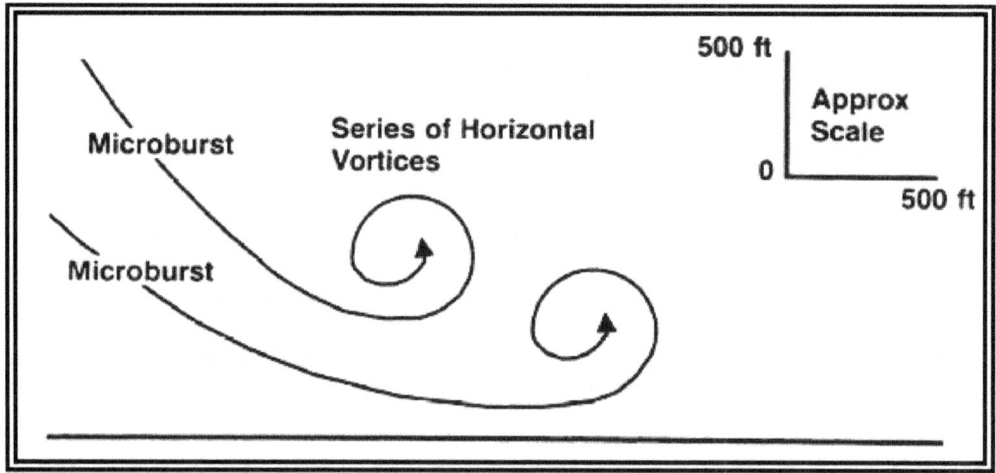

Figure 22-12. Illustration of a Series of Vortices Due to Microbursts Being Embedded in One Another

A downburst or microburst is dangerous to aircraft when climbing from takeoff or approaching to land. During these phases, the aircraft is operating at slow speeds. A major change of wind velocity can lead to loss of lift and a crash. During landing, if the pilot has reduced power and lowered the nose in response to the headwind shear (see Figure 22-13) this leaves the aircraft in a nose-low, power-low configuration when the tailwind shear occurs, which makes recovery more difficult. It can cause the airplane to stall or land short of the runway.

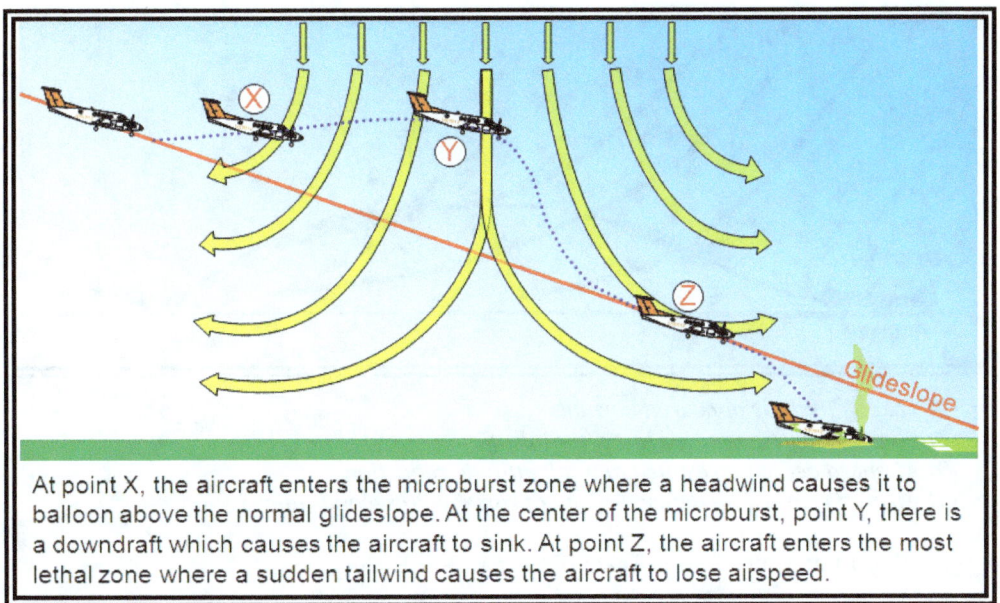

Figure 22-13. Landing in a Microburst

Pilots should be alert for indications of a microburst early in the approach phase, and ready to initiate a missed approach at the first indication. However, it may be impossible to recover from a microburst encounter at low altitude.

Pilots should be aware of asymmetrical microbursts (see Figure 22-10), since a significant airspeed increase may not occur upon entering the outflow, or may be much less than the subsequent airspeed loss experienced when exiting the microburst.

It is vital for pilots to recognize that some microbursts cannot be successfully escaped with any known techniques. Some wind shears that are within the performance capability of the aircraft have caused accidents.

22.7.3.1 Encounter During Takeoff—After Lift-Off

In previous accidents studied, the airplane encountered an increasing tailwind shear shortly after lifting off the runway (see Figure 22-14). For the first 5 seconds after lift-off, the takeoff appeared normal, but the airplane crashed off the end of the runway about 20 seconds after lift-off.

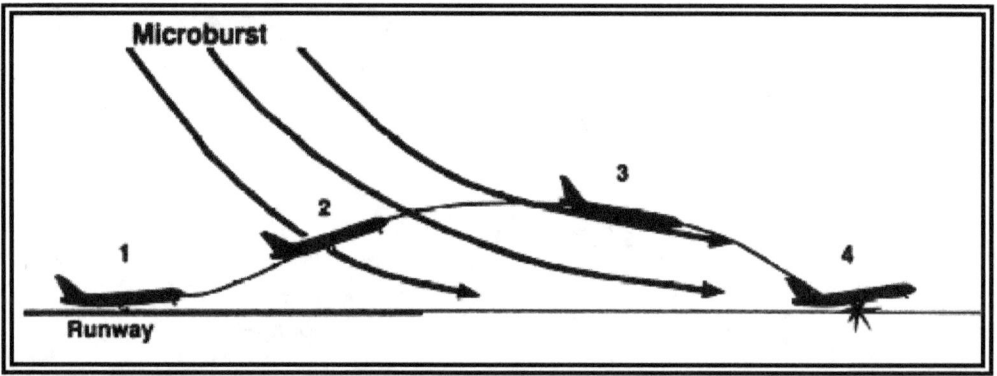

(1) Takeoff initially appeared normal.
(2) Wind shear encountered just after lift-off.
(3) Airspeed decrease resulted in pitch attitude reduction.
(4) Aircraft crashed off departure end of runway 20 seconds after lift-off.

Figure 22-14. Wind Shear Encounter During Takeoff After Lift-Off

In many events involving after-lift-off wind shear encounters, early trends in airspeed, pitch attitude, vertical speed, and altitude appeared normal. In this example, the airplane encountered wind shear before stabilized climb was established, which caused difficulty in detecting onset of shear. As the airspeed decreased, pitch attitude was reduced to regain trim airspeed (see Figure 22-15). By reducing pitch attitude, available performance capability was not utilized and the airplane lost altitude. As terrain became a factor, recovery to initial pitch attitude was initiated. This required unusually high stick force (up to 30 lb of pull may be needed on some airplanes). Corrective action, however, was too late to prevent ground contact since the downward flightpath was well established.

Reducing pitch attitude to regain lost airspeed, or allowing attitude to decrease in response to lost airspeed, is the result of past training emphasis on airspeed control. Successful recovery from an inadvertent wind shear encounter necessitates maintaining or increasing pitch attitude and accepting lower-than-usual airspeed. Unusual and unexpected stick forces may be needed to counter natural airplane pitching tendencies due to airspeed and lift loss.

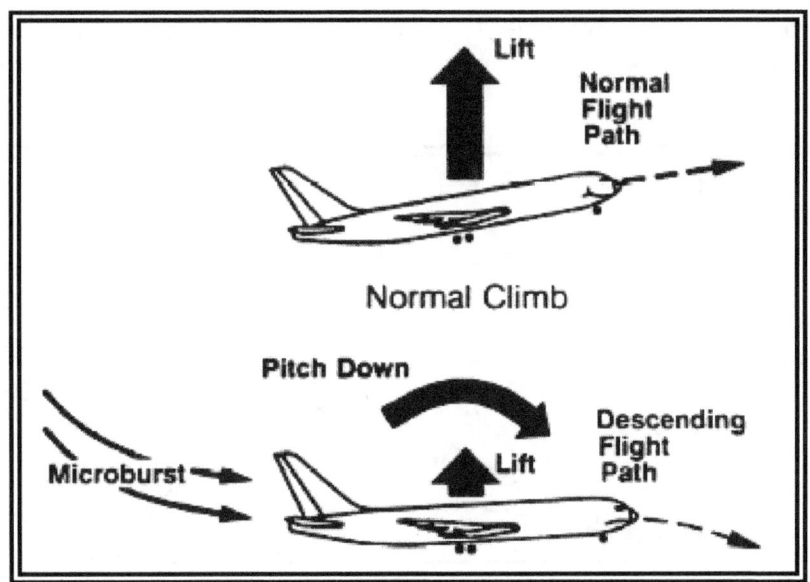

Microburst reduces airspeed and lift at normal attitude, which results in pitch-down tendency to regain airspeed.

Figure 22-15. Wind Shear Effects on Flightpath

To counter the loss of airspeed and lift resulting from wind shear, pitch attitude must not be allowed to fall below the normal range. Only by properly controlling pitch attitude and accepting reduced airspeed can flightpath degradation be prevented (see Figure 22-16). Once the airplane begins to deviate from the intended flightpath and high descent rates develop, it takes additional time and altitude to change flightpath direction.

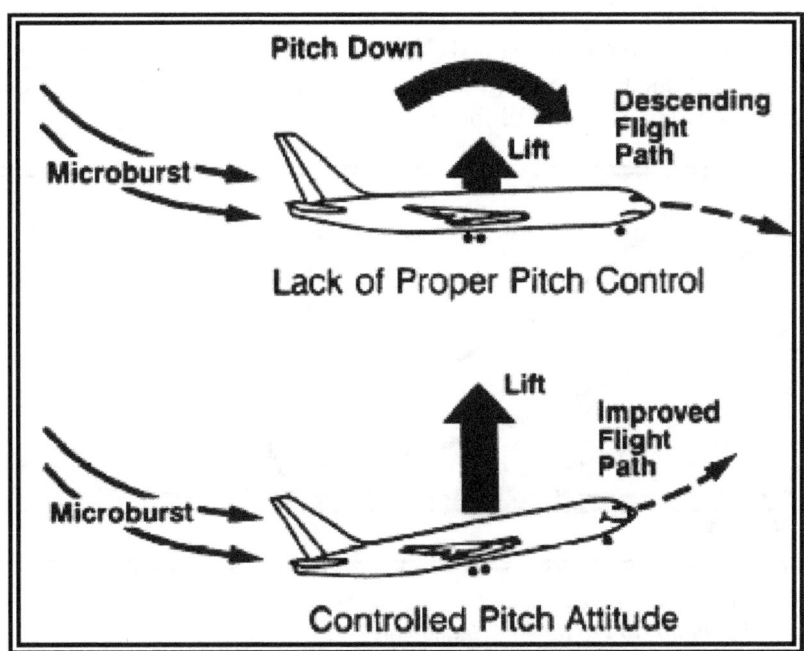

Control of pitch attitude and acceptance of reduced airspeed results in improved flightpath.

Figure 22-16. Pitch Control Effects on Flightpath

Chapter 22, Thunderstorms

Only 5 to 15 seconds may be available to recognize and respond to a wind shear encounter (see Figure 22-17). Therefore, it is of great importance that a wind shear encounter be recognized as soon as possible.

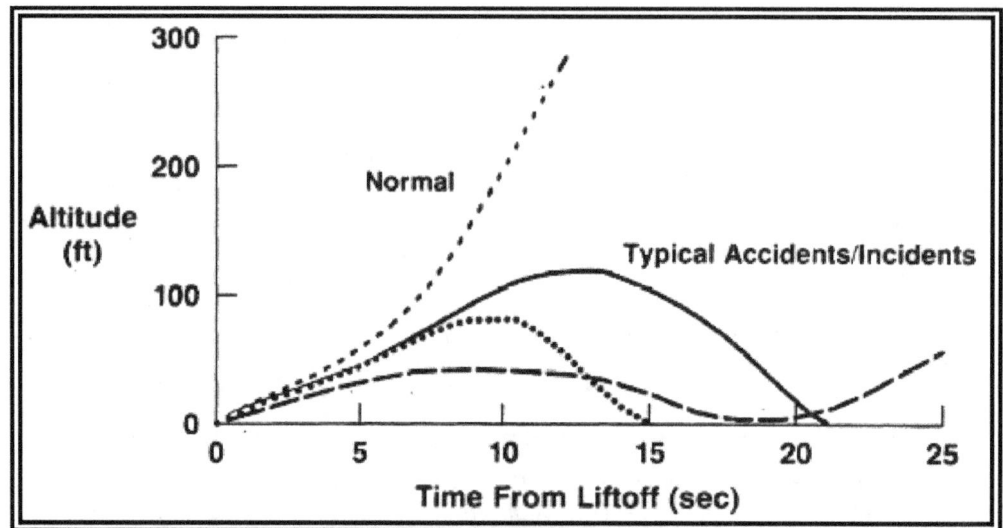

Takeoff initially appeared normal. Additional time is needed to arrest descent. Result: Only 5 to 15 seconds may be available for recognition and recovery.

Figure 22-17. Time Available to Respond to Wind Shear Encounter

22.7.3.2 Encounter During Takeoff—On Runway

Analysis of a typical accident where an increasing tailwind shear was encountered during takeoff ground roll showed that initial indications appeared normal (see Figure 22-18). Due to the increasing tailwind shear, however, the airplane did not reach rotation speed (V_R) until nearing the end of the runway. As the airplane lifted off, the tailwind continued increasing, preventing any further airspeed increase. The airplane contacted an obstacle off the departure end of the runway.

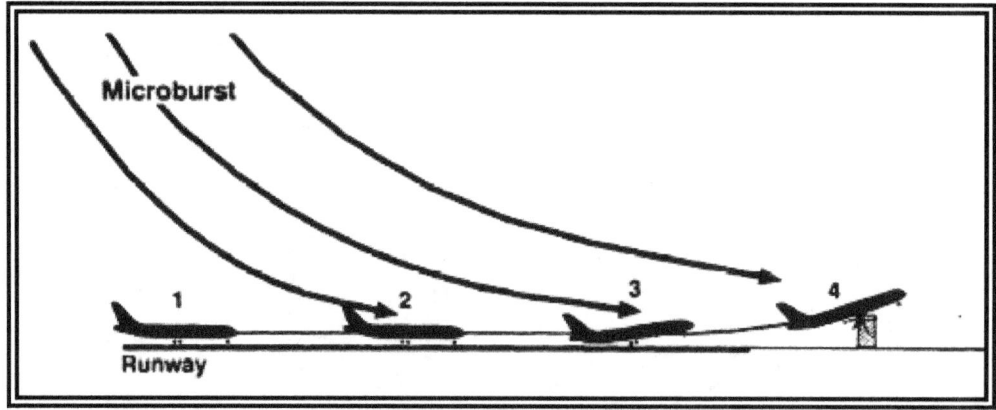

(1) Takeoff initially appeared normal.
(2) Airspeed buildup slowed due to wind shear.
(3) Airplane reached V_R near end of runway, lifted off but failed to climb.
(4) Airplane contacted obstacle off departure end of runway.

Figure 22-18. Wind Shear Encounter During Takeoff on Runway

Less-than-normal airspeed, due to wind shear encounter, resulted in reduced available lift at normal takeoff attitude (see Figure 22-19). In turn, the inability to lift off soon enough to clear obstacles resulted.

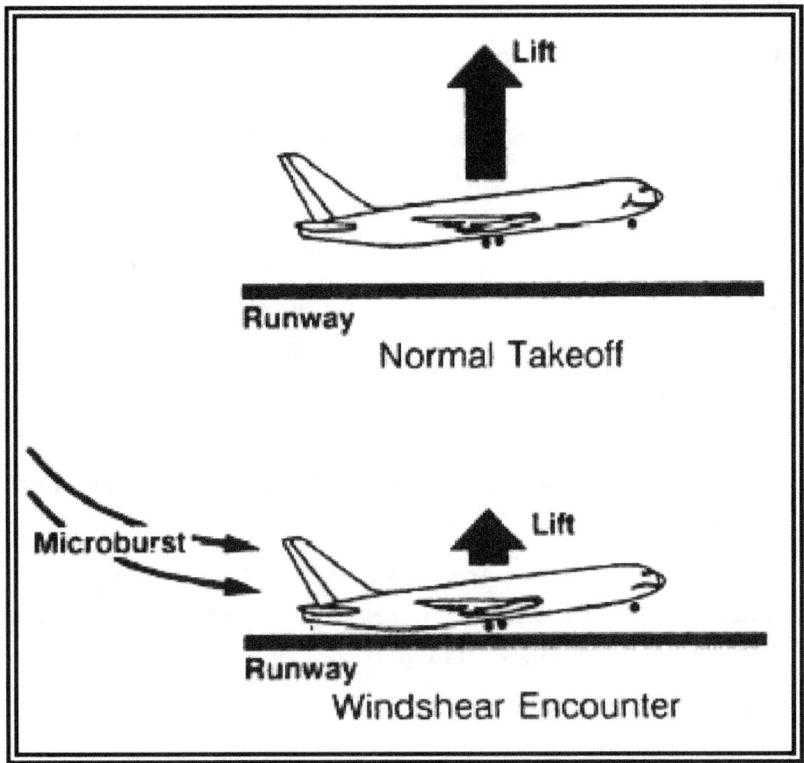

Microburst reduces airspeed and lift at normal attitude that results in inability to lift off.

Figure 22-19. Wind Shear Effects on Lift-Off

An additional factor is the difficulty of recognizing deteriorating airplane performance. Timely recognition of a wind shear encounter on the runway may be difficult since the only indication may be a slower-than-normal airspeed increase. The presence of gusts may mask abnormal airspeed buildup. Time available to respond effectively to a wind shear may be as little as 5 seconds from the initial encounter.

If there is insufficient runway left to accelerate to normal takeoff speed, and inadequate runway to stop, lift-off and safe climb may require rotation at speeds less than V_R. In this case, additional pitch attitude may be needed to achieve sufficient lift (see Figure 22-20). In traditional training, crews are frequently cautioned not to rotate at speeds less than V_R to avoid high pitch attitudes that could result in aft body contact. In a wind shear encounter, rotation toward normal takeoff pitch attitude at lower-than-normal airspeed may be needed to lift off in the remaining runway. This may result in aft body contact. To deal with an inadvertent wind shear encounter, the pilot should be prepared to apply techniques that differ from those ordinarily used.

Chapter 22, Thunderstorms

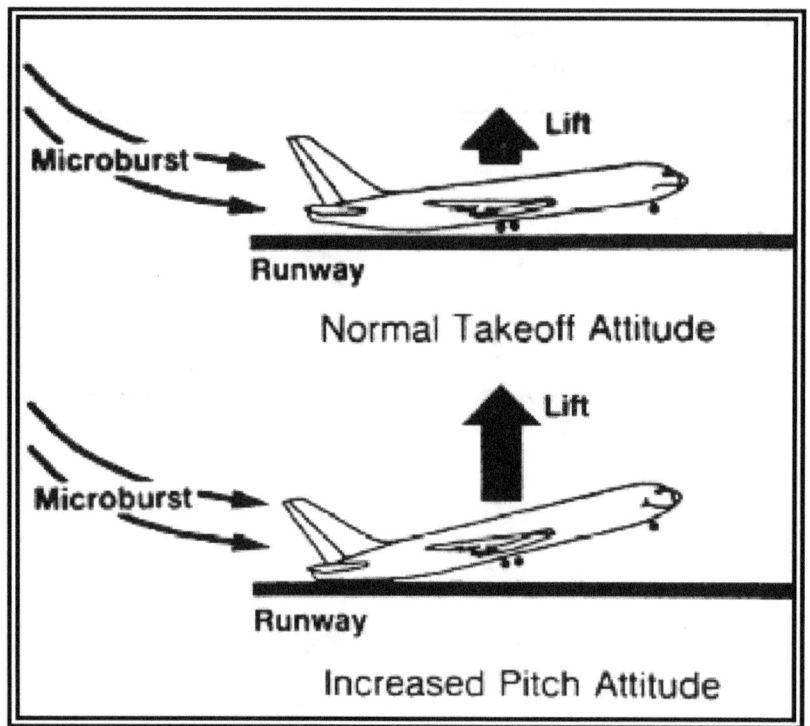

Increased pitch attitude generates lift needed for lift-off.

Figure 22-20. Pitch Attitude Effects on Lift-Off

22.7.3.3 Encounter on Approach

Analysis of a typical wind shear encounter on approach provided evidence of an increasing downdraft and tailwind along the approach flightpath (see Figure 22-21). The airplane lost airspeed, dropped below the target glidepath, and contacted the ground short of the runway threshold.

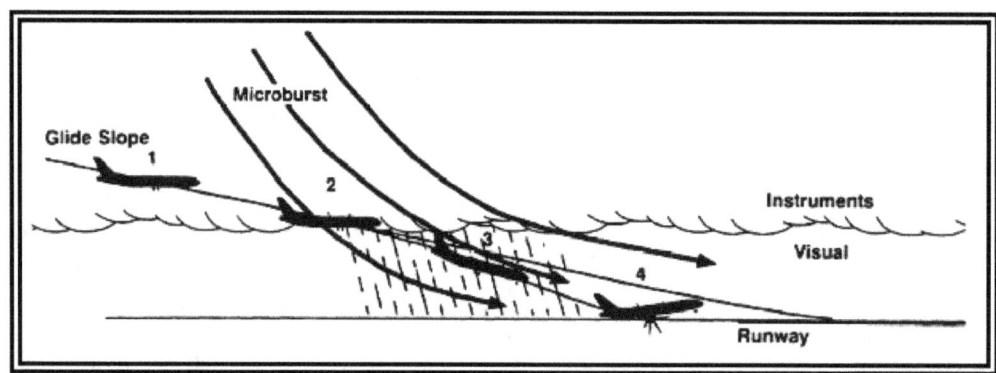

(1) Approach initially appeared normal.
(2) Increasing downdraft and tailwind encountered at transition.
(3) Airspeed decrease combined with reduced visual cues resulted in pitch attitude reduction.
(4) Airplane crashed short of approach end of runway.

Figure 22-21. Wind Shear Encounter During Approach

Chapter 22, Thunderstorms

Reduced airspeed, as the airplane encountered the wind shear, resulted in decreased lift. This loss of lift increased the descent rate (see Figure 22-22). The natural nose-down pitch response of the airplane to low airspeed caused additional altitude loss. Pitch attitude increase and recovery initiation were not used soon enough to prevent ground contact.

Lack of timely and appropriate response—affected by weather conditions, inadequate crew coordination, and limited recognition time—was a significant factor in delaying recovery initiation. Gradual application of thrust during approach may have masked the initial decreasing airspeed trend. Poor weather conditions caused increased workload and complicated the approach. Transition from instruments to exterior visual references may have detracted from instrument scan. Inadequate crew coordination may have resulted in a failure to be aware of flightpath degradation. A stabilized approach with clearly defined callouts is essential to aid in the recognition of unacceptable flightpath trends and the need to initiate recovery.

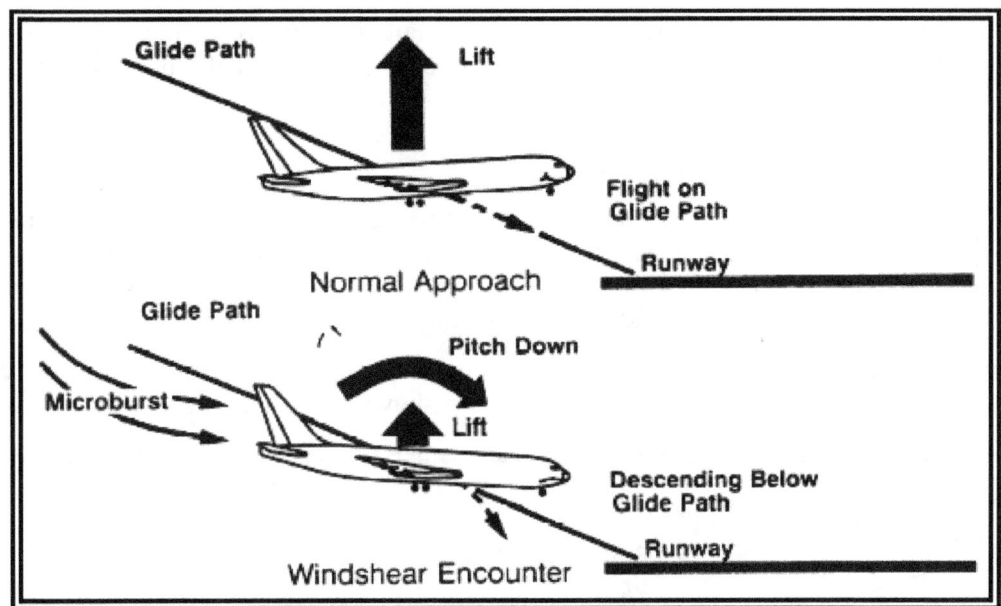

Microburst reduces airspeed and lift at normal attitude that results in pitch-down tendency to regain airspeed.

Figure 22-22. Wind Shear Effects on Flightpath on Approach

22.7.3.4 Wind Shear Effects on Airplanes and Systems

Several terms are used when discussing low-altitude wind variations with respect to aviation. These terms are defined as follows:

- Increasing Headwind Shear: Wind shear in which headwind increases, causing an airspeed increase.
- Decreasing Headwind Shear: Wind shear in which headwind decreases, causing an airspeed decrease.
- Decreasing Tailwind Shear: Wind shear in which tailwind decreases, causing an airspeed increase.
- Increasing Tailwind Shear: Wind shear in which tailwind increases, causing an airspeed decrease.

22.7.3.4.1 Headwind/Tailwind Shear Response

The various components of wind shear have unique effects on airplane performance. In addition, the magnitude of the shear depends on the flightpath through the microburst.

An increasing headwind (or decreasing tailwind) shear increases indicated airspeed and thus increases performance. The airplane will tend to pitch up to regain trim airspeed. An additional consideration is that this type of shear may reduce normal deceleration during flare, which could cause overrun.

Any rapid or large airspeed increase, particularly near convective weather conditions, should be viewed as a possible indication of a forthcoming airspeed decrease. Thus, a large airspeed increase may be reason for discontinuing the approach. However, since microbursts are often asymmetric and the headwind may not always be present, headwind shears are not to be relied upon to provide early indications of subsequent tailwind shears.

In contrast to shears that increase airspeed, an increasing tailwind (or decreasing headwind) shear will decrease indicated airspeed and performance capability. Due to airspeed loss, the airplane may tend to pitch down to regain trim speed.

22.7.3.4.2 Vertical Wind Shear Response

Vertical winds exist in every microburst and increase in intensity with altitude. Such winds usually reach peak intensity at heights greater than 500 ft above the ground. Downdrafts with speeds greater than 3,000 fpm can exist in the center of a strong microburst. The severity of the downdraft the airplane encounters depends on both the altitude and lateral proximity to the center of the microburst.

Perhaps more critical than sustained downdrafts, short duration reversals in vertical winds can exist due to the horizontal vortices associated with microbursts. This is shown in Figure 22-23 below.

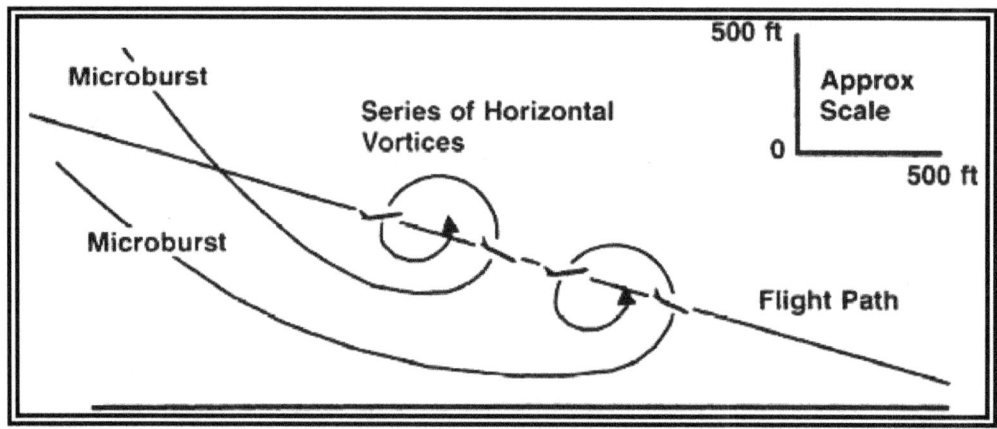

Rapid updraft/downdraft variations due to horizontal vortices can cause uncommanded pitch changes and may result in momentary stick shaker activation well above normal stick shaker speeds.

Figure 22-23. Illustration of an Encounter with Microburst Horizontal Vortices

An airplane flying through horizontal vortices as shown in Figure 22-23 experiences alternating updrafts and downdrafts causing pitch changes without pilot input. These vertical winds result in airplane angle-of-attack fluctuations which, if severe enough, may result in momentary stick shaker actuation or airframe shudder at speeds well above normal.

22.7.4 Convective Turbulence

See Section 19.2.1 for information on convective turbulence.

22.7.5 Convective Icing

See Section 20.3.9 for information on convective icing.

22.7.6 Hail

See Section 14.4.5 for information on hail.

22.7.7 Rapid Altimeter Changes

Pressure usually falls rapidly with the approach of a thunderstorm. Pressure then usually rises sharply with the onset of the first gust and arrival of the cold downdraft and heavy rain, falling back to normal as the thunderstorm passes. This cycle of pressure change may occur in 15 minutes. If the pilot does not receive a corrected altimeter setting, the altimeter may be more than 100 ft in error.

22.7.8 Static Electricity

Static electricity (a steady, high level of noise in radio receivers) is caused by intense corona discharges from sharp metallic points and edges of flying aircraft. It is encountered often in the vicinity of thunderstorms. When an aircraft flies through clouds, precipitation, or a concentration of solid particles (e.g., ice, sand, or dust), it accumulates a charge of static electricity. The electricity discharges onto a nearby surface or into the air, causing a noisy disturbance at lower frequencies.

The corona discharge is weakly luminous and may be seen at night. Although it has a rather eerie appearance, it is harmless. It was named "St. Elmo's Fire" by Mediterranean sailors, who saw the brushy discharge at the top of ship masts.

22.7.9 Tornado

A tornado is a violently rotating column of air in contact with the ground, either pendant from a cumuliform cloud or underneath a cumuliform cloud, and often (but not always) visible as a funnel cloud. The most violent thunderstorms draw air into their cloud bases with great force. If the incoming air has any initial rotating motion, it often forms an extremely concentrated vortex from the surface well into the cloud. Meteorologists have estimated that wind in such a vortex can exceed 200 kt; pressure inside the vortex is quite low. The strong winds gather dust and debris, and the low pressure generates a funnel-shaped cloud extending downward from the cumulonimbus base. If the cloud does not reach the surface, it is a "funnel cloud;" if it touches a land surface, it is a "tornado;" and if it touches water, it is a "waterspout." When tornadoes do occur without any visible funnel cloud, debris at the surface is usually the indication of the existence of an intense circulation in contact with the ground.

Tornadoes can occur almost anywhere in the world, but are most common in the central and eastern United States during spring and autumn months. They typically last only a few minutes and travel a few miles, but can persist much longer (e.g., more than 90 minutes) and track much farther (e.g., more than 100 mi), in extreme cases.

On a local scale, the tornado is the most intense of all atmospheric circulations. Its vortex is typically a few hundred yards in diameter, but can range in width from less than 10 yards (yd) to over 2 mi. Wind speeds are typically estimated on the basis of wind damage using the Enhanced Fujita (EF) Scale (see Table 22-1).

Table 22-1. Enhanced Fujita Scale for Tornado Damage

EF-Rating	Class	3-Second Wind Gust		Description	Relative Frequency
		mph	km/h		
EF-0	Weak	65–85	105–137	Gale	53.5%
EF-1	Weak	86–110	138–177	Weak	31.6%
EF-2	Strong	111–135	178–217	Strong	10.7%
EF-3	Strong	136–165	218–266	Severe	3.4%
EF-4	Violent	166–200	267–322	Devastating	0.7%
EF-5	Violent	>200	>322	Incredible	<0.1%

Note: The EF Scale is a set of wind estimates (not measurements) based on damage. The 3-second gust is not the same wind as in METAR/SPECI surface observations, which is a 2-minute average.

Note: Confirmed tornadoes with no reported damage (i.e., those that remain in open fields) are always rated EF-0.

Tornadoes occur with both isolated and squall line thunderstorms. However, over 80 percent of all tornadoes in the United States are produced by supercell thunderstorms. Multiple tornado occurrences associated with a particular large-scale weather system is termed a "tornado outbreak." On rare occasions, one supercell can produce multiple tornadoes over many hours. In addition, families of tornadoes have also been observed as appendages of the main cloud extending several miles outward from the area of lightning and precipitation. Thus, any cloud connected to a severe thunderstorm may contain hidden vortices.

An aircraft entering a tornado vortex is almost certain to suffer loss of control and structural damage. Since the vortex extends well into the cloud, any pilot inadvertently caught on instruments in a thunderstorm could encounter a hidden vortex.

22.7.10 Engine Water Ingestion

Turbine engines have a limit on the amount of water they can ingest. Updrafts are present in many thunderstorms, particularly those in the developing stages. If the updraft velocity in the thunderstorm approaches or exceeds the velocity of the falling raindrops, very high concentrations of water may occur. It is possible that these concentrations can be in excess of the quantity of water that turbine engines are designed to ingest. Therefore, severe thunderstorms may contain areas of high water concentration, which could result in flameout and/or structural failure of one or more engines.

22.8 Thunderstorm Avoidance

22.8.1 Airborne Weather Avoidance Radar (Aircraft Radar)

Airborne weather avoidance radar is, as the name implies, for avoiding severe weather—not for penetrating it. Whether to fly into an area of radar echoes depends on echo intensity, spacing between the echoes, and the capabilities of the pilot and the aircraft. The ability of airborne weather radar to detect weather phenomena is limited in both direction and range. Some airborne radars are fitted with a turbulence display mode, which is based on the Doppler effect. These Doppler radars can detect turbulence associated with precipitation (sometimes referred to as wet precipitation), but these radars are unable to detect CAT. The radar display also does not provide assurance of avoiding instrument weather conditions from clouds and fog. A phenomenon called attenuation (see Section 15.2.5) may exist when a cell absorbs or reflects all of the radio signals sent by the radar system (see Figure 15-5). Attenuation may prevent the radar from detecting additional cells that might lie behind the first cell. This is sometimes referred to as a radar "shadow." For aircraft equipped with airborne weather radar, pilots are expected to be familiar with the operating techniques and limitations of the specific system.

It is important to note that while hail always gives a radar echo, it may fall several miles from the nearest visible cloud, and hazardous turbulence may extend to as much as 20 mi from the echo edge.

22.8.2 Thunderstorm Avoidance Guidance

Never regard any thunderstorm lightly, even when radar observers report the echoes are of light intensity. Avoiding thunderstorms is the best policy. The following is guidance for avoiding thunderstorms:

1. Do not land or take off in the face of an approaching thunderstorm. A sudden gust front of low-level turbulence could cause loss of control.

2. Do not attempt to fly under a thunderstorm, even if you can see through to the other side. Turbulence and wind shear under the storm could be hazardous.

3. Do not attempt to fly under the anvil of a thunderstorm. There is a potential for severe and extreme CAT.

4. Do not fly without airborne radar into a cloud mass containing scattered embedded thunderstorms. Scattered thunderstorms not embedded usually can be visually circumnavigated.

5. Do not trust the visual appearance to be a reliable indicator of the turbulence inside a thunderstorm.

6. Do not assume that ATC will offer radar navigation guidance or deviations around thunderstorms.

7. Use data-linked weather radar (i.e., NEXRAD) mosaic imagery as the sole means for negotiating a path through a thunderstorm area (tactical maneuvering).

8. Remember that the data-linked NEXRAD mosaic imagery shows where the weather was, not where the weather is. The weather conditions may be 15 to 20 minutes older than the age indicated on the display.

9. Listen to chatter on the ATC frequency for PIREPs and other aircraft requesting to deviate or divert.

10. Ask ATC for radar navigation guidance or to approve deviations around thunderstorms, if needed.

11. Use data-linked weather NEXRAD mosaic imagery (e.g., FIS-B) for route selection to avoid thunderstorms entirely (strategic maneuvering).

12. Advise ATC, when switched to another controller, that you are deviating for thunderstorms before accepting to rejoin the original route.

13. Ensure that after an authorized weather deviation, before accepting to rejoin the original route, that the route of flight is clear of thunderstorms.

14. Avoid by at least 20 mi any thunderstorm identified as severe or giving an intense, heavy, or extreme radar echo. This is especially true under the anvil of a large cumulonimbus. Such echoes should be separated by at least 40 mi before flying between echoes. Separation distances may be reduced for avoiding weaker echoes.

15. Circumnavigate the entire area if more than half the area is covered by thunderstorms.

16. Vivid and frequent lightning indicates the probability of a severe thunderstorm.
17. Regard as extremely hazardous any thunderstorm with tops 35,000 ft or higher, whether the top is visually sighted or determined by radar.
18. Give a PIREP for the flight conditions.
19. Divert and wait out the thunderstorms on the ground if unable to navigate around an area of thunderstorms.

If unable to avoid penetrating a thunderstorm, the following is guidance for before entering the storm:

1. Tighten the safety belt, put on the shoulder harness (if installed), and secure all loose objects.
2. Plan and hold the course to take the aircraft through the storm in a minimum time.
3. To avoid the most critical icing, establish a penetration altitude below the freezing level or above the level of 15 °C.
4. Verify that pitot heat is on and turn on carburetor heat or jet engine anti-ice. Icing can be rapid at any altitude and cause almost instantaneous power failure and/or loss of airspeed indication.
5. Establish power settings for turbulence penetration airspeed recommended in the aircraft manual.
6. Turn up cockpit lights to highest intensity to lessen temporary blindness from lightning.
7. If using automatic pilot, disengage Altitude Hold Mode and Speed Hold Mode. The automatic altitude and speed controls will increase maneuvers of the aircraft; thus, increasing structural stress.
8. If using airborne radar, tilt the antenna up and down occasionally. This will permit the detection of other thunderstorm activity at altitudes other than the one being flown.
9. Keep eyes on the flight instruments. Looking outside the cockpit can increase danger of temporary blindness from lightning.
10. Do not change power settings; maintain settings for the recommended turbulence penetration airspeed.
11. Maintain constant attitude. Allow the altitude and airspeed to fluctuate.
12. Do not turn back once in the thunderstorm. A straight course through the storm most likely will get the aircraft out of the hazards most quickly. In addition, turning maneuvers increase stress on the aircraft.

23 Space Weather

23.1 Introduction

The term "space weather" is used to designate processes occurring on the Sun or in the Earth's magnetosphere, ionosphere, and thermosphere that could have a potential impact to the near-Earth environment. Space weather phenomena such as solar flares, radiation storms, and geomagnetic storms are some potential concerns for aviation.

This chapter provides an overall introduction to space weather general theory.

This chapter also outlines the potential effects of space weather on the aircraft, including communications, navigation (e.g., Global Positioning System (GPS)), radiation exposure, and radiation effects on avionics.

23.2 The Sun—Prime Source of Space Weather

The Sun is the dominant source of the conditions commonly described as space weather. Emissions from the Sun are both continuous (e.g., solar luminescence and solar wind) and eruptive (e.g., coronal mass ejections (CME) and flares). These solar eruptions may cause radio blackouts, magnetic storms, ionospheric storms, and radiation storms at Earth.

Similar to the charged particles that come from the Sun, Galactic Cosmic Rays (GCR) are charged particles that originate in more distant supernovae and contribute to the space weather conditions near Earth. Essentially, these charged particles comprise a steady drizzle of radiation at Earth.

The sum of the solar and nonsolar components equals the full extent of the potential radiation dose received. The size of the GCR flux varies inversely with the sunspot cycle (sunspots are described in Section 23.4); that is, during sunspot minimums when the interplanetary environment near Earth is laminar and steady, the GCR component is large due to its easier access to the near-Earth environment. At sunspot maximum, the turbulence and energetics associated with solar eruptions reduce GCR access to the vicinity of the Earth.

23.3 The Sun's Energy Output and Variability

The Sun is a variable star. That means the balance between the continuous emissions and the eruptive emissions changes with time. One metric that is commonly used to track this variability is the occurrence of sunspots. Astronomers have made sunspot observations continuously for hundreds, maybe even thousands, of years. Though the underlying physics is still not well understood, on average, sunspots come and go in an 11-year period. The magnitude and duration of individual cycles varies, but typically more eruptive events occur near the solar maximum, while few are observed near solar minimum. All solar electromagnetic emissions, from radio to x rays, are also stronger during solar maximum and less intense near solar minimum.

23.4 Sunspots and the Solar Cycle

Because space weather activity varies with sunspot activity, they are often used as a proxy index for changing space weather conditions. This is because sunspots, by their very nature, exist due to strong local magnetic fields. When these fields erupt, severe space weather can occur. While sunspots are easily seen, other events such as GCR, CMEs, and increased solar wind are more difficult to observe from the ground and may not be related to long historical records of sunspots.

23.5 Solar Wind

The solar wind is the continuous flow away from the Sun of charged particles and magnetic field, called plasma. Solar wind is a consequence of the very high temperature of the solar corona and the resultant expansion of the plasma into space.

The solar wind carries the energy from most solar eruptions that affect the near-Earth environment. The sole exception, solar flare photons consisting of light and x rays carry the energy released in solar flares. Even in the absence of an eruption, the constant flow of plasma fuels Earth's geomagnetic field. The solar wind may be fast and energetic if an eruption occurs, or can gradually increase due to a coronal-hole structure, which allows unimpeded high-speed solar wind to escape from the corona. As seen from the Earth, the Sun rotates on approximately a 27-day period, so well-established coronal-hole structures that persist for several months will swing by Earth on schedule, roughly every 27 days.

23.6 Solar Eruptive Activity

Most solar eruptions originate in areas that have strong magnetic fields. Usually marked with sunspots, these areas are commonly called active regions. Active regions are numerous and common during solar maximum and scarce during solar minimum.

Flares and CMEs are the two major types of solar eruptions. They may occur independently or at the same time. Solar flares have been recognized for more than 100 years, as they can be seen from the ground. In the past 50 years, Hydrogen-Alpha (656.3 nanometer (nm) wavelength) filter-equipped ground-based telescopes have been used to observe flares.

Flares are characterized by a very bright flash phase, which may last for a few minutes to a few hours during the largest flares. Flares can emit at all frequencies across the electromagnetic emission spectrum, from gamma rays to radio.

CMEs, in contrast to solar flares, are difficult to detect; they are not particularly bright, and may take hours to fully erupt from the Sun. CMEs literally are an eruption of a large volume of the solar outer atmosphere, the corona. Prior to the satellite era, they were very difficult to observe. The energy released in a large solar flare is on par with that released in a CME; however, CMEs are far more effective in perturbing Earth's magnetic field and are known to cause the strongest magnetic storms. A typical travel time for a CME from the Sun to Earth may range from less than 1 day to more than 4 days. The travel time of the electromagnetic emission produced during flares, by comparison, is at the speed of light. They instantaneously affect the day side of Earth upon observation.

The frequency of solar flares and CMEs tracks with the solar cycle. As many as 25 solar flares may occur per day during the maximum phase of the solar cycle. At solar minimum, it may take 6 months or more for 25 flares to occur. CME frequency varies from about five per day near solar maximum to one per week or longer at solar minimum.

Many CMEs observed lifting off the Sun miss Earth due to the CME's direction of travel.

23.7 Geospace

Geospace is the volume of space that surrounds Earth, influenced by the Earth's magnetic field in the solar wind. If Earth did not have a magnetic field, the solar wind would blow past unimpeded, affected only by the mass of Earth and its atmosphere. Earth's magnetic field extends outward in all directions. This forms a cocoon for the planet, protecting it from the flow of the solar wind. The cocoon is called the magnetosphere. The magnetosphere typically extends towards the Sun about 10 Earth radii on the day side and stretches away from the Sun many times more on the night side. The shape is similar to a comet tail, with it being extended during strong solar wind conditions and less so during more quiet times. On its flanks, the magnetosphere extends outward roughly 20 Earth radii in the dawn and dusk sectors.

The magnetosphere deflects most of the energy carried by the solar wind, while making a fraction of it available to be absorbed by the near-Earth system. When the Sun is active and CMEs interact with Earth, the additional energy disrupts the magnetosphere, resulting in a magnetic storm. Then, over time, the magnetosphere adjusts through various processes and once more returns to normal.

The most visible manifestation of the energy being absorbed from the solar wind into the magnetosphere is the aurora, both in the Northern and Southern Hemispheres. The aurora occurs when accelerated electrons from the Sun follow the magnetic field of Earth down to the polar regions, where they collide with oxygen and nitrogen atoms and molecules in Earth's upper atmosphere. In these collisions, the electrons transfer their energy to the atmosphere, thus exciting the atoms and molecules to higher energy states. When they relax to lower energy states, they release their energy in the form of light. Simply put, the more energy in the solar wind, the brighter and more widespread the aurora glow becomes.

Nearer to Earth is another region called the ionosphere. It is a shell of weak plasma, where electrons and ions exist embedded in the neutral atmosphere. The ionosphere begins at roughly 80 km in altitude and extends out many Earth radii, at the topside.

Extreme Ultraviolet (EUV) solar emissions create the ionosphere by ionizing the neutral atmosphere. The electrons and ions created by this process then engage in chemical reactions that progress faster in the lower ionosphere. The ionosphere changes significantly from day to night. When the Sun sets, chemical processes, together with other dynamic processes, allow some of the ionization to remain until the new day brings the solar EUV once again. An important point is that the energy that comes from the Sun in the solar wind makes its way to the ionosphere, where it alters the ambient conditions during space weather storms.

23.8 Galactic Cosmic Rays (GCR)

Galactic Cosmic Rays, more commonly known as GCR, is a consequence of distant supernovae raining charged particles, heavy ions, protons, and electrons onto the inner heliosphere. The abundance of GCR is inversely rated to the solar cycle. At solar maximum, when the solar wind flow is turbulent and strong, the GCR flux is inhibited and therefore low. At solar minimum, the GCR flux increases by about 30 percent in the near-Earth environment. When high-energy GCR enter Earth's atmosphere, it creates a cascade of interactions resulting in a range of secondary particles, including neutrons that make their way to Earth's surface.

23.9 Geomagnetic Storms

Geomagnetic storms are strong disturbances to Earth's magnetic field in the solar wind. These storms pose problems for many activities, technological systems, and critical infrastructure. The topology of Earth's magnetic field changes in the course of a storm, as the near-Earth system attempts to adjust to the jolt of energy from the Sun. CMEs and the shocks they drive are often the causative agent and can send the geomagnetic field into a disturbed state.

The most obvious and probably the only pleasing attribute of an energized geomagnetic field is the auroras. Geomagnetic storms tend to brighten auroras and allow them to move equatorward.

The duration of geomagnetic storms is usually on the order of days. The strongest storms may persist for almost 1 week. A string of CMEs may cause prolonged disturbed periods related to the additional energy being pumped toward the Earth.

Although the frequency of geomagnetic storms reflects the solar cycle, a closer look shows a bimodal distribution. Large numbers of storms cluster at solar maximum resulting from frequent CMEs, and again in the declining phase due to high-speed solar wind streams. Typically, the most intense storms occur near solar maximum, with weaker storms occurring during the declining phase.

23.10 Solar Radiation Storms

Solar radiation storms occur when large quantities of charged particles, primarily protons, are accelerated by processes at or near the Sun, then bathe the near-Earth environment with these charged particles. These particles cause an increase in the radiation dose to humans and increase the possibility of single-event upsets in electronics. Earth's magnetic field and atmosphere offer some protection from this radiation, but protection decreases with altitude, latitude, magnetic field strength, and direction. The polar regions on Earth are the most open to these charged particles. The magnetic field lines at the poles extend vertically downwards, intersecting Earth's surface. This allows the particles to spiral down the field lines and penetrate into the atmosphere and increase the ionization.

A significant factor related to the criticality of the radiation increase at Earth is the energy distribution of the solar protons. Protons of varying energies will bathe Earth as a function of the site of the eruption at the Sun and the magnetic connection between the Sun and Earth. High-energy protons cause radiation dose increases that are of concern to human beings. Lower energy protons have little effect on humans, but have a severe impact on the polar ionosphere.

The duration of solar radiation storms is a function of the magnitude of the solar eruption as well as the energy level of protons. For events that are of a large magnitude but low energy, the duration may last for 1 week. Events that are of high energy may last for only a few hours. There is a great diversity in the duration of solar radiation storms, as there are many factors that contribute to the acceleration and propagation of the charged particles near Earth.

Solar radiation storms can occur at any point in the solar cycle, but tend to be most common during the years around solar maximum.

23.11 Ionospheric Storms

Ionospheric storms arise from large influxes of solar particle and electromagnetic radiation. There is a strong coupling between the ionosphere and the magnetosphere, which means both regimes can be disturbed concurrently.

The symptoms of an ionospheric storm include enhanced currents, turbulence and wave activity, and a nonhomogeneous distribution of free electrons. This clustering of electrons, which leads to scintillation of signals passing through the cluster, is particularly problematic for the Global Navigation Satellite System (GNSS), which includes the United States' GPS.

The duration of the ionospheric storm impact may range from a few minutes to days-long prolonged events. As a general rule, these ionospheric storms mimic the duration of geomagnetic storms.

The intensity of ionospheric storms varies significantly as a function of local time, season, and time within the solar cycle.

The frequency of occurrence of ionospheric storms is also similar to geomagnetic storms with one important caveat. The near-equatorial ionosphere, a band extending approximately ±10° in latitude on either side of the magnetic equator, can be very disturbed in the post-sunset to near-midnight hours, even in the absence of a geomagnetic storm. This behavior is related to the internal electrodynamics of the ionosphere rather than external stimulation from the Sun.

23.12 Solar Flare Radio Blackouts

Radio blackouts primarily affect high frequency (HF) (3 to 30 megahertz (MHz)), although detrimental effects may spill over to VHF (30 to 300 MHz) and beyond, resulting in fading and diminished ability for reception. The blackouts are a consequence of enhanced electron densities caused by the emissions from solar flares that ionize the sunlit side of Earth.

The process consists of x ray and EUV bursts from a solar flare, increasing the number of free electrons in the atmosphere below 90 km, which in turn increases their interaction with the neutral atmosphere that increases the amount of radio energy lost as radio waves pass through this region. During a large flare event, the amount of radio energy lost is sufficient to make the return signal from the ionosphere too small to be useful with normal radio receivers. The net effect of this process is a blackout for HF transmissions.

The duration of dayside solar flare radio blackouts closely follows the duration of the solar flares that cause them beginning with the arrival of the x ray and EUV photons, and abating with their diminution. Usually, the radio blackouts last for several minutes, but they can last for hours.

23.13 Effects of Space Weather on Aircraft Operations

23.13.1 Communications

High frequency communications (HF COM) at low to mid-latitudes are used by aircraft during transoceanic flights and routes where line-of-sight VHF communication is not an option. HF enables a skip mode to send a signal around the curvature of Earth. HF COM on the Earth's day side can be adversely affected when a solar flare occurs and its photons rapidly alter the electron density of the lower altitudes of the ionosphere, causing fading, noise, or a total blackout. Usually these disruptions are short-lived (tens of minutes to a few hours), so the outage ends fairly quickly.

HF COM at high latitudes and polar regions are adversely affected for longer periods, sometimes days, due to some space weather events. The high latitude and polar ionosphere is a sink for charged particles, which alter the local ionization and provide steep local ionization gradients to deflect HF radio waves, as well as increase local absorption.

Satellite communication (SATCOM) signals pass through the bulk of the ionosphere and are a popular means of communicating over a wide area. The frequencies normally used for SATCOM are high enough for the ionosphere to appear transparent. However, when the ionosphere is turbulent and nonhomogeneous, an effect called scintillation (a twinkling in both amplitude and phase) is imposed upon the transmitted signal. Scintillations can result in loss-of-lock and the inability for the receiver to track a Doppler-shifted radio wave.

23.13.2 Navigation and GPS

Space weather adversely affects GPS in three ways: it increases the error of the computed position, it causes a loss-of-lock for receivers, and it overwhelms the transmitted signal with solar radio noise.

23.13.3 Radiation Exposure to Flightcrews and Passengers

Solar radiation storms occurring under particular circumstances cause an increase in radiation dose to flightcrews and passengers. As high polar latitudes and high altitudes have the least shielding from the particles, the threat is the greatest for higher altitude polar flights. The increased dose is much less of an issue for low- and mid-latitude flights.

23.13.4 Radiation Effects on Avionics

The electronic components of aircraft avionic systems are susceptible to damage from the highly ionizing interactions of cosmic rays, solar particles, and the secondary particles generated in the atmosphere. As these components become increasingly smaller, and therefore more susceptible, the risk of damage also increases.

Part 3: Technical Details Relating to Weather Products and Aviation Weather Tools

Types of Aviation Weather Information

This handbook groups aviation weather information into five types:

- Observations,
- Analysis,
- Advisories,
- Forecasts, and
- Aviation Weather Tools.

Observations

Observations are raw weather data collected automatically by sensor(s), manually by trained weather observers, or by a combination of both. The observations can either be in situ (i.e., surface or airborne) or remote (e.g., weather radar, satellite, and lightning).

Analysis

Analysis is the representation of an atmospheric variable (e.g., temperature, pressure) derived from a finite set of irregularly distributed observations onto a regular grid. Analyses of weather information are an enhanced depiction and/or interpretation of observed weather data.

Advisories

Aviation weather advisories, including warnings and alerts, described in this handbook are primarily produced by the NWS. They inform the pilot and/or operator about aviation weather that may be a safety of flight risk, or that may need mitigation or avoidance depending on the aircraft's and/or pilot's certification, current operating limits, or capabilities.

Forecasts

Forecasts are the predictions of the development and/or movement of weather phenomena based on meteorological observations and various mathematical models. This handbook describes the many forecasts primarily produced by the NWS that are either specific to aviation or are public products of interest to aviation users.

Aviation Weather Tools

Aviation weather tools are web-based applications that incorporate multiple weather products into a web-based interactive display.

24 Observations

24.1 Introduction

The first of five types of aviation weather information discussed in this handbook is observations. Observations are weather data collected automatically by sensor(s), manually by trained weather observers, or by a combination of both, and are the basic information upon which forecasts and advisories are made in support of a wide range of weather-sensitive activities within the public and private sectors, including aviation.

For this handbook, observations include the following:

- Surface observations,
- Trained weather observers,
- Aircraft observations,
- Radar observations,
- Satellite observations,
- Upper air observations (e.g., weather balloon),
- Aviation weather cameras.

24.2 Surface Observations

Surface weather observations are fundamental to all meteorological services. Observations are the basic information upon which forecasts and warnings are made in support of a wide range of weather-sensitive activities within the public and private sectors, including aviation.

There are three general types of surface weather observations:

- Manual,
- Automated, and
- Augmented.

24.2.1 Manual Observation

Manual surface weather observations are made by a human weather observer who is certified by the FAA. While manual observations were the primary type prior to the mid-1990s, automated and augmented observations make up the vast majority of today's surface observations in the United States.

24.2.2 Automated Observation

Automated observations are derived from instruments and algorithms without human input or oversight. In the United States, there are two main kinds of automated observing systems: ASOS and AWOS. Automated observations contain "AUTO" in the report unless they are augmented by a human weather observer.

24.2.3 Augmented Observation

At select airports in the United States, the automated observing system will have input and oversight by human weather observers or tower controllers certified in weather observing. These are referred to as augmented observations. Human observers report weather elements that are beyond the capabilities of the automated system and/or are deemed operationally significant. The weather elements observed and reported by the human observer vary, depending on the airport. "AUTO" is not used in augmented reports.

24.2.4 Recency of Observed Elements at Automated Stations

For those elements that the human observer evaluates using spatial averaging techniques (e.g., sky cover and visibility), the automated station substitutes time averaging of sensor data. Therefore, in an automated observation, sky condition is an evaluation of sensor data gathered during the 30-minute period ending at the actual time of the observation. All other elements are based on sensor data that is within 10 minutes or less of the actual time of the observation.

24.3 Automated Surface Observing System (ASOS) and Automated Weather Observing System (AWOS)

24.3.1 Automated Surface Observing System (ASOS)

The ASOS program is a joint effort of the NWS, the FAA, and the DOD. ASOS serves as the nation's primary surface weather observing network.

ASOS detects significant changes, disseminating hourly and special observations. Additionally, ASOS routinely and automatically provides computer-generated voice observations directly to aircraft in the vicinity of airports, using FAA ground-to-air radio. These messages are also available via a telephone. ASOS observes, formats, archives, and transmits observations automatically. ASOS transmits a special report when conditions exceed preselected weather element thresholds (e.g., the visibility decreases to less than 3 mi).

All ASOS locations prepare and disseminate METARs/SPECIs in accordance with the format shown in Section 24.4.3.

24.3.1.1 ASOS One-Minute Observations (OMO)

In addition to the ASOS METARs/SPECIs are ASOS OMOs, which are updated once a minute. OMOs can be in various formats and are sometimes encoded in the METAR format.

ASOS broadcasts can be different from the METAR seen on the internet or FIS-B, since the broadcast is the OMO ASOS data.

The OMOs are not "instant weather;" rather, the clouds and visibility are time averaged (30 minutes for clouds, 10 minutes for visibility). The averaging algorithms are designed to report deteriorating conditions much quicker than improving conditions.

24.3.1.2 ASOS Reporting

ASOS reports the following basic weather elements:

- Sky condition: cloud height and amount (e.g., clear, few, scattered, broken, overcast) up to 12,000 ft (future upgrade may raise height limit).

- Visibility (to at least 10 sm).

- Basic present weather information: type and intensity for rain (**RA**), snow (**SN**), freezing rain (**FZRA**), and unknown precipitation (**UP**).

- Thunderstorms on site (**TS**) or in the vicinity (**VCTS**).

- Obstructions to vision: fog (**FG**), mist (**BR**), and haze (**HZ**).

  Note: **FG** is reported when visibility is less than 5/8 sm. Freezing fog (**FZFG**) is reported when temperature is below 0 °C. **BR** or **HZ** is reported for visibilities from 5/8 sm to less than 7 sm, depending on the difference between the temperature and dewpoint. If the difference is 4 °F (about 2 °C) or less, then **FG** is reported; otherwise, **HZ** is reported.

- Pressure: sea level pressure and altimeter setting.

- Temperature: ambient temperature and dewpoint temperature.

- Wind: direction from which the wind is blowing, speed, and character (e.g., gusts, squalls).

  Note: National network distribution (e.g., FSS, internet, and FIS-B) of wind direction is in true degrees, while local dissemination (e.g., radio and telephone) is in magnetic degrees.

- Precipitation: accumulation.

- Selected significant remarks, including: variable cloud height, variable visibility, precipitation beginning/ending times, rapid pressure changes, pressure change tendency, wind shift, and peak wind, and may include density altitude.

24.3.2 Automated Weather Observing System (AWOS)

AWOS is a system similar to ASOS. Generally, AWOS does not report all the elements as ASOS and may not have the same level of backup sensors or maintenance response levels. Regardless, AWOS provides pilots with the necessary weather information to conduct 14 CFR part 91 flight operations as well as others, depending on their operations specifications (OpSpecs).

AWOS automatically provides computer-generated voice observations directly to aircraft in the vicinity of airports, using FAA ground-to-air radio. AWOS reports are also available via a telephone.

AWOS may be located on airports, at or near ground-based or rooftop-based heliports, as well as on offshore platforms and drill ships.

AWOS are either Federal or non-Federal. Federal AWOS units are owned, operated, and maintained by the FAA. Non-Federal AWOS are owned, operated, and maintained by the site owner.

AWOS generates a METAR at 20-minute intervals and does not report SPECIs. AWOS also provides OMOs available by phone or radio.

The OMOs are not "instant weather;" rather, the clouds and visibility are time averaged (30 minutes for clouds, 10 minutes for visibility). The averaging algorithms are designed to report deteriorating conditions much quicker than improving conditions. For example, if dense fog had been reported and then suddenly dissipated, it might take up to 10 minutes for the OMOs to report VFR conditions.

There are six types of AWOS systems:

- AWOS-A: The AWOS-A system measures and reports altimeter only.
- AWOS-AV: The AWOS-AV consists of an AWOS-A with a visibility sensor.
- AWOS-1: The AWOS-1 system measures and reports wind data (e.g., speed, direction, and gusts; temperature; dewpoint; altimeter; and density altitude).
- AWOS-2: The AWOS-2 system measures and reports all of the parameters of an AWOS-1 system plus visibility.
- AWOS-3: The AWOS-3 system measures and reports all of the parameters of an AWOS-2 system plus precipitation accumulation (rain gauge) and cloud height. AWOS-3 can have optional sensors such as precipitation type/intensity (present weather, P) and/or thunderstorm/lightning (T). The addition of an optional sensors will change the designation to AWOS-3P or AWOS-3PT.
- AWOS-4: The AWOS-4 system measures and reports all of the AWOS-3PT parameters plus freezing rain.

Depending on the type of AWOS unit, the following parameters may be measured:

- Altimeter.
- Wind speed.
- Wind direction (from which the wind is blowing).

    Note: National network distribution (e.g., FSS, internet, and FIS-B) of wind direction is in true degrees, while local dissemination (e.g., radio and telephone) is in magnetic degrees.

- Gusts.
- Temperature.
- Dewpoint.
- Density altitude.
- Visibility.
- Precipitation accumulation.

- Cloud height.
- Precipitation type.
- Precipitation intensity.
- Present weather.
- Thunderstorm/lightning.
- Freezing rain.
- Runway surface condition.

24.4 Aviation Routine Weather Report (METAR) and Aviation Selected Special Weather Report (SPECI)

The METAR and SPECI are the code form used for aviation surface observations (reports) to satisfy World Meteorological Organization (WMO) and ICAO instructions for reporting surface meteorological data. Although the METAR and SPECI code is used worldwide, there are some code differences among countries. Each country is allowed to make modifications to the code for use in their particular country, as long as they notify ICAO. These sections will focus on the METAR/SPECI code as used in the United States.

Traditionally, it was critical for pilots to know how to decode the METAR and SPECI. The majority of current preflight briefing services, including many weather applications, provide the decoded report in a plain language format in addition to the coded report. Table 24-1 provides an example of a decoded METAR. While the decoded version has been welcomed by many pilots, it is still important for all pilots to know how to decode the METAR/SPECI in case the source does not provide a decoded version.

Table 24-1. An Example of a Decoded METAR for an Aviation Weather Website

+	KPNS (Pensacola Intl, FL, US)
Text:	**KPNS 030053Z 36005KT 10SM CLR 30/23 A2992 RMK AO2 SLP132 T03000233**
Temperature:	30.0 °C (86 °F)
Dewpoint:	23.3 °C (74 °F) [RH = 67%]
Pressure (altimeter):	29.92 inHg (1013.3 mb) [Sea level pressure: 1013.2 mb]
Winds:	from the N (360 degrees) at 6 MPH (5 knots; 2.6 m/s)
Visibility:	10 or more sm (16+ km)
Ceiling:	at least 12,000 ft AGL
Clouds:	sky clear below 12,000 ft AGL

24.4.1 Aviation Routine Weather Report (METAR)

A METAR includes the airport identifier, time of observation, wind, visibility, Runway Visual Range (RVR), present weather phenomena, sky conditions, temperature, dewpoint, and altimeter setting. Excluding the airport identifier and the time of observation, this information is collectively referred to as the "body" of the report. Coded and/or plain language information elaborating on data in the body may be appended to the end of the METAR as "remarks." The contents of the remarks section vary with manual, automated, and augmented surface observations. At some designated stations, the METAR may be abridged to include only a few of the mentioned elements. METARs are sometimes referred to as "hourly" reports since they are routinely produced near the top of the hour.

24.4.2 Aviation Selected Special Weather Report (SPECI)

A SPECI is an unscheduled report taken when any of the criteria given in Table 24-2 are observed during the period between hourly reports. SPECIs contain all data elements found in a METAR. All SPECIs are issued as soon as possible when relevant criteria are observed.

Whenever SPECI criteria are met at the time of the routine METAR, a METAR is issued.

Table 24-2. SPECI Criteria

1	Wind Shift	Wind direction changes by 45° or more, in less than 15 minutes, and the wind speed is 10 knots or more throughout the wind shift.
2	Visibility	Visibility as reported in the body of the report decreases to less than, or if below, increases to equal or exceed: • 3 miles. • 2 miles. • 1 mile. • ½ mile. • ¼ mile. • The lowest standard IAP minimum as published in the U.S. Terminal Procedures, if not listed above.
3	RVR	The highest value from the designated RVR runway decreases to less than 2,400 ft during the preceding 10 minutes; or, if the RVR is below 2,400 ft, increases to equal to or exceed 2,400 ft during the preceding 10 minutes. U.S. military stations may not report a SPECI based on RVR.
4	Tornado, Funnel Cloud, or Waterspout	• Is observed. • Disappears from sight, or ends.
5	Thunderstorm	• Begins (a SPECI is not required to report the beginning of a new thunderstorm if one is currently reported). • Ends.
6	Precipitation	• Hail begins or ends. • Freezing precipitation begins, ends, or changes intensity. • Ice pellets begin, end, or change intensity. • Snow begins, ends, or changes intensity.
7	Squalls	When a squall occurs.

8	Ceiling	The ceiling changes[1] through: • 3,000 ft. • 1,500 ft. • 1,000 ft. • 500 ft. • The lowest standard IAP minimum.[2] [1] "Ceiling change" means that it forms, dissipates below, decreases to less than, or, if below, increases to equal or exceed the values listed. [2] As published in the U.S. Terminal Procedures. If none published, use 200 ft.
9	Sky Condition	A layer of clouds or obscurations aloft is present below 1,000 ft and no layer aloft was reported below 1,000 ft in the preceding METAR or SPECI.
10	Volcanic Eruption	When an eruption is first noted.
11	Aircraft Mishap	Upon notification of an aircraft mishap,[1] unless there has been an intervening observation. [1] "Aircraft mishap" is an inclusive term to denote the occurrence of an aircraft accident or incident.
12	Miscellaneous	Any other meteorological situation designated by the responsible agency of which, in the opinion of the observer, is critical.

24.4.3 METAR/SPECI Format

A U.S. METAR/SPECI has two major sections: the body (consisting of a maximum of 11 groups) and the remarks (consisting of 2 categories). When an element does not occur, or cannot be observed, the corresponding group is omitted from that particular report. See Figure 24-1 for the format.

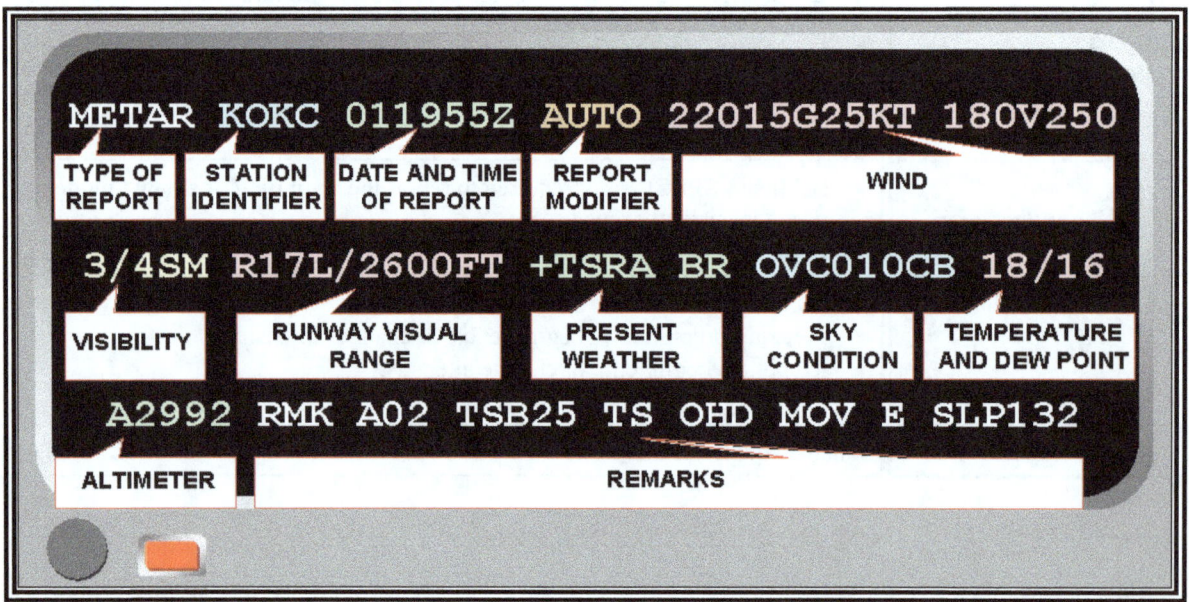

Figure 24-1. METAR/SPECI Coding Format

Chapter 24, Observations

24.4.3.1 Type of Report

METAR KOKC 011955Z AUTO 22015G25KT 180V250 3/4SM R17L/2600FT +TSRA BR OVC010CB 18/16 A2992 RMK AO2 TSB25 TS OHD MOV E SLP132

The type of report, **METAR** or **SPECI**, precedes the body of all reports, but may not be shown or displayed on all aviation weather websites.

24.4.3.2 Station Identifier

METAR **KOKC** 011955Z AUTO 22015G25KT 180V250 3/4SM R17L/2600FT +TSRA BR OVC010CB 18/16 A2992 RMK AO2 TSB25 TS OHD MOV E SLP132

The station identifier, in ICAO format, is included in all reports to identify the station to which the coded report applies.

The ICAO airport code is a four-letter alphanumeric code designating each airport around the world. The ICAO codes are used for flight planning by pilots and airline operation departments. These codes are not the same as the International Air Transport Association (IATA) codes encountered by the general public used for reservations, baggage handling, and in airline timetables.

Unlike the IATA codes, the ICAO codes have a regional structure. The first letter identifies the region and country (see Figure 24-2). In some regions, the second letter identifies the country. ICAO station identifiers in Alaska begin with "PA," Hawaii begins with "PH," Guam begins with "PG," and Puerto Rico begins with "TS." For example, the San Juan, Puerto Rico, IATA identifier "SJU" becomes the ICAO identifier "TSJU." The remaining letters are used to identify each airport.

In the CONUS, ICAO station identifiers are coded **K**, followed by the three-letter IATA identifier. For example, the Seattle, WA, IATA identifier "SEA" becomes the ICAO identifier "KSEA."

ICAO station identifiers in Alaska, Hawaii, and Guam begin with the continent code **P**.

For a list of all U.S. identifiers, refer to FAA Order JO 7350.9, Location Identifiers. For a complete worldwide listing, refer to ICAO Doc 7910, Location Indicators. Both are available online.

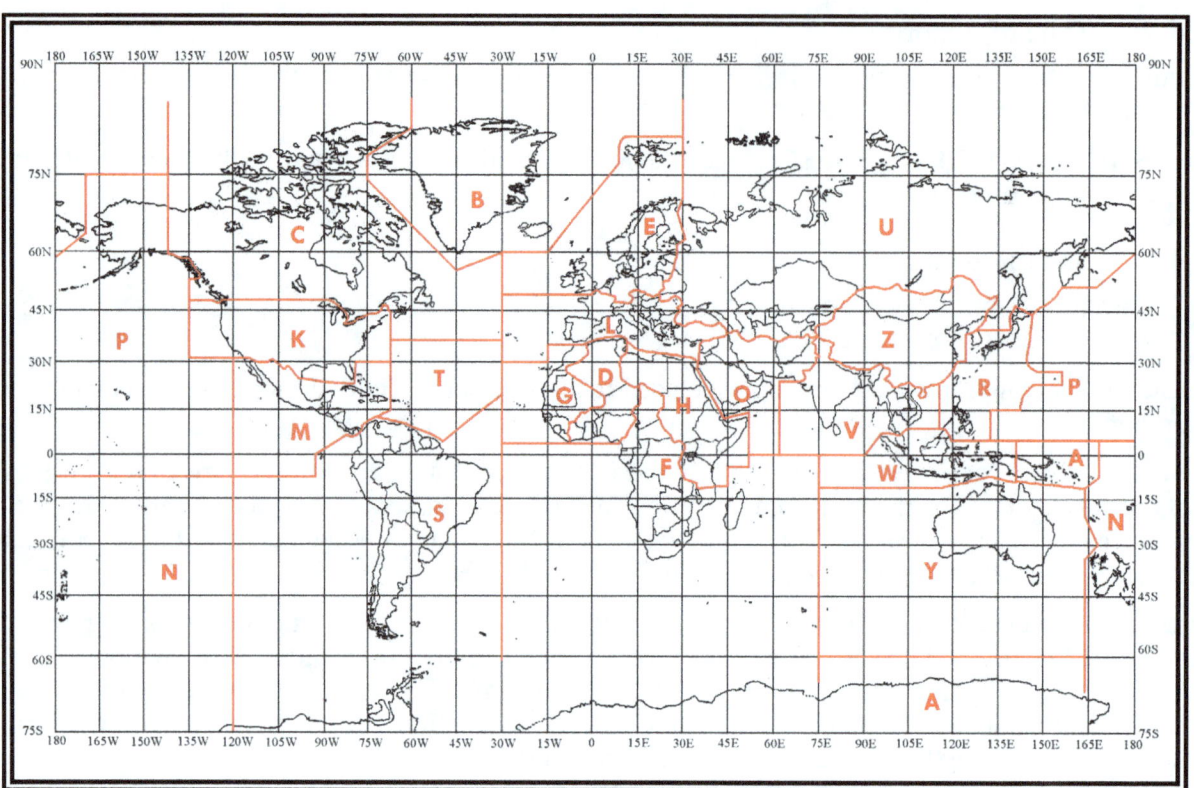

Figure 24-2. ICAO Continental Codes

24.4.3.3 Date and Time of Report

```
METAR KOKC 011955Z AUTO 22015G25KT 180V250 3/4SM R17L/2600FT +TSRA BR OVC010CB
18/16 A2992 RMK AO2 TSB25 TS OHD MOV E SLP132
```

The date and time are coded in all reports as follows: the day of the month is the first two digits (**01**), followed by the hour (**19**), and the minutes (**55**).

The coded time of observations is the actual time of the report, or when the criteria for a SPECI is met or noted.

If the report is a correction to a previously disseminated report, the time of the corrected report is the same time used in the report being corrected.

The date and time group always ends with a **Z**, indicating Zulu time (or Coordinated Universal Time (UTC)).

For example, **METAR KOKC 011955Z** would be disseminated as the 2,000-hour routine report for station KOKC, taken on the 1st of the month at 1955 UTC.

24.4.3.4 Report Modifier (as appropriate)

```
METAR KOKC 011955Z AUTO 22015G25KT 180V250 3/4SM R17L/2600FT +TSRA BR OVC010CB
18/16 A2992 RMK AO2 TSB25 TS OHD MOV E SLP132
```

The report modifier **AUTO** identifies the METAR/SPECI as a fully automated report with no human intervention or oversight. In the event of a corrected METAR or SPECI, the report modifier "COR" is substituted for "AUTO."

Chapter 24, Observations

24.4.3.5 Wind Group

```
METAR KOKC 011955Z AUTO 22015G25KT 180V250 3/4SM R17L/2600FT +TSRA BR OVC010CB
18/16 A2992 RMK AO2 TSB25 TS OHD MOV E SLP132
```

Wind is the horizontal motion of air past a given point. It is measured in terms of velocity, which is a vector that includes direction and speed. It indicates the direction the wind is coming from.

In the wind group, the wind direction is coded as the first three digits (**220**) and is determined by averaging the recorded wind direction over a 2-minute period. It is coded in tens of degrees relative to true north using three figures. Directions less than 100° are preceded with a 0. For example, a wind direction of 90° is coded as **090**. A wind from the north is coded as **360**.

Immediately following the wind direction is the wind speed coded in two or three digits (**15**). Wind speed is determined by averaging the speed over a 2-minute period and is coded in whole knots using the units, tens digits, and, when appropriate, the hundreds digit. When wind speeds are less than 10 kt, a leading 0 is used to maintain at least a two-digit wind code. For example, a wind speed of 8 kt will be coded **08KT**. The wind group is always coded with a **KT** to indicate wind speeds are reported in knots. Other countries may use km/h or meters per second (m/s) instead of knots.

Examples:

`05008KT`	Wind 50° at 8 kt
`15014KT`	Wind 150° at 14 kt
`340112KT`	Wind 340° at 112 kt

24.4.3.5.1 Wind Gust

Wind speed data for the most recent 10 minutes is examined to evaluate the occurrence of gusts. Gusts are defined as rapid fluctuations in wind speed with a variation of 10 kt or more between peaks and lulls. The coded speed of the gust is the maximum instantaneous wind speed.

Wind gusts are coded in two or three digits immediately following the wind speed. Wind gusts are coded in whole knots using the units, tens, and, when appropriate, the hundreds digit. For example, a wind out of the west at 20 kt with gusts to 35 kt would be coded **27020G35KT**.

24.4.3.5.2 Variable Wind Direction (speed 6 kt or less)

Wind direction may be considered variable when, during the previous 2-minute evaluation period, the wind speed was 6 kt or less. In this case, the wind may be coded as **VRB** in place of the three-digit wind direction. For example, if the wind speed was recorded as 3 kt, it would be coded **VRB03KT**.

24.4.3.5.3 Variable Wind Direction (speed greater than 6 kt)

Wind direction may also be considered variable when, during the 2-minute evaluation period, it varies by 60° or more and the speed is greater than 6 kt. In this case, a variable wind direction group immediately follows the wind group. The directional variability is coded in a clockwise direction and consists of the extremes of the wind directions separated by a **V**. For example, if the wind is variable from 180 to 240° at 10 kt, it would be coded **21010KT 180V240**.

24.4.3.5.4 Calm Wind

When no motion of air is detected, the wind is reported as calm. A calm wind is coded as **00000KT**.

24.4.3.6 Visibility Group

```
METAR KOKC 011955Z AUTO 22015G25KT 180V250 3/4SM R17L/2600FT +TSRA BR OVC010CB
18/16 A2992 RMK AO2 TSB25 TS OHD MOV E SLP132
```

Visibility is a measure of the opacity of the atmosphere. It is defined as the greatest horizontal distance at which selected objects can be seen and identified, or its equivalent derived from instrumental measurements.

Prevailing visibility is the reported visibility considered representative of recorded visibility conditions at the manual station during the time of observation. It is the greatest distance that can be seen throughout at least half of the horizon circle, not necessarily continuous.

Surface visibility is the prevailing visibility from the surface at manual stations or the visibility derived from sensors at automated stations.

The visibility group is coded as the surface visibility in statute miles. A space is coded between whole numbers and fractions of reportable visibility values. The visibility group ends with **SM** to indicate that the visibility is in statute miles. For example, a visibility of 1½ sm is coded **1 1/2SM**. Most other countries use meters.

U.S. automated stations use an **M** to indicate "less than." For example, **M1/4SM** means a visibility of less than ¼ sm.

24.4.3.7 Runway Visual Range (RVR) Group

```
METAR KOKC 011955Z AUTO 22015G25KT 180V250 3/4SM R17L/2600FT +TSRA BR OVC010CB
18/16 A2992 RMK AO2 TSB25 TS OHD MOV E SLP132
```

The RVR is an instrument-derived value representing the horizontal distance a pilot may see down the runway.

RVR is reported whenever the station has RVR equipment and prevailing visibility is 1 sm or less, and/or the RVR for the designated instrument runway is 6,000 ft or less. Otherwise, the RVR group is omitted.

RVR is coded in the following format: The initial **R** is code for runway and is followed by the runway number. When more than one runway is defined with the same runway number, a directional letter is coded on the end of the runway number. Next is a solidus (/) followed by the visual range in feet, and then **FT** completes the RVR report. For example, an RVR value for Runway 01L of 800 ft would be coded **R01L/0800FT**. Most other countries use meters.

In the United States, RVR values are coded in increments of 100 ft up to 1,000 ft, increments of 200 ft from 1,000 to 3,000 ft, and increments of 500 ft from 3,000 to 6,000 ft. Manual RVR is not reported below 600 ft.

For U.S. airports only, the touchdown zone's (TDZ) RVR is reported. For U.S. airports with multiple runways, the operating runway with the lowest touchdown RVR is reported. RVR may be reported for up to four designated runways in other countries.

When the RVR varies by more than one reportable value, the lowest and highest values will be shown with **V** between them, indicating variable conditions. For example, the 10-minute RVR for Runway 01L varying between 600 and 1,000 ft would be coded **R01L/0600V1000FT**.

If RVR is less than its lowest reportable value, the visual range group is preceded by **M**. For example, an RVR for Runway 01L of less than 600 ft is coded **R01L/M0600FT**.

If RVR is greater than its highest reportable value, the visual range group is preceded by a **P**. For example, an RVR for Runway 27 of greater than 6,000 ft will be coded **R27/P6000FT**.

24.4.3.8 Present Weather Group

```
METAR KOKC 011955Z AUTO 22015G25KT 180V250 3/4SM R17L/2600FT +TSRA BR OVC010CB
18/16 A2992 RMK AO2 TSB25 TS OHD MOV E SLP132
```

Present weather includes precipitation, obscurations, and other weather phenomena. The appropriate notations found in Table 24-3 are used to code present weather.

Table 24-3. METAR/SPECI Notations for Reporting Present Weather[1]

Qualifier		Weather Phenomena		
Intensity or Proximity **1**	**Descriptor** **2**	**Precipitation** **3**	**Obscuration** **4**	**Other** **5**
- Light Moderate[2] + Heavy VC In the Vicinity[3]	MI Shallow PR Partial BC Patches DR Low Drifting BL Blowing SH Shower(s) TS Thunderstorms FZ Freezing	DZ Drizzle RA Rain SN Snow SG Snow Grains IC Ice Crystals (Diamond Dust) PL Ice Pellets GR Hail GS Snow Pellets UP Unknown Precipitation	BR Mist FG Fog FU Smoke VA Volcanic Ash DU Widespread Dust SA Sand HZ Haze PY Spray	PO Dust/Sand Whirls SQ Squalls FC Funnel Cloud, Tornado, or Waterspout[4] SS Sandstorm DS Dust Storm

1. The weather groups are constructed by considering columns 1 through 5 in Table 24-3 in sequence (i.e., intensity followed by description, followed by weather phenomena). For example, heavy rain shower(s) is coded as +SHRA.
2. To denote moderate intensity, no entry or symbol is used.
3. See text for vicinity definitions.
4. Tornadoes and waterspouts are coded as +FC.

Separate groups are used for each type of present weather. Each group is separated from the other by a space. METARs/SPECIs contain no more than three present weather groups.

When more than one type of present weather is reported at the same time, present weather is reported in the following order:

- Tornadic activity (tornado, funnel cloud, or waterspout).
- Thunderstorm(s) (with and without associated precipitation).
- Present weather in order of decreasing dominance (i.e., the most dominant type reported first).
- Left to right in Table 24-3 (columns 1 through 5).

Qualifiers may be used in various combinations to describe weather phenomena. Present weather qualifiers fall into two categories:

- Intensity or proximity, and
- Descriptors.

24.4.3.8.1 Intensity Qualifier

The intensity qualifiers are light, moderate, and heavy. They are coded with precipitation types, except ice crystals (**IC**) and hail (**GR**), including those associated with a thunderstorm (**TS**) and those of a showery nature (**SH**). Tornadoes and waterspouts are coded as heavy (**+FC**). No intensity is ascribed to the obscurations of blowing dust (**BLDU**), blowing sand (**BLSA**), and blowing snow (**BLSN**). Only moderate or heavy intensity is ascribed to sandstorm (**SS**) and dust storm (**DS**).

When more than one form of precipitation is occurring at a time, or precipitation is occurring with an obscuration, the reported intensities are not cumulative. The reported intensity will not be greater than the intensity for each form of precipitation. For example, **-FZRAPL** is light freezing rain and light ice pellets, *not* light freezing rain and moderate ice pellets.

24.4.3.8.2 Proximity Qualifier

Weather phenomena occurring beyond the point of observation (between 5 and 10 sm) are coded as in the vicinity (**VC**). **VC** can be coded in combination with thunderstorm (**TS**), fog (**FG**), shower(s) (**SH**), well-developed dust/sand whirls (**PO**), blowing dust (**BLDU**), blowing sand (**BLSA**), blowing snow (**BLSN**), sandstorm (**SS**), and dust storm (**DS**). Intensity qualifiers are not coded in conjunction with **VC**.

For example, **VCFG** can be decoded as meaning some form of fog is between 5 and 10 sm of the point of observation. If **VCSH** is coded, showers are occurring between 5 and 10 sm of the point of observation.

Weather phenomena occurring at the point of observation (at the station) or in the vicinity of the point of observation are coded in the body of the report. Weather phenomena observed beyond 10 sm from the point of observation (at the station) is not coded in the body, but may be coded in the remarks section.

24.4.3.8.3 Descriptor Qualifier

Descriptors are qualifiers that further amplify weather phenomena and are used in conjunction with some types of precipitation and obscurations. The descriptor qualifiers are: shallow (**MI**), partial (**PR**), patches (**BC**), low drifting (**DR**), blowing (**BL**), shower(s) (**SH**), thunderstorm (**TS**), and freezing (**FZ**).

Only one descriptor is coded for each weather phenomena group (e.g., **FZDZ**).

The descriptors shallow (**MI**), partial (**PR**), and patches (**BC**) are only coded with fog (**FG**) (e.g., **MIFG**). Mist (**BR**) is not coded with any descriptor.

The descriptors low drifting (**DR**) and blowing (**BL**) will only be coded with dust (**DU**), sand (**SA**), and snow (**SN**) (e.g., **BLSN** or **DRSN**). **DR** is coded with **DU**, **SA**, or **SN** for raised particles drifting less than 6 ft above the ground.

When blowing snow is observed with snow falling from clouds, both phenomena are reported (e.g., **SN BLSN**). If blowing snow is occurring and the observer cannot determine whether or not snow is also falling, then **BLSN** is reported. Spray (**PY**) is coded only with blowing (**BL**).

The descriptor for showery-type precipitation (**SH**) is coded only with one or more of the precipitation qualifiers for rain (**RA**), snow (**SN**), ice pellets (**PL**), or hail (**GR**). When any type of precipitation is coded with **VC**, the intensity and type of precipitation is not coded.

The descriptor for thunderstorm (**TS**) may be coded by itself when the thunderstorm is without associated precipitation. A thunderstorm may also be coded with the precipitation types of rain (**RA**), snow (**SN**), ice pellets (**PL**), snow pellets (**GS**), or hail (**GR**). **TS** is not coded with **SH**.

The descriptor freezing (**FZ**) is only coded in combination with fog (**FG**), drizzle (**DZ**), or rain (**RA**) (e.g., **FZRA**). **FZ** is not coded with **SH**.

24.4.3.8.4 Precipitation

Precipitation is any form of water particle, whether liquid or solid, that falls from the atmosphere and reaches the ground. The precipitation types are: drizzle (**DZ**), rain (**RA**), snow (**SN**), snow grains (**SG**), ice crystals (**IC**), ice pellets (**PL**), hail (**GR**), snow pellets (**GS**), and unknown precipitation (**UP**). **UP** is reported if an automated station detects the occurrence of precipitation, but the precipitation sensor cannot recognize the type.

Up to three types of precipitation may be coded in a single present weather group. They are coded in order of decreasing dominance based on intensity.

24.4.3.8.5 Obscuration

Obscurations are any phenomenon in the atmosphere, other than precipitation, that reduces the horizontal visibility in the atmosphere. The obscuration types are: mist (**BR**), fog (**FG**), smoke (**FU**), volcanic ash (**VC**), widespread dust (**DU**), sand (**SA**), haze (**HZ**), and spray (**PY**). Spray (**PY**) is coded only as **BLPY**.

With the exception of volcanic ash, low drifting dust, low drifting sand, low drifting snow, shallow fog, partial fog, and patches (of) fog, an obscuration is coded in the body of the report if the surface visibility is less than 7 mi, or considered operationally significant. Volcanic ash is always reported when observed.

24.4.3.8.6 Other Weather Phenomena

Other weather phenomenon types include: well-developed dust/sand whirls (**PO**), sandstorms (**SS**), dust storms (**DS**), squalls (**SQ**), funnel clouds (**FC**), and tornados and waterspouts (**+FC**).

Examples:

`-DZ`	Light drizzle.
`-RASN`	Light rain and (light) snow.
`SN BR`	(Moderate) snow, mist.
`-FZRA FG`	Light freezing rain, fog.
`SHRA`	(Moderate) rain shower.
`VCBLSA`	Blowing sand in the vicinity.
`-RASN FG HZ`	Light rain and (light) snow, fog, haze.
`TS`	Thunderstorm (without precipitation).

+TSRA   Thunderstorm, heavy rain.

+FC TSRAGR BR   Tornado, thunderstorm, (moderate) rain, hail, mist.

24.4.3.9 Sky Condition Group

```
METAR KOKC 011955Z AUTO 22015G25KT 180V250 3/4SM R17L/2600FT +TSRA BR OVC010CB
18/16 A2992 RMK AO2 TSB25 TS OHD MOV E SLP132
```

Sky condition is a description of the appearance of the sky. It includes cloud cover, vertical visibility, or clear skies.

The sky condition group is based on the amount of cloud cover (the first three letters) followed by the height of the base of the cloud cover (final three digits). No space is between the amount of cloud cover and the height of the layer. The height of the layer is recorded in feet AGL.

Sky condition is coded in ascending order and ends at the first overcast layer. At mountain stations, if the layer is below station level, the height of the layer will be coded with three solidi (///).

Vertical visibility is coded as **VV**, followed by the vertical visibility into the indefinite ceiling. An "indefinite ceiling" is a ceiling classification applied when the reported ceiling value represents the vertical visibility upward into surface-based obscuration. No space is between the group identifier and the vertical visibility. Figure 24-3 illustrates the effect of an obscuration on the vision from a descending aircraft.

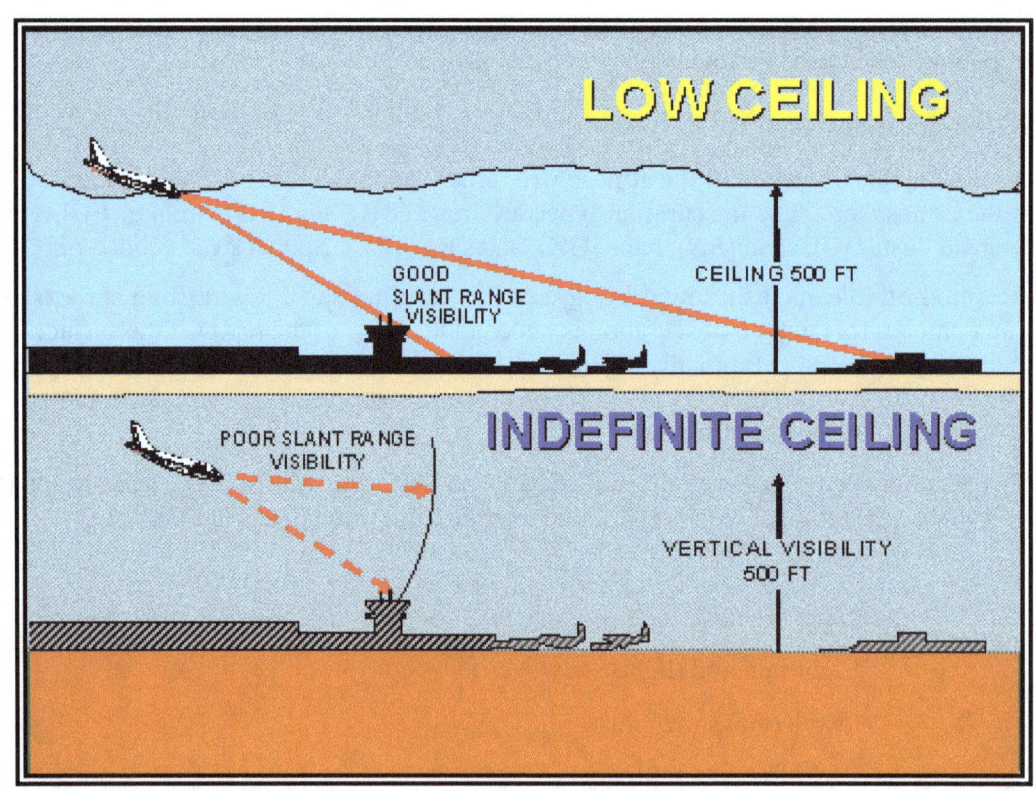

The ceiling is 500 ft in both examples, but the indefinite ceiling example (bottom) produces a more adverse impact to landing aircraft. This is because an obscuration (e.g., fog, blowing dust, snow) limits runway acquisition due to reduced slant range visibility. This pilot would be able to see the ground, but not the runway. If the pilot was at approach minimums, the approach could not be continued and a missed approach would need to be executed.

Figure 24-3. Obscuration Effects on Slant Range Visibility

Clear skies are coded in the format **SKC** or **CLR**. When **SKC** is used, an observer indicates no layers are present; **CLR** is used by automated stations to indicate no layers are detected at or below 12,000 ft.

Each coded layer is separated from the others by a space. Each layer reported is coded by using the appropriate reportable contraction seen in Table 24-4. A report of clear skies (**SKC** or **CLR**) is a complete layer report within itself. The abbreviations **FEW**, **SCT**, **BKN**, and **OVC** will be followed (without a space) by the height of the layer.

Table 24-4. METAR/SPECI Contractions for Sky Cover

Reportable Contraction	Meaning	Summation Amount of Layer
VV	Vertical Visibility	8/8
SKC or CLR[1]	Clear	0
FEW[2]	Few	1/8 – 2/8
SCT	Scattered	3/8 – 4/8
BKN	Broken	5/8 – 7/8
OVC	Overcast	8/8

1. *The abbreviation **CLR** will be used at automated stations when no layers at or below 12,000 ft are reported; the abbreviation **SKC** will be used at manual stations when no layers are reported.*
2. *Any layer amount less than 1/8 is reported as **FEW**.*

The height is coded in hundreds of feet above the surface using three digits in accordance with Table 24-5.

Table 24-5. METAR/SPECI Increments of Reportable Values of Sky Cover Height

Range of Height Values (ft)	Reportable Increment (ft)
Less than or equal to 5,000	To nearest 100
5,001 to 10,000	To nearest 500
Greater than 10,000	To nearest 1,000

The ceiling is the lowest layer aloft reported as broken or overcast. If the sky is totally obscured with ground-based clouds, the vertical visibility is the ceiling.

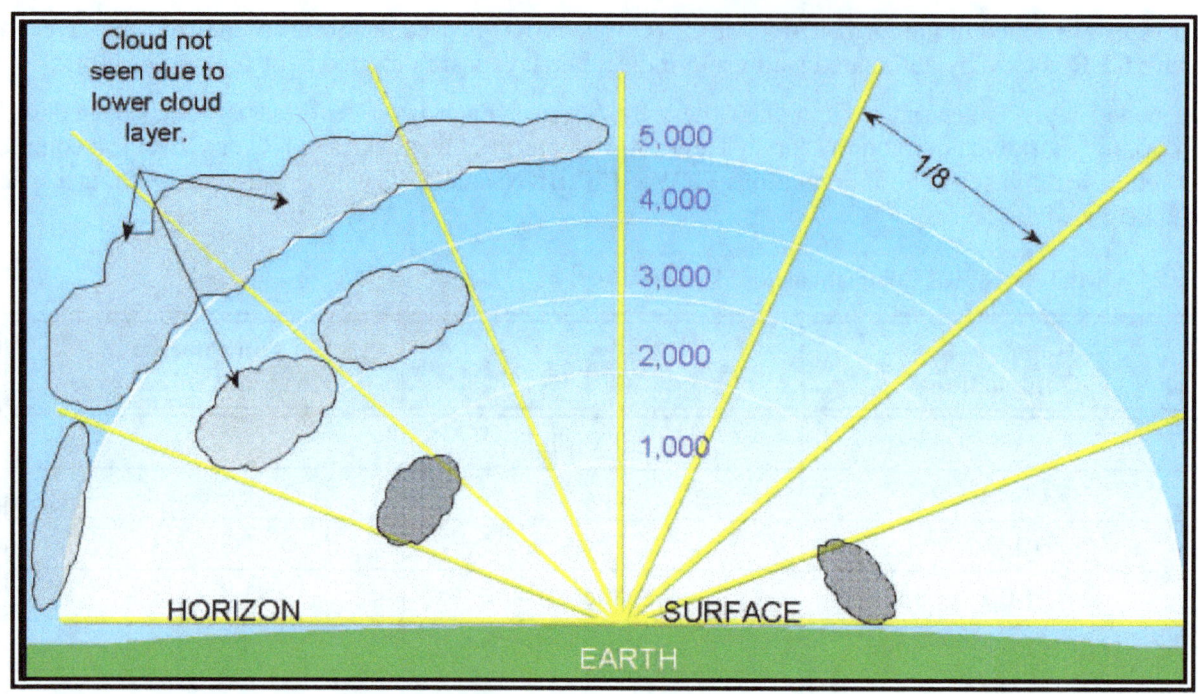

Clouds at 1,200 ft obscure 2/8 of the sky (FEW). Higher clouds at 3,000 ft obscure an additional 1/8 of the sky, and because the observer cannot see above the 1,200-ft layer, he is to assume that the higher 3,000-ft layer also exists above the lower layer (SCT). The highest clouds at 5,000 ft obscure 2/8 of the sky, and again since the observer cannot see past the 1,200 and 3,000-ft layers, he is to assume the higher 5,000-ft layer also exists above the lower layers (BKN). The sky condition group would be coded as: FEW012 SCT030 BKN050.

Figure 24-4. METAR/SPECI Sky Condition Coding

At manual stations, cumulonimbus (**CB**) or towering cumulus (**TCU**) is appended to the associated layer. For example, a scattered layer of towering cumulus at 1,500 ft would be coded **SCT015TCU**, and would be followed by a space if there were additional higher layers to code.

Examples:

SKC	No layers are present.
CLR	No layers are detected at or below 12,000 ft AGL.
FEW004	Few at 400 ft AGL.
SCT023TCU	Scattered layer of towering cumulus at 2,300 ft AGL.
BKN100	Broken layer (ceiling) at 10,000 ft AGL.
OVC250	Overcast layer (ceiling) at 25,000 ft AGL.
VV001	Indefinite ceiling with a vertical visibility of 100 ft AGL.
FEW012 SCT046	Few clouds at 1,200 ft, scattered layer at 4,600 ft AGL.
SCT033 BKN085	Scattered layer at 3,300 ft, broken layer (ceiling) at 8,500 ft AGL.

```
SCT018 OVC032CB          Scattered layer at 1,800 ft AGL, overcast layer (ceiling) of cumulonimbus
                         at 3,200 ft AGL.

SCT009 SCT024 BKN048     Scattered layer at 900 ft AGL, scattered layer at 2,400 ft AGL, broken layer
                         (ceiling) at 4,800 ft AGL.
```

24.4.3.10 Temperature/Dewpoint Group

```
METAR KOKC 011955Z AUTO 22015G25KT 180V250 3/4SM R17L/2600FT +TSRA BR OVC010CB
18/16 A2992 RMK AO2 TSB25 TS OHD MOV E SLP132
```

Temperature is the degree of hotness or coldness of the ambient air, as measured by a suitable instrument. Dewpoint is the temperature to which a given parcel of air must be cooled at constant pressure and constant water vapor content for the air to become fully saturated.

Temperature and dewpoint are coded as two digits rounded to the nearest whole degree Celsius. For example, a temperature of 0.3 °C would be coded at **00**. Sub-zero temperatures and dewpoints are prefixed with an **M**. For example, a temperature of 4 °C with a dewpoint of -2 °C would be coded as **04/M02**; a temperature of -2 °C would be coded as **M02**.

If temperature is not available, the entire temperature/dewpoint group is not coded. If dewpoint is not available, temperature is coded followed by a solidus (/) and no entry is made for dewpoint. For example, a temperature of 1.5 °C and a missing dewpoint would be coded as **02/**.

24.4.3.11 Altimeter Group

```
METAR KOKC 011955Z AUTO 22015G25KT 180V250 3/4SM R17L/2600FT +TSRA BR OVC010CB
18/16 A2992 RMK AO2 TSB25 TS OHD MOV E SLP132
```

The altimeter setting group codes the current pressure at elevation. This setting is then used by aircraft altimeters to determine the true altitude above a fixed plane of MSL.

The altimeter group always starts with an **A** and is followed by the four-digit group representing the pressure in tens, units, tenths, and hundredths of inches of mercury. The decimal point is not coded. For example, an altimeter setting of 29.92 inHg would be coded as **A2992**.

24.4.3.12 Remarks (RMK)

```
METAR KOKC 011955Z AUTO 22015G25KT 180V250 3/4SM R17L/2600FT +TSRA BR OVC010CB
18/16 A2992 RMK AO2 TSB25 TS OHD MOV E SLP132
```

Remarks are included in METAR and SPECI, when appropriate.

Remarks are separated from the body of the report by the contraction **RMK**. When no remarks are necessary, the contraction **RMK** is not used.

METAR/SPECI remarks fall into two categories (see Table 24-6):

- Automated, Manual, and Plain Language; and
- Additive and Automated Maintenance Data.

Table 24-6. METAR/SPECI Order of Remarks

Automated, Manual, and Plain Language				Additive and Automated Maintenance Data	
1.	Volcanic Eruptions	14.	Hailstone Size	27.	Precipitation Amount Within a Specified Time Period*
2.	Funnel Cloud	15.	Virga	28.	Cloud Types*
3.	Type of Automated Station	16.	Variable Ceiling Height	29.	Duration of Sunshine*
4.	Peak Wind	17.	Obscurations	30.	Hourly Temperature and Dewpoint
5.	Wind Shift	18.	Variable Sky Condition	31.	6-Hourly Maximum Temperature*
6.	Tower or Surface Visibility	19.	Significant Cloud Types	32.	6-Hourly Minimum Temperature*
7.	Variable Prevailing Visibility	20.	Ceiling Height at Second Location	33.	24-Hour Maximum and Minimum Temperature*
8.	Sector Visibility	21.	Pressure Rising or Falling Rapidly	34.	3-Hourly Pressure Tendency*
9.	Visibility at Second Location	22.	Sea Level Pressure	35.	Sensor Status Indicators
10.	Lightning	23.	Aircraft Mishap	36.	Maintenance Indicator
11.	Beginning and Ending of Precipitation	24.	No SPECI Reports Taken	Note: Additive data is primarily used by the NWS for climatological purposes. *These groups should have no direct impact on the aviation community and will not be discussed in this document.	
12.	Beginning and Ending of Thunderstorms	25.	Snow Increasing Rapidly		
13.	Thunderstorm Location	26.	Other Significant Information		

Remarks are made in accordance with the following:

- Time entries are made in minutes past the hour if the time reported occurs during the same hour the observation is taken. Hours and minutes are used if the hour is different.

- Present weather coded in the body of the report as **VC** may be further described (e.g., direction from the station, if known). Weather phenomena beyond 10 sm of the point(s) of observation are coded as distant (**DSNT**) followed by the direction from the station. For example, precipitation of unknown intensity within 10 sm east of the station would be coded as **VCSH E**; lightning 25 sm west of the station would be coded as **LTG DSNT W**.

- Distance remarks are in statute miles, except for automated lightning remarks, which are in nautical miles.

- Movement of clouds or weather, when known, is coded with respect to the direction toward which the phenomena are moving. For example, a thunderstorm moving toward the northeast would be coded as **TS MOV NE**.

- Directions use the eight points of the compass coded in a clockwise order.
- Insofar as possible, remarks are entered in the order they are presented in the following sections (and Table 24-6).

24.4.3.13 Automated, Manual, and Plain Language Remarks

These remarks generally elaborate on parameters reported in the body of the report. An automated station or observer may generate automated and manual remarks. Only an observer can provide plain language remarks.

24.4.3.13.1 Volcanic Eruptions

Volcanic eruptions are coded in plain language and contain the following, when known:

- **Name** of volcano;
- **Latitude and longitude**, or the direction and approximate distance from the station;
- **Date/time** (UTC) of the eruption;
- Size **description**, approximate height, and direction of movement **of the ash cloud**; and
- Any **other pertinent data** about the eruption.

For example, a remark on a volcanic eruption would look like the following:

```
RMK MT. AUGUSTINE VOLCANO 70 MILES SW ERUPTED AT 231505 LARGE ASH CLOUD EXTENDING
TO APRX 30000 FEET MOVING NE.
```

Pre-eruption volcanic activity is not coded. Pre-eruption refers to unusual and/or increasing volcanic activity that could presage a volcanic eruption.

24.4.3.13.2 Funnel Cloud

At manual stations, tornadoes, funnel clouds, and waterspouts are coded in the following format: tornadic activity, **TORNADO, FUNNEL CLOUD**, or **WATERSPOUT**, followed by the beginning and/or ending time, followed by the location and/or direction of the phenomena from the station, and/or movement, when known. For example, **TORNADO B13 6 NE** would indicate that a tornado began at 13 minutes past the hour and was 6 sm northeast of the station.

24.4.3.13.3 Type of Automated Station

AO1 or **AO2** is coded in all METARs/SPECIs from automated stations. Automated stations without a precipitation discriminator are identified as **AO1**; automated stations with a precipitation discriminator are identified as **AO2**.

24.4.3.13.4 Peak Wind

Peak wind is coded in the following format: the remark identifier **PK WND**, followed by the direction of the wind (first three digits), peak wind speed (next two or three digits) since the last METAR, and the time of occurrence. A space is between the two elements of the remark identifier and the wind direction/speed group; a solidus (/), without spaces, separates the wind direction/speed group and the time. For example, a peak wind of 45 kt from 280° that occurred at 15 minutes past the hour is coded **PK WND 28045/15**.

24.4.3.13.5 Wind Shift

Wind shift is coded in the following format: the remark identifier **WSHFT**, followed by the time the wind shift began. The contraction **FROPA** is entered following the time if there is reasonable data to consider the wind shift was the result of a frontal passage. A space is between the remark identifier and the time and, if applicable, between the time and the frontal passage contraction. For example, a remark reporting a wind shift accompanied by a frontal passage that began at 30 minutes past the hour would be coded **WSHFT 30 FROPA**.

24.4.3.13.6 Tower or Surface Visibility

Tower or surface visibility is coded in the following format: tower (**TWR VIS**) or surface (**SFC**), followed by the observed tower/surface visibility value. A space is coded between each of the remark elements. For example, the control tower visibility of 1½ sm would be coded **TWR VIS 1 1/2**.

24.4.3.13.7 Variable Prevailing Visibility

Variable prevailing visibility is coded in the following format: the remark identifier **VIS**, followed by the lowest and highest visibilities evaluated, separated by the letter **V**. A space follows the remark identifier, and no spaces are between the letter **V** and the lowest/highest values. For example, a visibility that was varying between ½ and 2 sm would be coded **VIS 1/2V2**.

24.4.3.13.8 Sector Visibility

Sector visibility is coded at manual stations in the following format: the remark identifier **VIS**, followed by the sector referenced to eight points of the compass, and the sector visibility in statute miles. For example, a visibility of 2½ sm in the northeastern octant is coded **VIS NE 2 1/2**.

24.4.3.13.9 Visibility at Second Location

At designated automated stations, the visibility at a second location is coded in the following format: the remark identifier **VIS**, followed by the measured visibility value and the specific location of the visibility sensor(s) at the station. This remark will only be generated when the condition is lower than that contained in the body of the report. For example, a visibility of 2½ sm measured by a second sensor located at Runway 11 is coded **VIS 2 1/2 RWY11**.

24.4.3.13.10 Lightning

When lightning is observed at a manual station, the frequency, type of lightning, and location are reported. The contractions for the type and frequency of lightning are based on Table 24-7 (e.g., **OCNL LTGICCG NW**, **FRQ LTG VC**, or **LTG DSNT W**).

When lightning is detected by an automated system:

- Within 5 NM of the Airport Location Point (ALP), it is reported as **TS** in the body of the report with no remark.

- Between 5 and 10 NM of the ALP, it is reported as **VCTS** in the body of the report with no remark.

- Beyond 10 but less than 30 NM of the ALP, it is reported in remarks only as **LTG DSNT**, followed by the direction from the ALP.

Table 24-7. METAR/SPECI Type and Frequency of Lightning

Type of Lightning		
Type	**Contraction**	**Definition**
Cloud-Ground	CG	Lightning occurring between cloud and ground.
In-Cloud	IC	Lightning that takes place within the cloud.
Cloud-Cloud	CC	Streaks of lightning reaching from one cloud to another.
Cloud-Air	CA	Streaks of lightning that pass from a cloud to the air, but do not strike the ground.
Frequency of Lightning		
Frequency	**Contraction**	**Definition**
Occasional	OCNL	Less than 1 flash/minute.
Frequent	FRQ	About 1 to 6 flashes/minute.
Continuous	CONS	More than 6 flashes/minute.

24.4.3.13.11 Beginning and Ending of Precipitation

At designated stations, the beginning and ending times of precipitation are coded in the following format: the type of precipitation, followed by either a **B** for beginning or an **E** for ending, and the time of occurrence. No spaces are coded between the elements. The coded times of the precipitation start and stop times are found in the remarks section of the next METAR. The times are not required to be in the SPECI. The intensity qualifiers are coded. For example, if rain began at 0005 and ended at 0030, and then snow began at 0020 and ended at 0055, the remark would be coded **RAB05E30SNB20E55**. If the precipitation was showery, the remark is coded **SHRAB05E30SHSNB20E55**. If rain ended and snow began at 0042, the remark would be coded as **RAESNB42**.

24.4.3.13.12 Beginning and Ending of Thunderstorms

The beginning and ending times of thunderstorms are coded in the following format: the thunderstorm identifier **TS**, followed by either a **B** for beginning or an **E** for ending, and the time of occurrence. No spaces are between the elements. For example, if a thunderstorm began at 0159 and ended at 0230, the remark is coded **TSB0159E30**.

24.4.3.13.13 Thunderstorm Location

Thunderstorm locations are coded in the following format: the thunderstorm identifier **TS**, followed by the location of the thunderstorm(s) from the station, and the direction of movement, when known. For example, a thunderstorm southeast of the station and moving toward the northeast is coded **TS SE MOV NE**.

24.4.3.13.14 Hailstone Size

At designated stations, the hailstone size is coded in the following format: the hail identifier **GR**, followed by the size of the largest hailstone. The hailstone size is coded in ¼ in increments. For example, **GR 1 3/4** would indicate that the largest hailstones were 1¾ in in diameter. When small hail with a size less than ¼ in is occurring, the hailstone size is reported in the remarks as **GR LESS THAN 1/4 in**.

24.4.3.13.15 Virga

At designated stations, virga is coded in the following format: the identifier **VIRGA**, followed by the direction from the station. The direction of the phenomena from the station is optional (e.g., **VIRGA** or **VIRGA SW**).

24.4.3.13.16 Variable Ceiling Height

The variable ceiling height is coded in the following format: the identifier **CIG**, followed by the lowest ceiling height recorded, **V** denoting variability between two values, and ending with the highest ceiling height. A single space follows the identifier with no other spaces between the letter **V** and the lowest/highest ceiling values. For example, **CIG 005V010** would indicate a ceiling is variable between 500 and 1,000 ft.

24.4.3.13.17 Obscurations

Obscurations, surface-based or aloft, are coded in the following format: the weather identifier causing the obscuration at the surface or aloft, followed by the sky cover of the obscuration aloft (**FEW, SCT, BKN, OVC**) or at the surface (**FEW, SCT, BKN**), and the height. Surface-based obscurations have a height of 000. A space separates the weather causing the obscuration and the sky cover; no space is between the sky cover and the height. For example, fog hiding 3/8 to 4/8 of the sky is coded **FG SCT000**; a broken layer at 2,000 ft composed of smoke is coded **FU BKN020**.

24.4.3.13.18 Variable Sky Condition

Variable sky condition remarks are coded in the following format: the two operationally significant sky conditions (**FEW, SCT, BKN,** and **OVC**), separated by spaces, and **V** denoting the variability between the two ranges. If several layers have the same condition amount, the layer height of the variable layer is coded. For example, a cloud layer at 1,400 ft varying between broken and overcast is coded **BKN014 V OVC**.

24.4.3.13.19 Significant Cloud Types

At manual stations, significant cloud type remarks are coded in all reports.

24.4.3.13.19.1 Cumulonimbus

Cumulonimbus not associated with thunderstorms is coded as **CB**, followed by the direction from the station, and the direction of movement, when known. The location, direction, and direction of movement entries are separated from each other by a space. For example, a cumulonimbus up to 10 sm west of the station moving toward the east would be coded **CB W MOV E**. If the cumulonimbus was more than 10 sm to the west, the remark is coded **CB DSNT W**.

24.4.3.13.19.2 Towering Cumulus

Towering cumulus clouds are coded in the following format: the identifier **TCU**, followed by the direction from the station. The cloud type and direction entries are separated by a space. For example, a towering cumulus cloud up to 10 sm west of the station is coded **TCU W**.

24.4.3.13.19.3 Standing Lenticular or Rotor Clouds

Stratocumulus (**SCSL**), altocumulus (**ACSL**), cirrocumulus (**CCSL**), or rotor clouds are coded in the following format: the cloud type followed by the direction from the station. The cloud type and direction entries are separated by a space. For example, altocumulus standing lenticular clouds observed southwest through west of the station are coded **ACSL SW-W**.

24.4.3.13.20 Ceiling Height at Second Location

At designated stations, the ceiling height at a second location is coded in the following format: the identifier **CIG**, followed by the measured height of the ceiling and the specific location of the ceilometer(s) at the station. This remark is only generated when the ceiling is lower than that contained in the body of the report. For example, if the ceiling measured by a second sensor located at Runway 11 is broken at 200 ft, the remark would be coded **CIG 002 RWY11**.

24.4.3.13.21 Pressure Rising or Falling Rapidly

At designated stations, the reported pressure is evaluated to determine if a pressure change is occurring. If the pressure is rising or falling at a rate of at least 0.06 in per hour and the pressure change totals 0.02 in or more at the time of the observation, a pressure change remark is reported. When the pressure is rising or falling rapidly at the time of observation, the remark pressure rising rapidly (**PRESRR**) or pressure falling rapidly (**PRESFR**) is included in the remarks.

24.4.3.13.22 Sea Level Pressure

At designated stations, the sea level pressure is coded in the following format: the identifier **SLP**, immediately followed by the sea level pressure in millibars. The hundreds and thousands units are not coded and must be inferred. For example, a sea level pressure of 998.2 mb is coded **SLP982**. A sea level pressure of 1,013.2 mb would be coded **SLP132**. For a METAR, if sea level pressure is not available, it is coded **SLPNO**.

24.4.3.13.23 Aircraft Mishap

If a SPECI is taken to document weather conditions when notified of an aircraft mishap, the remark **ACFT MSHP** is coded in the report, but the SPECI is not transmitted.

24.4.3.13.24 No SPECI Reports Taken

At manual stations where SPECIs are not taken, the remark **NOSPECI** is coded to indicate that no changes in weather conditions will be reported until the next METAR.

24.4.3.13.25 Snow Increasing Rapidly

At designated stations, the snow increasing rapidly remark is reported in the next METAR whenever the snow depth increases by 1 in or more in the past hour. The remark is coded in the following format: the remark indicator **SNINCR**, the depth increase in the past hour, and the total depth of snow on the ground at the time of the report. The depth of snow increase in the past hour and the total depth on the ground are separated from each other by a solidus (/). For example, a snow depth increase of 2 in in the past hour with a total depth on the ground of 10 in is coded **SNINCR 2/10**.

24.4.3.13.26 Other Significant Information

Agencies may add to a report other information significant to their operations, such as information on fog dispersal operations, runway conditions, **FIRST** or **LAST** reports from station, etc.

24.4.3.14 Additive and Automated Maintenance Data

Additive data groups (see Table 24-6) are only reported at designated stations and are primarily used by the NWS for climatological purposes. Most have no direct impact on the aviation community, but a few are discussed below.

24.4.3.14.1 Hourly Temperature and Dewpoint

At designated stations, the hourly temperature and dewpoint group are further coded to the tenth of a degree Celsius. For example, a recorded temperature of +2.6 °C and dewpoint of -1.5 °C would be coded **T00261015**.

The format for the coding is as follows:

T	Group indicator
0	Indicates the following number is positive; a 1 would be used if the temperature was reported as negative at the time of observation
026	Temperature disseminated to the nearest tenth and read as 02.6
1	Indicates the following number is negative; a 0 would be used if the number was reported as positive at the time of observation
015	Dewpoint disseminated to the nearest tenth and read as 01.5

No spaces are between the entries. For example, a temperature of 2.6 °C and dewpoint of -1.5 °C is reported in the body of the report as **03/M01** and the hourly temperature and dewpoint group as **T00261015**. If the dewpoint is missing, only the temperature is reported; if the temperature is missing, the hourly temperature and dewpoint group are not reported.

24.4.3.14.2 Maintenance Data Groups

The following maintenance data groups, sensor status indicators and the maintenance indicator, are only reported from automated stations.

24.4.3.14.2.1 Sensor Status Indicators

Sensor status indicators are reported as indicated below:

- If the RVR is missing and would normally be reported, **RVRNO** is coded.
- When automated stations are equipped with a present weather identifier and the sensor is not operating, the remark **PWINO** is coded.
- When automated stations are equipped with a tipping bucket rain gauge and the sensor is not operating, **PNO** is coded.
- When automated stations are equipped with a freezing rain sensor and the sensor is not operating, the remark **FZRANO** is coded.
- When automated stations are equipped with a lightning detection system and the sensor is not operating, the remark **TSNO** is coded.
- When automated stations are equipped with a secondary visibility sensor and the sensor is not operating, the remark **VISNO LOC** is coded.
- When automated stations are equipped with a secondary ceiling height indicator and the sensor is not operating, the remark **CHINO LOC** is coded.

24.4.3.14.2.2 Maintenance Indicator

A maintenance indicator ($) is coded when an automated system detects that maintenance is needed on the system.

24.5 Aircraft Observations and Reports

There are three kinds of aircraft observations: PIREPs, Aircraft Reports (AIREP), and Volcanic Activity Reports (VAR). Both PIREPs and AIREPS have two types:

- Routine PIREPs and Urgent PIREPs.
- Routine AIREPs and Special AIREPs.

PIREPs are reported by the pilot (or aircrew), while AIREPs can either be reported by the pilot or generated from sensors onboard the aircraft (automated AIREPs). PIREPs and AIREPs are coded differently. The PIREP format is a U.S.-only format. The AIREP format is used worldwide. Automated AIREPs are common over the United States.

The VAR is a report for aircraft encounters with volcanic ash and/or sulfur dioxide (SO_2).

24.5.1 Pilot Weather Reports (PIREP)

Pilots can report any observation, good or bad, to assist other pilots with flight planning and preparation. If conditions were forecasted to occur but not encountered, a pilot can also report the observed condition. This will help the NWS verify forecast products and create more accurate products for the aviation community.

A PIREP is prepared using a prescribed format (see Figure 24-5). Elements for all PIREPs are: message type, location, time, altitude/FL, type aircraft, and at least one other element to describe the reported phenomena. The other elements are omitted when no data is reported. All altitude references are mean sea level unless otherwise noted. Distance for visibility is in statute miles and all other distances are in nautical miles. Time is reported in Coordinated Universal Time.

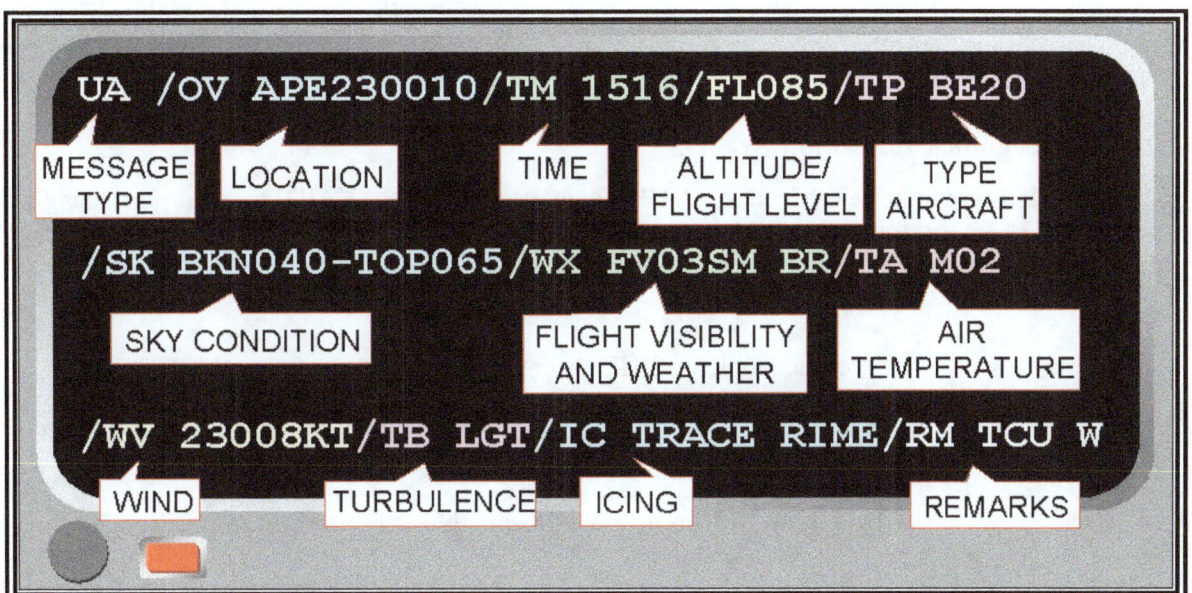

Figure 24-5. PIREP Coding Format

24.5.1.1 Message Type (UUA/UA)

The two types of PIREPs are Urgent (**UUA**) and Routine (**UA**).

24.5.1.1.1 Urgent PIREPs

Urgent (**UUA**) PIREPs contain information about:

- Tornadoes, funnel clouds, or waterspouts;
- Severe or extreme turbulence (including CAT);
- Severe icing;
- Hail;
- LLWS within 2,000 ft of the surface (LLWS PIREPs are classified as **UUA** if the pilot reports air speed fluctuations of 10 kt or more; or if air speed fluctuations are not reported but LLWS is reported, the PIREP is classified as **UUA**);
- Volcanic ash clouds; and/or
- Any other weather phenomena reported that are considered by the air traffic controller or Flight Service specialist receiving the report as being hazardous, or potentially hazardous, to flight operations.

24.5.1.1.2 Routine PIREPs

Routine (**UA**) PIREPs are issued after receiving a report from a pilot that does not contain any urgent information as listed in Section 24.5.1.1.1.

24.5.1.2 Location (/OV)

The location (**/OV**) is the position reference where the phenomenon occurred. *It is not the location of the aircraft when the report is submitted.* Location can be referenced either by geographical position or by route segment. A position reference is preferred by meteorologists to aid forecast precision, monitoring, and verification.

24.5.1.2.1 Geographical Position

Geographical position can be referenced to a VHF Navigational Aid (NAVAID) or an airport, using either the three-letter IATA or four-letter ICAO identifier. If appropriate, the PIREP is encoded using the identifier, then three digits to define a radial and three digits to define the distance in nautical miles.

Examples:

`/OV APE`	Over the Appleton VHF omni-directional range station (VOR).
`/OV KJFK`	Over John F. Kennedy International Airport, New York, NY.
`/OV APE230010`	230° at 10 NM from the Appleton VOR.
`/OV KJFK107080`	107° at 80 NM from John F. Kennedy International Airport, New York, NY.

24.5.1.2.2 Route Segment

A PIREP can also be referenced using two or more fixes to describe a route.

Examples:

`/OV KSTL-KMKC`	From St. Louis Lambert International Airport, St. Louis, MO, to Charles B. Wheeler Downtown Airport, Kansas City, MO.
`/OV KSTL090030-KMKC045015`	From 90° at 30 NM from St. Louis Lambert International Airport, St. Louis, MO, to 45° at 15 NM from Charles B. Wheeler Downtown Airport, Kansas City, MO.

24.5.1.3 Time (/TM)

Time (**/TM**) is the time when the reported phenomenon occurred or was encountered. It is coded in four digits UTC.

Example:

`/TM 1315`   1315 UTC.

24.5.1.4 Altitude/Flight Level (/FL)

The altitude/flight level (**/FL**) is the altitude in hundreds of feet MSL where the phenomenon was first encountered. If not known, **UNKN** is entered. If the aircraft was climbing or descending, the appropriate contraction (**DURC** or **DURD**) is entered in the remarks (**/RM**). If the condition was encountered within a layer, the altitude range is entered within the appropriate element that describes the condition.

Examples:

`/FL085`	8,500 ft MSL.
`/FL310`	Flight level 310.
`/FLUNKN /RM DURC`	Flight level unknown, remarks, during climb.

24.5.1.5 Aircraft Type (/TP)

Aircraft type (**/TP**) is entered. If not known, **UNKN** is entered. Icing and turbulence reports always include aircraft type.

Examples:

`/TP BE20`	Super King Air 200.
`/TP SR22`	Cirrus 22.
`/TP P28R`	Piper Arrow.
`/TP UNKN`	Type unknown.

24.5.1.6 Sky Condition (/SK)

The sky condition (**/SK**) group is used to report height of cloud bases, tops, and cloud cover. The height of base of a layer of clouds is coded in hundreds of feet MSL. The top of a layer is entered in hundreds of feet MSL preceded by the word **TOP**. If reported as clear above the highest cloud layer, **SKC** is coded following the reported level.

Examples:

`/BKN040-TOP065`	Base of broken layer 4,000 ft MSL, top 6,500 ft MSL.
`/SK OVC100-TOP110/ SKC`	Base of an overcast layer 10,000 ft MSL, top 11,000 ft MSL, clear above.
`/SK OVC015-TOP035/OVC230`	Base of an overcast layer 1,500 ft MSL, top 3,500 ft MSL, base of an overcast layer 23,000 ft MSL.
`/SK OVC-TOP085`	Overcast layer, top 8,500 ft MSL.

Cloud cover amount ranges are entered with a hyphen separating the amounts (e.g., **BKN-OVC**).

Examples:

`/SK SCT-BKN050-TOP100`	Base of a scattered to broken layer 5,000 ft MSL, top 10,000 ft MSL.
`/SK BKN-OVCUNKN-TOP060/BKN120-TOP150/ SKC`	Base of a broken to overcast layer unknown, top 6,000 ft MSL, base of a broken layer 12,000 ft MSL, top 15,000 ft MSL, clear above.

Unknown heights are indicated by the contraction **UNKN**.

Examples:

`/SK OVC065-TOPUNKN`	Base of an overcast layer 6,500 ft MSL, top unknown. If a pilot indicates being in the clouds, **IMC** is entered.
`/SK OVC065-TOPUNKN /RM IMC`	Base of an overcast layer 6,500 ft MSL, top unknown, remark, in the clouds. When more than one layer is reported, layers are separated by a solidus (/).

24.5.1.7 Flight Visibility and Weather (/WX)

The pilot reports the weather conditions encountered as follows: flight visibility, when reported, is entered first in the **/WX** field. It is coded **FV** followed by a two-digit visibility value rounded down, if necessary, to the nearest whole statute mile and appended with **SM** (e.g., **FV03SM**). If visibility is reported as unrestricted, **FV99SM** is entered.

Flight weather types are entered using one or more of the surface weather reporting phenomena contained in Table 24-3.

Example:

`/WX FV01SM +DS000-TOP083/SKC /RM DURC`	Flight visibility 1 sm, base heavy dust storm layer at the surface, top 8,300 ft MSL, clear above, remarks, during climb.

When more than one form of precipitation is combined in the report, the dominant type is reported first.

Examples:

`/WX FV00SM +TSRAGR`	Flight visibility 0 sm, thunderstorm, heavy rain, hail.
`/WX FV02SM BRHZ000-TOP083`	Flight visibility 2 sm, base of a haze and mist layer at the surface, top 8,300 ft MSL.

If a funnel cloud is reported, it is coded **FC** following the **/WX** group and is spelled out as **FUNNEL CLOUD** after the **/RM** group. If a tornado or waterspout is reported, it is coded **+FC** following the **/WX** group, and **TORNADO** or **WATERSPOUT** is spelled out after the **/RM** group.

Examples:

`/WX FC /RM FUNNEL CLOUD`	Funnel cloud, remarks, funnel cloud.
`/WX +FC /RM TORNADO`	Tornado, remarks, tornado.

When more than one type of weather is reported, they are reported in the following order:

- **TORNADO**, **WATERSPOUT**, or **FUNNEL CLOUD**.
- Thunderstorm with or without associated precipitation.
- Weather phenomena in order of decreasing predominance.
- No more than three groups are used in a single PIREP.
- Weather layers are entered with the base and/or top of the layer when reported. The same format as in the sky condition (**/SK**) group is used.

Example:

`/WX FU002-TOP030`   Base of a smoke layer, 200 ft MSL, top 3,000 ft MSL.

24.5.1.8 Air Temperature (/TA)

Outside air temperature (**/TA**) is reported using two digits in degrees Celsius. Negative temperatures are prefixed with an **M** (e.g., **/TA 08** or **/TA M08**).

24.5.1.9 Wind Direction and Speed (/WV)

Wind direction and speed are encoded using three digits to indicate wind direction, relative to magnetic north, and two or three digits to indicate reported wind speed. When the reported speed is less than 10 kt, a leading **0** is used. The wind group will always have **KT** appended to represent the units in knots.

Examples:

`/WV 02009KT`	Wind 20° at 9 kt.
`/WV 28057KT`	Wind 280° at 57 kt.
`/WV 350102KT`	Wind 350° at 102 kt.

24.5.1.10 Turbulence (/TB)

Turbulence intensity, type, and altitude are reported after wind direction and speed.

Duration (intermittent (**INTMT**), occasional (**OCNL**), or continuous (**CONS**)) is coded first (if reported by the pilot), followed by the intensity (light (**LGT**), moderate (**MOD**), severe (**SEV**), or extreme (**EXTRM**)). Range or variation of intensity is separated with a hyphen (e.g., **MOD-SEV**). If turbulence was forecast, but not encountered, negative (**NEG**) is entered.

Type is coded second. **CAT** or **CHOP** is entered if reported by the pilot. High-level turbulence not associated with clouds (including thunderstorms) is reported as **CAT**.

Altitude is reported last, and only if it differs from the value reported in the altitude/flight level (**/FL**) group. When a layer of turbulence is reported, height values are separated with a hyphen. If lower or upper limits are not defined, below (**BLO**) or above (**ABV**) is used.

Examples:

/TB LGT	Light turbulence.
/TB LGT 040	Light turbulence at 4,000 ft MSL.
/TB OCNL MOD-SEV BLO 080	Occasional moderate to severe turbulence below 8,000 ft MSL.
/TB MOD-SEV CAT 350	Moderate to severe CAT at 35,000 ft MSL.
/TB NEG 120-180	Negative turbulence between 12,000 and 18,000 ft MSL.
/TB CONS MOD CHOP 220/NEG 230-280	Continuous moderate chop at 22,000 ft MSL, negative turbulence between 23,000 and 28,000 ft MSL.
/TB MOD CAT ABV 290	Moderate CAT above 29,000 ft MSL.

Turbulence reports should include location, altitude (or range of altitudes), and aircraft type, as well as, when reported, whether in clouds or clear air. The pilot determines the degree of turbulence, intensity, and duration (occasional, intermittent, or continuous). Reports should be obtained and disseminated, when possible, in conformance with the Turbulence Reporting Criteria Table in the AIM, Chapter 7, Section 1, Meteorology.

24.5.1.11 Icing (/IC)

Icing intensity, type, and altitude are reported after turbulence.

Intensity is coded first using contractions **TRACE**, light (**LGT**), moderate (**MOD**), or severe (**SEV**). Reports of a range or variation of intensity is separated with a hyphen. If icing was forecast but not encountered, negative (**NEG**) is coded. Icing type is reported second. Reportable types are **RIME**, clear (**CLR**), or mixed (**MX**).

The AIM, Chapter 7, Section 1, Meteorology provides classification of icing intensity, according to its operational effects on aircraft, as well as tables of icing types and icing conditions.

The reported icing/altitude is coded last, and only if different from the value reported in the altitude/flight level (**/FL**) group. A hyphen is used to separate reported layers of icing. Above (**ABV**) or below (**BLO**) is coded when a layer is not defined.

Pilot reports of icing should also include air temperature (**/TA**).

Examples:

/IC LGT-MOD MX 085	Light to moderate mixed icing, 8,500 ft MSL.
/IC LGT RIME	Light rime icing.
/IC MOD RIME BLO 095	Moderate rime icing below 9,500 ft MSL.
/IC SEV CLR 035-062	Severe clear icing 3,500 to 6,200 ft MSL.

24.5.1.12 Remarks (/RM)

The remarks (**/RM**) group is used to report a phenomenon that is considered important but does not fit in any of the other groups. This includes, but is not limited to, LLWS reports, thunderstorm lines, coverage and movement, lightning, SO₂ gas smell, clouds observed but not encountered, and geographical or local descriptions of where the phenomenon occurred. Hazardous weather is reported first. LLWS is described to the extent possible.

24.5.1.12.1 Wind Shear

Fluctuations in wind speed 10 kt or more (+/-10 kt), within 2,000 ft of the surface, are issued as an Urgent (**UUA**) PIREP. When LLWS is entered in a PIREP, **LLWS** is entered as the first remark in the remarks (**/RM**) group.

Example:

/RM LLWS +/-15 KT SFC-008 DURC RY22 JFK	Remarks, LLWS, air speed fluctuations of plus or minus 15 kt, surface to 800 ft during climb, Runway 22, John F. Kennedy International Airport, New York, NY.

24.5.1.12.2 Funnel Cloud, Tornado, and Waterspout

FUNNEL CLOUD, **TORNADO**, and **WATERSPOUT** are entered with the direction of movement, when reported.

Example:

/RM TORNADO W MOV E   Remarks, tornado west moving east.

24.5.1.12.3 Thunderstorm

Thunderstorm coverage is coded as isolated (**ISOL**), few (**FEW**), scattered (**SCT**), or numerous (**NMRS**), followed by a description as line (**LN**), broken line (**BKN LN**), or solid line (**SLD LN**), when reported. This is followed with **TS**, the location and movement, and the type of lightning, when reported.

Example:

/RM NMRS TS S MOV E   Remarks, numerous thunderstorms south moving east.

24.5.1.12.4 Lightning

Lightning frequency is coded as occasional (**OCNL**) or frequent (**FRQ**), followed by type as lightning in cloud (**LTGIC**), lightning cloud to cloud (**LTGCC**), lightning cloud to ground (**LTGCG**), lightning cloud to air (**LTGCA**), or combinations, when reported.

Example:

/RM OCNL LTGICCG   Remarks, occasional lighting in cloud, cloud to ground.

24.5.1.12.5 Electrical Discharge

For an electrical discharge, **DISCHARGE** is coded followed by the altitude.

Example:

/RM DISCHARGE 120   Remarks, discharge, 12,000 ft MSL.

24.5.1.12.6 Clouds

Remarks are used when clouds can be seen but were not encountered and reported in the sky condition group (**/SK**).

Examples:

```
/RM CB E MOV N        Remarks, cumulonimbus east moving north.
/RM OVC BLO           Remarks, overcast below.
```

24.5.1.12.7 Other Remarks

Remarks that do not fit in other groups, like during climb (**DURC**), during descent (**DURD**), reach cruising altitude (**RCA**), or top of climb (**TOP** or **TOC**), may be included. If specific phraseology or contractions are not adequate, plain language is used to describe the phenomena or local geographic locations.

Example:

```
/RM DONNER SUMMIT PASS
```

24.5.1.12.8 Volcanic Eruptions

Volcanic ash alone is an Urgent PIREP. A report of volcanic activity includes as much information as possible, including the name of the mountain, ash cloud and movement, height of the top and bottom of the ash, etc.

Example:

```
/UUA/OV ANC240075/TM 2110/FL370/TP DC10/WX VA/RM VOLCANIC ERUPTION 2008Z MT
AUGUSTINE ASH 40S MOV SSE
```

Urgent PIREP, 240° at 75 NM from Anchorage International Airport, Anchorage, AK, 2110 UTC, FL370, a DC-10 reported volcanic ash, remarks, volcanic eruption occurred at 2008 UTC Mount Augustine, ash 40 NM south moving south-southeast.

24.5.1.12.9 Skyspotter

The **SKYSPOTTER** program is a result of a recommendation from the "Safer Skies" FAA/Industry Joint Safety Analysis and Implementation Teams. The term **SKYSPOTTER** indicates a pilot has received specialized training in observing and reporting in-flight weather phenomena or PIREPs.

When a PIREP is received from a pilot identifying themselves as a **SKYSPOTTER** aircraft, the additional comment "**/AWC**" is added at the end of the remarks section of the PIREP.

An **AWC-WEB/xxxx** in the remarks indicates the PIREP was submitted by an airline dispatcher or CWSU meteorologist directly to the AWC. The "xxxx" represents the airline abbreviation or ARTCC of the CWSU that submitted the PIREP.

Example:

```
PIREP TEXT/RM REMARKS/AWC
PIREP TEXT/RM REMARKS/AWC-WEB/KZFW
```

24.5.2 Aircraft Reports (AIREP)

AIREPs are messages from an aircraft to a ground station. AIREPs are normally comprised of the aircraft's position, time, FL, ETA over its next reporting point, destination ETA, fuel remaining, and meteorological information. It is beyond the scope of this document to describe the details of all the elements in the AIREP, but this section will focus on the meteorological information.

The AWC's website provides AIREPs over the CONUS and portions of the Atlantic and Pacific Oceans.

24.5.2.1 AIREP Types and Content

There are two types of AIREPs: routine or position report (**ARP**) and special (**ARS**). AIREPs can be reported by the pilot, but the majority of routine AIREPs are automated and downlinked from the aircraft to a service provider (e.g., a flight planning company) for processing and forwarding to an airline and the NWS.

The majority of AIREPs report wind and temperature at selected intervals along the flight route, derived from onboard sensors and probes. Some aircraft are equipped with sensors and probes to measure humidity/water vapor, turbulence, and icing data.

The format for the AIREP is governed by the WMO and ICAO. The AWC's website includes AIREPs on their PIREP web page that is formatted for web display, with some weather elements decoded.

24.5.2.2 AIREP Examples

The following examples are from the AWC's website. The actual airline's call sign was replaced with a fictitious call sign, and the Special AIREP was created from the routine report.

24.5.2.2.1 Routine AIREP Example

```
ARP XXX836 2443N 15516W 2229 F350 M43 315/128 TB LGT
```

ARP	Routine report.
XXX836	Aircraft call sign.
2423N 15516W	Location in latitude and longitude, 24° and 23 minutes north, 155° and 16 minutes west.
F350	Flight level or altitude, FL350.
M43	Temperature in Celsius, minus 43 °C.
315/128	Wind direction (reference to true north) and speed, 315° and 128 kt.
TB LGT	Light turbulence.

24.5.2.2.2 Special AIREP Example

```
ARS XXX836 2443N 15516W 2229 F350 M43 315/128 TB SEV
```

ARS	Special AIREP.

Same as the routine example except:

TB SEV	Severe turbulence.

24.5.3 Volcanic Activity Reports (VAR)

The VAR (see Figure 24-6) is a report for aircraft encounters with volcanic ash and/or SO_2 clouds. The first part of the VAR is reported to ATC (as an Urgent PIREP or Special AIREP) as soon as practical. The second part of the VAR is submitted postflight. The VAR is used by volcano scientists and forecasters to better understand the characteristics of volcanic eruptions, including their volcanic ash and/or SO2 clouds. Additional information can be found in FAA Order 8900.1, Volume 3, Chapter 26, Section 7, Safety Assurance System: Volcanic Ash Avoidance, Concepts, Policies, and Guidance.

```
                              VOLCANIC ACTIVITY REPORT
   Air-reports are critically important in assessing the hazards which volcanic ash cloud presents to aircraft operations.

   OPERATOR:                                    A/C IDENTIFICATION: (as indicated on flight plan)
   PILOT-IN-COMMAND:
   DEP FROM:       DATE:      TIME; UTC:        ARR AT:          DATE:        TIME; UTC:
   ADDRESSEE                                    AIREP SPECIAL

   Items 1–8 are to be reported immediately to the ATS unit that you are in contact with.
   1)  AIRCRAFT IDENTIFICATION                  2)   POSITION
   3)  TIME                                     4)   FLIGHT LEVEL OR ALTITUDE
   5)  VOLCANIC ACTIVITY OBSERVED AT
       (position or bearing, estimated level of ash cloud and distance from aircraft)
   6)  AIR TEMPERATURE                          7)   SPOT WIND

                                          Other _____
   8)  SUPPLEMENTARY INFORMATION
       SO₂ detected    Yes ☐    No ☐
       Ash encountered Yes ☐    No ☐     (Brief description of activity especially vertical and lateral extent of ash cloud
                                          and, where possible, horizontal movement, rate of growth, etc.)

   After landing complete items 9–16 then fax form to: (Fax number to be provided by the meteorological authority based on local arrangements between
   the meteorological authority and the operator concerned.)
   9)  DENSITY OF ASHCLOUD        ☐ (a) Wispy        ☐ (b) Moderate dense    ☐ (c) Very dense
   10) COLOUR OF ASH CLOUD        ☐ (a) White        ☐ (b) Light grey        ☐ (c) Dark grey
       ☐                          (d) Black          ☐ (e) Other _____
   11) ERUPTION                   ☐ (a) Continuous   ☐ (b) Intermittent      ☐ (c) Not visible
   12) POSITION OF ACTIVITY       ☐ (a) Summit       ☐ (b) Side              ☐ (c) Single
                                  ☐ (d) Multiple     ☐ (e) Not observed
   13) OTHER OBSERVED             ☐ (a) Lightning    ☐ (b) Glow              ☐ (c) Large rocks
       FEATURES OF ERUPTION       ☐ (d) Ash fallout  ☐ (e) Mushroom cloud    ☐ (f) All
   14) EFFECT ON AIRCRAFT         ☐ (a) Communication ☐ (b) Navigation systems ☐ (c) Engines
                                  ☐ (d) Pitot static ☐ (e) Windscreen        ☐ (f) Windows
   15) OTHER EFFECTS              ☐ (a) Turbulence   ☐ (b) St. Elmo's Fire   ☐ (c) Other fumes
   16) OTHER INFORMATION
       (Any information considered useful.)
   Date: 07/19/2010
```

Figure 24-6. VAR Form

24.5.4 Turbulence Observations

Since the 1990s, several innovations have improved the quality and availability of turbulence reports. Automated turbulence reporting systems are common on many commercial aircraft using the Aircraft Meteorological Data Relay (AMDAR) system.

Modern commercial aircraft are equipped with meteorological sensors and associated sophisticated data acquisition and processing systems. These systems continuously record meteorological information on the aircraft and send these observations at selected intervals to ground stations via satellite or radio links where they are processed and disseminated.[11] Participating airlines add turbulence information with these reports. See Figure 24-7 for a plot of AMDAR reports from 2019.

---

[11] In 2019, it was estimated that NOAA was receiving about 600,000 wind and temperature observations per day, with about 80 percent of the reports over the CONUS. This data comes from more than 9,000 aircraft.

AMDAR reports turbulence in terms of Eddy Dissipation Rate (EDR). EDR is the ICAO standard dimension for automated turbulence reporting. EDR is a state of the atmosphere measure rather than a state of the aircraft measure, and so it is independent of aircraft type.

Note: This information is restricted.

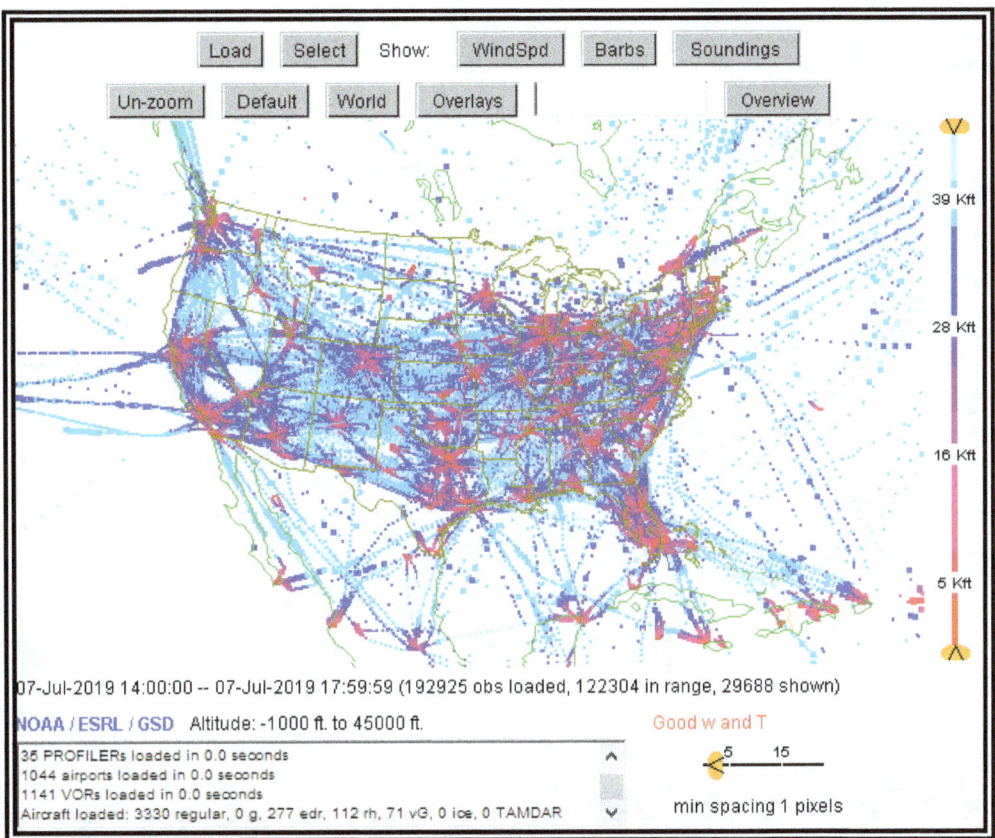

Figure 24-7. A Plot of AMDAR Reports Received During a 24-Hour Period in 2019

24.6 Radar Observations

24.6.1 Weather Surveillance Radar—1988 Doppler (WSR-88D) Description

Weather radar observations and their resultant images are graphical displays of precipitation and nonprecipitation targets detected by weather radars. WSR-88D, also known as NEXRAD, displays these targets on a variety of products, which can be found on the websites of all NWS WFOs, the AWC, the SPC, and websites and phone applications of various flight planning and weather service providers.

For information on radar basics, see Chapter 15, Weather Radar.

24.6.1.1 Issuance

WSR-88D radars are continuously generating radar observations. Each radar observation, called a volume scan, consists of 5 to 14 separate elevation "tilts," and takes between 4 and 11 minutes to generate, depending on the radar's mode of operation. Once one observation is complete, the next one begins. Radar observation times are not standard, nor are they synchronized with other radars. The valid time of the observation is the time assigned to the product, which is the end of the last radar scan.

24.6.1.1.1 WSR-88D Radar (NEXRAD) Network

The WSR-88D radar network consists of 160 radars operated by the NWS, FAA, and DOD. Figure 24-8, Figure 24-9, and Figure 24-10 show the locations of the radars.

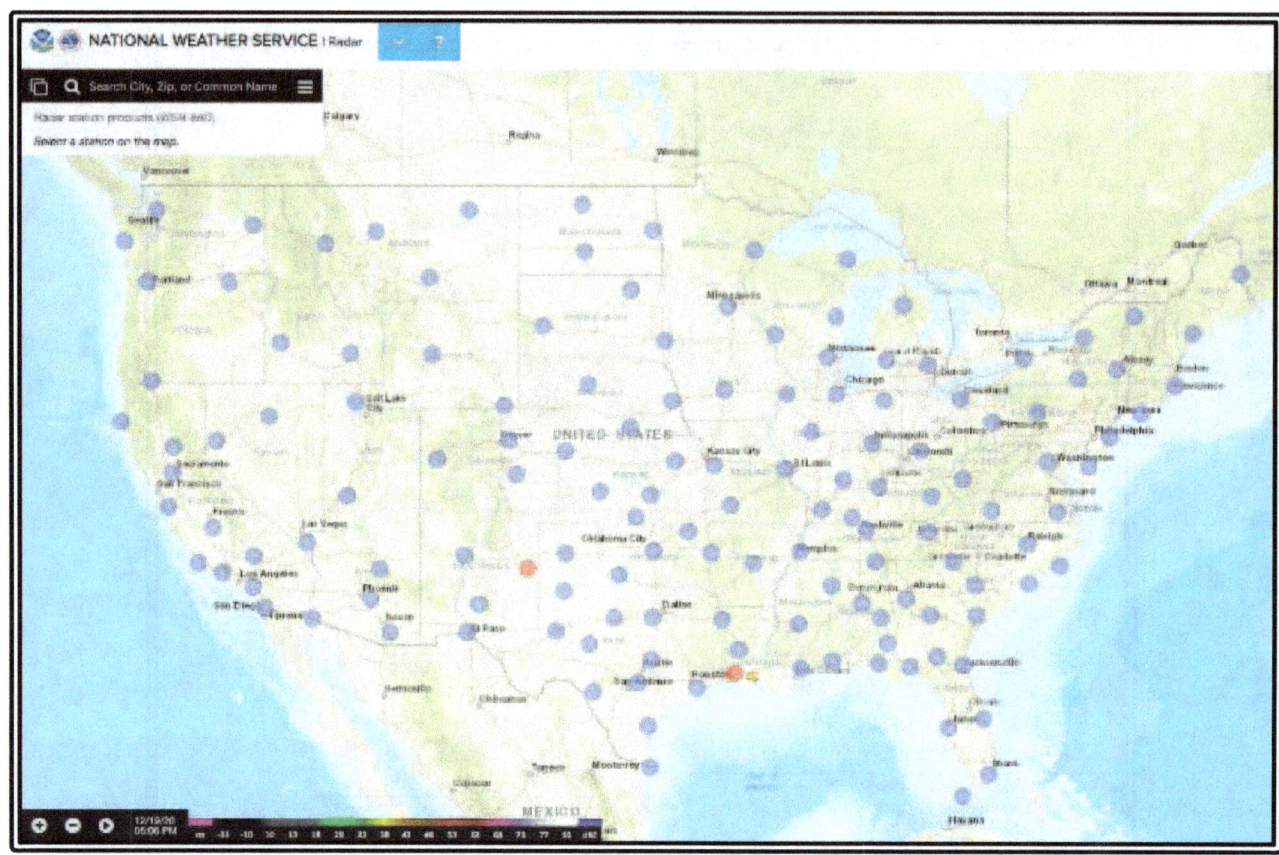

Locations of WSR-88D weather radar are indicated by gray circles. Red circles indicate radars that are temporarily out of service.

Figure 24-8. Locations of WSR-88D Weather Radar in the CONUS

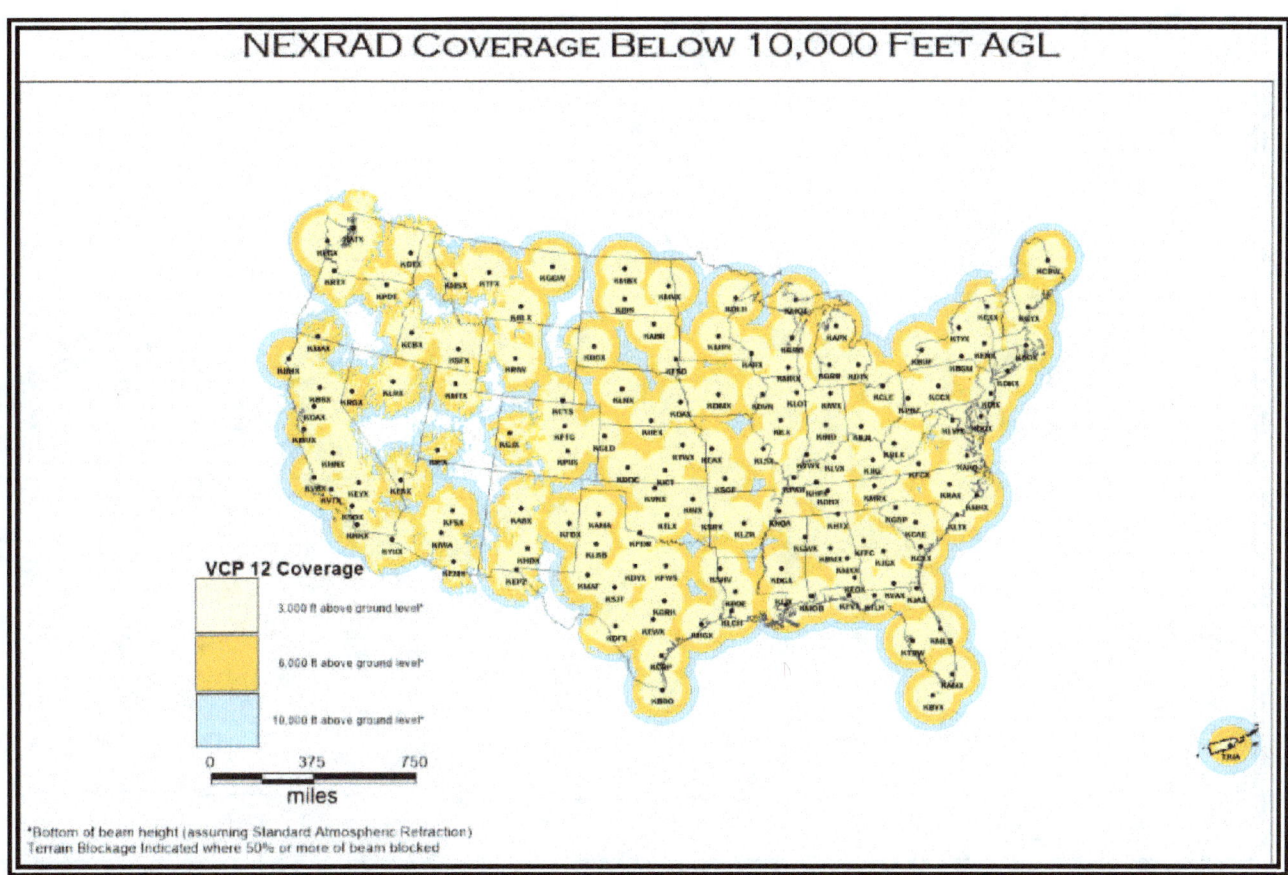

Figure 24-9. WSR-88D Weather Radar Coverage at 3,000 ft AGL, 6,000 ft AGL, and 10,000 ft AGL over the CONUS and Puerto Rico

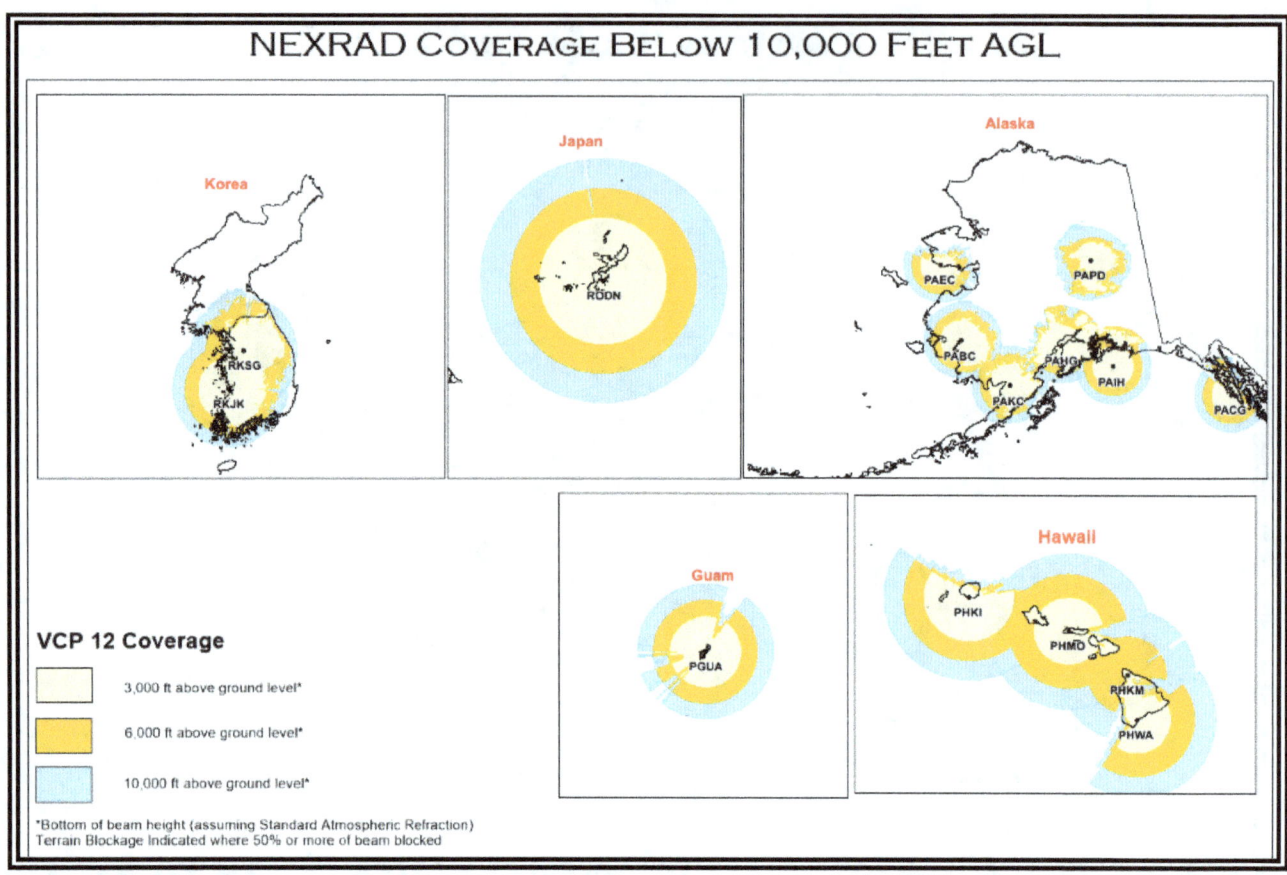

Figure 24-10. Additional Locations of WSR-88D Weather Radar and Coverage Outside of the CONUS

24.6.1.2 Coverage

Figure 24-9 and Figure 24-10 depict the radar coverage at 3,000 ft AGL, 6,000 ft AGL, and 10,000 ft AGL (i.e., above the height of the radar). Several WSR-88D radars are located in mountainous areas, such as the western United States. For example, the radar in southern Utah (near Cedar City) is on top of a 10,000-ft mountain. This means that the coverage begins at 10,000 ft AGL in that area. Any precipitation from low-topped clouds would not be detected by this radar due to overshooting of the radar beam (see Section 15.2.8).

24.6.1.3 Modes of Operation

The WSR-88D employs scanning strategies in which the antenna automatically raises to higher and higher preset angles (or elevation scans) as it rotates. These elevation scans comprise a volume coverage pattern (VCP) that NWS forecasters utilize to help analyze the atmosphere around the radar. These different VCPs have varying numbers of elevation tilts and rotation speeds of the radar itself. Therefore, each VCP can provide a different perspective of the atmosphere. Once the radar sweeps through all elevation slices, a volume scan is complete. The WSR-88D radar can use several VCPs.

There are two main classes of VCPs, which are commonly referred to as Clear Air Mode and Precipitation Mode.

24.6.1.3.1 Clear Air Mode

In Clear Air Mode, the radar is in its most sensitive operation. The NWS uses Clear Air Mode when there is no rain within the range of the radar. This mode has the slowest antenna rotation rate, which permits the radar to sample the atmosphere longer. This slower sampling increases the radar's sensitivity and ability to detect smaller objects in the atmosphere. The term "clear air" does not imply "no precipitation" mode. Even in Clear Air Mode, the WSR-88D can detect light, stratiform precipitation (e.g., snow) due to the increased sensitivity. Generally, the only returned energy to the radar will be very close to the radar's location.

Many of the radar returns in Clear Air Mode are airborne dust and particulate matter. The WSR-88D images are updated approximately every 10 minutes when operating in this mode.

24.6.1.3.2 Precipitation Mode

Precipitation targets typically provide stronger return signals to the radar than nonprecipitation targets. Therefore, the WSR-88D is operated in Precipitation Mode when precipitation is present, although some nonprecipitation echoes can still be detected in this operating mode. The NWS uses Precipitation Mode to see higher into the atmosphere when precipitation is occurring to analyze the vertical structure of the storms.

The faster rotation of the WSR-88D in Precipitation Mode allows images to update at a faster rate, approximately every 4 to 6 minutes.

24.6.1.4 Echo Intensities

The colors on radar images represent the reflective power of the precipitation target. In general, the amount of radar power received is proportional to the intensity of the precipitation. This reflective power, commonly referred to by meteorologists as "reflectivity," is measured in terms of dBZ. A decibel is a unit that describes the change of power emitted versus the power received. Since the power emitted is constant, the power received is related to the intensity of the precipitation target. Each reflectivity image includes a color scale that describes the relationship among reflectivity value, color on the radar image, and precipitation intensity (see Figure 24-11). The color scale and decibel scale can vary depending on the service provider and website.

Reflectivity is correlated to intensity of precipitation. For example, in Precipitation Mode, when the decibel value reaches 15, light precipitation is present. The higher the indicated reflectivity value, the higher the rainfall rate. The interpretation of reflectivity values is the same for both Clear Air and Precipitation Modes.

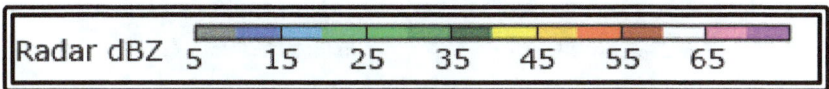

Figure 24-11. Example WSR-88D (NEXRAD) Weather Radar Echo Intensity Legend

Reflectivity is also correlated with intensity terminology (phraseology) for ATC purposes. Table 24-8 defines this correlation.

Table 24-8. WSR-88D Weather Radar Precipitation Intensity Terminology

Reflectivity (dBZ) Ranges	Weather Radar Echo Intensity Terminology
<30 dBZ	Light
30–40 dBZ	Moderate
>40–50 dBZ	Heavy
50+ dBZ	Extreme

Note: En route ATC radar's weather and radar processor (WARP) does not display light precipitation.

Values below 15 dBZ are typically associated with clouds. However, they may also be caused by atmospheric particulate matter such as dust, insects, pollen, or other phenomena. The scale *cannot* reliably be used to determine the intensity of snowfall. However, snowfall rates generally increase with increasing reflectivity.

24.6.1.5   Radar Products

The NWS produces many radar products that serve a variety of users. Some of these products are of interest to the aviation community. This section will discuss radar mosaics, Composite Reflectivity, Base Reflectivity, and Echo Tops products.

24.6.1.5.1   Radar Mosaic

A radar mosaic consists of multiple single-site radar images combined to produce a radar image on a regional or national scale. Regional and national mosaics can be found on the websites of the NWS, AWC, and all NWS WFOs, as well as commercial aviation weather providers. Radar mosaics can be assembled from Composite Reflectivity (see Figure 24-12), Base Reflectivity (see Figure 24-13), and Echo Tops (see Figure 24-14), depending on the website or data provider.

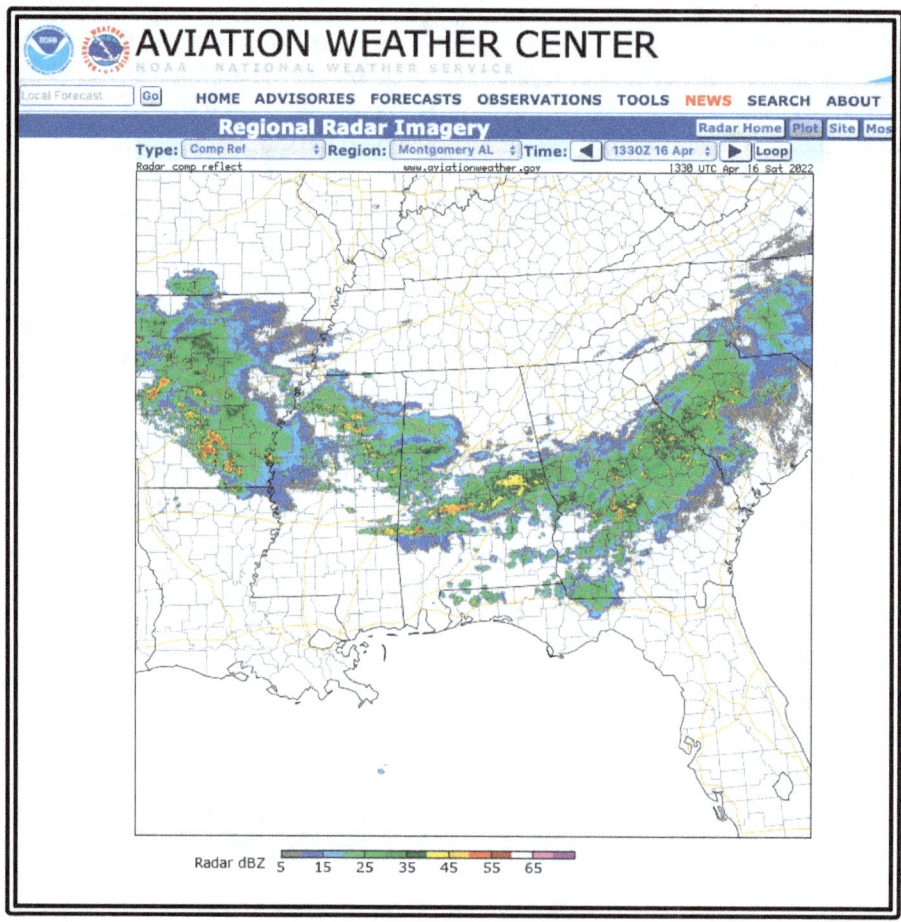

Figure 24-12. Regional Radar Mosaic Example Using Composite Reflectivity

Chapter 24, Observations

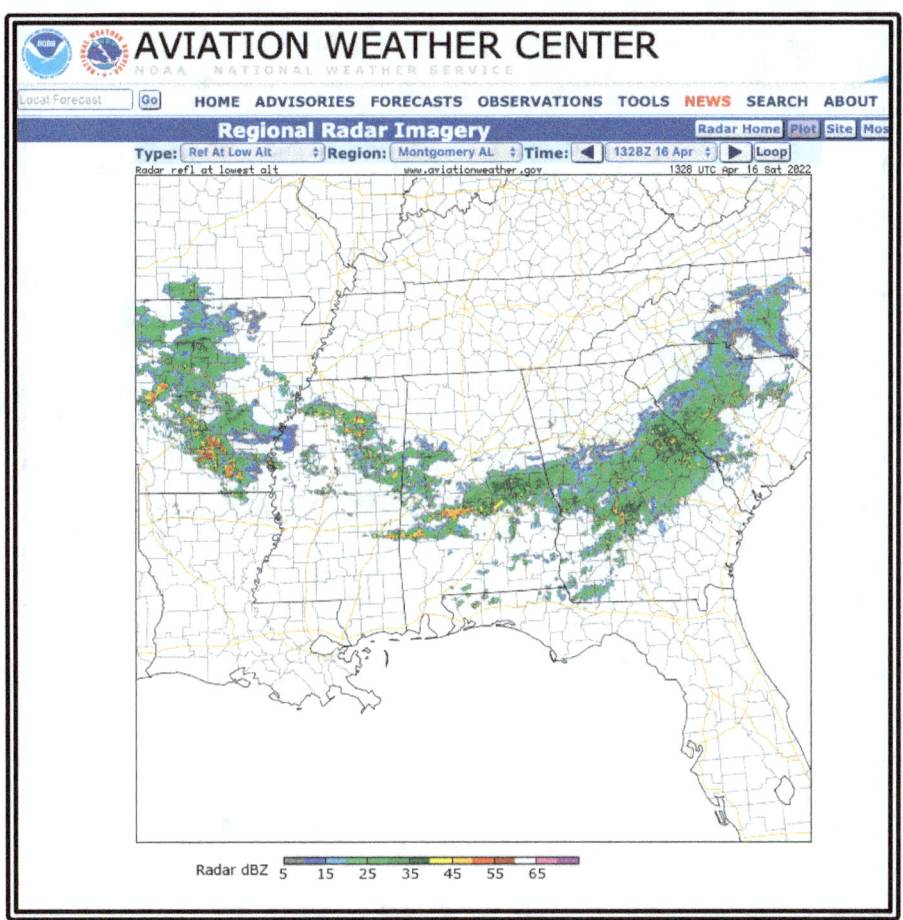

Figure 24-13. Regional Radar Mosaic Example Using Base Reflectivity (Reflectivity at Lowest Altitude)

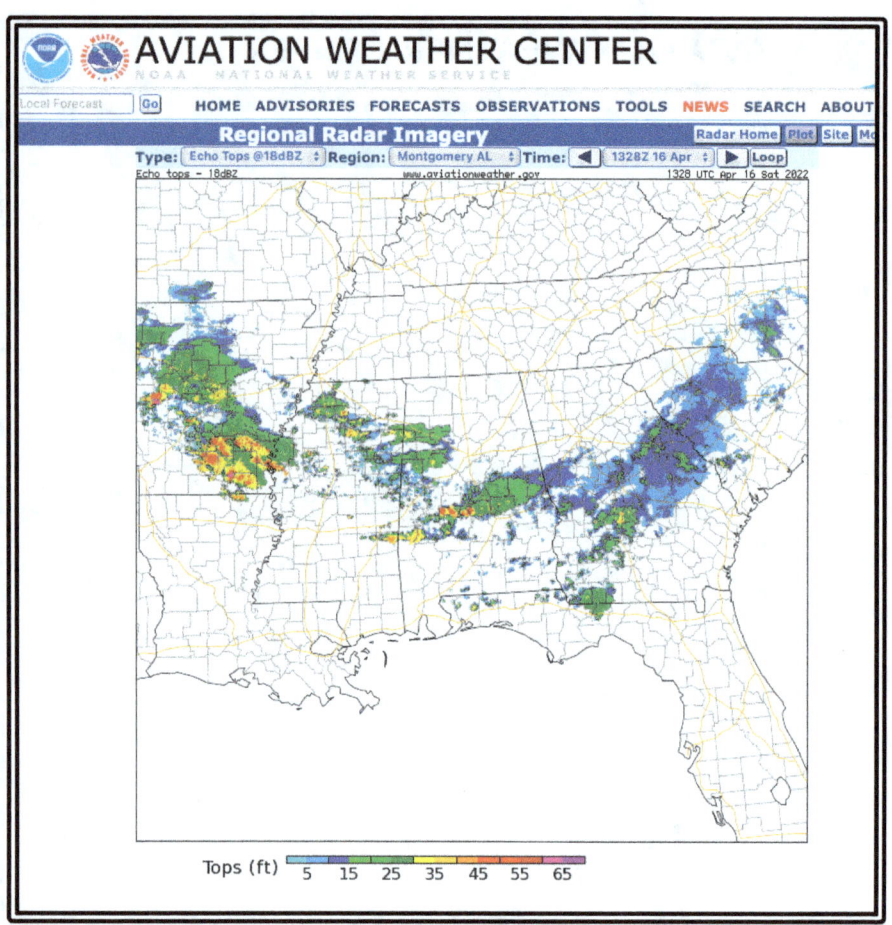

Figure 24-14. Regional Radar Mosaic Example Using Echo Tops (Tops of the 18 dBZ Radar Echo in Thousands of Feet Above Sea Level)

24.6.1.5.2   Composite Reflectivity

Because the highest precipitation intensity can be at any altitude, the Composite Reflectivity product (see Figure 24-15) is needed. Composite Reflectivity is the maximum echo intensity (reflectivity) detected within a column of the atmosphere above a location. During its tilt sequence, the radar scans through all of the elevation slices to determine the highest decibel value in the vertical column (see Figure 24-16), then displays that value on the product. When compared with Base Reflectivity, the Composite Reflectivity can reveal important storm structure features and intensity trends of storms (see Figure 24-17).

NEXRAD radar displays on airplane avionics use the Composite Reflectivity data for their radar mosaics.

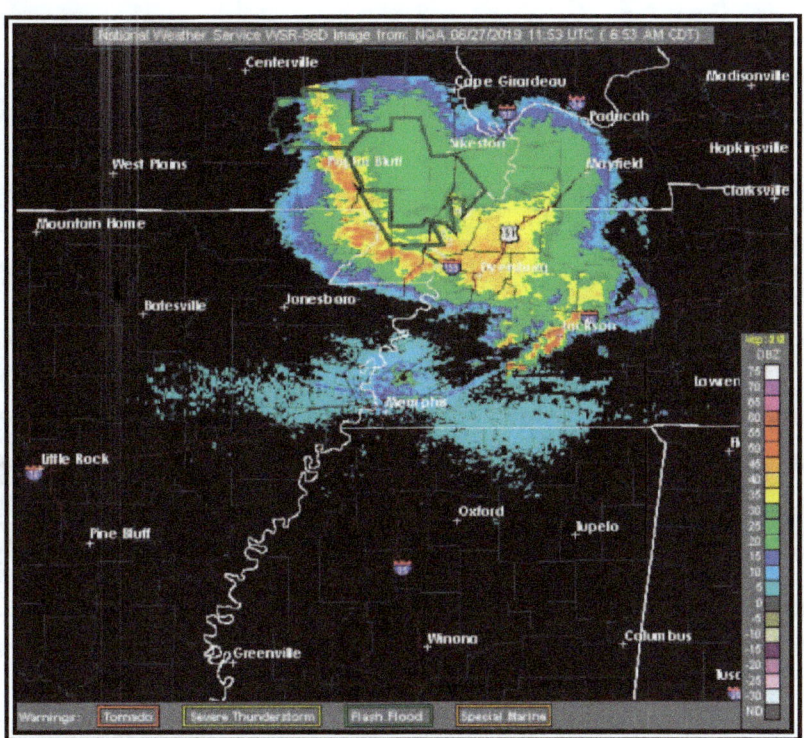

Figure 24-15. WSR-88D Weather Radar Composite Reflectivity, Single-Site Product Example

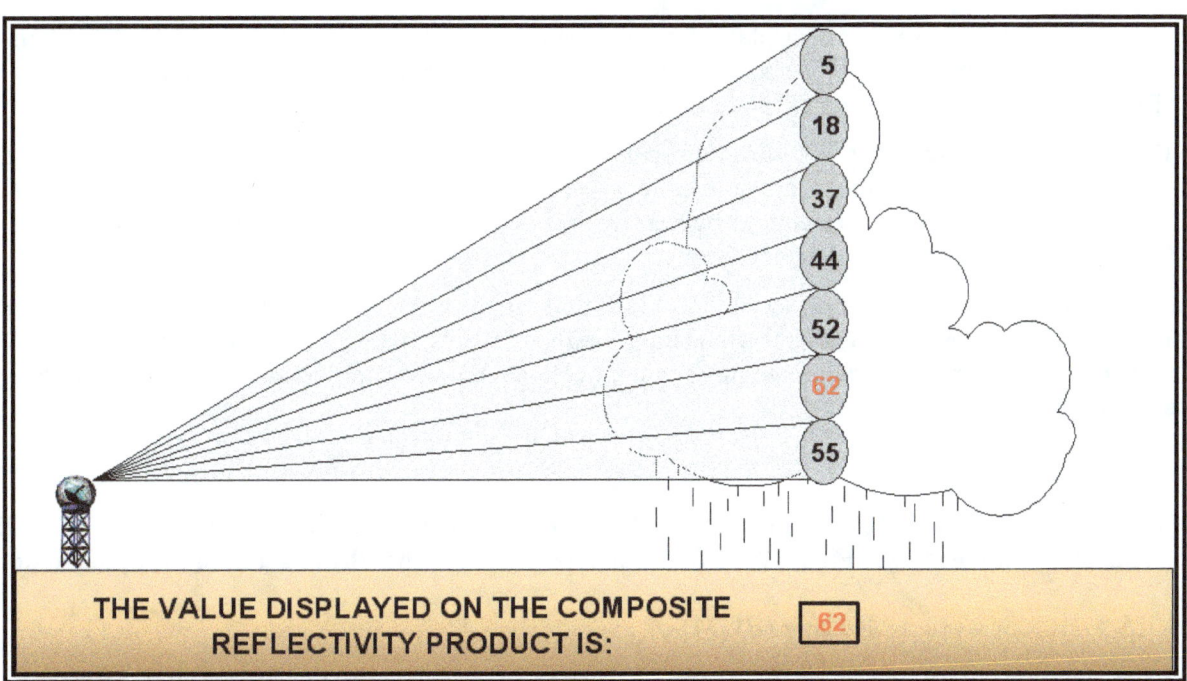

The Composite Reflectivity product displays the highest reflectivity of all elevation scans.

Figure 24-16. Creation of a Composite Reflectivity, Single-Site Product

Chapter 24, Observations

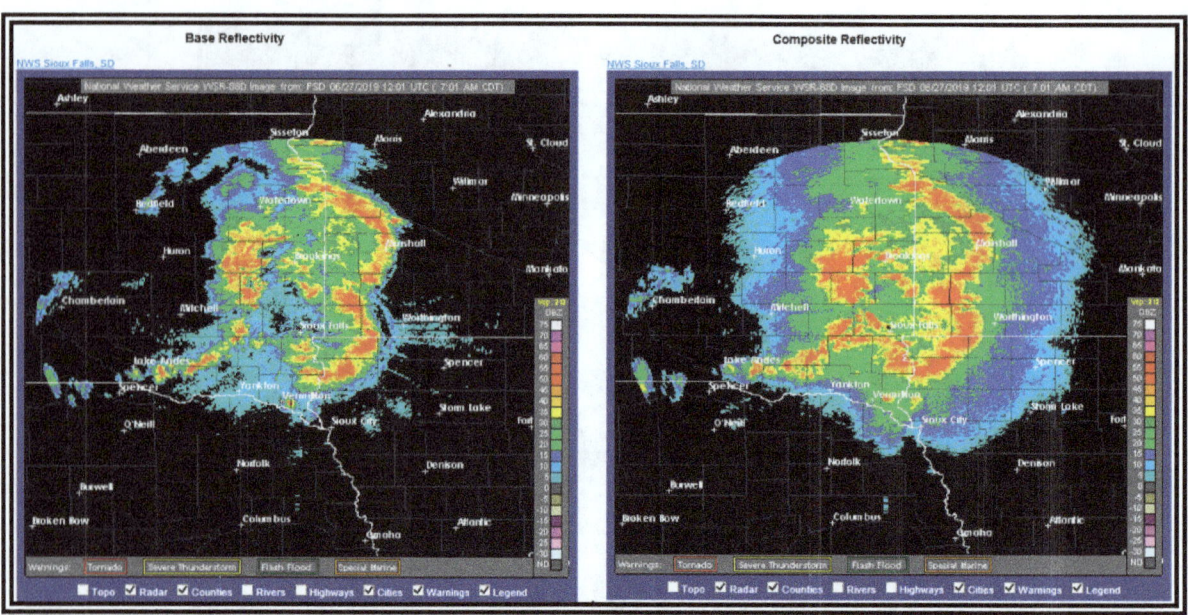

This Composite Reflectivity shows that in many locations the highest precipitation intensity occurs at an altitude higher than precipitation detected at the elevation of the base elevation angle.

Figure 24-17. Weather Radar 0.5° Base Reflectivity (left) versus Composite Reflectivity (right) Comparison

24.6.1.5.3 Base Reflectivity

The Base Reflectivity product is a display of both the location and intensity of reflectivity data from the lowest elevation angle scan, or 0.5° above the horizon. Base Reflectivity is also known as "Lowest Tilt" and "Reflectivity at Lowest Altitude," depending on the website or weather data service provider.

The Base Reflectivity product is one elevation scan, whereas Composite Reflectivity looks at all elevation scans. Base Reflectivity products are available several minutes sooner than Composite Reflectivity products. Precipitation at any location may be heavier than depicted on the Base Reflectivity image because it is occurring above the lowest elevation angle.

Depending on the radar website or service provider, the range of the Base Reflectivity single-site radar product is either 124 NM or 248 NM. When using a single-site radar (i.e., not using a radar mosaic), echoes farther than 124 NM from the radar site might not be displayed, even if precipitation may be occurring at these greater distances.

24.6.1.5.4 Echo Tops

An Echo Tops product provides an estimation of the top of the precipitation by using the height of 18 dBZ radar echo above sea level (see Figure 24-18). Cloud tops will be higher than the top of the precipitation.

24.6.1.5.5 Radar Coded Message (RCM)

With the deployment of the WSR-88D (NEXRAD) radar network in the 1990s, the manual radar observations, known by the acronyms "ROB," "SD," and "RAREP," were replaced by the RCM. The RCM is the encoded and transmitted report of radar features observed by a WSR-88D radar. The RCM is generated automatically and has no human input or oversight.

The actual RCM is highly detailed, complicated, and not intended for use by the pilot or operator in raw format. However, the RCM is used to generate certain radar charts or displays, such as the current "RCM

Radar Plot," available on the AWC's website (see Figure 24-18), as well as other weather service providers' websites.

The RCM display is updated every 30 minutes. The display is an image representation of the NEXRAD radar Composite Reflectivity overlaid with cloud and echo tops and centroid movement. The RCM display includes the maximum echo top for each radar's area of coverage. The other tops shown on the display are derived from the satellite images.

The echo tops shown on the display can be erroneous in some cases. The algorithm computes echo tops (the altitude of echoes greater than or equal to 18 dBZ) within the range of the radar. This can lead to spurious values when precipitation is far from the site. Values greater than 50,000 ft (500 ft on the plot) can be disregarded, especially if their locations do not correspond to any precipitation.

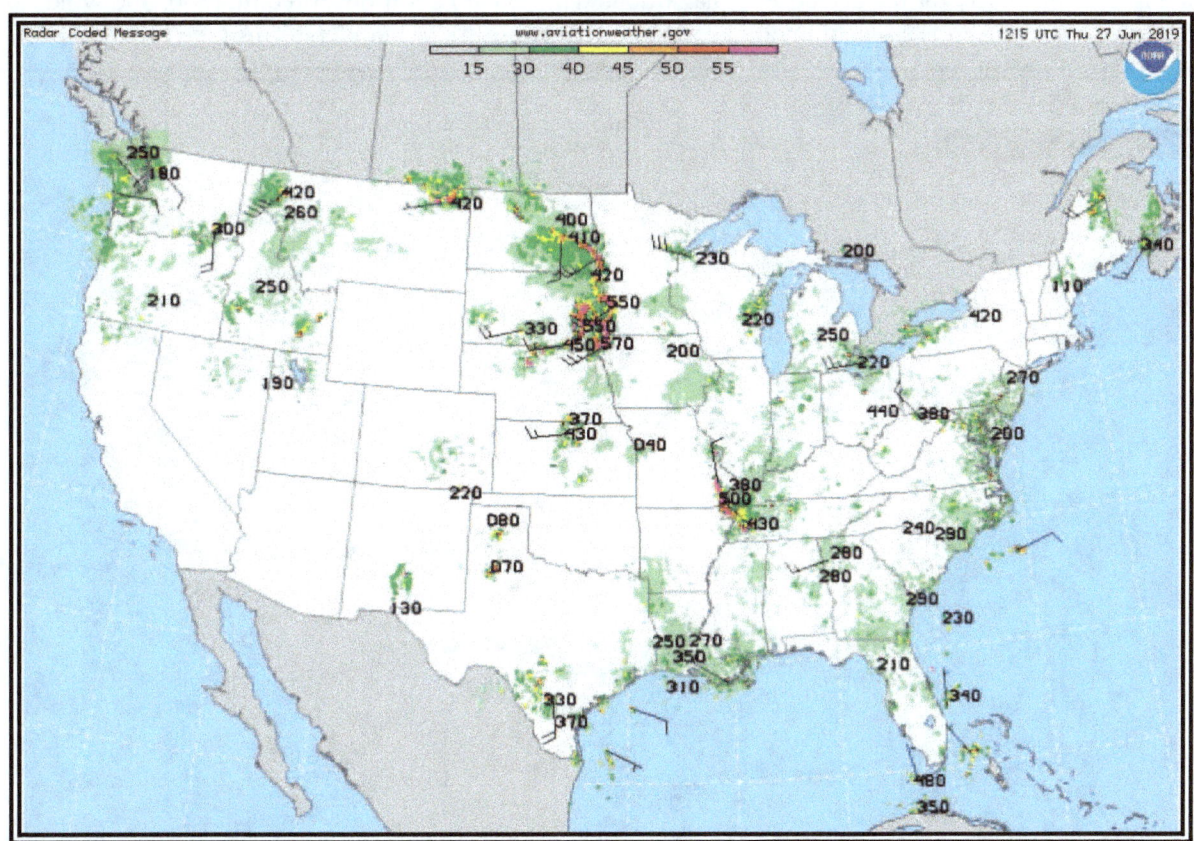

Figure 24-18. RCM Display

24.6.1.6 Limitations

Limitations associated with Composite Reflectivity and Base Reflectivity images include:

- The radar beam may overshoot or undershoot targets (see Section 15.2.8); and
- The image may be contaminated by (see Sections 15.2.9 through 15.2.14):
    - Beam blockage,
    - Ground clutter,
    - AP,
    - Ghosts,

- Angels, and/or
- Other nonmeteorological phenomena.

Limitations associated with mosaics include:

- Datalinked mosaic radar imagery shows where the precipitation *was*, not where the precipitation *is* occurring. The displayed radar precipitation may be 15 to 20 minutes older than the age indicated on the display.

24.6.2 Terminal Doppler Weather Radar (TDWR)

The terminal Doppler weather radar (TDWR) is a Doppler weather radar system operated by the FAA, which is used primarily for the detection of hazardous wind shear conditions, precipitation, and winds aloft on and near major airports situated in climates with great exposure to thunderstorms (see Figure 24-19).

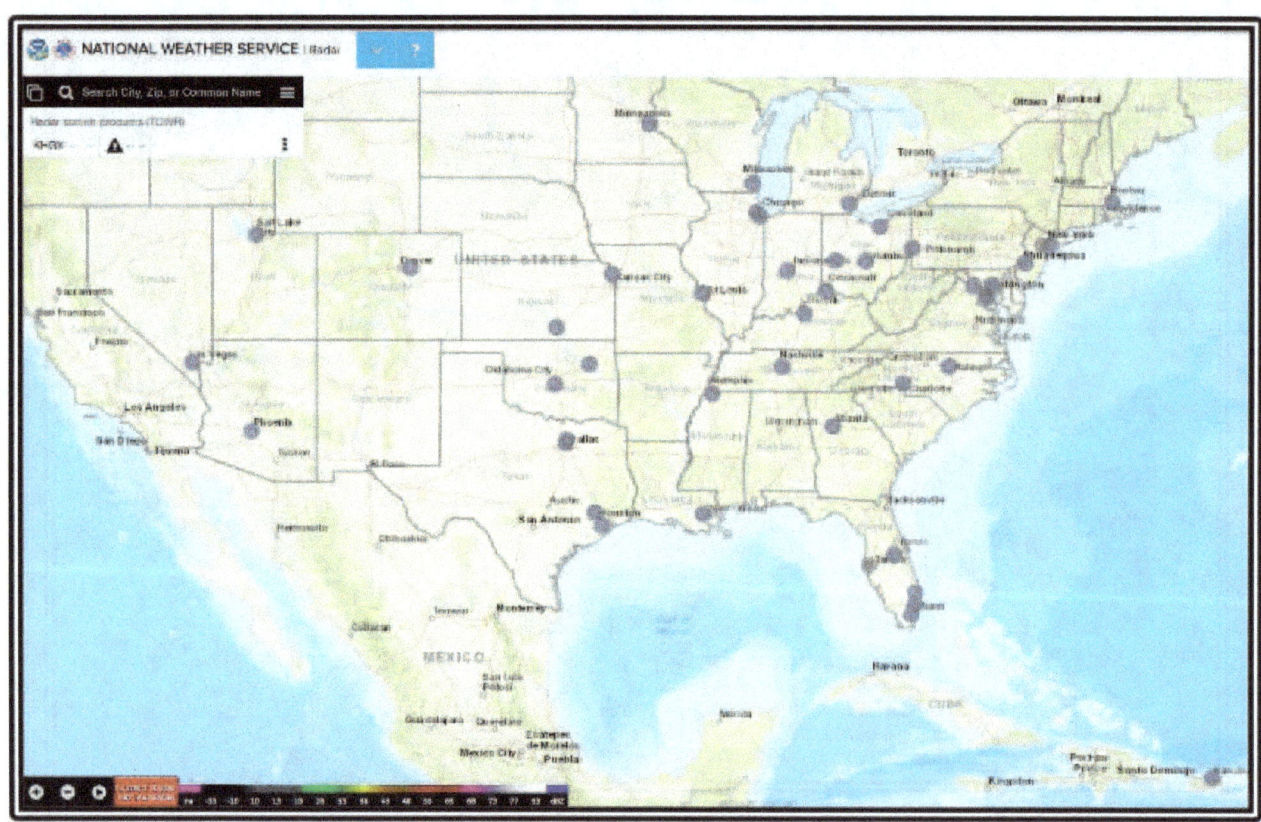

TWDR locations are indicated by gray circles.

Figure 24-19. TDWR Locations in the CONUS and Puerto Rico

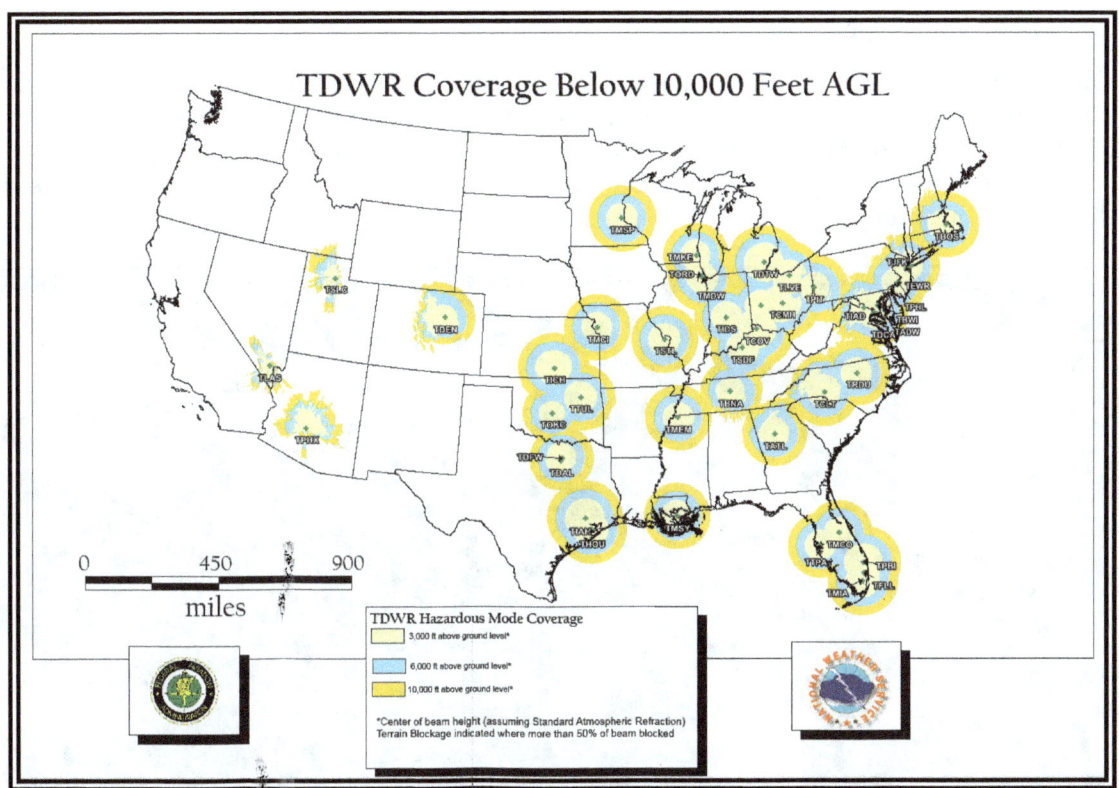

Figure 24-20. TDWR Coverage

TDWR has long- and short-range scans as well as monitor and hazardous weather modes. Update times vary from around 5 minutes in monitor mode to 1 minute in hazardous weather mode. Select TDWR products are available on NWS radar websites.

24.7 Satellite Observations

24.7.1 Description

Satellite is perhaps the single most important source of weather data worldwide, particularly over data-sparse regions, such as countries without organized weather data collection and the oceans.

There are two kinds of weather satellite: geostationary and polar-orbiting. Geostationary satellites are located about 22,000 mi above the equator and orbit the Earth at a very fast speed that effectively makes them appear stationary. Polar-orbiting satellites orbit the Earth at a much lower altitude. Their track is such that the Earth rotates underneath the satellite as it orbits from the North Pole to the South Pole, then back to the North Pole.

The vast majority of weather satellite imagery for aviation comes from NOAA's Geostationary Operational Environmental Satellites (GOES).

24.7.2 Imagery Types

Four types of satellite imagery are commonly used: GeoColor, visible, infrared (IR), and water vapor. Visible imagery is only useful during daylight hours. IR and water vapor imagery are useful day or night.

24.7.2.1 GeoColor Imagery

GeoColor imagery (see Figure 24-21) is a multispectral product composed of true color (using a simulated green component) during the daytime, and an IR product at night. During the day, the imagery looks approximately as it would appear when viewed with human eyes from space. At night, the blue colors represent liquid water clouds such as fog and stratus, while gray to white indicate higher ice clouds, and the city lights come from a static database.

Figure 24-21. GeoColor Satellite Image—U.S. Example

24.7.2.2 Visible Imagery

Visible imagery (see Figure 24-22) displays reflected sunlight from the Earth's surface, clouds, and particulate matter in the atmosphere. Visible satellite images, which look like black and white photographs, are derived from the satellite signals. Clouds usually appear white, while land and water surfaces appear in shades of gray or black.

The visible channel senses reflected solar radiation. Clouds, the Earth's atmosphere, and the Earth's surface all absorb and reflect incoming solar radiation. Since visible imagery is produced by reflected sunlight (radiation), it is only useful during daylight.

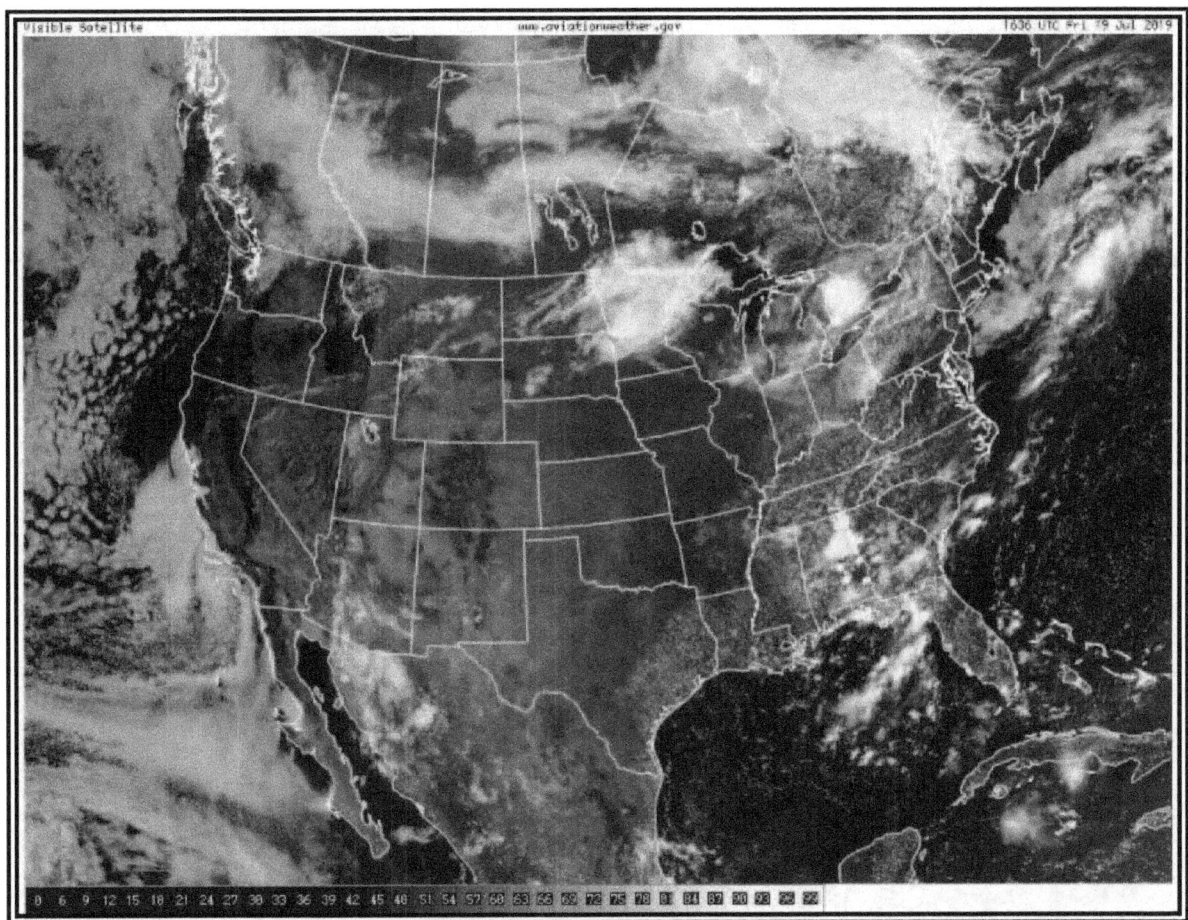

Figure 24-22. Visible Satellite Image—U.S. Example

24.7.2.2.1 Visible Image Data Legend

The data legend on a visible image displays albedo, or reflectance, expressed as a percentage (see Figure 24-23). For example, an albedo of 72 means 72 percent of the sunlight that struck a feature was reflected back to space.

The gray shades (values) represent albedo, or reflectance, expressed as a percentage.

Figure 24-23. Visible Satellite Image Data Legend

24.7.2.3 Infrared (IR) Imagery

IR images (see Figure 24-24 and Figure 24-25) display temperatures of the Earth's surface, clouds, and particulate matter. Generally speaking, the warmer an object, the more IR energy it emits. The satellite sensor measures this energy and calibrates it to temperature using a very simple physical relationship.

Clouds that are very high in the atmosphere are generally quite cold (e.g., -50 °C), whereas clouds very near the Earth's surface are generally quite warm (e.g., +5 °C). Likewise, land may be even warmer than the lower clouds (e.g., +20 °C). Those colder clouds emit much less IR energy than the warmer clouds, and the land emits more than those warm clouds.

The data measured by satellite is calibrated and colorized according to the temperature. If the temperature of the atmosphere decreases with height (which is typical), cloud-top temperature can be used to roughly determine which clouds are high-level and which are low-level.

When clouds are present, the temperature displayed on the IR images is that of the tops of clouds. When clouds are not present, the temperature is that of the ground or the ocean. A major advantage of the IR channel is that it can sense energy at night, so this imagery is available 24 hours a day.

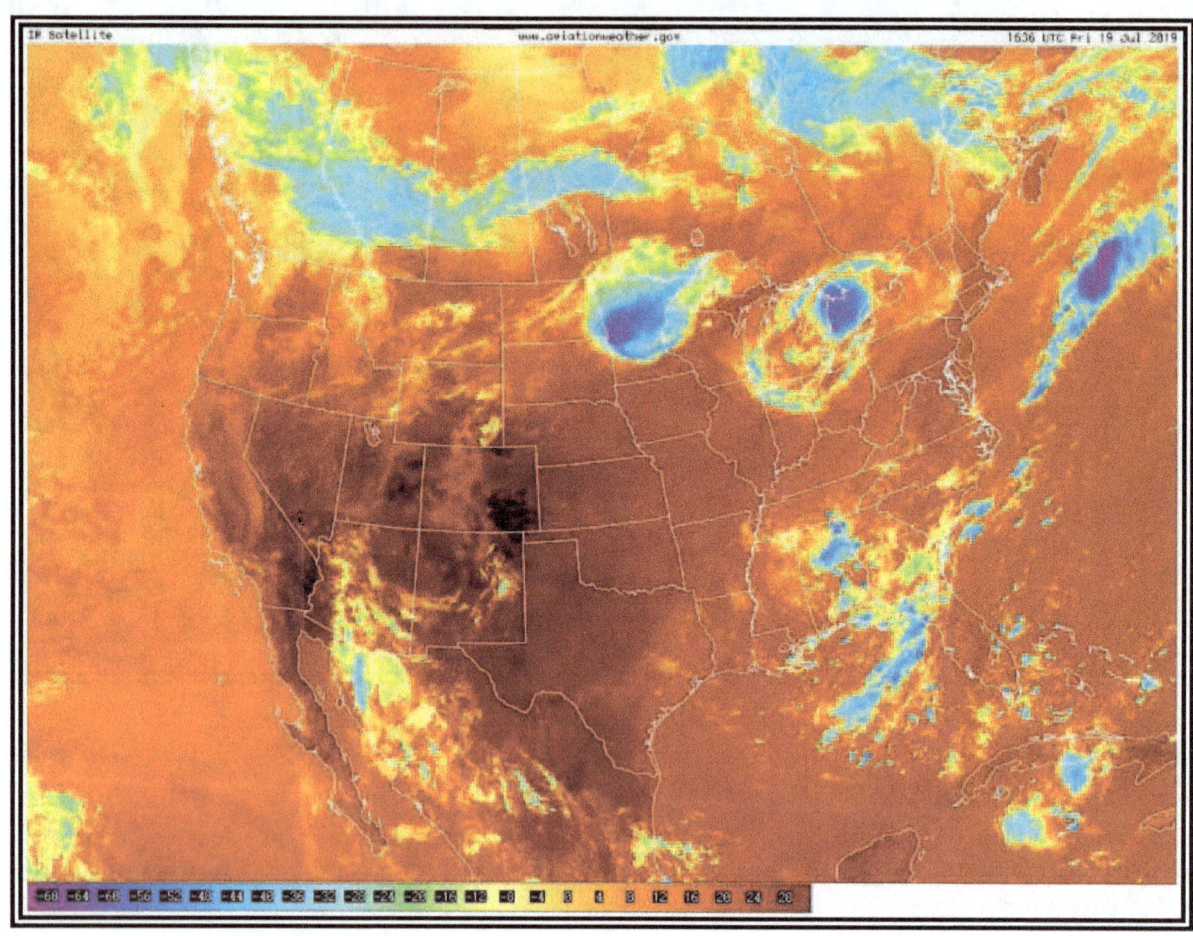

The scale is in degrees Celsius. Blue/purple colors indicate colder temperatures, while orange/red colors indicate warmer temperatures.

Figure 24-24. Infrared (Color) Satellite Image—U.S. Example

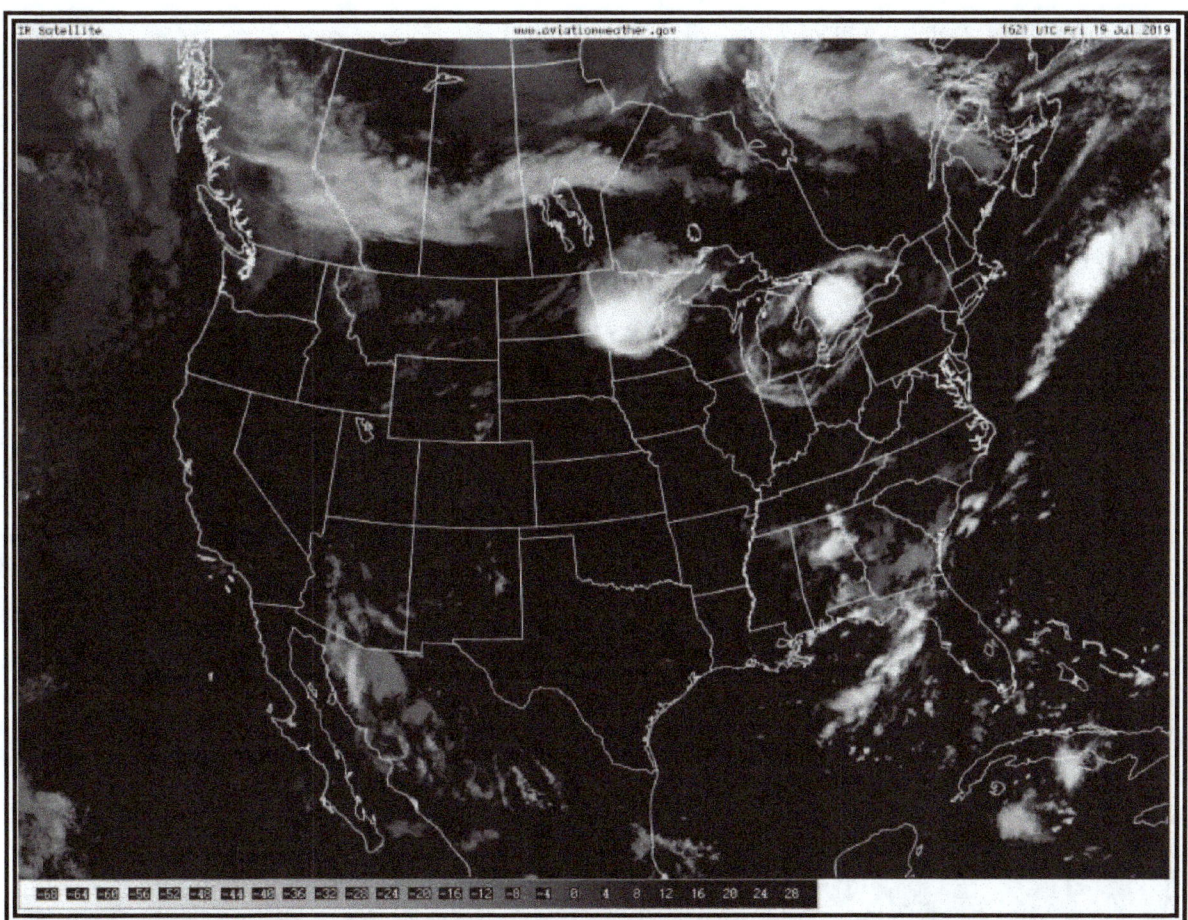

The scale is in degrees Celsius. Lighter gray shades indicate colder temperatures, while darker gray shades indicate warmer temperatures.

Figure 24-25. Unenhanced Infrared (Black and White) Satellite Image—U.S. Example

24.7.2.3.1 Infrared Image Data Legends

The data legend on an IR image is calibrated to temperature expressed in degrees Celsius (see Figure 24-26 and Figure 24-27). The legend may vary based on the satellite image provider.

The colors (values) represent temperature in degrees Celsius.

Figure 24-26. Infrared (Color) Satellite Image Data Legend

The gray shades (values) represent temperature in degrees Celsius.

Figure 24-27. Unenhanced Infrared (Black and White) Satellite Image Data Legend

24.7.2.4 Water Vapor Imagery

Water vapor imagery (see Figure 24-28) displays the quantity of water vapor generally located in the middle and upper troposphere within the layer between 700 mb (approximately 10,000 ft MSL) and 200 mb (approximately FL390). The actual numbers displayed on the water vapor images correspond to temperature in degrees Celsius. No direct relationship exists between these values and the temperatures of clouds, unlike IR imagery. Water vapor imagery does not really "see" clouds, but "sees" high-level water vapor instead.

The most useful information to be gained from the water vapor images is the locations and movements of weather systems, jet streams, and thunderstorms. Another useful tidbit is aided by the color scale used on the images. In general, regions displayed in shades of red are *very* dry in the upper atmosphere and *may* correlate to crisp, blue skies from a ground perspective. On the contrary, regions displayed in shades of blue or green are indicative of a lot of high-level moisture and may also indicate cloudiness. This cloudiness could simply be high-level cirrus types or thunderstorms. That determination cannot be ascertained from this image by itself, but could easily be determined when used in conjunction with corresponding visible and IR satellite images. A major advantage of the water vapor channel is that it can sense energy at night, so this imagery is available 24 hours a day.

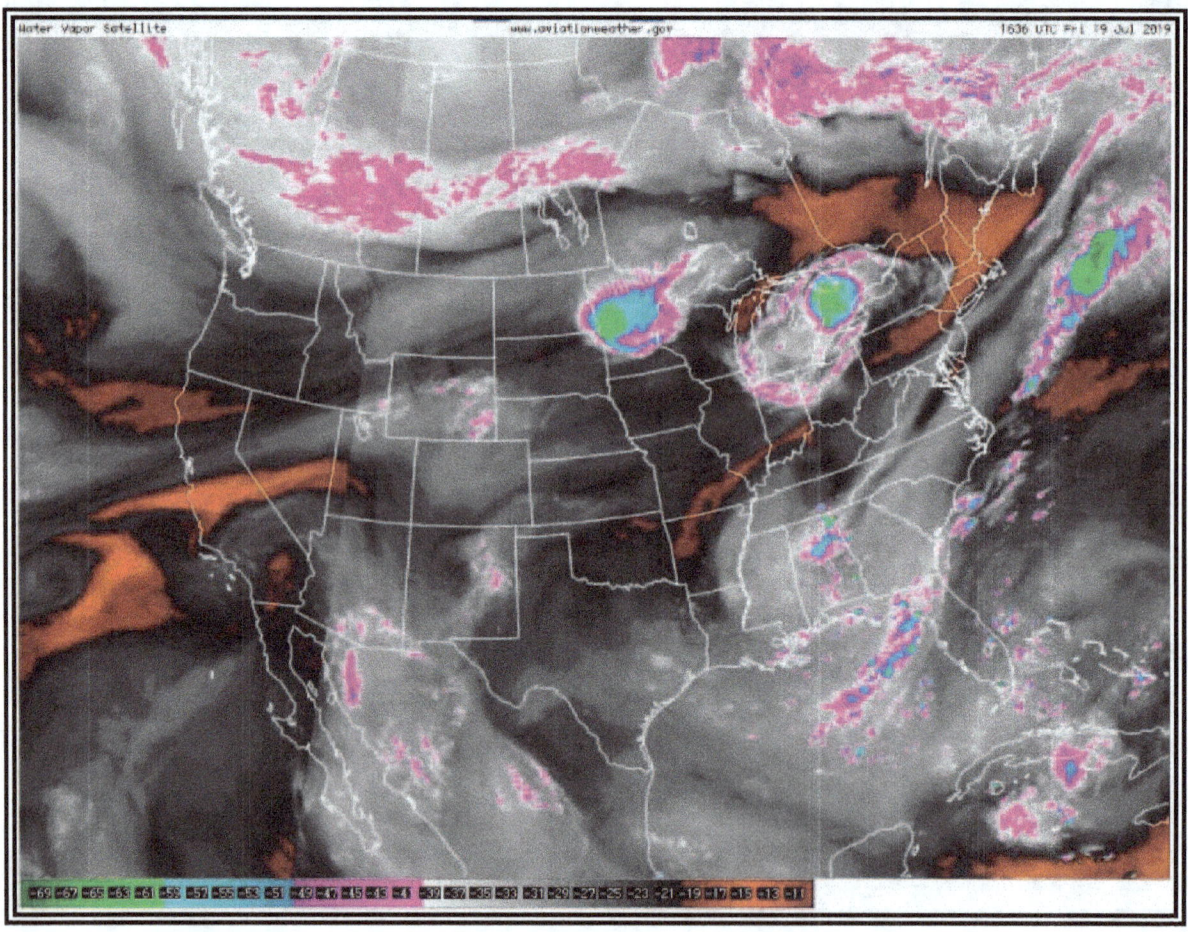

The scale is in degrees Celsius. Blue/green colors indicate moisture and/or clouds in the mid/upper troposphere, while dark gray/orange/red colors indicate dry air in the mid/upper troposphere.

Figure 24-28. Water Vapor Satellite Image—U.S. Example

24.7.2.4.1 Water Vapor Image Data Legend

The data legend on water vapor images is calibrated to temperature expressed in degrees Celsius (see Figure 24-29). The actual data values on the water vapor images are not particularly useful. Interpretation of the patterns and how they change over time is more important. The legend may vary depending on the satellite image provider.

The colors (values) represent temperature in degrees Celsius.

Figure 24-29. Water Vapor Satellite Image Data Legend

24.7.3 Polar Operational Environment Satellites (POES)

"POES" stands for the Polar Operational Environment Satellites. Polar satellites are not stationary. They track along various orbits around the poles. Typically, they are somewhere between 124 and 1,240 mi above the Earth's surface. The satellites scan the Earth in swaths as they pass by on their tracks.

The NWS AAWU posts POES images on their website.

24.7.3.1 Benefits

Because polar satellites are so much closer to Earth, you can get very high resolution (i.e., better than 0.5 km (or about 5/8 mi)). This allows for weather and surface features to be seen in much greater detail.

This is particularly useful over the poles and arctic areas. The quality of geostationary satellite data degrades the closer you get to the poles, while polar satellite data provides high resolution in those areas.

24.7.3.2 Shortfalls

By far the most significant shortfall is the latency, or the time between the satellite scanning the area and the time that the data is available to a user. Because polar satellites are moving, they cannot continuously transmit to a single station. Instead, there is a series of stations around the globe through which the data is collected. Data is then transmitted from those stations to other locations. At times, it can be several hours old (or more) by the time it reaches operational users in the United States (in polar areas it is much quicker). There are some direct ground stations closer to the United States, which can cut the latency to about 45 minutes when utilizing the newer polar satellites.

24.8 Upper Air Observations

24.8.1 Radiosonde Observations (Weather Balloon)

Since the late 1930s, the NWS has taken routine scheduled upper air observations with radiosondes attached to weather balloons, usually referred to as soundings. Weather data from the radiosondes are foundational to all computer model forecasts produced by the NWS and others.

The radiosonde is a small, expendable instrument package (weighing 100 to 500 g), which consists of radio gear and sensing elements, that is suspended below a large balloon inflated with hydrogen or helium gas (see Figure 24-30). As the radiosonde rises at about 300 m per minute (about 1,000 ft per minute), sensors on the radiosonde measure profiles of pressure, temperature, and moisture. These sensors are linked to a battery-powered radio transmitter that sends the sensor measurements to a ground tracking antenna. Wind speed and direction aloft are also obtained by tracking the position of the radiosonde in flight using the GPS. Most stations around the world take rawinsonde observations. However, meteorologists and other data users frequently refer to a rawinsonde observation as a radiosonde observation.

Figure 24-30. Weather Balloon and Radiosonde

24.8.1.1 Issuance

Weather balloons with radiosondes are launched twice a day worldwide from designated locations at around 1100 UTC and 2300 UTC (see Figure 24-31). It takes approximately 90 minutes for the balloon to reach an altitude of 100,000 ft. The weather data collected is assigned the observation times of 1200 UTC and 0000 UTC. Special radiosondes may be launched at select times for various reasons, including when severe weather is expected in a region.

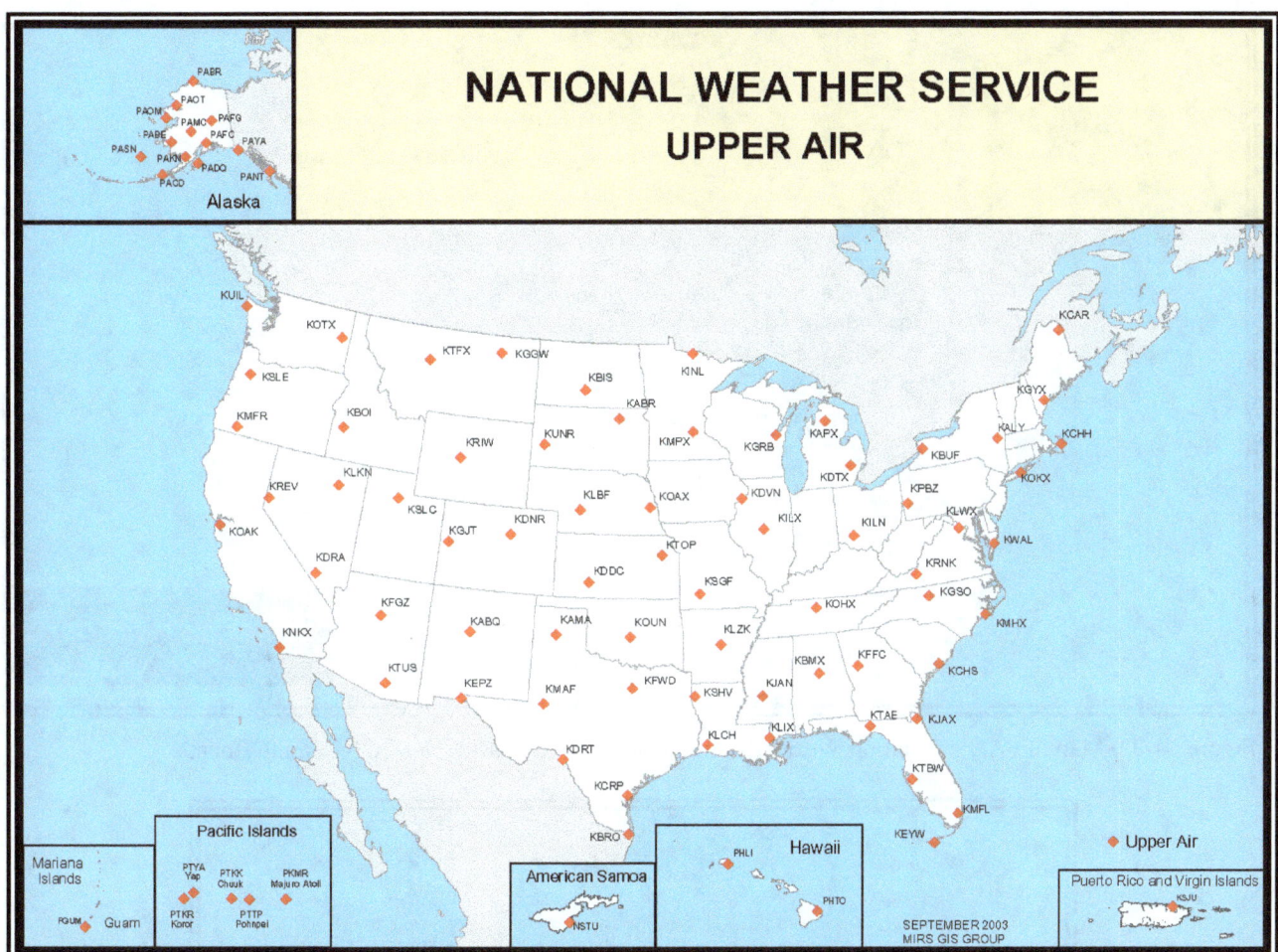

Figure 24-31. U.S. Radiosonde Network

24.9 Aviation Weather Cameras

24.9.1 FAA Aviation Weather Camera Network

In July 2007 the FAA, under the leadership of Walter Combs, began a program to assist Alaska pilots with their significant needs for weather information. Combs' passion for safe flight in Alaska provided the fuel for much improved situational awareness, seeing the weather picture in mountain passes. His work, along with management support, created the Aviation Weather Camera Network.

Flight Service delivers the FAA's Aviation Weather Camera Network (see Figure 24-32 and Figure 24-33), which consists of nearly 500 camera facilities in remote or mountainous locations. Additional sites will be added to the network as they are installed and become operational. Each site has up to four cameras. The direction of each camera (see Figure 24-34) is provided with reference to a Sectional Chart. A "clear day" image (see Figure 24-35) is provided for reference and comparison to the latest image (see Figure 24-36).

In Alaska, locations with cameras are marked on Sectional Charts and listed in Chart Supplement Alaska, Section 2, Airport/Facility Directory, as well as under Section 4, Associated Data, FAA Aviation Camera Locations.

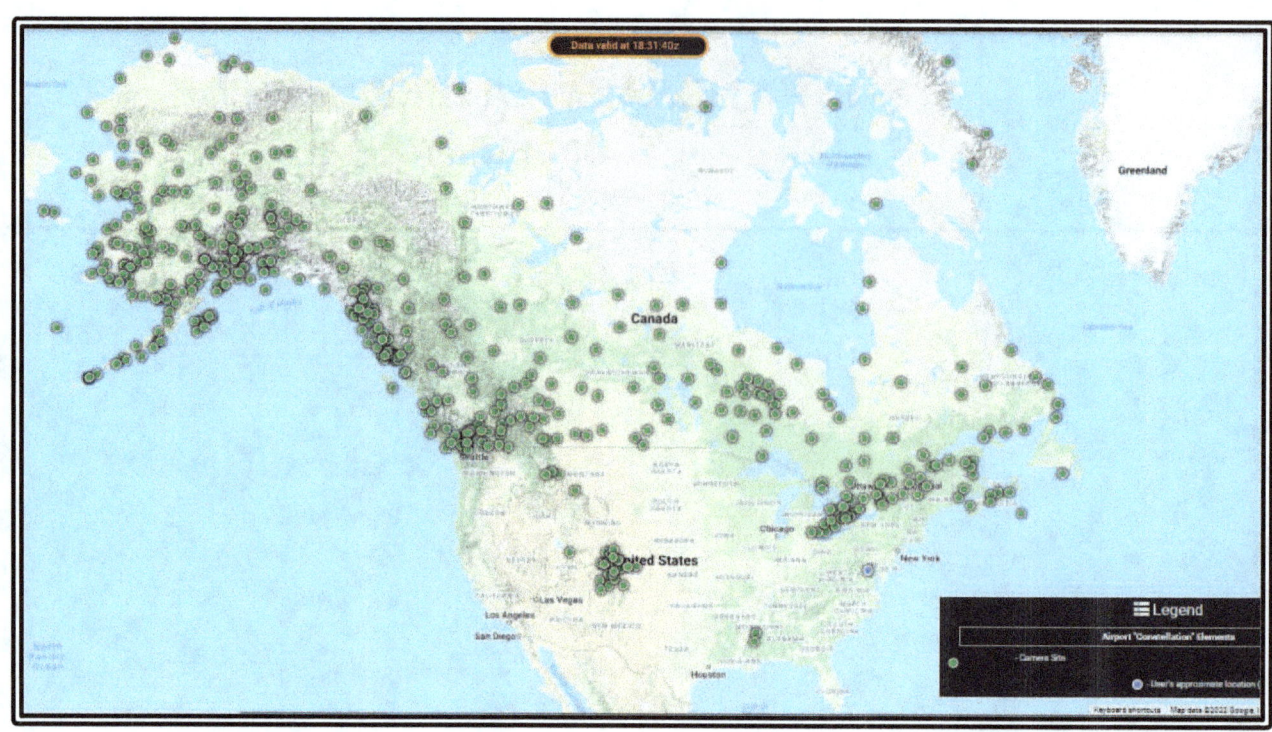

Figure 24-32. Map of FAA's Aviation Weather Camera Network in Alaska, the CONUS, and Canada[12]

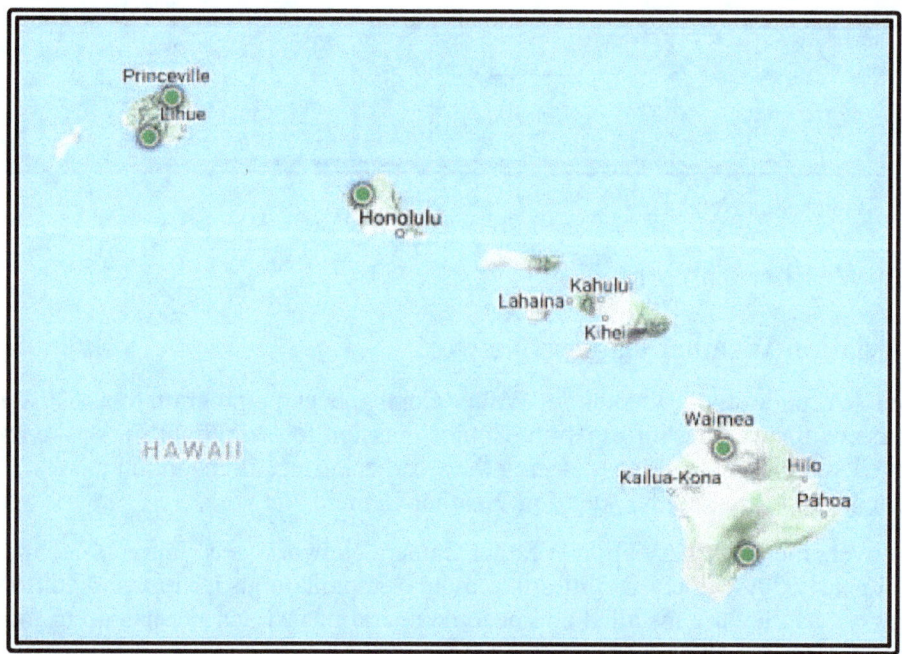

Figure 24-33. Map of the FAA's Aviation Weather Camera Network in Hawaii

---

[12] Canadian camera sites are owned and operated by NAV CANADA, Canada's civilian air navigation service provider.

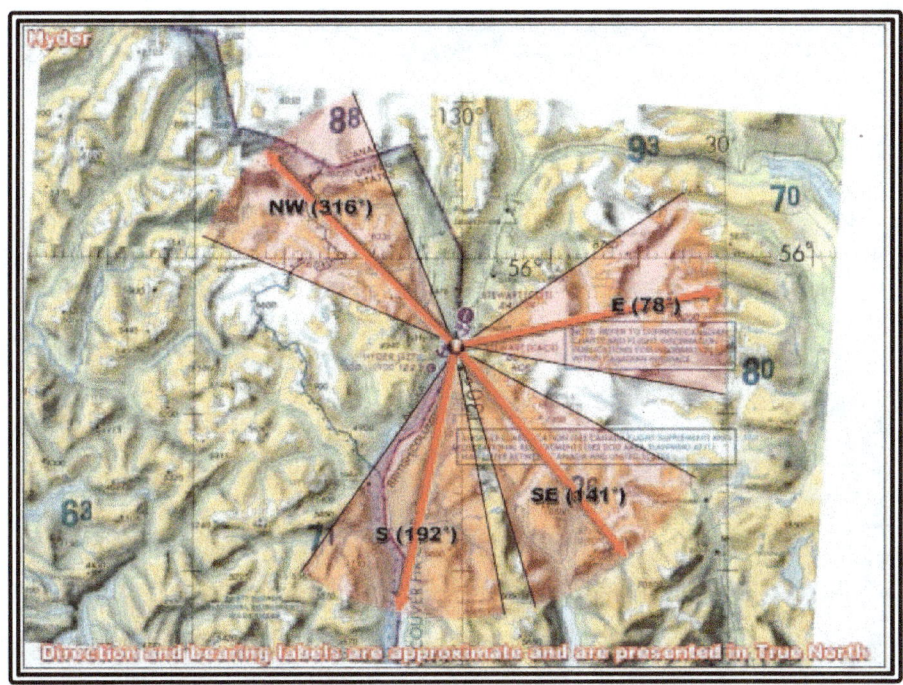

Figure 24-34. Sectional Chart Depicting Hyder (4Z7), Alaska Camera Orientations

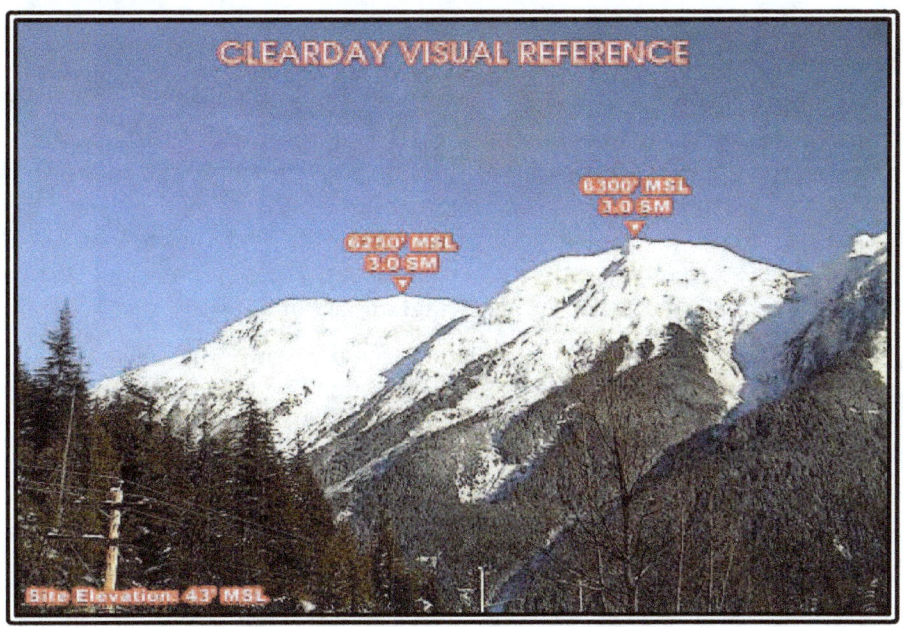

Figure 24-35. "Clear Day" Image from the Aviation Weather Camera at Hyder (4Z7), Alaska

Figure 24-36. Aviation Weather Camera Image at Hyder (4Z7), Alaska

24.9.2 Visibility Estimation through Image Analytics (VEIA)

VEIA, which is available on the FAA's Aviation Weather Camera Network, is an algorithm that uses the FAA's Aviation Weather Camera Network to produce visibility estimates (in statute miles) by analyzing the views in the camera imagery. VEIA uses all of the cameras at a single site to produce one prevailing visibility estimate for that location by identifying and measuring the strength of the edges of permanent features in the landscape/scene and comparing them to a clear day reference. VEIA visibility estimates allow users to quickly access visibility estimates at the departure location, along the route, and at the destination between sunrise and sunset (not during twilight hours or at night).

VEIA visibility estimates are supplemental information, for use with other meteorological information, displayed in conjunction with the specific site's camera images. The visibility estimates are accessible on the FAA's Aviation Weather Camera website. See Figure 24-37 for an example VEIA visibility estimate.

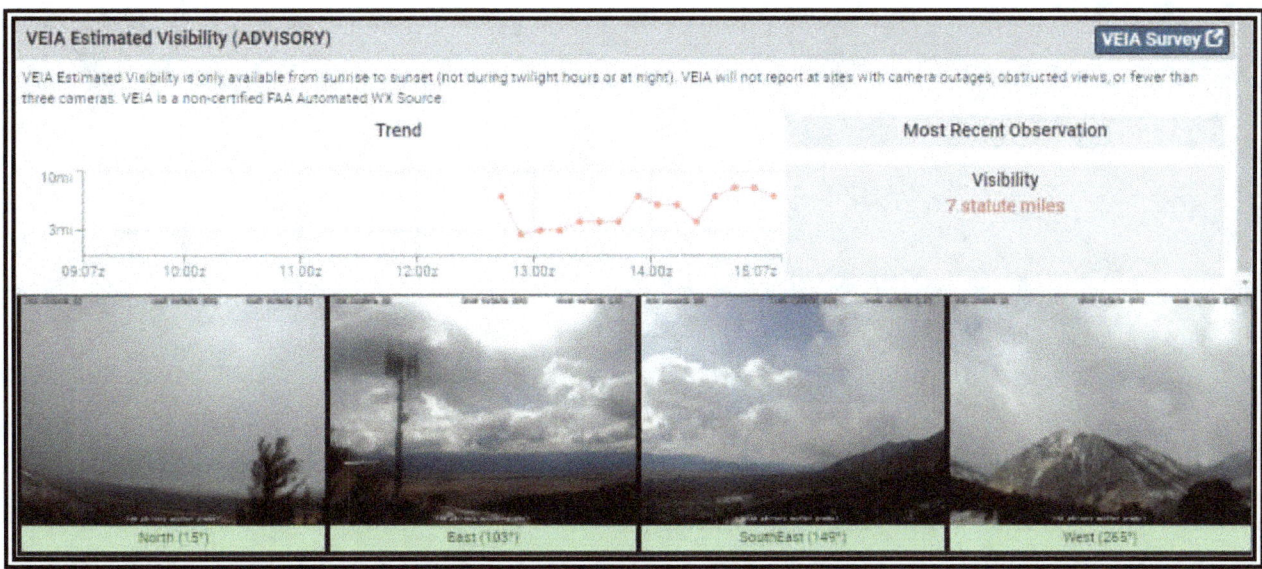

Figure 24-37. VEIA Example at Bald Mountain (K7BM) in Colorado

24.9.3 Visual Weather Observation System (VWOS)

VWOS, which is available on the FAA's Aviation Weather Camera Network, is an advanced camera system that is comprised of a suite of weather sensors and 360-degree camera images that collectively observe and report several critical data fields, including winds, cloud height, visibility, present weather, temperature, dewpoint, and pressure. VWOS uses automated processes to self-check and validate its operations and data outputs. It provides operators with visual and textual weather observations as guidance, for use with other weather information, to make flight decisions into airports that do not possess an AWOS/METAR.

In the future, VEIA will be a core component of VWOS.

24.9.4 Issuance

Camera images are available on the FAA's Aviation Weather Camera website, which can be found in Appendix A. Images are generally updated every 10 minutes. The time of the last update is indicated on each image. Actual site conditions may differ from displayed images due to a variety of reasons (e.g., rapidly changing weather conditions, image update frequency, or optical distortion).

Camera images are to be used to improve situational awareness. They are not to be used to determine weather minimums for VFR or IFR flight, unless authorized by the Administrator of the FAA.

In addition to the camera images, the website also delivers a variety of safety of flight information, including adverse conditions (e.g., AIRMETs and SIGMETs), current conditions (e.g., METARs, radar, satellite imagery, and weather trends (see Figure 24-38), TAFs, PIREPs, and other aeronautical information (e.g., RCOs, TFRs, and charts).

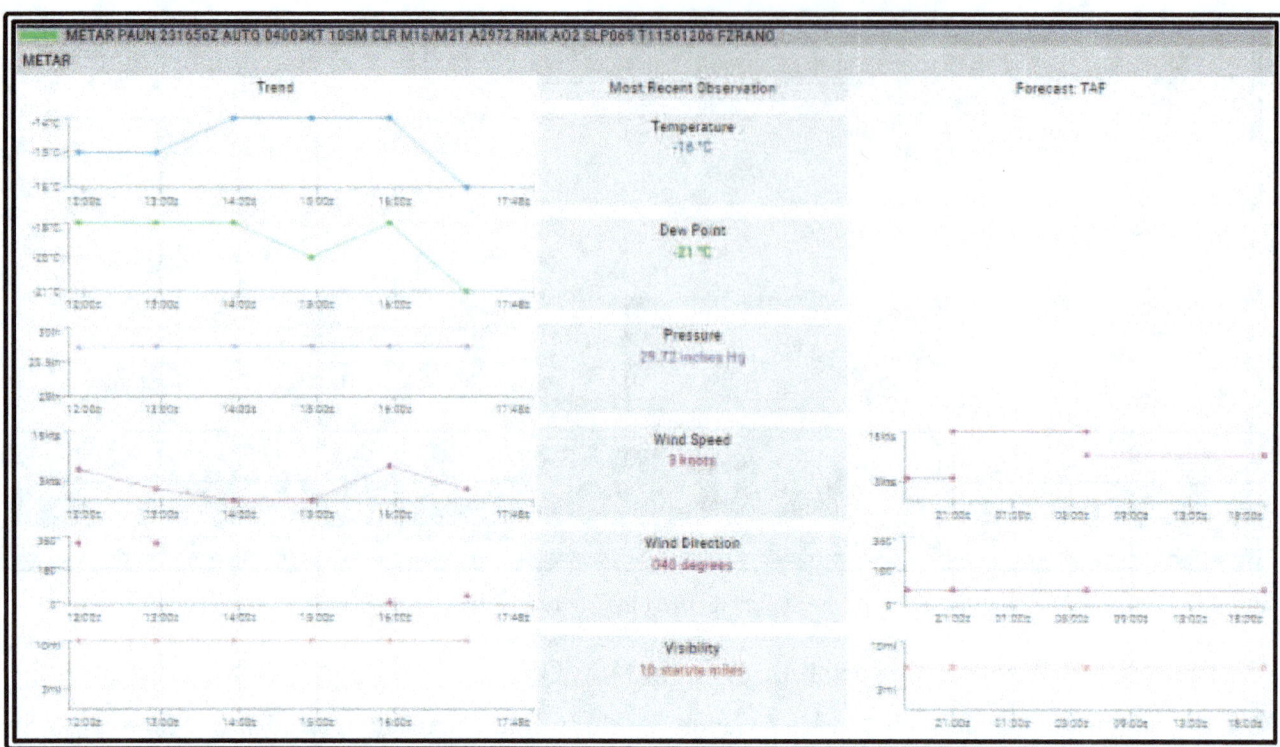

Figure 24-38. Weather Trends Example at Unalakleet (PAUN), Alaska

25 Analysis

25.1 Introduction

The second of five types of aviation weather information discussed in this handbook are analyses. Analyses of weather information are an enhanced depiction and/or interpretation of observed weather data. Prior to the 1990s, most analysis charts were hand-drawn by forecasters. Today's analyses are automated, and depending on the weather information provider (e.g., the NWS, commercial weather services, and flight planning services), the appearance and content of these analyses will vary.

This chapter will only focus on those analyses produced by the NWS and made available on various websites, including the AWC, the WPC, the Ocean Prediction Center (OPC), and the AAWU.

For this handbook, analyses include the following:

- Surface Chart Analysis,
- Upper-Air Analysis,
- Freezing Level Analysis,
- Icing Analysis (Current Icing Product (CIP)),
- Turbulence (Graphical Turbulence Guidance (GTG)) Analysis, and
- Real-Time Mesoscale Analysis (RTMA).

25.2 Weather Charts

A weather chart is a map on which data and analyses are presented that describe the state of the atmosphere over a large area at a given moment in time.

The possible variety of such charts is enormous, but in meteorological history there has been a more or less standard set of charts, including surface charts and the constant-pressure charts of the upper atmosphere. Because weather systems are three dimensional (3D), both surface and upper air charts are needed. Surface weather charts depict weather on a constant-altitude (usually sea level) surface, while upper air charts depict weather on constant-pressure surfaces.

The NWS produces many weather charts that support the aviation community.

25.2.1 Weather Observation Sources

Weather analysis charts can be based on observations from a variety of data sources (see Figure 25-1), including:

- Land surface (e.g., ASOS, AWOS, and mesonet);
- Marine surface (e.g., ship, buoy, Coastal-Marine Automated Network (C-MAN), and tide gauge);
- Sounding (e.g., radiosonde, dropsonde, pibal, profiler, and Doppler weather radar Velocity Azimuth Display (VAD) wind profile);
- Aircraft (e.g., AIREPs and PIREPs), AMDAR, and Aircraft Communications Addressing and Reporting System (ACARS)); and
- Satellite (e.g., GOES sensors that provide temperature, moisture, and wind (through cloud movement)).

Note: Human observers can augment automated reports.

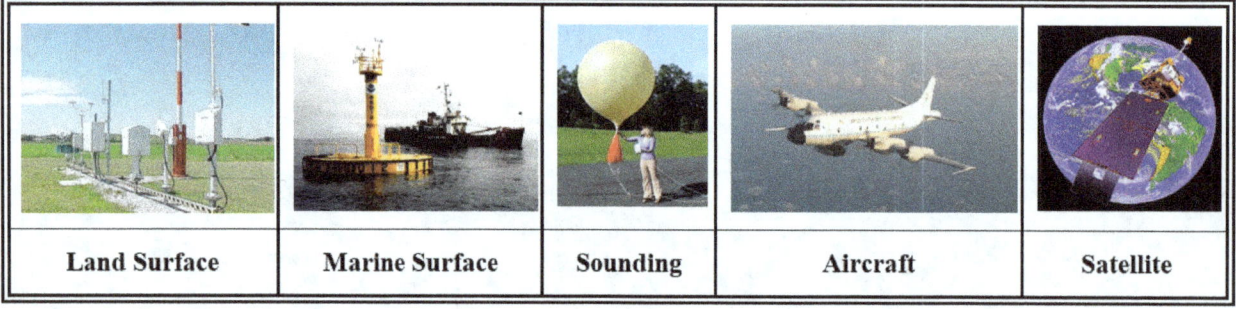

Figure 25-1. Weather Observation Sources

25.2.2 Analysis

Analysis is the drawing and interpretation of the patterns of various elements on a weather chart. It is an essential part of the forecast process. If meteorologists do not know what is currently occurring, it is nearly impossible to predict what will happen in the future. Computers have been able to analyze weather charts for many years and are commonly used in the process. However, computers cannot interpret what they analyze. Thus, many meteorologists still perform a subjective analysis of weather charts when needed.

25.2.2.1 Analysis Procedure

The analysis procedure is similar to drawing in a dot-to-dot coloring book. Just as one would draw a line from one dot to the next, analyzing weather charts is similar in that lines of equal values, or isopleths, are drawn between dots representing various elements of the atmosphere. An isopleth is a broad term for any line on a weather map connecting points with equal values of a particular atmospheric variable. See Table 25-1 for common isopleths.

Table 25-1. Common Isopleths

Isopleth	Variable	Definition
Isobar	Pressure	A line connecting points of equal or constant pressure.
Contour Line (also called Isoheight)	Height	A line of constant elevation above MSL of a defined surface, typically a constant-pressure surface.
Isotherm	Temperature	A line connecting points of equal or constant temperature.
Isotach	Wind Speed	A line connecting points of equal wind speed.
Isohume	Humidity	A line drawn through points of equal humidity.
Isodrosotherm	Dewpoint	A line connecting points of equal dewpoint.

The weather chart analysis procedure begins with a map of the plotted data, which is to be analyzed (see Figure 25-2). It is assumed that bad or obviously incorrect data has been removed before beginning the analysis process. At first, the chart will appear to be a big jumble of numbers. However, when the analysis procedure is complete, patterns will appear, and significant weather features will be revealed.

25.2.2.1.1 Step 1: Determine the Optimal Contour Interval and Values to be Analyzed

The first step in the weather chart analysis procedure is to identify the maxima and minima data values and their ranges to determine the optimal contour interval and values to be analyzed. The best contour interval will contain enough contours to identify significant weather features, but not so many that the chart becomes cluttered. Each weather element has a standard contour interval on NWS weather charts, but these values can be adjusted in other analyses as necessary.

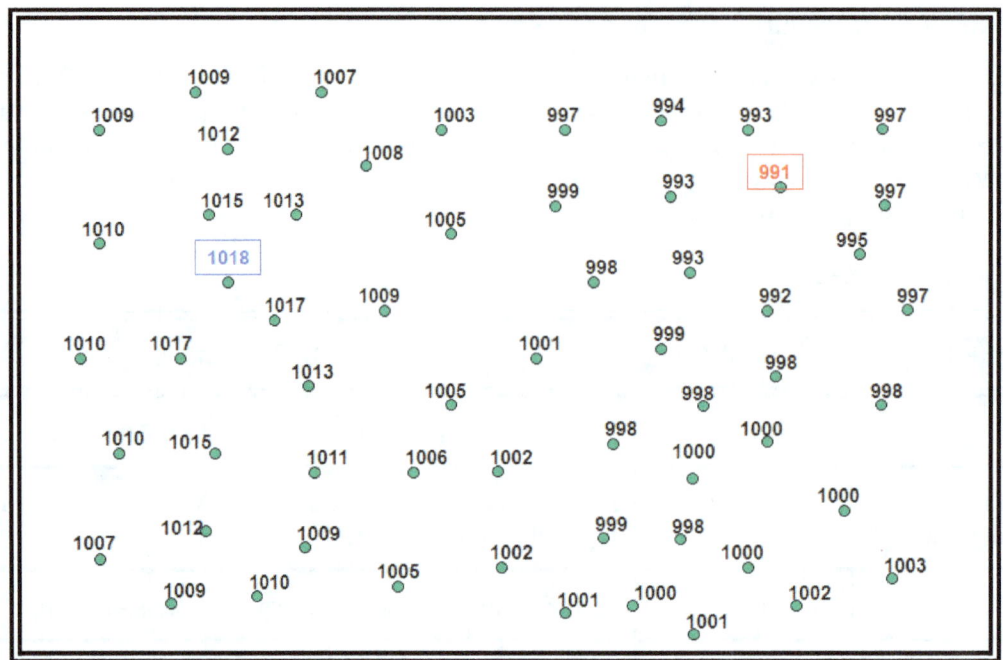

Figure 25-2. Analysis Procedure Step 1: Determine the Optimal Contour Interval and Values to be Analyzed

Every contour value must be evenly divisible by the contour interval. So, for example, if the contour interval is every 4 units, a 40-unit contour is all right, but a 41-unit contour is not. In the surface pressure analysis example shown in Figure 25-2, an isobar analysis will be performed beginning at a value of 992 mb and using a contour interval of 4 mb, which is standard on the NWS Surface Analysis Chart.

25.2.2.1.2 Step 2: Draw the Isopleths and Extrema

The second step is to draw the isopleths and extrema (maxima and minima) using the beginning contour value and contour interval chosen in the first step (see Figure 25-3). It is usually easiest to begin drawing an isopleth either at the edge of the data domain (edge of the chart) or at a data point that matches the isopleth value being drawn. Interpolation must often be used to draw isopleths between data points and determine the extrema. "Interpolate" means to estimate a value within an interval between known values.

When drawing isopleths and extrema on a weather chart, certain rules must be followed:

1. The analysis must remain within the data domain. Analysis must never be drawn beyond the edge of the chart where there are no data points. That would be guessing.

2. Isopleths must not contain waves and kinks between two data points. This would indicate a feature too small to be supported by the data. Isopleths should be smooth and drawn generally parallel to each other.

3. When an isopleth is complete, all data values must be higher than the isopleth's value on one side of the line and lower on the other.

4. A closed-loop isopleth must contain an embedded extremum (maximum or minimum).

5. When a maximum (minimum) is identified, data values must decrease (increase) in all directions away from it.

6. Isopleths can never overlap, intersect, or cross over extrema. It is impossible for one location to have more than one data value simultaneously.

7. Each isopleth must be labeled. A label must be drawn wherever an isopleth exits the data domain. For closed-loop isopleths, a break in the loop must be created where a label can be drawn. For very long and/or complex isopleths, breaks should be created where additional labels can be drawn, as necessary.

8. Extrema must be labeled. Extrema are often denoted by an "x" embedded within a circle. Beneath the label, the analyzed value of the field must be written and underlined.

9. Isopleths and labels should not be drawn over the data point values. If necessary, breaks in the isopleths should be created so that the data point values can still be read.

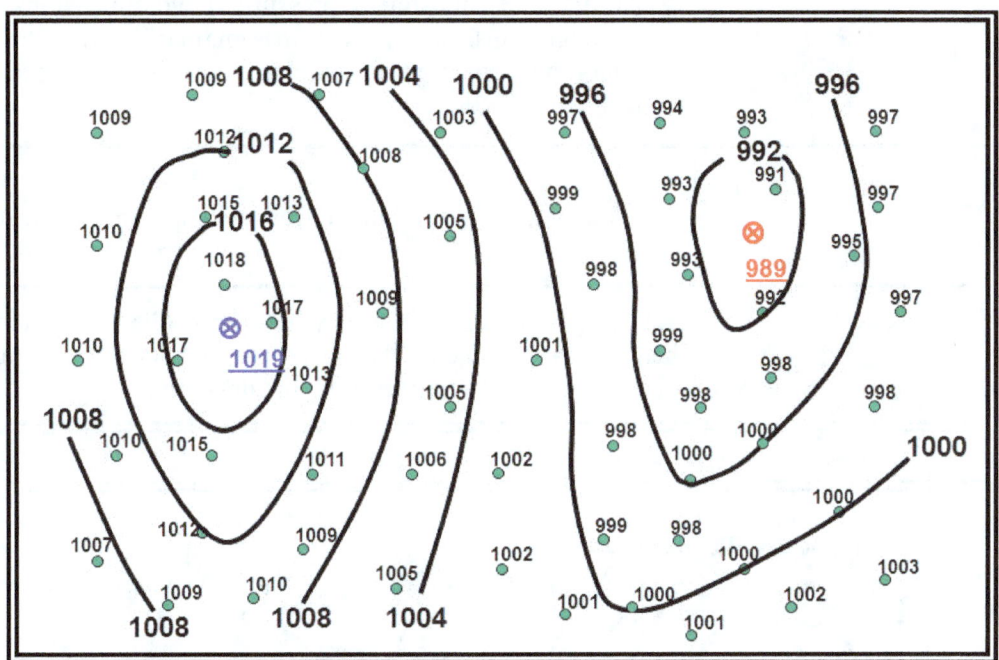

Figure 25-3. Analysis Procedure Step 2: Draw the Isopleths and Extrema

25.2.2.1.3 Step 3: Identify Significant Weather Features

The third (and final) step is to interpret significant weather features. The conventional labels for extrema are H (high) and L (low) for pressure and height, W (warm) and K (cold) for temperature (they stand for the German words for warm and cold), and X (maxima) and N (minima) for all other elements. Tropical storms, hurricanes, and typhoons are low-pressure systems with their names and central pressures denoted. Troughs, ridges, and other significant features are often identified as well.

For surface analysis charts, positions and types of fronts are shown by symbols in Figure 25-4. The symbols on the front indicate the type of front and point in the direction toward which the front is moving. Two short lines across a front indicate a change in front type.

Table 25-2 provides the most common weather chart symbols.

In the surface pressure analysis in Figure 25-4, a high, low, trough, and ridge have been identified.

Table 25-2. Common Weather Chart Symbols

Feature	Symbol	Definition
Low	L	A minimum of atmospheric pressure in two dimensions (closed isobars) on a surface chart, or a minimum of height (closed contours) on a constant-pressure chart. Also known as a cyclone.
High	H	A maximum of atmospheric pressure in two dimensions (closed isobars) on a surface chart, or a maximum of height (closed contours) on a constant-pressure chart. Also known as an anticyclone.
Trough	— — —	An elongated area of relatively low atmospheric pressure or height.
Ridge	⋀⋀⋀	An elongated area of relatively high atmospheric pressure or height. May also be used as reference to other meteorological quantities, such as temperature and dewpoint.

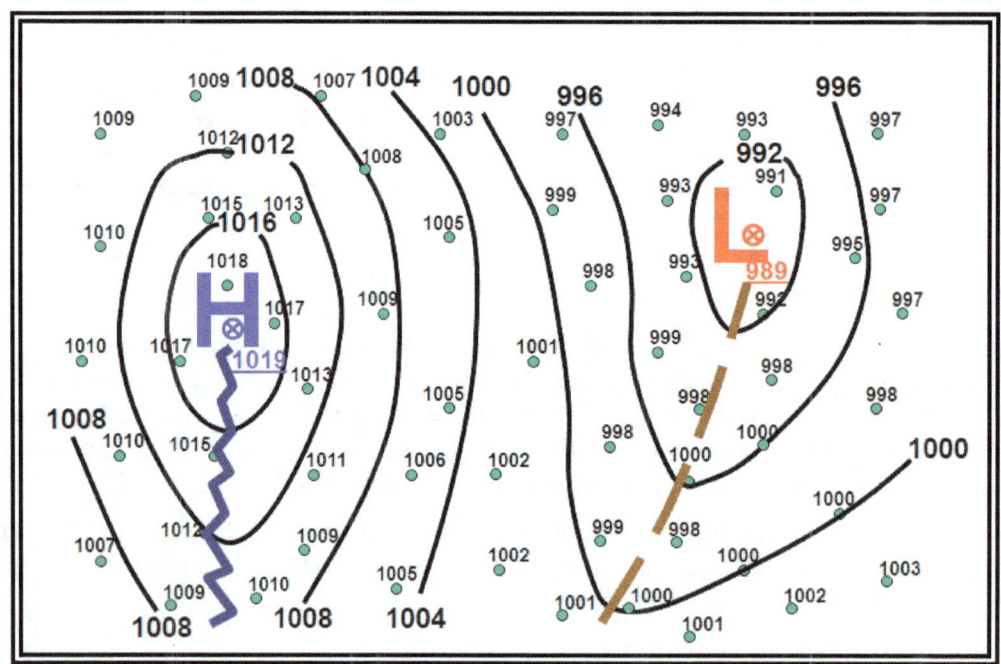

Figure 25-4. Analysis Procedure Step 3: Interpret Significant Weather Features

25.2.3 Surface Analysis Chart

The WPC in College Park, MD, produces a variety of surface analysis charts for North America that are available on their website. The WPC's surface analysis is also available on the AWC's and other providers' websites.

A surface chart (also called surface map or sea level pressure chart) is an analyzed chart of surface weather observations. Essentially, a surface chart shows the distribution of sea level pressure (lines of equal pressure are isobars). Hence, the surface chart is an isobaric analysis showing identifiable, organized pressure patterns. The chart also includes the positions of highs, lows, ridges, and troughs, and the location and character of fronts and various boundaries, such as drylines, outflow boundaries, and sea breeze fronts. Although the pressure is referred to as MSL, all other elements on this chart are presented as they occur at the surface point of observation. A chart in this general form is the one commonly referred to as the weather map. See Figure 25-5 for a schematic example of a surface chart.

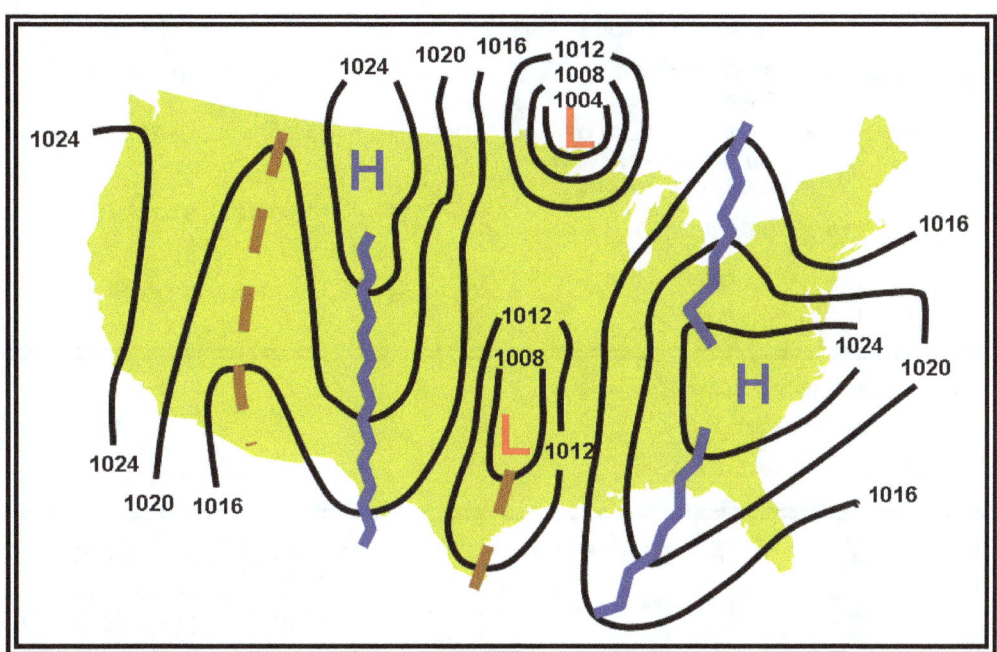

Figure 25-5. Schematic of Surface Chart Pressure Patterns

Figure 25-7 is one example of several surface analysis products available on the WPC's website.

Some of the WPC's surface analysis charts are combined with radar or satellite imagery (see Figure 25-8 and Figure 25-9) as well as having different background features (e.g., terrain).

25.2.3.1 Issuance

The WPC issues surface analysis charts for North America eight times daily, valid at 00, 03, 06, 09, 12, 15, 18, and 21 UTC.

25.2.3.2 Analysis Symbols

Figure 25-6 shows analysis symbols used on NWS surface analysis charts.

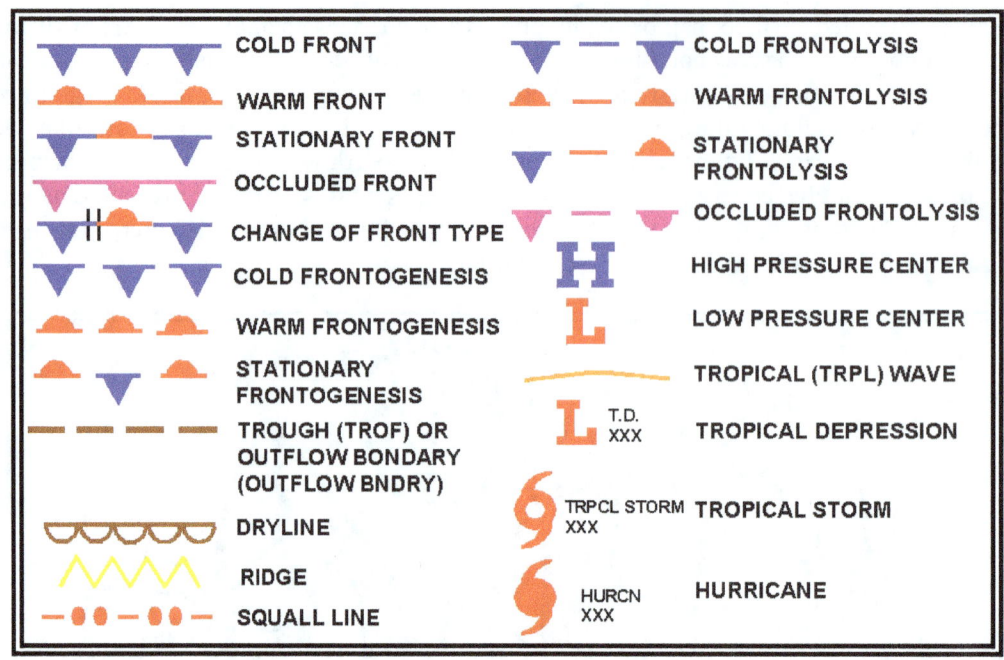

Figure 25-6. NWS Surface Analysis Chart Symbols

25.2.3.3 Examples

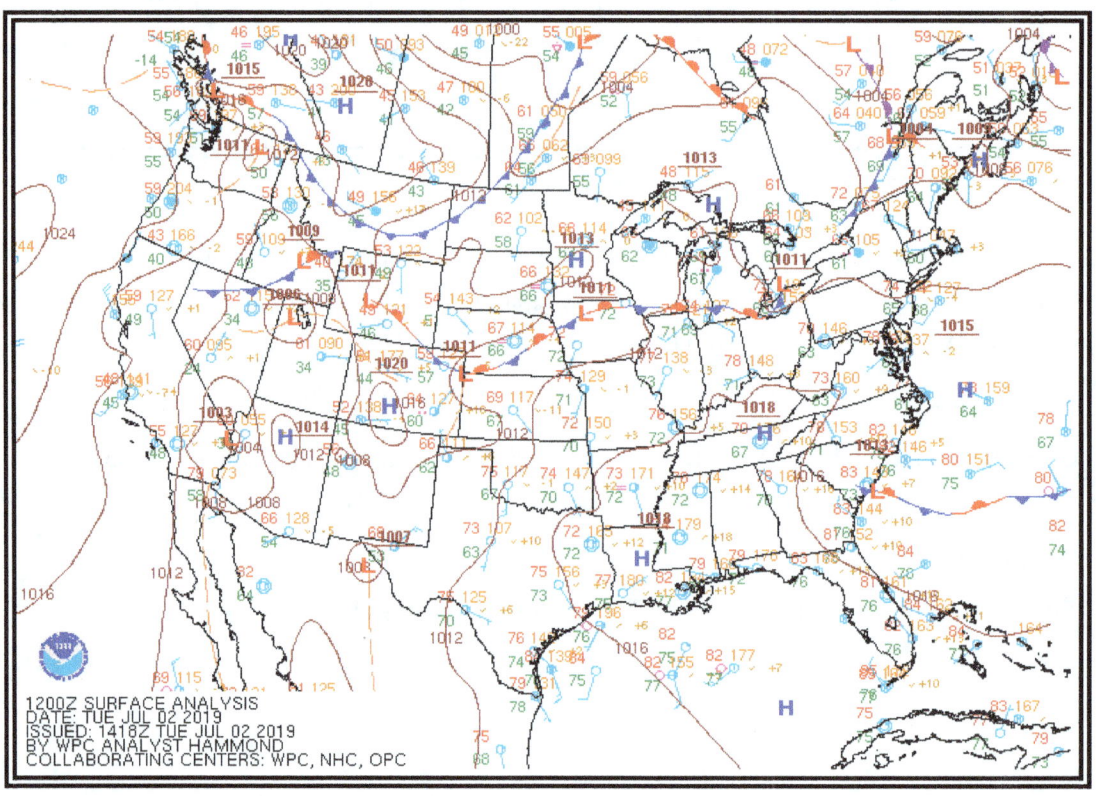

Figure 25-7. Example of a Surface Chart with Surface Observations

Chapter 25, Analysis

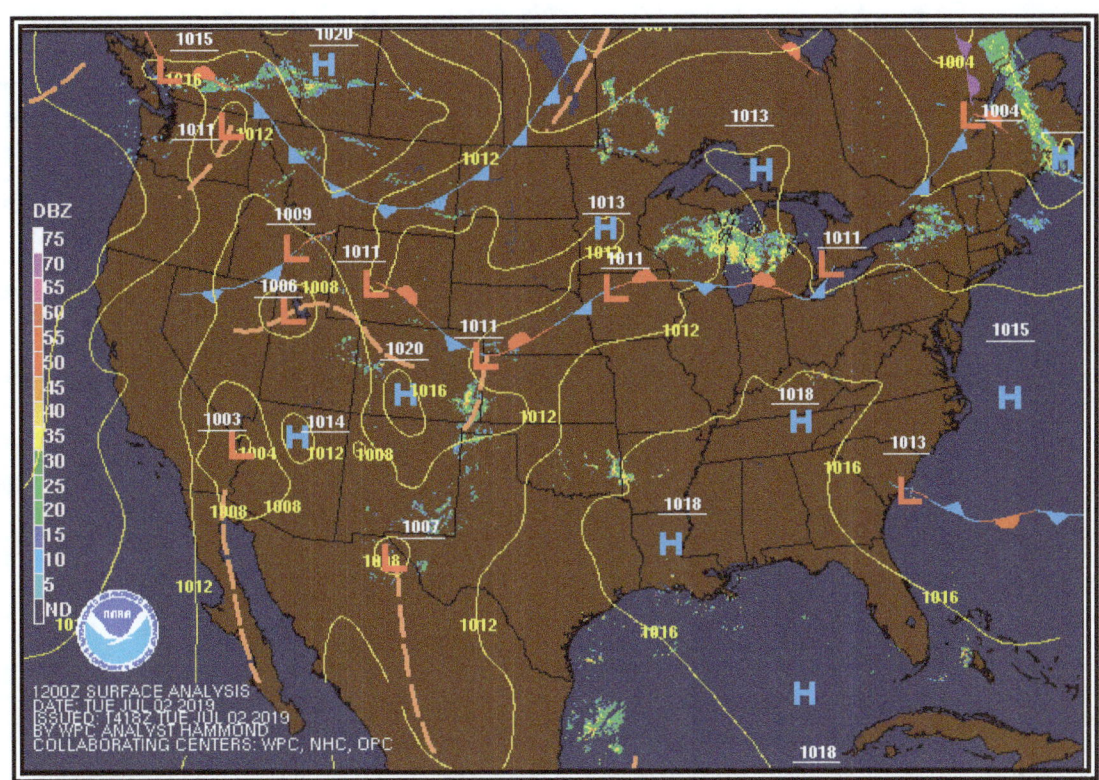

Figure 25-8. Surface Analysis with Radar Composite Example

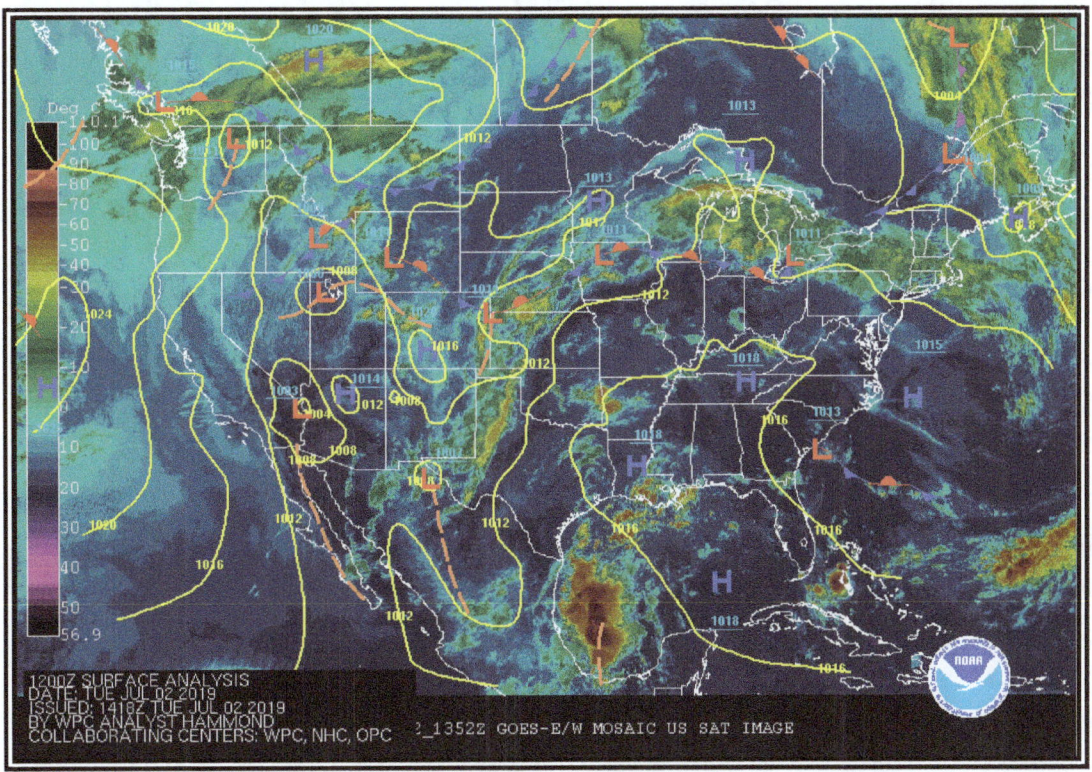

Figure 25-9. Surface Analysis with Satellite Composite Example

Chapter 25, Analysis

25.2.3.4 Station Plot Models

Land, ship, buoy, and C-MAN stations are plotted on the chart to aid in analyzing and interpreting the surface weather features. These plotted observations are referred to as station models. Some stations may not be plotted due to space limitations. However, all reporting stations are used in the analysis.

Figure 25-10 and Figure 25-11 contain the most commonly used station plot models used in surface analysis charts.

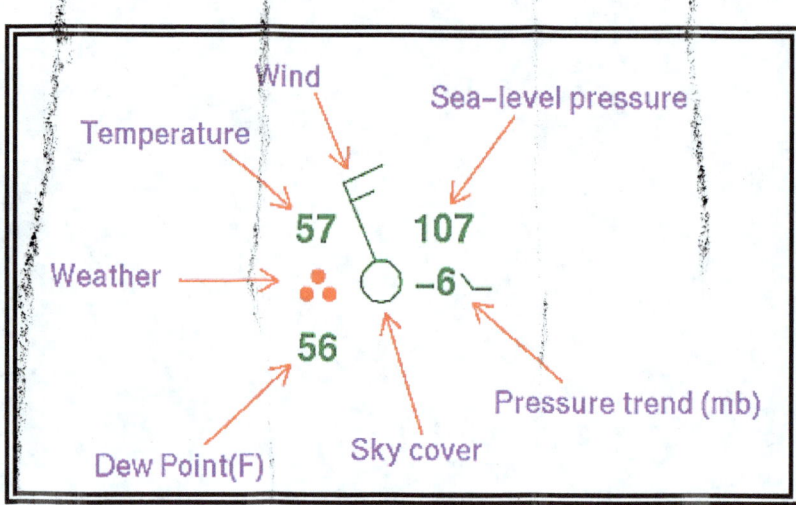

Figure 25-10. NWS Surface Analysis Chart Station Plot Model

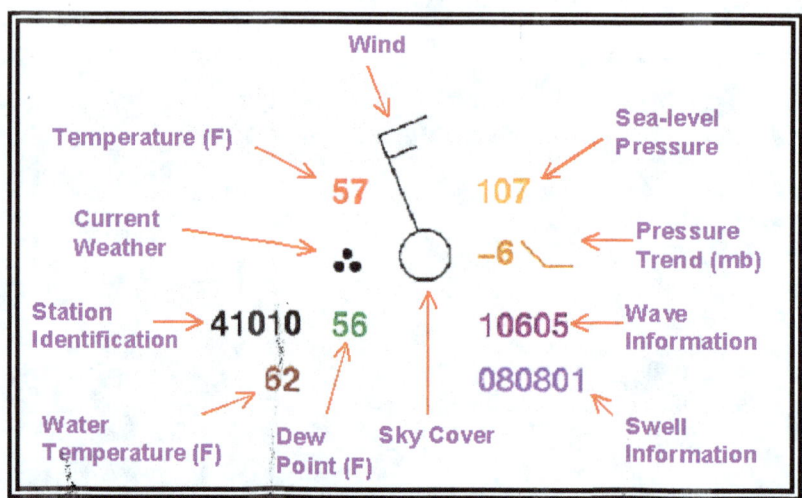

Figure 25-11. NWS Surface Analysis Chart Ship/Buoy Plot Model

The WPC also produces surface analysis charts specifically for the aviation community. Figure 25-12 contains the station plot model for these charts.

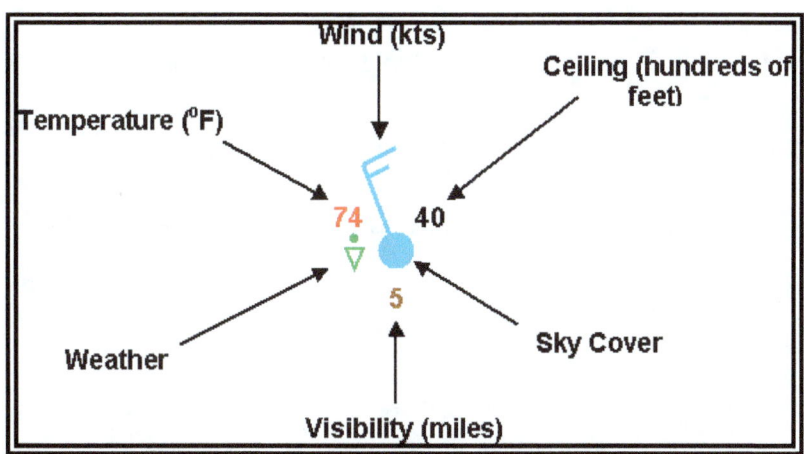

Figure 25-12. NWS Surface Analysis Chart for Aviation Interests Station Plot Model

25.2.3.4.1 Station Identifier

The format of the station identifier depends on the observing platform:

- Ship: Typically four or five characters. If five characters, then the fifth will usually be a digit.
- Buoy: Whether drifting or stationary, a buoy will have a five-digit identifier. The first digit will always be a **4**.
- C-MAN: Usually located close to coastal areas. Their identifier will appear like a five-character ship identifier; however, the fourth character will identify off which state the platform is located.
- Land: Land stations will always be three characters, making them easily distinguishable from ship, buoy, and C-MAN observations.

25.2.3.4.2 Temperature

The air temperature is plotted in whole degrees Fahrenheit.

25.2.3.4.3 Dewpoint

The dewpoint temperature is plotted in whole degrees Fahrenheit.

25.2.3.4.4 Weather

A weather symbol is plotted if, at the time of observation, precipitation is either occurring or a condition exists causing reduced visibility.

Figure 25-13 contains a list of the most common weather symbols.

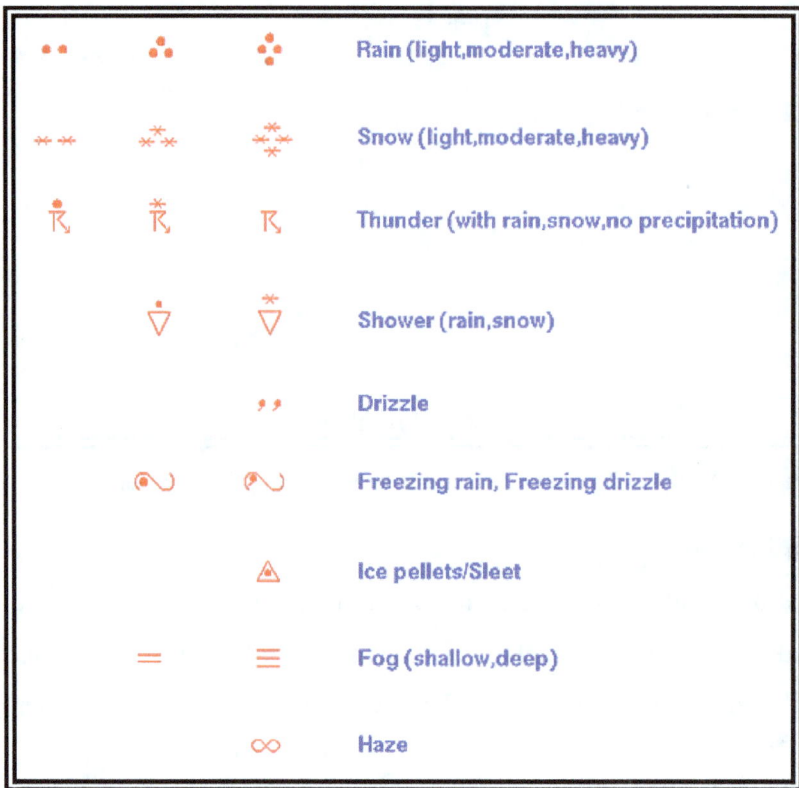

Figure 25-13. NWS Surface Analysis Chart Common Weather Symbols

25.2.3.4.5 Wind

Wind is plotted in increments of 5 kt. The wind direction is referenced to true north and is depicted by a stem (line) pointed in the direction from which the wind is blowing. Wind speed is determined by adding the values of the flags (50 kt), barbs (10 kt), and half-barbs (5 kt) found on the stem.

If the wind is calm at the time of observation, only a single circle over the station is depicted.

Figure 25-14 includes some sample wind symbols.

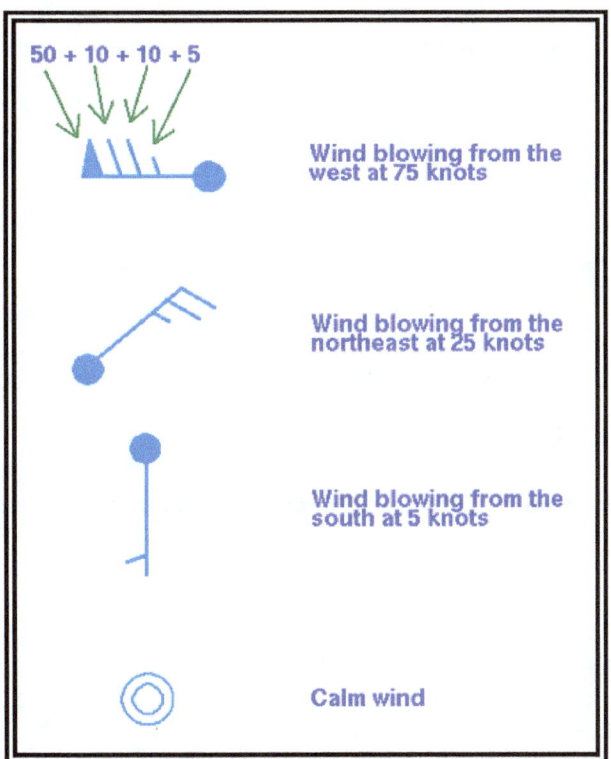

Figure 25-14. NWS Surface Analysis Chart Sample Wind Symbols

25.2.3.4.6 Ceiling

Ceiling is plotted in hundreds of feet AGL.

25.2.3.4.7 Visibility

Surface visibility is plotted in whole statute miles.

25.2.3.4.8 Pressure

Sea level pressure is plotted in tenths of millibars, with the first two digits (generally 10 or 9) omitted. For reference, 1013 mb is equivalent to 29.92 inHg. Below are some sample conversions between plotted and complete sea level pressure values.

```
410     1041.0 mb
103     1010.3 mb
987     998.7 mb
872     987.2 mb
```

25.2.3.4.9 Pressure Trend

The pressure trend has two components, a number and a symbol, to indicate how the sea level pressure has changed during the past 3 hours. The number provides the 3-hour change in tenths of millibars, while the symbol provides a graphic illustration of how this change occurred.

Figure 25-15 contains the meanings of the pressure trend symbols.

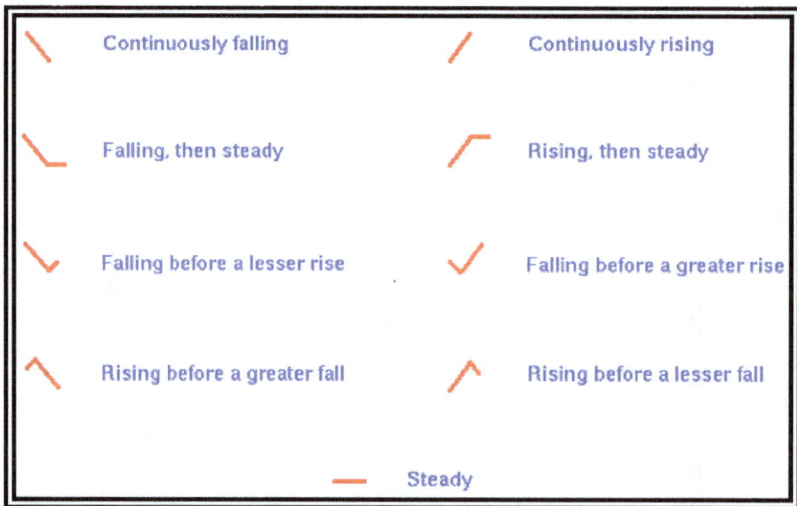

Figure 25-15. NWS Surface Analysis Chart Pressure Trends

25.2.3.4.10 Sky Cover

The approximate amount of sky cover can be determined by the circle at the center of the station plot. The amount that the circle is filled reflects the amount of sky covered by clouds. Figure 25-16 contains the common cloud cover depictions.

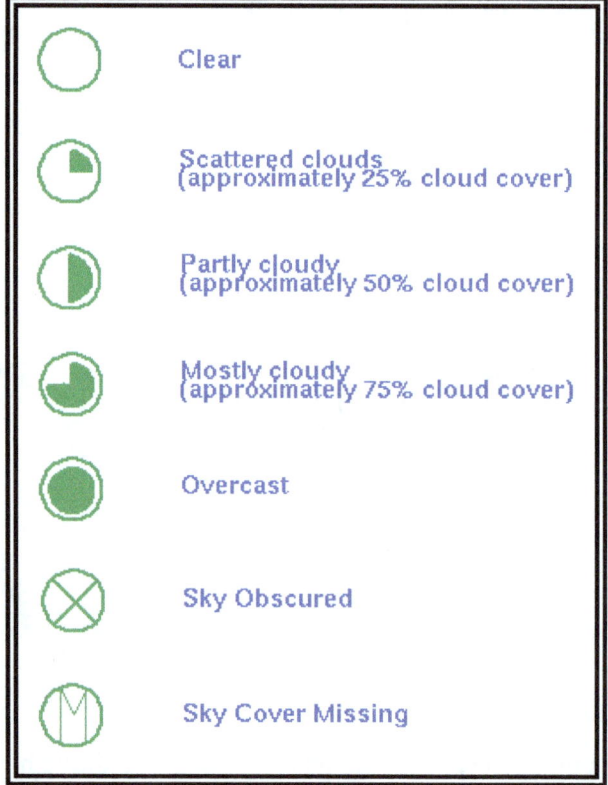

Figure 25-16. NWS Surface Analysis Chart Sky Cover Symbols

25.2.3.4.11 Water Temperature

Water temperature is plotted in whole degrees Fahrenheit.

25.2.3.4.12 Swell Information

Swell direction, period, and height are represented in the surface observations by a six-digit code. The first two digits represent the swell direction, the middle digits describe the swell period (in seconds), and the last two digits are the swell's height (in half meters).

`090703`:

- 09: The swell direction is from 90° (i.e., it is coming from due east).
- 07: The period of the swell is 7 seconds.
- 03: The height of the swell is 3 half m.

`271006`:

- 27: The swell direction is from 270° (due west).
- 10: The period is 10 seconds.
- 06: The height of the swell is 6 half m.

25.2.3.4.13 Wave Information

Period and height of waves are represented by a five-digit code. The first digit is always **1**. The second and third digits describe the wave period (in seconds), and the final two digits give the wave height (in half meters).

`10603`:

- 1: A group identifier. The first digit will always be 1.
- 06: The wave period is 6 seconds.
- 03: The wave height is 3 half m.

`10515`:

- 1: A group identifier.
- 05: The wave period is 5 seconds.
- 15: The wave height is 15 half m.

In some charts by the OPC, only the wave height (in feet) is plotted.

25.2.4 Unified Surface Analysis Chart

The NWS Unified Surface Analysis Chart is a surface analysis product produced collectively and collaboratively by the NWS WPC, the OPC, the NHC, and WFO Honolulu. The chart contains an analysis of isobars, pressure systems, and fronts.

This chart is available on the OPC's website as well as the AWC's website. Users can zoom in by clicking an area on the map to enlarge (see Figure 25-17 and Figure 25-18) and show station plot models (see Section 25.2.3.4).

25.2.4.1 Issuance

The Unified Surface Analysis Chart is issued four times daily for valid times 00, 06, 12, and 18 UTC.

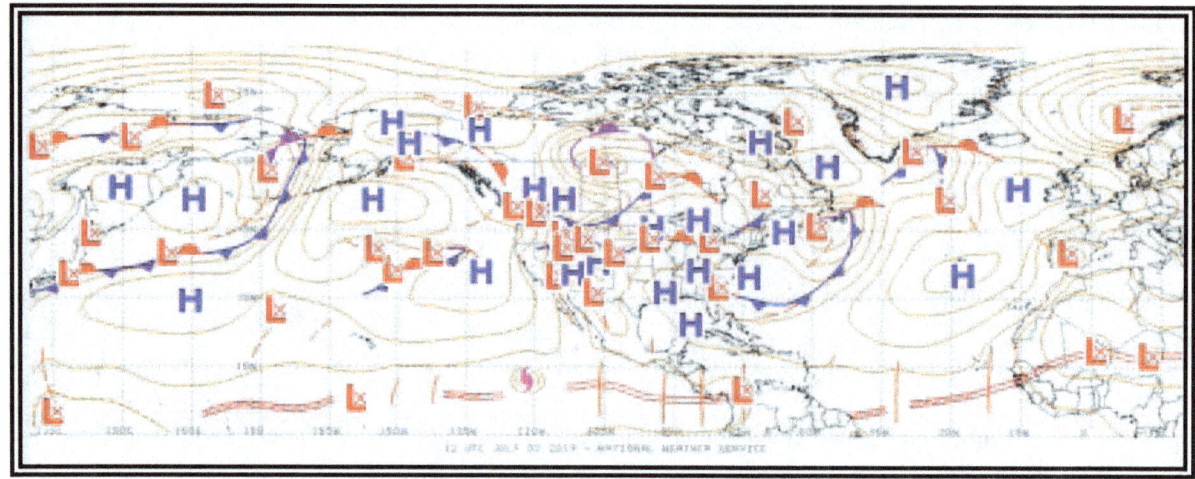

Figure 25-17. Unified Surface Analysis Chart Example

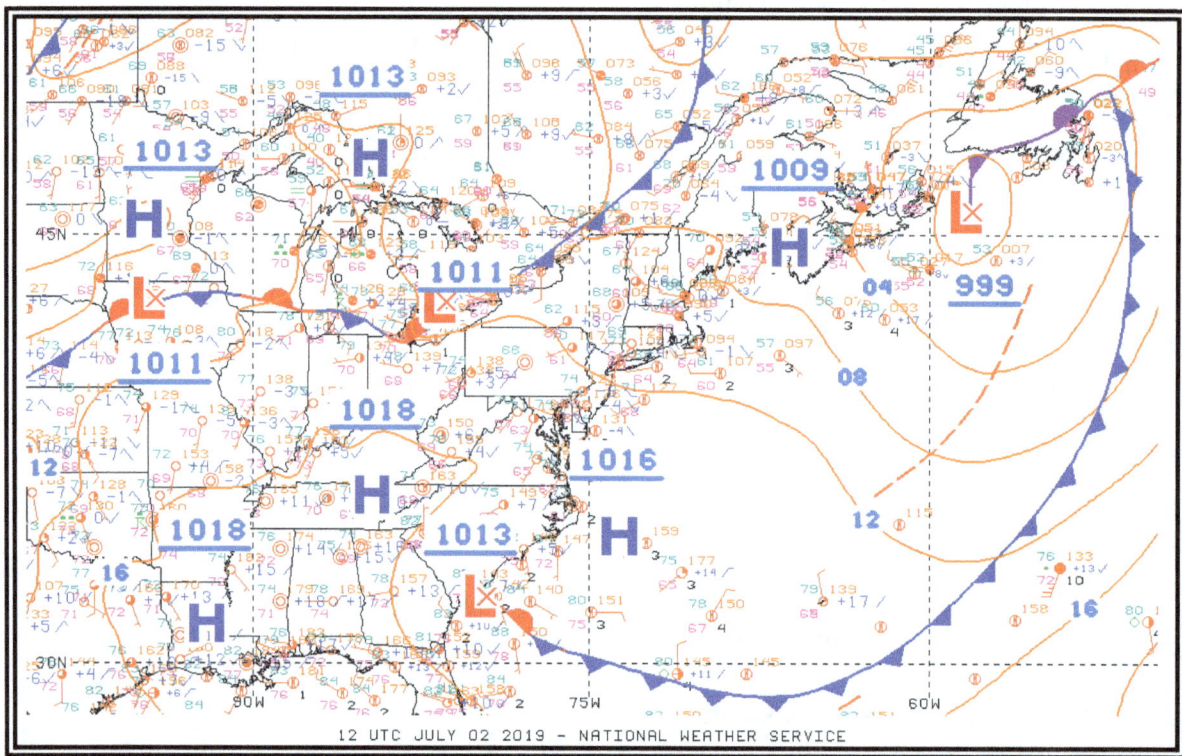

Figure 25-18. Unified Surface Analysis Chart Example (Enlarged Area)

25.2.4.2 Analysis Symbols

Unified Surface Analysis Charts use the symbols shown in Figure 25-6.

25.2.5 AAWU Surface Chart

The NWS Unified Surface Analysis Chart covers the Alaska area. The AAWU also provides a fixed area image of the Unified Surface Analysis Chart centered over Alaska (see Figure 25-19). This chart is available from the AAWU's website.

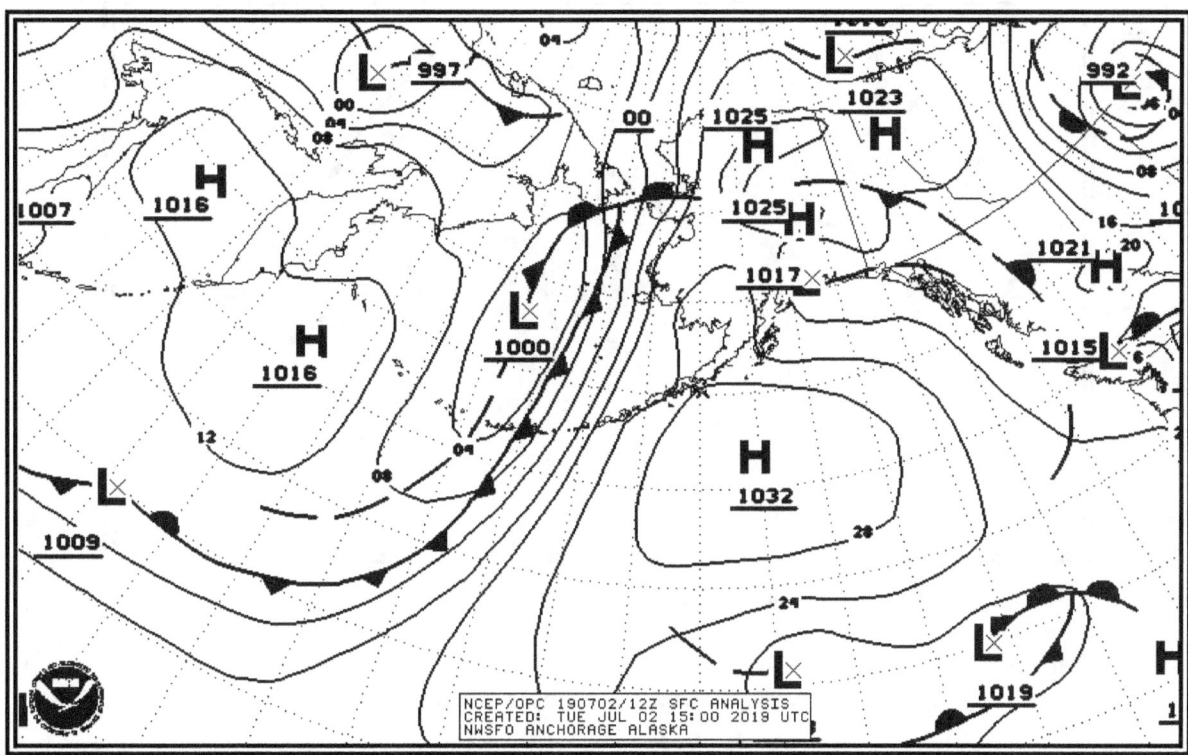

Figure 25-19. Unified Surface Analysis Chart Example with Fixed Area Coverage over Alaska

25.3 Upper-Air Analysis

An upper air chart (also known as a constant-pressure chart or an isobaric chart) is a weather map representing conditions on a surface of equal atmospheric pressure. A constant-pressure surface is a surface along which the atmospheric pressure is everywhere equal at a given instant. For instance, the 500 mb constant-pressure surface has a pressure of 500 mb everywhere on it. For example, a 500 mb chart will display conditions at the level of the atmosphere at which the atmospheric pressure is 500 mb.

Constant-pressure charts usually contain plotted data and analyses of the distribution of height of the surface (contours), wind (isotachs), temperature (isotherms), and sometimes humidity (isohumes). The height above sea level at which the pressure is that particular value may vary from one location to another at any given time, and also varies with time at any one location, so it does not represent a surface of constant altitude/height (i.e., the 500 mb level may be at a different height above sea level over Dallas than over New York at a given time, and may also be at a different height over Dallas from one day to the next). The height (altitude) of a constant-pressure surface varies primarily due to temperature; these heights can be measured by a rawinsonde.

Constant-pressure charts are most commonly known by their pressure value (see Table 25-3 for common constant pressure levels). For example, the 1,000 mb chart (which closely corresponds to the surface chart), the 850 mb chart, 700 mb chart, 500 mb chart, etc.

A contour analysis (see Figure 25-20 and Figure 25-21) can reveal highs, ridges, lows, and troughs aloft just as the surface chart shows such systems at the surface. These systems of highs/ridges and lows/troughs are called pressure waves. These pressure waves are similar to waves seen on bodies of water. They have crests (ridges) and valleys (troughs) and are in constant movement.

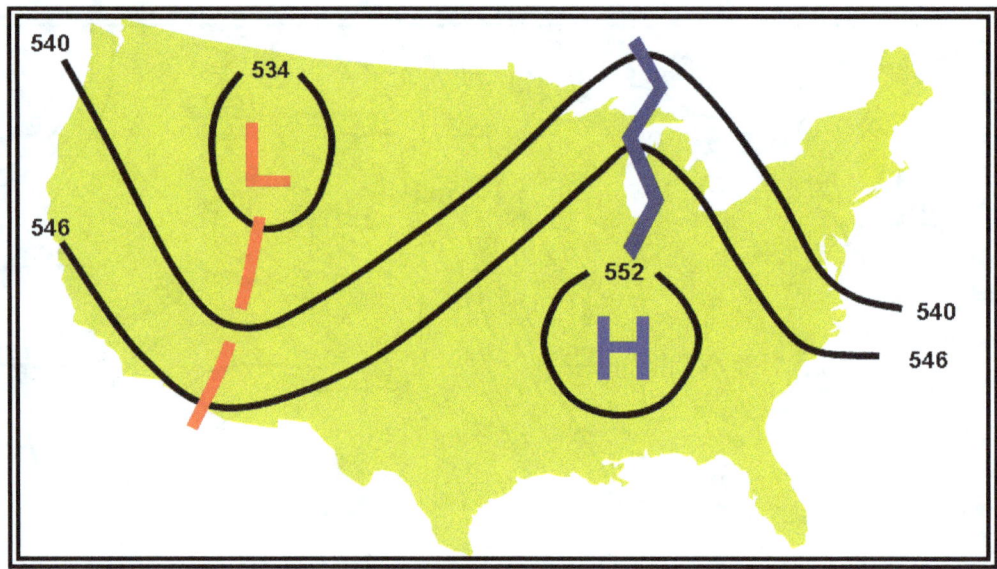

Figure 25-20. Schematic of 500 mb Constant-Pressure Chart

Table 25-3. Common Constant-Pressure Charts

Chart	Pressure Altitude (approximate)	
	Feet (ft)	Meters (m)
100 mb	53,000 ft	16,000 m
150 mb	45,000 ft	13,500 m
200 mb	39,000 ft	12,000 m
250 mb	34,000 ft	10,500 m
300 mb	30,000 ft	9,000 m
500 mb	18,000 ft	5,500 m
700 mb	10,000 ft	3,000 m
850 mb	5,000 ft	1,500 m
925 mb	2,500 ft	750 m

25.3.1 Issuance

The NWS provides data to produce upper-air analysis charts.

25.3.1.1 Examples

See Figure 25-21 for an example of a 500 mb chart.

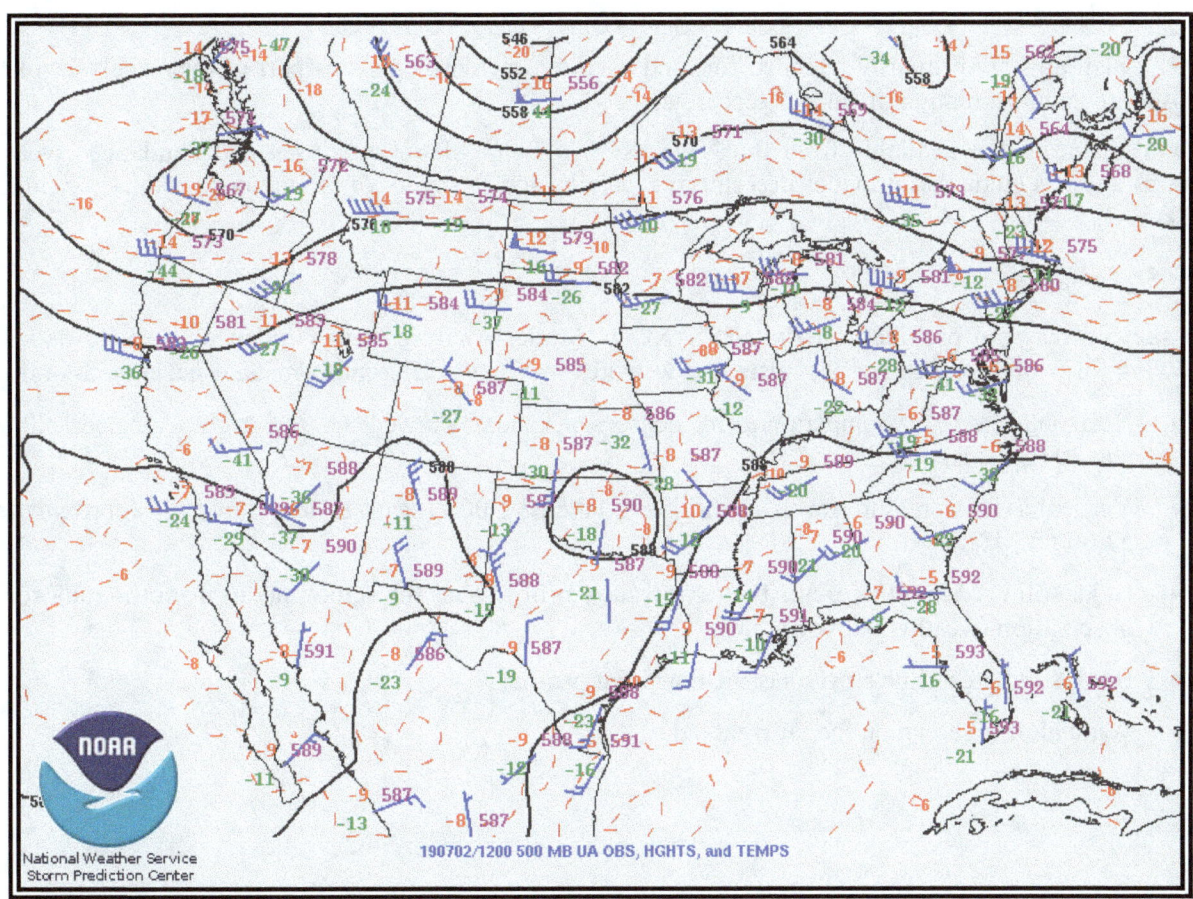

Figure 25-21. Example of a 500 mb Constant-Pressure Chart

Constant pressure level forecasts are used to provide an overview of weather patterns at specified times and pressure altitudes and are the source for wind and temperature aloft forecasts.

Pressure patterns cause and characterize much of the weather. Typically, lows and troughs are associated with clouds and precipitation, while highs and ridges are associated with fair weather, except in winter when valley fog may occur. The location and strength of the jet stream can be viewed at 300 mb, 250 mb, and 200 mb levels.

25.3.2 Radiosonde Observation (Weather Balloon) Analysis

A common means of analyzing radiosonde observations is the skew-T diagram (see Figure 25-22). Skew-T diagrams are primarily intended for, and used by, meteorologists as part of their analyses of the atmosphere. For example, the skew-T diagram can be used to:

- Determine the freezing level (or levels);
- Determine the stability of the atmosphere;

- Determine the potential for severe weather;
- Determine the height and depth of inversions;
- Infer cloud bases, tops, and layers; and
- Determine soaring conditions.

25.3.2.1 Issuance

Skew-T diagrams are primarily intended for, and used by, meteorologists as part of their analyses of the atmosphere and formulation of various forecasts.

Skew-T diagrams are available from the NWS NCO. Their "Model Analyses and Guidance" website contains a user's guide that provides descriptions, details, and examples of the various products, including the skew-T.

25.3.2.2 Format

The skew-T diagram provided on the NWS NCO "Model Analyses and Guidance" website uses the following format (other weather providers and websites may have different formats, especially colors):

- Horizontal axis is temperature in degrees Celsius, skewed to the right, labeled -20, 0, and 20 (Celsius).
- Vertical axis is pressure levels in millibars, labeled 1,000 (near sea level) to 100 (approximately 53,000 ft MSL).
- Bold, solid red line represents the temperature profile over the station taken from the radiosonde observation (weather balloon).
- Bold, solid green line represents the dewpoint profile.
- Wind aloft is shown on the far right side.

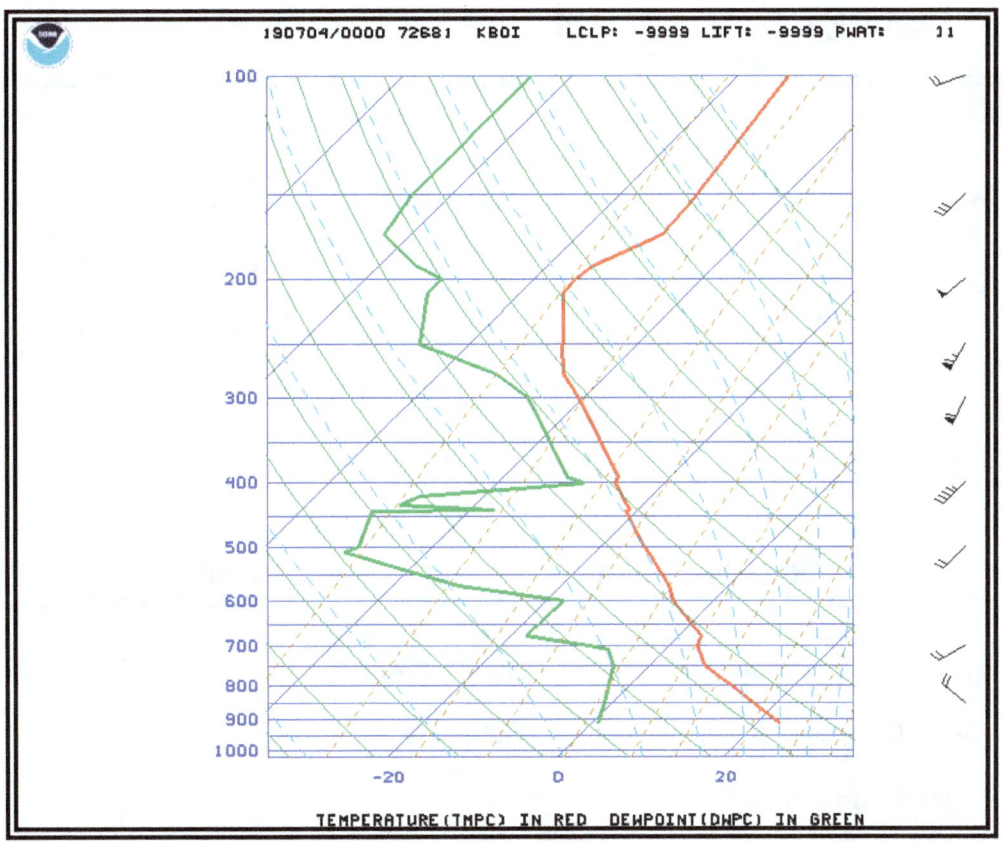

Figure 25-22. Skew-T Diagram Example

25.3.2.3 Examples

Two examples are provided below: a multiple freezing level example (see Figure 25-23) and a cloud top example (see Figure 25-24).

25.3.2.3.1 Multiple Freezing Level Example

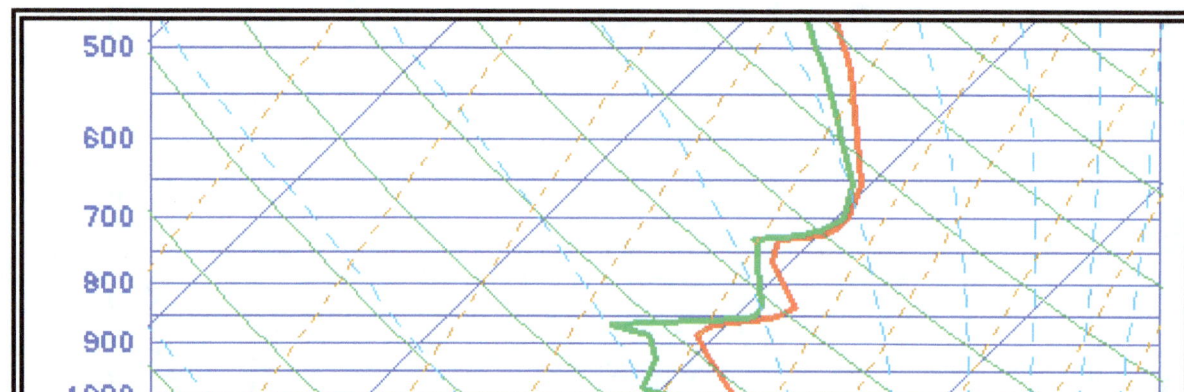

Note how the temperature profile (bold red line) crosses the 0 degree temperature line (also known as an isotherm) 5 times (near 900 mb, 860 mb, 775 mb, 725 mb, and 675 mb).

Figure 25-23. Skew-T Diagram—Multiple Freezing Level Example

25.3.2.3.2 Cloud Top Example

Figure 25-24 is the radiosonde observation from Vandenberg Air Force Base (VAFB), California, for 1200 UTC for a typical coastal stratus cloud.

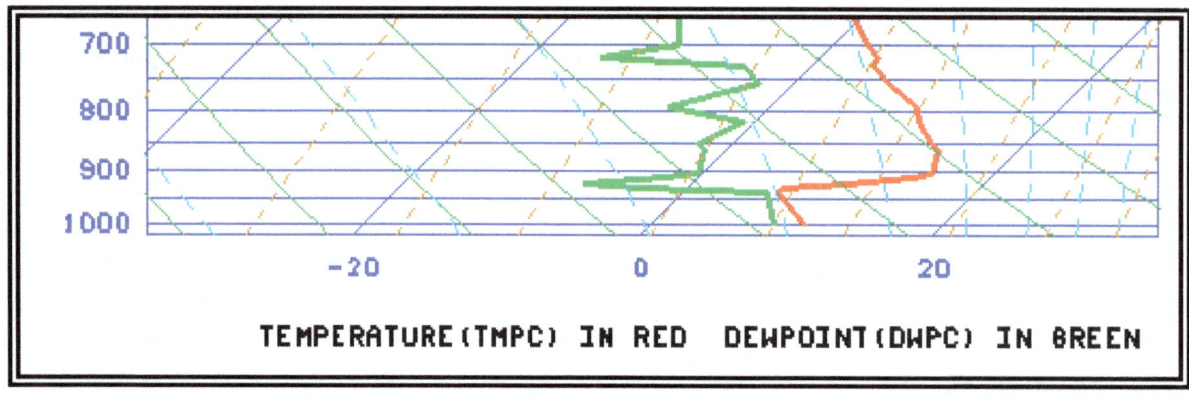

At about 950 mb, the temperature profile (bold red line) and dewpoint profile (bold green line) almost touch each other. This is the profile of a cloud top. The temperature and dewpoint quickly diverge, representing a change from the cool, moist air (and associated stratus cloud) to the dry and warmer air (cloud free) above.

Figure 25-24. Skew-T Diagram—Cloud Top Example

25.4 Freezing Level Analysis

The freezing level is the lowest altitude in the atmosphere over a given location at which the air temperature reaches 0 °C. This altitude is also known as the height of the 0 °C constant-temperature surface. A freezing level analysis graphic (see Figure 25-25) shows the height of the 0 °C constant-temperature surface.

The initial analysis is updated hourly. The colors represent the height in hundreds of feet above MSL of the lowest freezing level. Regions with white indicate the surface and the entire depth of the atmosphere are below freezing. Hatched or spotted regions (if present) represent areas where the surface temperature is below freezing with multiple freezing levels aloft.

More information on the freezing level forecast graphics is available on the AWC's website.

See Section 27.11 for additional information on freezing level forecast graphics.

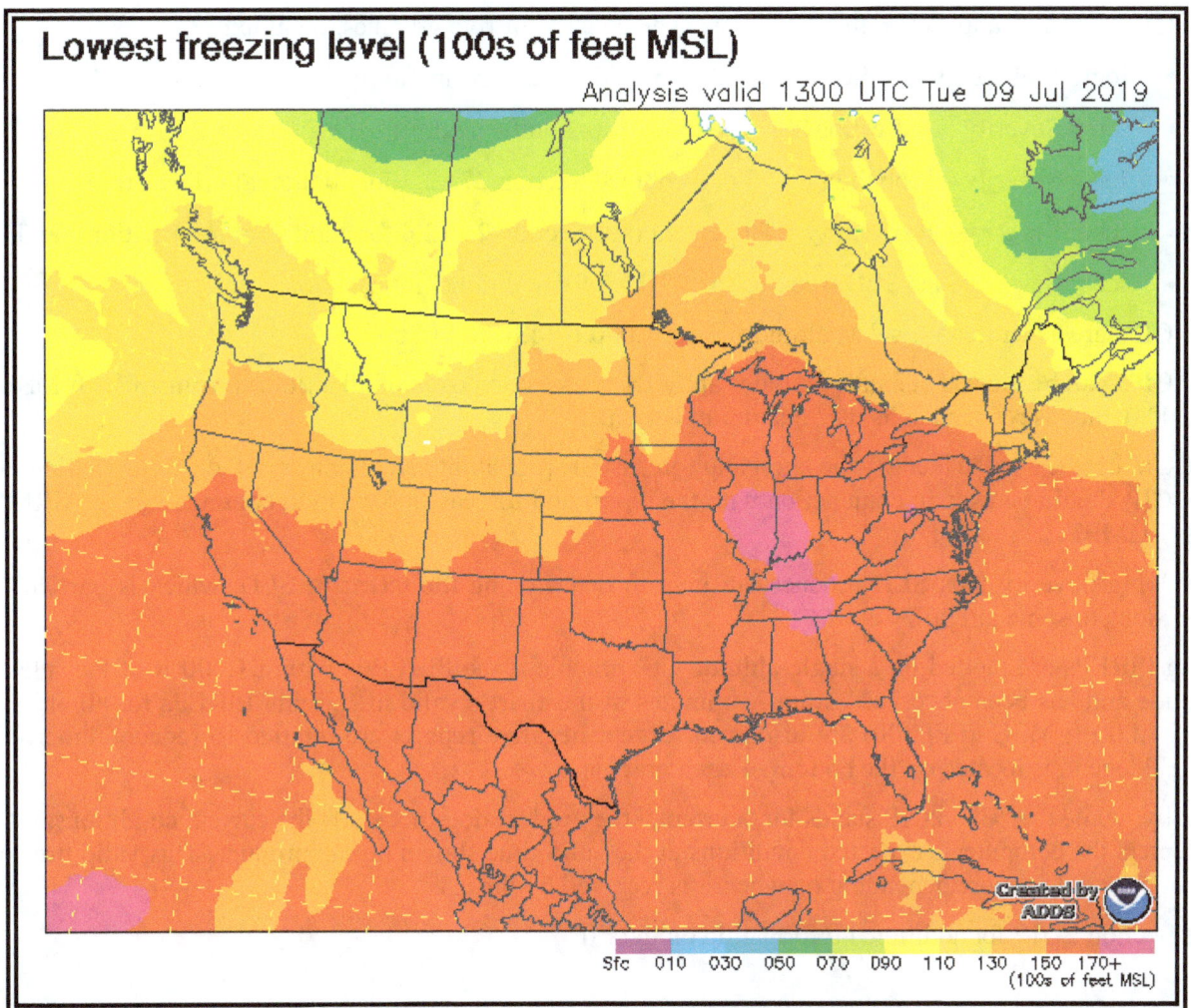

Figure 25-25. Freezing Level Analysis Graphic—Example

25.5 Icing Analysis (Current Icing Product (CIP))

The NWS produces icing products that are derived from NWS computer model data with no forecaster modifications. One of these products is the CIP.

The CIP combines sensor and NWS model data to provide an hourly 3D diagnosis of the icing environment. This information is displayed on a suite of graphics available for the CONUS, much of Canada and Mexico, and their respective coastal waters. See Figure 25-26 for an example CIP Analysis.

The CIP (and its forecast counterpart Forecast Icing Product (FIP) (see Section 27.12)) provide a broad-brush approach to describing icing intensity using estimated liquid water content, drop size, and temperature to depict ice accumulation rate. The intensity of ice accumulation rate varies by aircraft wing shape. Hence, the icing intensity categories depicted in CIPs and FIPs (e.g., none, light, moderate, heavy) is only a broad-brushed indication of ice accumulation rate, and not necessarily of aircraft performance. The icing terms used in SIGMETs and AIRMETs *do* refer to icing impact on aircraft.

CIPs will continue to evolve over the coming years with increased model resolutions, additional horizontal layers, and improvements to the algorithms and/or data sets used to produce the products. Along with these improvements may come a change in references to the product update version. Users can find additional information on these products and any changes on the AWC's icing website.

The CIP suite as it appears on the AWC's website consists of five graphics, including:

- Icing Probability (see Section 27.12.1 for additional information);
- Icing Severity (see Section 27.12.2 for additional information);
- Icing Severity—Probability > 25 percent (see Section 27.12.3 for additional information);
- Icing Severity—Probability > 50 percent (see Section 27.12.4 for additional information); and
- Icing Severity plus SLD (see Section 27.12.5 for additional information).

The CIPs are generated for select altitudes from 1,000 ft MSL to FL300.

The CIPs can be viewed at single altitudes and FLs or as a composite of all altitudes from 1,000 ft MSL to FL300, which is referred to as the "maximum" or "max."

The CIP can be used to identify the latest and forecast 3D probability and intensity of ice accumulation rate. The CIP should be used in conjunction with the report and forecast information contained in an AIRMET and SIGMET.

The Icing Severity plus SLD product can help in determining the threat of SLD, which is particularly hazardous to some aircraft.

Icing PIREPs are plotted on a single-altitude CIP graphic if the PIREP is within 1,000 ft of the selected altitude and has been observed within 75 minutes of the chart's valid time. Icing PIREPs for all altitudes (i.e., 1,000 ft MSL to FL300) are displayed, except negative reports are omitted to reduce clutter. The PIREP legend is located on the bottom of each graphic.

Finally, while the "C" in CIP stands for "current," the product does not show the current conditions, rather it depicts the computer's expected conditions at the valid time shown on the product. This valid time can be an hour old or more depending on when it is received by the user.

See Section 27.12 for additional information on the FIP.

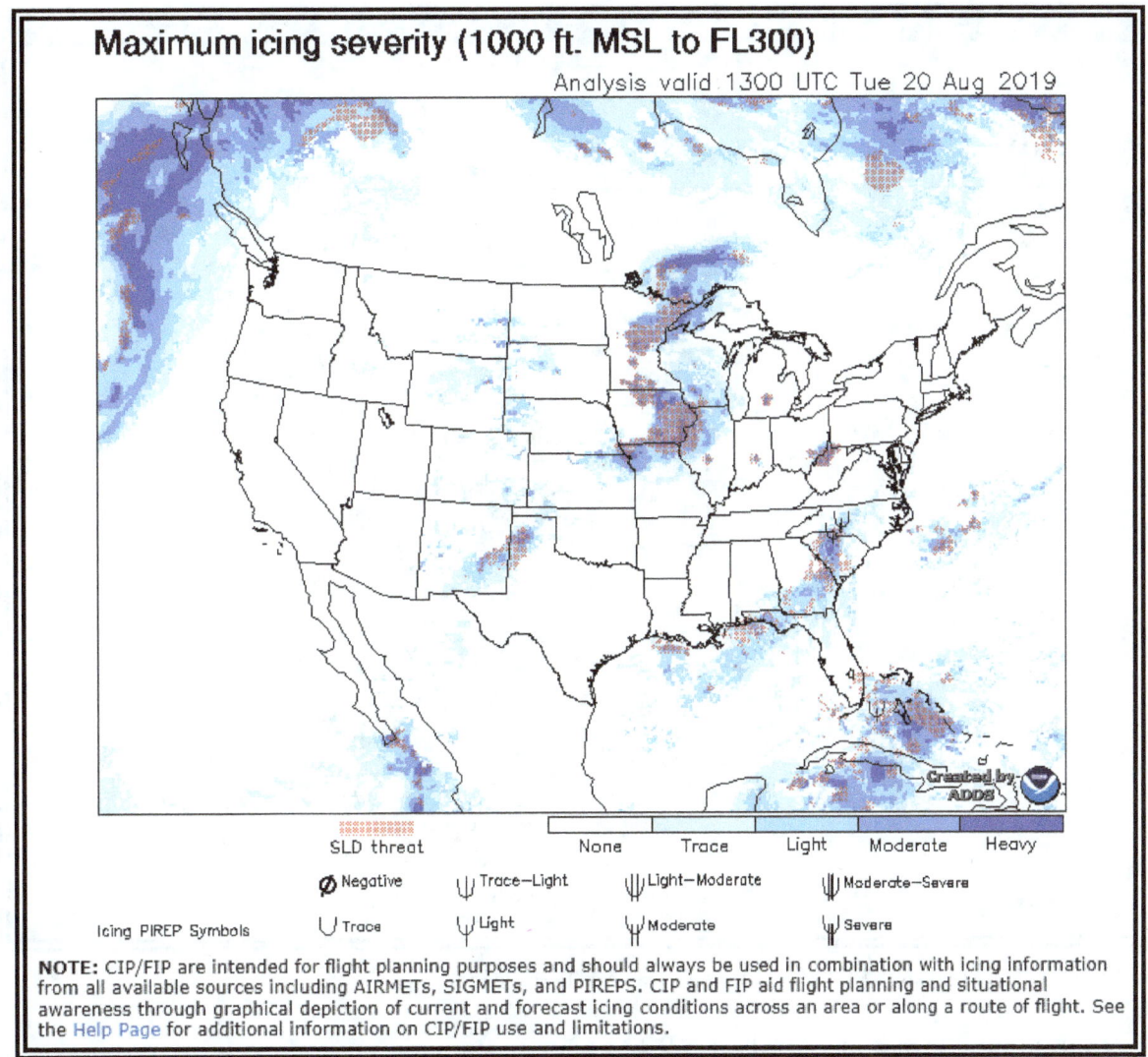

Figure 25-26. CIP Maximum Icing Severity Plus SLD—Max Example

25.6 Turbulence (Graphical Turbulence Guidance (GTG)) Analysis

The NWS produces a turbulence product that is derived from NWS model data with no forecaster modifications. This product is the GTG.

The GTG computes the results from more than 10 turbulence algorithms, then compares the results of each algorithm with turbulence observations from both PIREPs and AMDAR data to determine how well each algorithm matches reported turbulence conditions from these sources. See Figure 25-27 for an example of the GTG Analysis.

The GTG Analysis is essentially a 0-hour GTG forecast and is labeled accordingly. This product overlays turbulence PIREPs that correspond to the valid time of the product.

This turbulence product will continue to evolve over the coming years with increased model resolutions, additional horizontal layers, and improvements to the algorithms and/or data sets used to produce the product. Users can find additional information on these products and any changes on the AWC's turbulence web page.

See Section 27.13 for additional information on the GTG forecast.

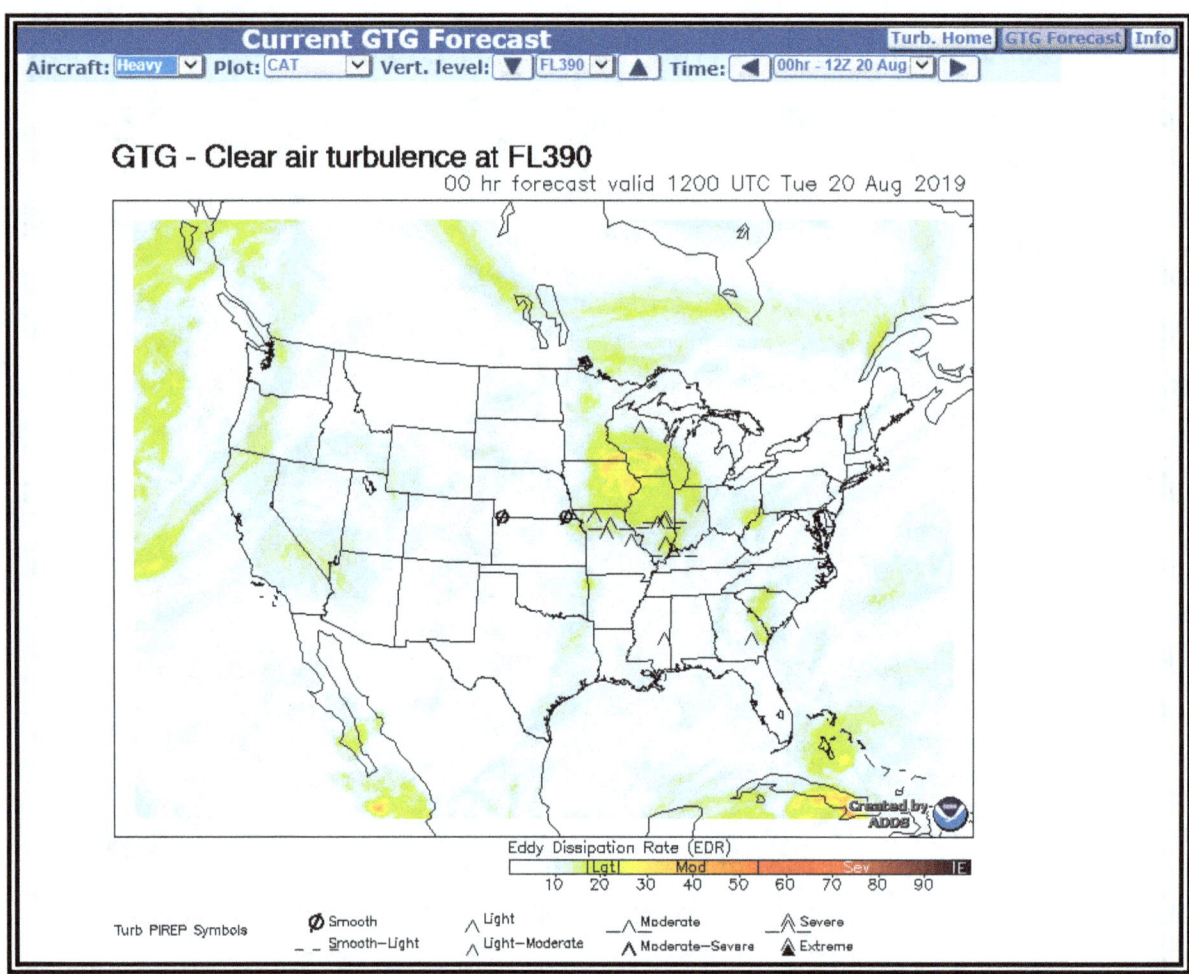

Figure 25-27. GTG Analysis—FL390 Example (Clear Air Turbulence)

25.7 Real-Time Mesoscale Analysis (RTMA)

RTMA is an hourly analysis system by the NWS's Environmental Modeling Center that produces analyses of surface weather elements. The FAA has determined that RTMA temperature information is a suitable replacement for missing temperature observations for a subset of airports. RTMA temperature is intended for use by operators, pilots, and aircraft dispatchers when an airport lacks a surface temperature report from an automated weather system (e.g., ASOS or AWOS sensor) or human observer. Airports with RTMA data available are located in Alaska, Guam, Hawaii, Puerto Rico, and the CONUS.

RTMA is issued by the NWS every hour, 24 hours a day. Temperatures are reported for an airport station including the latitude and the longitude. Temperatures are reported in degrees Celsius. See Figure 25-28 for an example RTMA temperature report.

```
*************************************************************
** RTMA temperature in degrees Celsius at select station
locations COMPUTED: 0743Z 18 May 2015
VALID: 0743Z 18 May 2015 to 0843Z 18 May 2015
*************************************************************

Station     Lat      Lon         T
KAVP        41.33   -75.73      17.67
KLAN        42.77   -84.60      21.37
KSEA        47.45  -122.30      14.54
KEFD        29.60   -95.17      26.08
KMMH        37.62  -118.83       2.26
KSPS        33.98   -98.50      18.56
KGGG        32.35   -94.65      21.09
KAIA        42.05  -102.80       6.46
KCAK        40.92   -81.43      20.07
```

Figure 25-28. RTMA Surface Temperature Example

26 Advisories

26.1 Introduction

The third of five types of aviation weather information contained in this handbook are advisories. For this handbook, advisories include the following:

1. Significant Meteorological Information (SIGMET):
    - Inside the CONUS:
        - Nonconvective.
        - Convective.
    - Outside the CONUS.
2. Airmen's Meteorological Information (AIRMET):
    - Inside the CONUS.
    - Outside the CONUS.
3. Center Weather Advisory (CWA).
4. Volcanic Ash Advisory (VAA).
5. Tropical Cyclone Advisory (TCA).
6. Space Weather Advisory.
7. Wind Shear Alerts.
8. Airport Weather Warning (AWW).

26.2 Significant Meteorological Information (SIGMET)

A SIGMET is a concise description of the occurrence or expected occurrence of specified en route weather phenomena that may affect the safety of aircraft operations. SIGMETs are issued in text format and intended for dissemination to all pilots in flight to enhance safety. SIGMETs are issued as soon as practical to give notice to operators and aircrews of potentially hazardous en route conditions.

Although the areas where the SIGMETs apply may be shown graphically, a graphical depiction of the SIGMET area is not the entire SIGMET. Additional information regarding the SIGMET may be contained in the text version.

26.2.1 SIGMET Issuance

SIGMETs are issued from MWOs. The United States has three MWOs: the AWC, the AAWU, and the WFO Honolulu.

The AWC is responsible for:

- Twenty domestic ARTCC FIRs covering the CONUS and adjacent coastal waters. CONUS SIGMETs, except for Convective SIGMETs, are grouped into six areas (see Figure 26-1).

- The New York, Houston, Miami, and San Juan Oceanic FIRs (see Figure 26-2).

- The Oakland Oceanic FIR north of 30° N latitude and the portion east of 140° W longitude, which is between the Equator and 30° N latitude (see Figure 26-3).

The AAWU is responsible for the Anchorage Continental FIR and the Anchorage Oceanic FIR (see Figure 26-3).

WFO Honolulu is responsible for the Oakland Oceanic FIR south of 30° N latitude and between 140° W and 130° E longitude (see Figure 26-3).

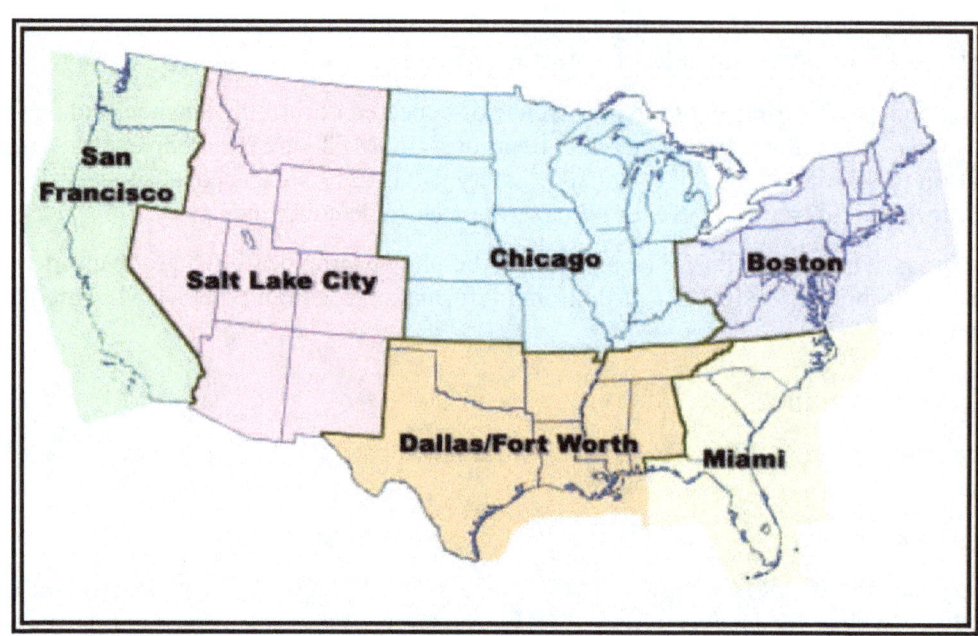

Figure 26-1. AWC SIGMET Areas of Responsibility—CONUS

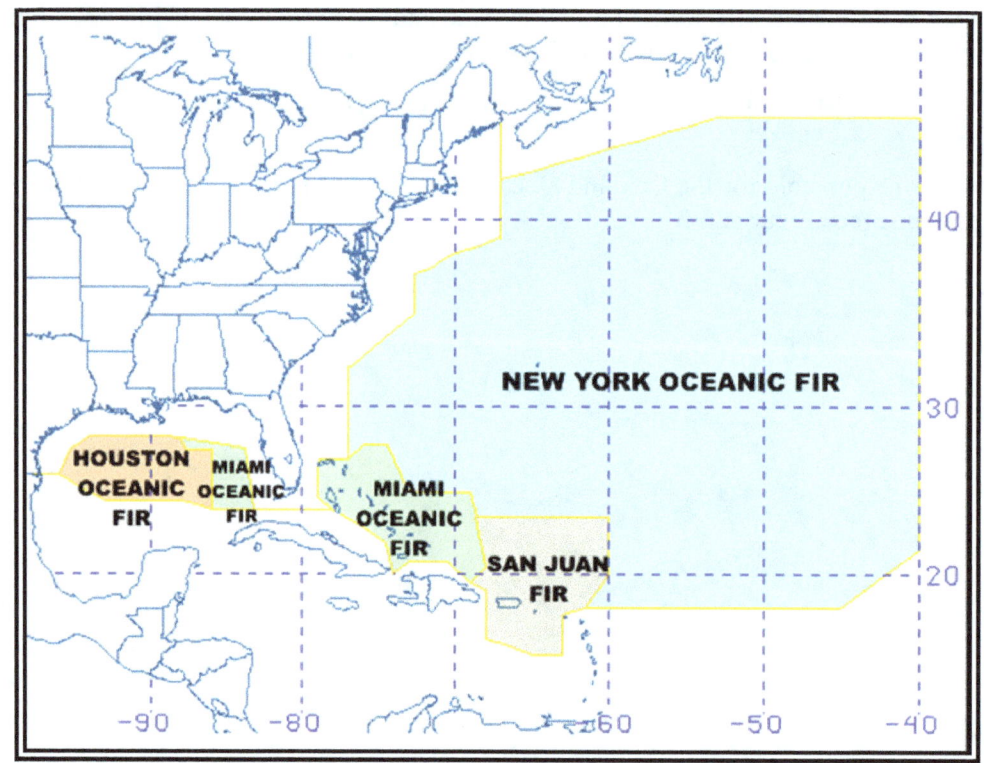

Figure 26-2. AWC SIGMET Areas of Responsibility—Atlantic Basin

Chapter 26, Advisories

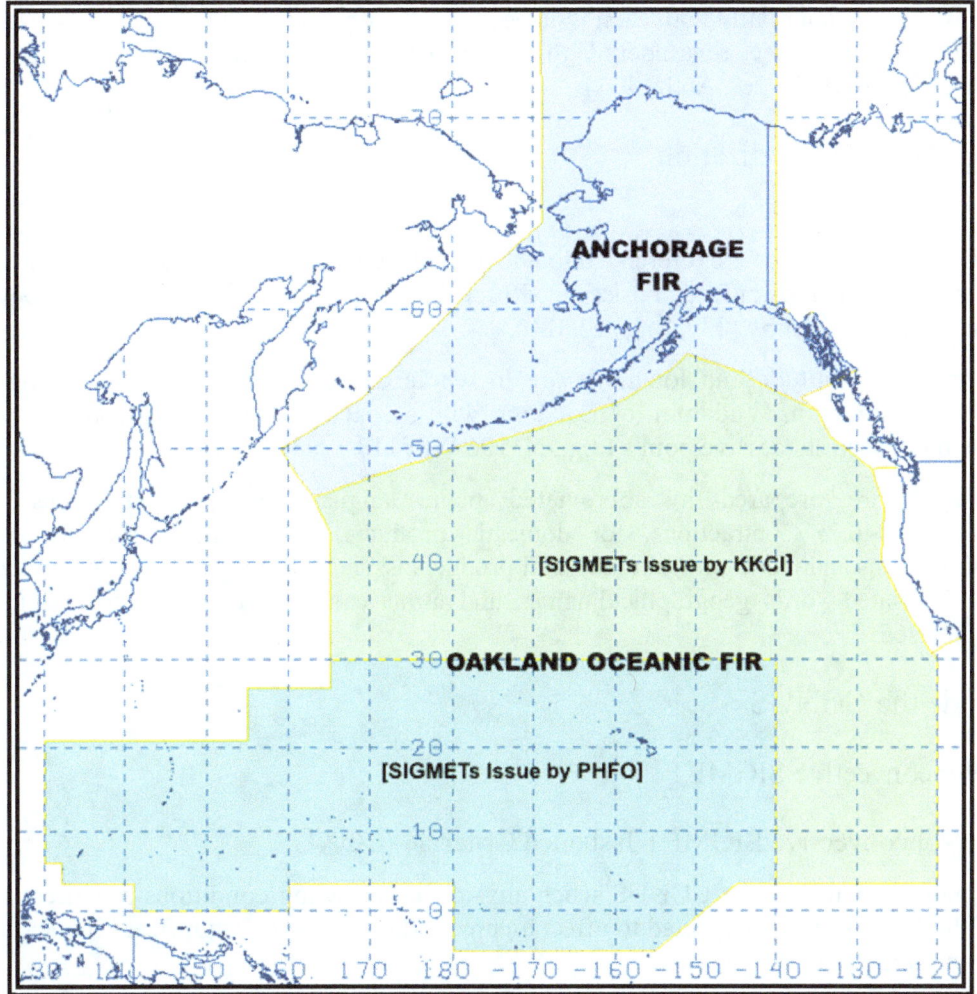

Note that KKCI refers to the AWC that is in Kansas City, MO.

Figure 26-3. SIGMET Areas of Responsibility—Pacific Basin

26.2.2 SIGMET Identification

U.S. SIGMETs (other than Convective SIGMETs) are assigned a series identifier:

- AWC for the CONUS:
  - NOVEMBER, OSCAR, PAPA, QUEBEC, ROMEO, UNIFORM, VICTOR, WHISKEY, XRAY, and YANKEE. Note that SIERRA and TANGO are excluded.
- AWC for the Oakland Oceanic FIR:
  - ALFA, BRAVO, CHARLIE, DELTA, ECHO, FOXTROT, GOLF, and HOTEL.
- Honolulu MWO for the Oakland Oceanic FIR:
  - NOVEMBER, OSCAR, PAPA, QUEBEC, ROMEO, SIERRA, TANGO, UNIFORM, VICTOR, WHISKEY, XRAY, YANKEE, and ZULU.
- AAWU for the Anchorage FIR:
  - INDIA, JULIET, KILO, LIMA, and MIKE.

A number is assigned sequentially with each issuance until the phenomenon ends. At 0000 UTC each day, all continuing SIGMETs are renumbered to 1, regardless of a continuation of the phenomena (e.g., YANKEE 1, YANKEE 2, YANKEE 3).

26.2.3 SIGMET Standardization

SIGMETs follow these standards:

- All heights or altitudes are referenced to above MSL, unless otherwise noted, and annotated using the height in hundreds of feet, consisting of three digits (e.g., 040). For heights at or above 18,000 ft, the level is preceded by "FL" (e.g., FL180).

- References to latitude and longitude are in whole degrees and minutes following the model: Nddmm or Sddmm, Wdddmm, or Edddmm with a space between latitude and longitude and a hyphen between successive points (e.g., N3106 W07118—N3011 W7209).

- Messages are prepared in abbreviated plain language using contractions from FAA Order JO 7340.2, Contractions, for domestic products, and from ICAO Doc 8400, ICAO Abbreviations and Codes, for international products issued for Oceanic FIRs. A limited number of nonabbreviated words, geographical names, and numerical values of a self-explanatory nature may also be used.

26.2.4 Inside the CONUS

26.2.4.1 Nonconvective SIGMET

26.2.4.1.1 Nonconvective SIGMET Issuance Criteria—CONUS

A SIGMET may be issued in the CONUS when any of the following conditions are affecting or, in the judgment of the forecaster, are expected to affect an area judged to have a significant impact on the safety of aircraft operations:

- Severe or greater turbulence (**SEV TURB**).
- Severe icing (**SEV ICE**).
- Widespread dust storm (**WDSPR DS**).
- Widespread sandstorm (**WDSPR SS**).
- Volcanic ash (**VA**).

26.2.4.1.2 Nonconvective SIGMET Issuance Time and Valid Period—CONUS

A SIGMET is an unscheduled product issued any time conditions reaching SIGMET criteria are occurring or expected to occur within a 4-hour period. A SIGMET can have a valid period up to, but not exceeding, 4 hours. SIGMETs for continuing phenomena will be reissued at least every 4 hours as long as SIGMET conditions continue to occur in the area of responsibility.

26.2.4.1.3 Nonconvective SIGMET Format and Example—CONUS

The content and order of elements in the SIGMET are as follows:

- Series name and number.
- Valid beginning and ending time (UTC).
- List of states affected by the phenomena.

- Location of phenomena delineated by high-altitude VOR coordinates covering the affected area during the SIGMET valid time.
- Phenomena description (e.g., **SEV ICE**).
- Vertical extent (base and top), if appropriate.
- Movement, if appropriate.
- Intensity change (**INTSF**—intensifying, **WKN**—weakening, **NC**—no change).
- Indication that the weather condition will continue during the 4 hours beyond the valid time of the SIGMET.

```
SFOR UWS 100130
SIGMET ROMEO 1 VALID UNTIL 100530
OR WA
FROM SEA TO PDT TO EUG TO SEA
SEV TURB BTN FL280 AND FL350. CONDS CONTG BYD 0530Z.
```

Figure 26-4. Nonconvective SIGMET for the CONUS—Example

Table 26-1. Decoding a Nonconvective SIGMET for the CONUS

Line	Content	Description
1	SFO	SIGMET area identifier
	R	SIGMET series
	UWS	Product identifier
	100130	Issuance date/time UTC
2	SIGMET	Product type
	ROMEO	SIGMET series name
	1	Series issuance number
	VALID UNTIL 100530	Ending valid date/time UTC
3	OR WA	Phenomenon location (states)
4	FROM SEA TO PDT TO EUG TO SEA	Phenomenon location (high-altitude VOR coordinates)
5	SEV TURB BTN FL280 AND FL350. CONDS CONTG BYD 1000Z	Phenomenon description

Chapter 26, Advisories

The SIGMET in Figure 26-4 is decoded as the following:

- (Line 1) SIGMET ROMEO series issued for the San Francisco area at 0130 UTC on the 10th day of the month.
- (Line 2) This is the first issuance of the SIGMET ROMEO series and is valid until the 10th day of the month at 0530 UTC.
- (Line 3) The affected states within the San Francisco area are Oregon and Washington.
- (Line 4) From Seattle, WA; to Pendleton, OR; to Eugene, OR; to Seattle, WA.
- (Line 5) Severe turbulence between FL280 and FL350. Conditions continuing beyond 0530Z.

26.2.4.1.4 Nonconvective SIGMET Cancellations—CONUS

A CONUS Nonconvective SIGMET is cancelled when the phenomenon is no longer occurring, no longer expected to occur, or has moved out of the area of responsibility.

26.2.4.1.5 Nonconvective SIGMET Amendments—CONUS

Amendments to CONUS Nonconvective SIGMETs are not issued. Instead, a new SIGMET is issued using the next series number.

26.2.4.1.6 SIGMET (Nonconvective) Corrections—CONUS

Corrections to CONUS Nonconvective SIGMETs are issued as necessary. The corrected SIGMET is identified by a "COR" located at the end of the first line after the issuance UTC date/time.

26.2.4.2 Convective SIGMET

Convective SIGMETs are issued for the CONUS instead of SIGMETs for thunderstorms. Any Convective SIGMET implies severe or greater turbulence, severe icing, and LLWS.

Although the areas where the Convective SIGMETs apply may be shown graphically, such a graphical depiction, the Convective SIGMET polygon is a "snapshot" that outlines the area (or line) of thunderstorms at the issuance time of 55 past each hour. During the valid time of the SIGMET, the area/line will move according to the movement vector given in the SIGMET. For fast-moving areas or lines, they will very likely end up outside the SIGMET polygon by the end of the hour. Slow-moving or stationary areas or lines will likely remain in or very close to the original polygon. For additional clarification, the movement "MOV FROM..." within the Convective SIGMET describes the current movement of the SIGMET area or line. In cases when cell movements are different within the area, the SIGMET will include an additional line that states "CELL MOV FROM." Detailed information regarding the Convective SIGMET depiction should be compared to the textual version for storm movement, velocity, cloud tops, and several other important elements. Users should exercise caution as areas of convection and their associated polygons can change and should only be used for strategic planning.

26.2.4.2.1 Convective SIGMET Routine Issuance Criteria

A Convective SIGMET will be issued when any of the following conditions are occurring or, in the judgment of the forecaster, are expected to occur:

- A line of thunderstorms at least 60 mi long with thunderstorms affecting at least 40 percent of its length.

- An area of active thunderstorms judged to have a significant impact on the safety of aircraft operations covering at least 40 percent of the area concerned and exhibiting a very strong radar reflectivity intensity or a significant satellite or lightning signature.

- Embedded or severe thunderstorm(s) expected to occur for more than 30 minutes during the valid period, regardless of the size of the area.

26.2.4.2.2 Convective SIGMET Special Issuance Criteria

A special Convective SIGMET may be issued when any of the following criteria are occurring or, in the judgment of the forecaster, are expected to occur for more than 30 minutes of the valid period:

- Tornado, hail greater than or equal to ¾ in (at the surface), or wind gusts greater than or equal to 50 kt (at the surface) are reported.

- Indications of rapidly changing conditions if, in the forecaster's judgment, they are not sufficiently described in existing Convective SIGMETs.

Special issuance is not required for a valid Convective SIGMET.

26.2.4.2.3 Convective SIGMET Issuance Time and Valid Period

Convective SIGMET bulletins for the eastern, central, and western regions of the CONUS (see Figure 26-5) are issued on a scheduled basis, hourly at 55 minutes past the hour. Each bulletin contains all valid Convective SIGMETs within the region. Convective SIGMETs are valid for 2 hours or until superseded by the next hourly issuance. A Convective SIGMET bulletin must be transmitted each hour for each region. When conditions do not meet or are not expected to meet Convective SIGMET criteria within a region at the scheduled time of issuance, a "CONVECTIVE SIGMET...NONE" message is transmitted.

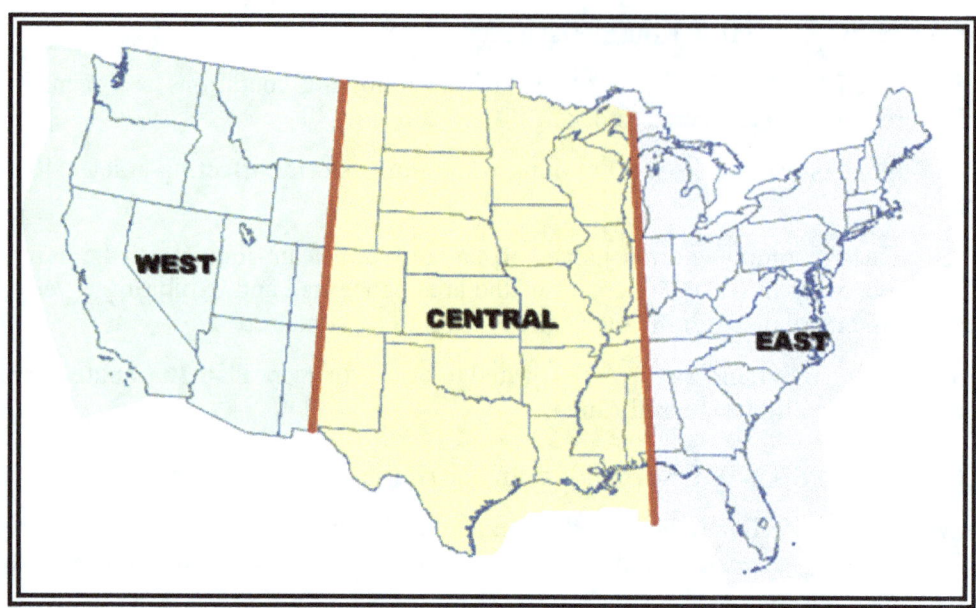

Figure 26-5. AWC Convective SIGMET Areas of Responsibility

26.2.4.2.4   Convective SIGMET Format and Example

Each Convective SIGMET bulletin includes one or more individually numbered Convective SIGMETs for the region. The content and order of each bulletin is as follows:

- Convective SIGMET series number and region letter (**E**, **W**, or **C**).
- Valid ending time (UTC).
- List of states affected by the phenomena.
- Location of phenomena delineated by high-altitude VOR coordinates covering the affected area during the SIGMET valid time.
- Phenomena description (e.g., **AREA SEV EMBD TS**).
- Movement (e.g., **MOV FROM 26030KT**).
- Cloud top (e.g., **TOPS ABV FL450**).
- Remarks (e.g., **TORNADOES...HAIL TO 2.5 IN...WIND GUSTS TO 70KT POSS**).

Note: Tropical cyclone information will be added to the remarks section of the CONUS Convective SIGMETs when appropriate.

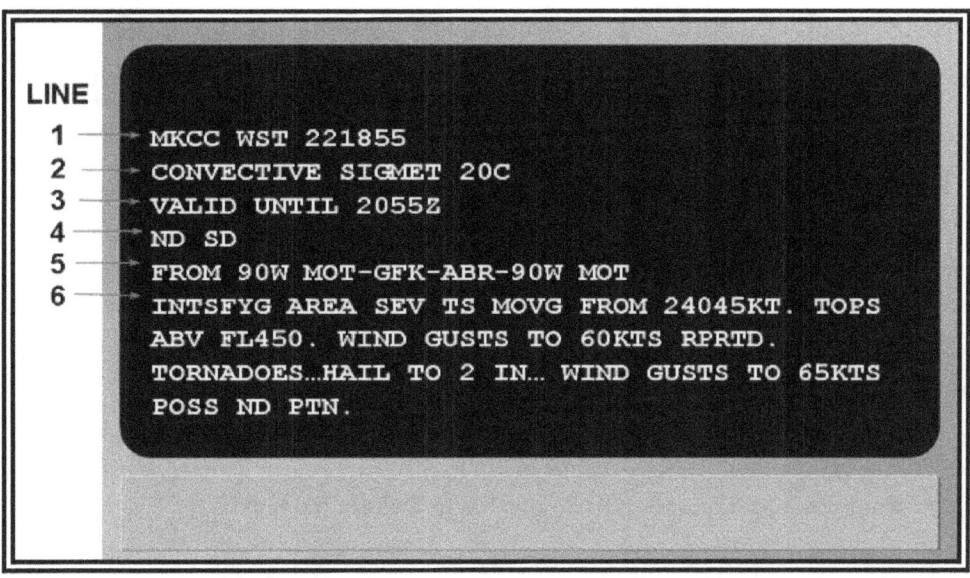

Figure 26-6. Convective SIGMET—Example

Table 26-2. Decoding a Convective SIGMET for the CONUS

Line	Content	Description
1	MKC	Issuing office (AWC)
	C	Region (East, Central, or West)
	WST	Product identifier
	221855	Issuance date/time (DDHHMM)
2	CONVECTIVE SIGMET	Product type
	20	Issuance number
	C	Region (East, Central, or West)
3	VALID UNTIL 2055Z	Valid ending time (UTC)
4	ND SD	States/areas affected
5	FROM 90W MOT-GFK-ABR-90W MOT	Phenomenon location (high-altitude VOR coordinates)
6	INTSFYG AREA SEV TS MOVG FROM 24045KT. TOPS ABV FL450. WIND GUSTS TO 60KTS RPRTD. TORNADOES...HAIL TO 2 IN... WIND GUSTS TO 65KTS POSS ND PTN	Phenomenon description, movement, cloud top, remarks

The Convective SIGMET in Figure 26-6 is decoded as the following:

- (Line 1) Convective SIGMET issued for the central portion of the United States on the 22nd day of the month at 1855Z.
- (Line 2) This is the 20th Convective SIGMET issued on the 22nd day of the month for the central United States as indicated by "20C."
- (Line 3) Valid until 2055Z.
- (Line 4) The affected states are North Dakota and South Dakota.
- (Line 5) From 90 NM west of Minot, ND; to Grand Forks, ND; to Aberdeen, SD; to 90 NM west of Minot, ND.
- (Line 6) An intensifying area of severe thunderstorms moving from 240° at 45 kt (to the northeast). Thunderstorm tops above FL450. Wind gusts to 60 kt reported. Tornadoes, hail with diameter of 2 in, and wind gusts to 65 kt possible in the North Dakota portion.

26.2.4.2.5 Convective SIGMET Outlook

Each Convective SIGMET bulletin includes a 2 to 6-hour outlook at the end of the bulletin. The content and order of each bulletin is as follows:

- Beginning and ending valid times.
- Location of expected Convective SIGMET issuances delineated by high-altitude VOR coordinates for the outlook valid time.

26.2.4.2.6 Convective SIGMET Cancellations

Convective SIGMETs are not cancelled but are superseded by the next Convective SIGMET in the series.

26.2.4.2.7 Convective SIGMET Amendments

Amended Convective SIGMETs are not issued. Instead, a new Convective SIGMET is issued for that region.

26.2.4.2.8 Convective SIGMET Corrections

Corrections to Convective SIGMETs are issued as necessary. The corrected Convective SIGMET is identified by a "COR" located at the end of the first line after the issuance UTC date/time.

26.2.5 Outside the CONUS

26.2.5.1 SIGMET Issuance Criteria—Outside the CONUS

U.S. SIGMETs outside the CONUS are issued when any of the following conditions are affecting or, in the judgment of the forecaster, are expected to affect an area judged to have a significant impact on the safety of aircraft operations:

- Thunderstorm—of type below:*
  - Obscured (**OBSC TS**).
  - Embedded (**EMBD TS**).
  - Widespread (**WDSPR TS**).

- o Squall line (**SQL TS**).
- o Isolated severe (**ISOL SEV TS**).
- Severe turbulence (**SEV TURB**).
- Severe icing (**SEV ICE**); with freezing rain (**SEV ICE (FZRA)**).
- Widespread dust storm (**WDSPR DS**).
- Widespread sandstorm (**WDSPR SS**).
- Volcanic ash (**VA**).
- Tropical cyclone (**TC**).

*Tornado (**TDO**), funnel cloud (**FC**), waterspout (**WTSPT**), and heavy hail (**HVY GR**) may be used as further descriptions of the thunderstorm, as necessary.

26.2.5.2   SIGMET Issuance Time and Valid Period—Outside the CONUS

A SIGMET is an unscheduled product issued any time conditions reaching SIGMET criteria are occurring or expected to occur within a 4-hour period. A SIGMET outside the CONUS can have a valid period up to, but not exceeding, 4 hours, except for volcanic ash (**VA**) and tropical cyclone (**TC**), which can be valid up to 6 hours. SIGMETs for continuing phenomena will be reissued at least every 4 (or 6) hours as long as SIGMET conditions continue to occur in the area of responsibility.

26.2.5.3   SIGMET Format and Example—Outside the CONUS

SIGMETs outside the CONUS contain the following information, related to the specific phenomena and in the order indicated:

- Phenomenon and its description (e.g., **SEV TURB**).
- An indication whether the information is observed, using "OBS" and/or "FCST." The time of observation will be given in UTC.
- Location of the phenomenon referring to, where possible, the latitude and longitude and FLs (altitude) covering the affected area during the SIGMET valid time. SIGMETs for volcanic ash clouds and tropical cyclones contain the positions of the ash cloud, tropical cyclone center, and radius of convection at the start of the valid time of the SIGMET.
- Movement toward or expected movement using 16 points of the compass, with speed in knots, or stationary, if appropriate.
- Thunderstorm maximum height as FL.
- Changes in intensity, using, as appropriate, the abbreviations for intensifying (**INTSF**), weakening (**WKN**), or no change (**NC**).
- Forecast position of the volcanic ash cloud or the center of the tropical cyclone at the end of the validity period of the SIGMET message.

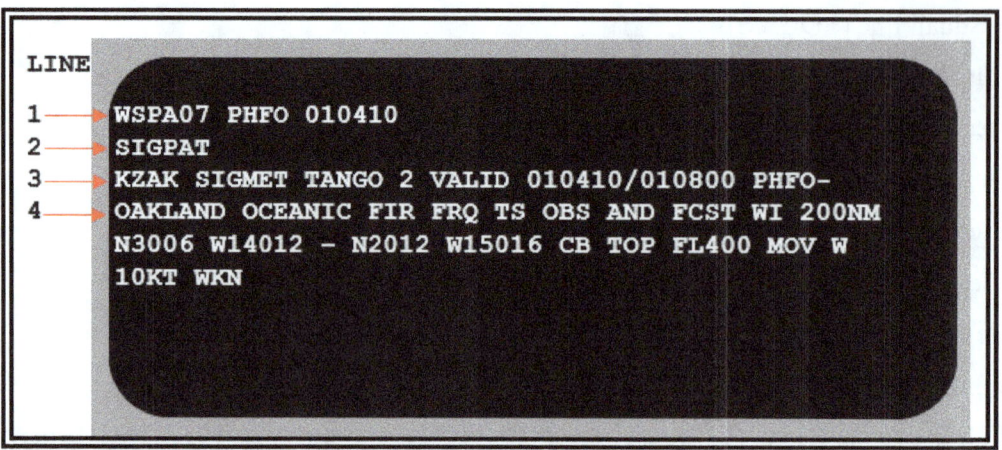

Figure 26-7. SIGMET Outside the CONUS—Example

Table 26-3. Decoding a SIGMET Outside of the CONUS

Line	Content	Description
1	WSPA07	ICAO communication header
	PHFO	Issuance MWO
	010410	Issuance UTC date/time
2	SIGPAT	NWS AWIPS communication header
3	KZAK	Area control center
	SIGMET	Product type
	TANGO	SIGMET series
	2	Issuance number
	VALID 010410/010800	Valid period UTC date/time
	PHFO	Issuance office
4	OAKLAND OCEANIC FIR	FIR
	FRQ TS OBS AND FCST WI 200NM N3006 W14012 - N2012 W15016 CB TOP FL400 MOV W 10KT WKN.	Phenomenon description

The SIGMET in Figure 26-7 is decoded as the following:

- (Line 1) The WMO product header is WSPA07. Issued by the Honolulu MWO on the 1st day of the month at 0410 UTC.

- (Line 2) The NWS Advanced Weather Interactive Processing System (AWIPS) communication header is SIGPAT.

- (Line 3) For the Oakland ARTCC (KZAK). This is the 2nd issuance of SIGMET Tango series, valid from the 1st day of the month at 0410 UTC until the 1st day of the month at 0800 UTC, issued by the Honolulu MWO.

- (Line 4) Concerning the Oakland Oceanic FIR, frequent thunderstorms observed and forecast within 200 NM of 30° and 6 minutes north; 140° and 12 minutes west; to 20° and 12 minutes north, 150° and 16 minutes west, cumulonimbus tops to FL400 moving west at 10 kt, weakening.

26.2.5.4 SIGMETs for Volcanic Ash (VA)—Outside the CONUS

A SIGMET for a volcanic ash cloud is issued for volcanic eruptions. A volcanic eruption is any volcanic activity, including the emission of volcanic ash, regardless of the eruption's magnitude. Initial volcanic ash SIGMETs may be issued based on credible PIREPs in the absence of a VAA, but are updated once a VAA is issued. Volcanic ash SIGMETs will continue to be issued until the ash cloud is no longer occurring or expected to occur.

SIGMETs for a volcanic ash cloud are valid up to 6 hours and provide an observed or forecast location of the ash cloud at the beginning of the SIGMET. A 6-hour forecast position for the ash cloud, valid at the end of the validity period of the SIGMET message, is also included.

26.2.5.5 SIGMETs for Tropical Cyclone (TC)—Outside the CONUS

SIGMETs for a tropical cyclone (which includes a hurricane, typhoon (western Pacific), and tropical storm) may be issued for nonfrontal synoptic-scale cyclones meeting the following criteria:

- Originates over tropical or subtropical waters with organized convection and definite cyclonic surface wind circulation.
- Wind speeds reach 35 kt independent of the wind averaging time used by the TCAC.

SIGMETs for tropical cyclones will be valid up to 6 hours. SIGMETs for tropical cyclones will include two positions. The first position will be the observed position of the center of the tropical cyclone, taken from the TCA. The second position will be the forecast position of the center of the tropical cyclone valid at the end of the SIGMET period, which will coincide with the 6-hour forecast position in the TCA.

In addition to the two storm positions, SIGMETs will include associated convection when applicable. SIGMETs will be reissued at least every 6 hours while the tropical cyclone winds remain or are expected to remain above 34 kt.

26.2.5.6 SIGMET Cancellation—Outside the CONUS

SIGMETs are cancelled when the phenomenon is no longer occurring or expected to occur.

26.2.5.7 SIGMET Amendments—Outside the CONUS

SIGMET amendments will not be issued. Instead, the next SIGMET in the series is issued to accomplish the update. The valid time of the new SIGMET is reset to reflect the new 4-hour valid period (6-hour period for volcanic ash and tropical cyclone SIGMETs).

26.2.5.8 SIGMET Corrections—Outside the CONUS

Corrections to SIGMETs are issued as necessary. This is done by issuing a new SIGMET in the series, which advances the SIGMET number and cancels the previous SIGMET.

26.2.5.8.1 SIGMET for Volcanic Ash Example—Outside the CONUS

```
WVNT06 KKCI 082030
TJZS SIGMET FOXTROT 2 VALID 082030/090230 KKCI-
SAN JUAN FIR VA FROM SOUFRIERE HILLS LOC 1642N06210W
VA CLD OBS AT 2030Z WI N1730 W06400 - N1700 W06300 - N1650 W06300 - N1710 W06400
- N1730 W06400. SFC/060. MOV W 15KT. FCST 0230Z VA CLD APRX N1730 W06500 - N1700
W06300 - N1650 W06300 - N1710 W06500 - N1730 W06500.
```

The ICAO communication header for this product is WVNT06. It is a SIGMET issued by the AWC (**KKCI**) in Kansas City, MO, on the 8th day of the month at 2030 UTC. This is the second (**2**) issuance of SIGMET series Foxtrot valid from the 8th day of the month at 2030 UTC until the 9th day of the month at 0230 UTC. Within the San Juan Oceanic FIR, volcanic ash from the Soufriere Hills volcano located at 16°/42 minutes north, 62°/10 minutes west. Volcanic ash cloud observed at 2030 UTC within an area bounded by 17°/30 minutes north, 64°/00 minutes west to 17°/00 minutes north, 63°/00 minutes west to 16°/50 minutes north, 63°/00 minutes west to 17°/10 minutes north, 64°/00 minutes west to 17°/30 minutes north, 64°/00 minutes west. From the surface to 6,000 ft MSL. Moving to the west at 15 kt. Forecast at 0230 UTC, volcanic ash cloud located approximately at 17°/30 minutes north, 65°/00 minutes west to 17°/00 minutes north, 63°/00 minutes west to 16°/50 minutes north, 63°/00 minutes west to 17°/10 minutes north, 65°/00 minutes west to 17°/30 minutes north, 65°/00 minutes west.

26.2.5.8.2 SIGMET for Tropical Cyclone Example—Outside the CONUS.

```
WSNT03 KKCI 081451
SIGA0C
KZNY SIGMET CHARLIE 11 VALID 081500/082100 KKCI-
NEW YORK OCEANIC FIR TC KYLE OBS N3106 W07118 AT 1500Z CB TOP FL500 WI 120NM OF
CENTER MOV WSW 5 KT NC FCST 2100Z TC CENTER N3142 W07012
```

The ICAO communication header for this product is WSNT03. It is a SIGMET issued by the AWC (**KKCI**) in Kansas City, MO, on the 8th day of the month at 1451 UTC. The NWS AWIPS communication header for this product is SIGA0C. This is the eleventh (**11**) issuance of SIGMET series Charlie valid from the 8th day of the month at 1500 UTC until the 8th day of the month at 2100 UTC. Within the New York Oceanic FIR, Tropical Cyclone Kyle observed at 31°/6 minutes north, 71°/18 minutes west at 1500 UTC, cumulonimbus tops to FL500 (approximately 50,000 ft MSL), within 120 NM of the center, moving from west-southwest at 5 kt, no change in intensity is forecast, at 2100 UTC the tropical cyclone center will be at 31°/42 minutes north, 70°/12 minutes west.

26.3 Airmen's Meteorological Information (AIRMET)

An AIRMET is a concise description of the occurrence or expected occurrence of specified en route weather phenomena which may affect the safety of aircraft operations. AIRMETs are intended to inform all pilots, especially VFR pilots and operators of sensitive aircraft, of potentially hazardous weather phenomena. AIRMETs contain details about IFR, extensive mountain obscuration, moderate turbulence, strong surface winds, moderate icing, and freezing levels.

AIRMETs are intended for dissemination to all pilots in the preflight and en route phases of flight to enhance safety. En route, AIRMETs are available over Flight Service frequencies. Over the CONUS, AIRMETs are also available on equipment intended to display weather and other non-ATC-related flight information to pilots using the FIS-B. In Alaska and Hawaii, AIRMETs are broadcast on air traffic frequencies.

AIRMETs are issued for the CONUS, Alaska, and Hawaii. No AIRMETs are issued for U.S. Oceanic FIRs in the Gulf of Mexico, Caribbean, and Western Atlantic and Pacific Oceans.

26.3.1 AIRMET Issuance

AIRMETs are issued from the three U.S. MWOs located at the AWC, the AAWU, and the WFO Honolulu. Their areas of responsibility are:

- AWC: The CONUS and adjacent coastal waters.
- AAWU: Alaska and adjacent coastal waters.
- WFO Honolulu: Hawaii and adjacent waters.

26.3.1.1 AIRMETs Over the CONUS

AIRMETs over the CONUS are displayed graphically on websites such as https://aviationweather.gov and https://www.1800wxbrief.com and equipment receiving FIS-B information. AIRMETs over the CONUS provide a higher forecast resolution than AIRMETs issued in text format. AIRMETs over the CONUS are valid at discrete times no more than 3 hours apart for a period of up to 12 hours into the future (e.g., 00, 03, 06, 09, and 12 hours). Additional forecasts may be inserted during the first 6 hours (e.g., 01, 02, 04, and 05). A forecast at 00-hour represents the initial conditions, and the subsequent graphics depict the area affected by the particular hazard at that valid time. Forecasts valid at 00 through 06 hours correspond to the text AIRMET bulletin.

AIRMETs over the CONUS depict the following en route aviation weather hazards:

- IFR conditions (ceiling <1,000 ft and/or surface visibility <3 mi).
- Widespread mountain obscuration.
- Moderate icing.
- Freezing levels.
- Moderate turbulence.
- Nonconvective LLWS potential below 2,000 ft AGL.
- Sustained surface winds greater than 30 kt.

Interpolation of time periods between AIRMETs over the CONUS valid times: Users should keep in mind when using the AIRMET over the CONUS that if a 00-hour forecast shows no significant weather and a 03-hour forecast shows hazardous weather, they must assume a change is occurring during the period between the two forecasts. It should be taken into consideration that the hazardous weather starts immediately after the 00-hour forecast, unless there is a defined initiation or ending time for the hazardous weather. The same would apply after the 03-hour forecast. The user should assume the hazardous weather condition is occurring between the snapshots unless informed otherwise. For example, if a 00-hour forecast shows no hazard, a 03-hour forecast shows the presence of hazardous weather, and a 06-hour forecast shows no hazard, the user should assume the hazard exists from the 0001 hour to the 0559 hour time period.

26.3.1.2 AIRMETs Over Alaska and Hawaii

AIRMETs over Alaska and Hawaii are in text format. The hazard areas are described using well-known geographical areas. AIRMETs over Alaska are issued for three Alaska regions corresponding to Alaska area forecasts.

AIRMETs over Alaska are valid up to 8 hours. AIRMETs over Hawaii are valid up to 6 hours. Unscheduled issuances contain an update number for easier identification.

There are three AIRMET categories: Sierra, Tango, and Zulu.

- AIRMET Sierra describes IFR conditions and/or extensive mountain obscurations.
- AIRMET Tango describes moderate turbulence, sustained surface winds of 30 kt or greater, and/or nonconvective LLWS.
- AIRMET Zulu describes moderate icing and provides freezing level heights.

26.3.1.3 AIRMET Issuance Criteria

An AIRMET may be issued when any of the following weather phenomena are occurring or are expected to occur:

- Ceiling less than 1,000 ft and/or visibility less than 3 sm (**IFR**).
    - Weather phenomena restricting the visibility including, but not limited to, precipitation (**PCPN**), smoke (**FU**), haze (**HZ**), mist (**BR**), fog (**FG**), and blowing snow (**BS**).
- Widespread mountain obscuration (**MTN OBSCN**).
    - Weather phenomena causing the obscuration are included, but not limited to, clouds (**CLDS**), precipitation (**PCPN**), smoke (**FU**), haze (**HZ**), mist (**BR**), and fog (**FG**).
- Moderate turbulence (**MOD TURB**).
    - Top and bottom of **MOD TURB** layer are specified.
- Sustained surface wind greater than 30 kt (**STG SFC WND**).
- Moderate icing (**MOD ICE**).
    - Top and bottom of **MOD ICE** are specified.
    - The range of freezing level altitudes is given when the bottom altitude of **MOD ICE** is the freezing level (**FRZLVL**).
    - Areas with multiple freezing levels are specified.
    - Range of freezing levels over the area is specified.
    - Lowest freezing levels AGL at intervals of 4,000 ft MSL (or surface (**SFC**) as appropriate) are specified.
- Nonconvective LLWS potential below 2,000 ft AGL (**LLWS POTENTIAL**).

26.3.2 AIRMET Standardization

All AIRMETs follow these standards:

- All heights or altitudes are referenced to above MSL, unless otherwise noted, and annotated using the height in hundreds of feet, consisting of three digits (e.g., **040**). For heights at or above 18,000 ft, the level is preceded by "FL" to represent flight levels (e.g., **FL180**).
- Messages are prepared in abbreviated plain language using contractions from Order JO 7340.2. A limited number of nonabbreviated words, geographical names, and numerical values of a self-explanatory nature may also be used.
- Weather and obstructions to visibility are described using the weather abbreviations for surface weather observations (METAR/SPECI). Refer to Federal Meteorological Handbook No. 1, Surface Weather Observations and Reports, or Section 24.3 of this handbook.

26.3.3 AIRMET Issuance Times and Valid Periods

AIRMETs are issued following the schedule listed in Table 26-4. AIRMETs are issued four times a day except for those in Alaska, which are issued three times a day. Unscheduled AIRMETs are issued when conditions are occurring or expected to occur but were not forecast.

Table 26-4. AIRMET Issuance Schedule

Product Type	Issuance Time	Issuance Frequency
AIRMETs over the CONUS	0245, 0845, 1445, 2045 UTC	Every 6 hours
AIRMETs over Alaska	0115, 1315, 2115 UTC (Standard Time) 0415, 1215, 2015 UTC (Daylight Saving Time)	Every 8 hours
AIRMETs over Hawaii	0400, 1000, 1600, 2200 UTC	Every 6 hours

AIRMETs are valid for 6 hours except for those in Alaska, which are valid for 8 hours. The valid period of an AIRMET message cannot exceed the valid time of the AIRMET bulletin. However, note that each AIRMET contains remarks concerning the continuance of the phenomenon during the 6 hours following the AIRMET ending time. Also, AIRMET bulletins can contain a separate outlook when conditions meeting AIRMET criteria are expected to occur during the 6-hour period (8-hour for Alaska) after the valid time of the AIRMET bulletin.

26.3.4 AIRMET Formats and Examples

The AWC's website provides several options to display AIRMETs over the CONUS. There is an interactive display, as well as snapshots that may be viewed as static (single), combined, and looped images using the tools provided on the website. Figure 26-8 is an example of the AIRMET snapshot static image.

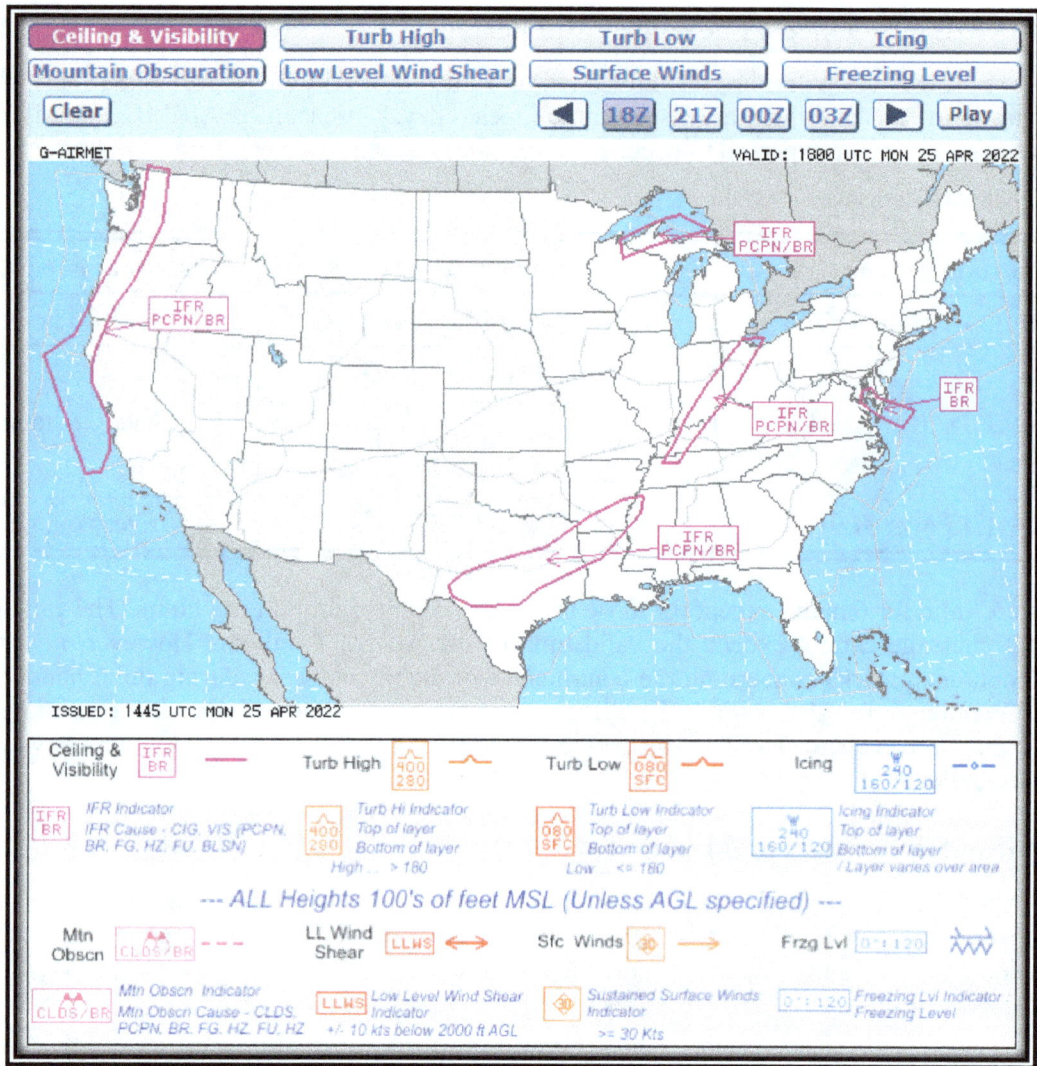

Figure 26-8. AIRMET—Ceiling and Visibility (IFR) Snapshot Example

An AIRMET message for Alaska and Hawaii includes the following information as appropriate and in the order indicated:

- Reference to appropriate active SIGMETs affecting the area at the time of AIRMET issuance (e.g., **SEE SIGMET BRAVO SERIES**).

- Beginning time of the AIRMET phenomenon if different from the AIRMET beginning valid time.

- AIRMET name (**SIERRA**, **TANGO**, or **ZULU**), update number, weather phenomenon, and ending valid time. (Note: The AIRMET number is reset to 1 after 0000 UTC each day.)
  - AIRMET Sierra describes IFR conditions and/or extensive mountain obscurations. Hawaii AIRMETs for mountain obscuration may be issued for an area less than 3,000 mi^2.
  - AIRMET Tango describes moderate turbulence, sustained surface winds of 30 kt or greater, and nonconvective LLWS.
  - AIRMET Zulu describes moderate icing and provides freezing-level heights.

- Fixed geographic locations (Alaska and Hawaii, see Figure 27-5 and Figure 27-6).

- Description of phenomenon for the AIRMET issuance.

- Vertical extent (bases and tops), as appropriate.

- Ending time of phenomenon if different from the AIRMET ending time.

- For Alaska AIRMETs: Intensity change remarks (**INTSF, WKN, IMPR**, or **NC**) concerning the continuance of the phenomenon following the valid period.

- For Hawaii AIRMETs: A separate AIRMET outlook is included in the AIRMET bulletin when conditions meeting AIRMET criteria are expected to occur during the 6-hour period after the valid time of the AIRMET bulletin.

26.3.4.1 Example of AIRMET Sierra Issued for the Southeast Alaska Area

```
WAAK47 PAWU 241324
WA7O
JNUS WA 241315
AIRMET SIERRA FOR IFR AND MT OBSC VALID UNTIL 242115
.
LYNN CANAL AND GLACIER BAY JB
MTS OBSC BY CLDS/ISOL PCPN. NC.
.
CNTRL SE AK JC
MTS OCNL OBSC IN CLDS. NC.
.
SRN SE AK JD
PAWG-PAKT LN W OCNL CIGS BLW 010/VIS BLW 3SM BR. IMPR.
.
SRN SE AK JD
MTS OCNL OBSC IN CLDS. NC.
.
ERN GLF CST JE
OCNL CIGS BLW 010/VIS BLW 3SM BR/-RA BR. DTRT.
.
ERN GLF CST JE
MTS OBSC IN CLDS/ISOL PCPN. DTRT.
.
SE AK CSTL WTRS JF
MTS OCNL OBSC IN CLDS. NC.
.
=JNUT WA 241315
AIRMET TANGO FOR TURB/STG SFC WINDS VALID UNTIL 242115
.
ERN GLF CST JE
OFSHR ICY BAY W SUSTAINED SFC WND 30 KTS
OR GTR. SPRDG E. INTSF.
.
=JNUZ WA 241315
AIRMET ZULU FOR ICING VALID UNTIL 242115
.
ERN GLF CST JE
16Z TO 19Z ALG CST W ICY BAY OCNL MOD ICEIC 080-160.
FZLVL 045 EXC 015 INLAND. WKN.
```

26.3.4.2 Example of AIRMET Tango Issued for Hawaii FA Area

```
WAHW31 PHFO 241529
```

```
WA0HI

HNLS WA 241600
AIRMET SIERRA UPDATE 2 FOR IFR VALID UNTIL 242200
.
NO SIGNIFICANT IFR EXP.

=HNLT WA 241600
AIRMET TANGO UPDATE 3 FOR TURB VALID UNTIL 242200
.
AIRMET TURB...HI
OVER AMD IMT S THRU W OF MTN.
TEMPO MOD TURB BLW 070.
COND CONT BEYOND 2200Z.

=HNLZ WA 241600
AIRMET ZULU UPDATE 2 FOR ICE AND FZLVL VALID UNTIL 242200
.
NO SIGNIFICANT ICE EXP.
```

26.3.5 AIRMET Updates and Amendments

AIRMETs are issued over the CONUS by the AWC every 6 hours and are updated/amended as necessary.

If an AIRMET issued for Alaska or Hawaii is amended, "AMD" is added after the date/time group. The issuance time of the AIRMET bulletin is updated to reflect the time of the amendment. The ending valid time remains unchanged.

26.4 Center Weather Advisory (CWA)

A CWA is a concise description of the occurrence or expected occurrence of specified weather phenomena that meet or approach in-flight advisory (e.g., AIRMET, SIGMET, or Convective SIGMET) criteria. A CWA is issued for hazardous weather when there is no existing in-flight advisory from the AWC or the AAWU.

26.4.1 CWA Issuance

CWAs are issued by the NWS CWSU. CWSU areas of responsibility are depicted in Figure 26-9.

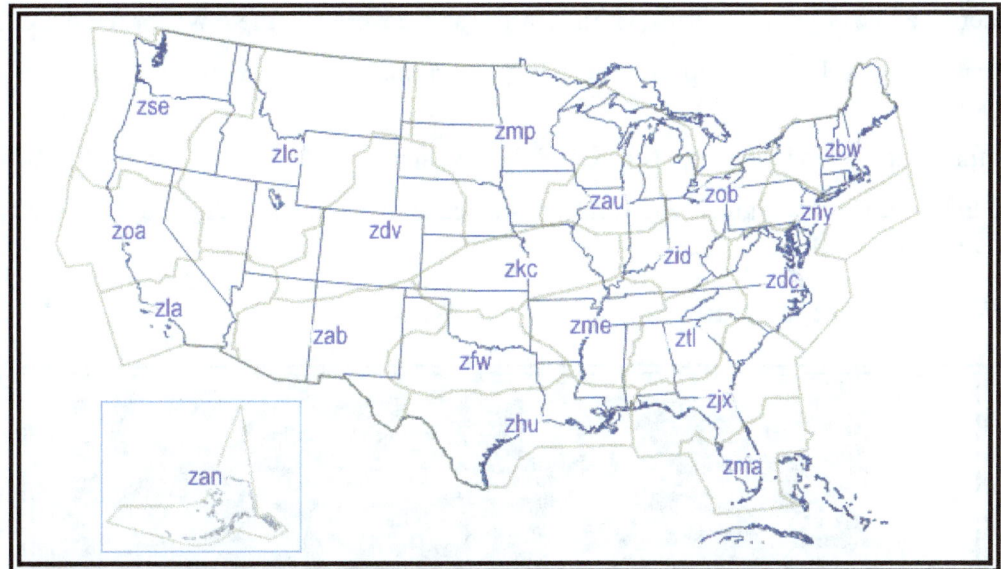

Figure 26-9. CWSU Areas of Responsibility

CWAs are valid for up to 2 hours and may include forecasts of conditions expected to begin within 2 hours of issuance. If conditions are expected to persist after the advisory's valid period, a statement to that effect is included in the last line of the text. Additional CWAs will subsequently be issued as appropriate. Notice of significant changes in the phenomenon described in a CWA is provided by a new CWA issuance for that phenomenon. If the forecaster deems it necessary, CWAs may be issued hourly for convective activity.

26.4.2 CWA Criteria

CWAs may be issued for the following situations:

- There is no existing AWC or AAWU advisory in effect.
- Any of the following conditions occur:
  - Conditions meeting Convective SIGMET criteria.
  - Icing—moderate or greater.
  - Turbulence—moderate or greater.
  - Heavy and extreme precipitation.
  - Freezing precipitation.
  - Conditions at or approaching LIFR.
  - Surface wind gust at or above 30 kt.
  - LLWS (surface–2,000 ft).
  - Volcanic ash, dust storms, or sandstorms.
- When a hazard has grown significantly outside of the boundary defined by the AWC or the AAWU advisory.
- To upgrade a thunderstorm advisory to include severe thunderstorms.
- To upgrade an AIRMET to include isolated severe turbulence or icing. If greater than isolated severe turbulence or icing is occurring, then a new CWA should be issued.

- To define a line of thunderstorms within a larger area covered by the AWC or the AAWU advisory.
- To better define hazards expected at a major terminal already within an AWC or an AAWU advisory.
- Anything that in the judgment of the CWSU forecaster will add value to an existing advisory.
- If in the forecaster's judgment the conditions listed above, or any others, may adversely impact the safe flow of air traffic.

26.4.3 CWA Format and Example

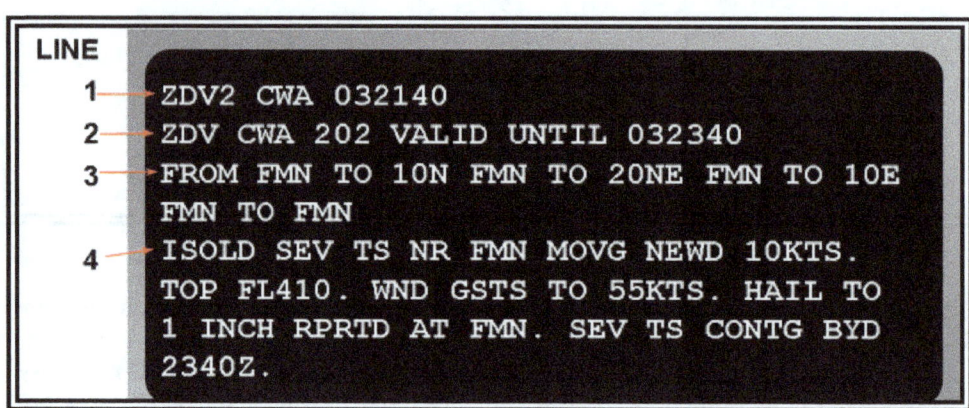

Figure 26-10. CWA—Example

Table 26-5. Decoding a CWA

Line	Content	Description
1	ZDV	ARTCC identification
	2	Phenomenon number (single digit, 1–6)
	CWA	Product type (CWA)
	032140	Beginning and/or issuance UTC date/time
2	ZDV	ARTCC identification
	CWA	Product type
	2	Phenomenon number (single digit, 1–6)
	02	Issuance number (issued sequentially for each Phenomenon Number)
	VALID TIL 032340Z	Ending valid UTC date/time
3	FROM FMN TO 10N FMN TO 20NE FMN TO 10E FMN TO FMN	Phenomenon location
4	ISOLD SEV TS NR FMN MOVG NEWD 10KTS. TOP FL410. WND GSTS TO 55KTS. HAIL TO 1 INCH RPRTD AT FMN. SEV TS CONTG BYD 2340Z	Phenomenon description

Chapter 26, Advisories

The CWA in Figure 26-10 is decoded as follows:

- (Line 1) CWA issued for the Denver ARTCC (**ZDV**) CWSU. The "2" after **ZDV** in the first line denotes this is the second meteorological event of the local calendar-day. This CWA was issued/begins on the 3rd day of the month at 2140 UTC.

- (Line 2) The Denver ARTCC (**ZDV**) is identified again. The "202" in the second line denotes the phenomena number again (**2**) and the issuance number (**02**) for this phenomenon. This CWA is the valid until the 3rd day of the month at 2340 UTC.

- (Line 3) From Farmington, NM, to 10 NM north of Farmington, NM, to 20 NM northeast of Farmington, NM, to 10 NM east of Farmington, NM, to Farmington, NM.

- (Line 4) Isolated severe thunderstorms near Farmington moving northeastward at 10 kt. Tops to FL410. Wind gusts to 55 kt. Hail to one inch reported at Farmington. Severe thunderstorms continuing beyond 2340 UTC.

26.5 Volcanic Ash Advisory (VAA)

The VAA is advisory information on volcanic ash cloud issued in abbreviated plain language, using approved ICAO abbreviations and numerical values of self-explanatory nature.

A graphical (e.g., chart) version of the VAA is also produced and is sometimes referred to as "VAG." The VAG contains the same information as in the VAA but in a four-panel chart format.

26.5.1 Volcanic Ash Advisory Center (VAAC)

A VAAC is a meteorological office designated by an ICAO regional air navigation agreement to provide advisory volcanic ash information to MWOs, WAFCs, area control centers, flight information centers, and international operational meteorological (OPMET) data banks regarding the lateral and vertical extent and forecast movement of volcanic ash in the atmosphere following a volcanic eruption. There are nine VAACs worldwide (see Figure 26-11). The duties of a VAAC include:

- Monitoring relevant geostationary and polar-orbiting satellite data to detect the existence and extent of volcanic ash in the atmosphere in the area concerned;

- Activating the volcanic ash numerical trajectory/dispersion model in order to forecast the movement of any ash cloud that has been detected or reported; and

- Issuing advisory information regarding the extent and forecast movement of the volcanic ash cloud.

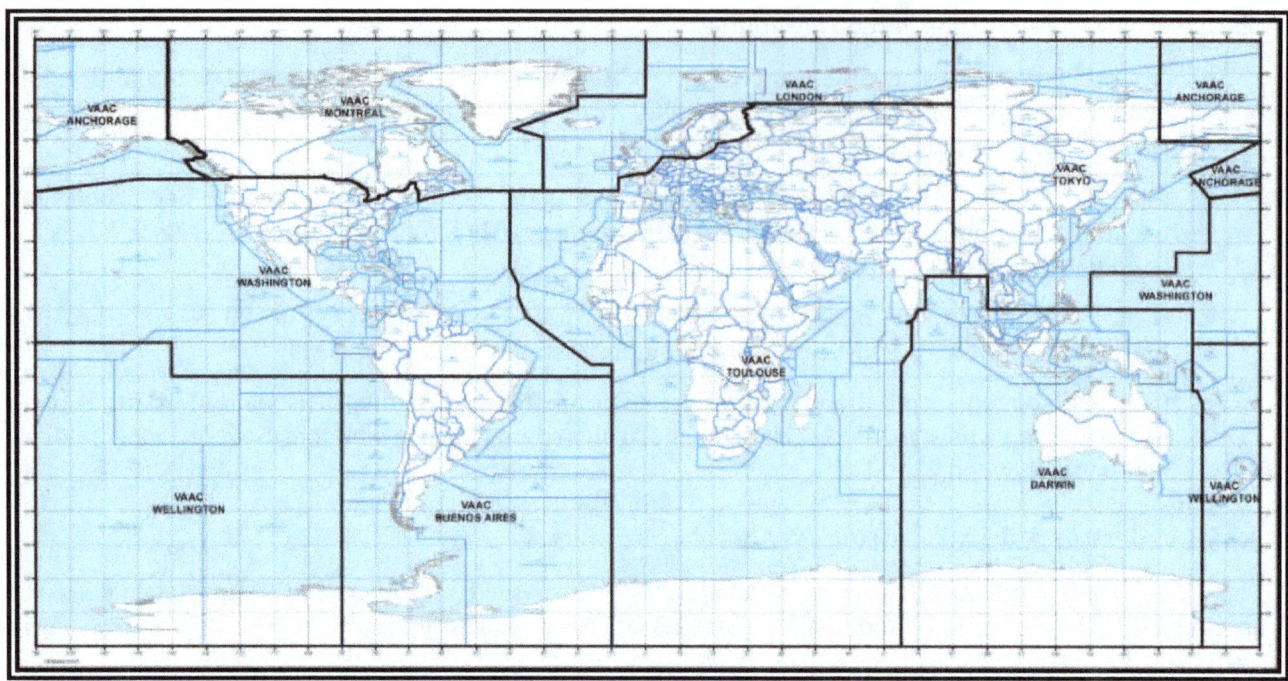

Areas of responsibility for VAACs include Anchorage, Montreal, Washington, Wellington (New Zealand), Buenos Aires (Argentina), London, Toulouse (France), Tokyo, and Darwin.

Figure 26-11. Volcanic Ash Advisory Centers

The United States has two VAACs with responsibilities defined in ICAO Annex 3, Meteorological Service for International Air Navigation. The Washington VAAC is jointly managed by the NESDIS SAB and the NWS NCO. The Anchorage VAAC is managed by the NWS AAWU.

The Washington VAAC areas of responsibility are:

- FIRs in the CONUS and adjacent coastal waters.
- The Oakland Oceanic FIR over the Pacific Ocean.
- The New York FIR over the western Atlantic Ocean.
- FIRs over and adjacent to the Caribbean and Central and South America north of 10° S latitude.

The Anchorage VAAC areas of responsibility are:

- The Anchorage FIR.
- Russian FIRs north of 60° N latitude and east of 150° E longitude.

26.5.1.1 VAA Issuance

Each VAAC issues VAAs to provide guidance to MWOs for SIGMETs involving volcanic ash.

VAAs are issued as necessary, but at least every 6 hours until such time as the volcanic ash cloud is no longer identifiable from satellite data, no further reports of volcanic ash are received from the area, and no further eruptions of the volcano are reported.

VAAs in both text and graphical format are available from the websites of the VAACs as well as other weather information sources.

26.5.1.2 VAA Format

The VAA format conforms to the template included in ICAO Annex 3, Table A2-1, Template for Advisory Message for Volcanic Ash.

26.5.1.3 VAA Text Example

```
FVAK21 PAWU 190615
VOLCANIC ASH ADVISORY
ISSUED:20030419/0615Z
VAAC:    ANCHORAGE
VOLCANO:   CHIKURACHKI, 900-36
LOCATION:  N5019 E15527
AREA:   KAMCHATKA NORTHERN KURIL ISLANDS
SUMMIT ELEVATION: 7674 FT (2339 M)
ADVISORY NUMBER:   2003-02
INFORMATION SOURCE:    SATELLITE
AVIATION COLOR CODE:  NOT GIVEN
ERUPTION DETAILS: NEW ERUPTION OCCURRED APPROX 190500 UTC.
HEIGHT IS ESTIMATED AT FL300. ESTIMATE IS BASED ON OBSERVED AND MODEL WINDS.
MOVEMENT APPEARS TO BE E AT 75 KTS.
OBS ASH DATA/TIME:    19/0500Z
OBS ASH CLOUD: VA EXTENDS FM NEAR VOLCANO EWD TO N50 E160.
FCST ASH CLOUD +6HR:  30NM EITHER SIDE OF LN FM NIPPI N49 E159 - N50 E175.
FCST ASH CLOUD +12HR: 30NM EITHER SIDE OF LN FM N50 E168 - N50 E180.
FCST ASH CLOUD +18HR: 30NM EITHER SIDE OF LN FM N51 E175 - N50 E185.
NEXT ADVISORY: 20030419/1500Z
REMARKS:    UPDATES AS SOON AS INFO BECOMES AVAILABLE.
```

26.5.1.4 VAA Graphic Example

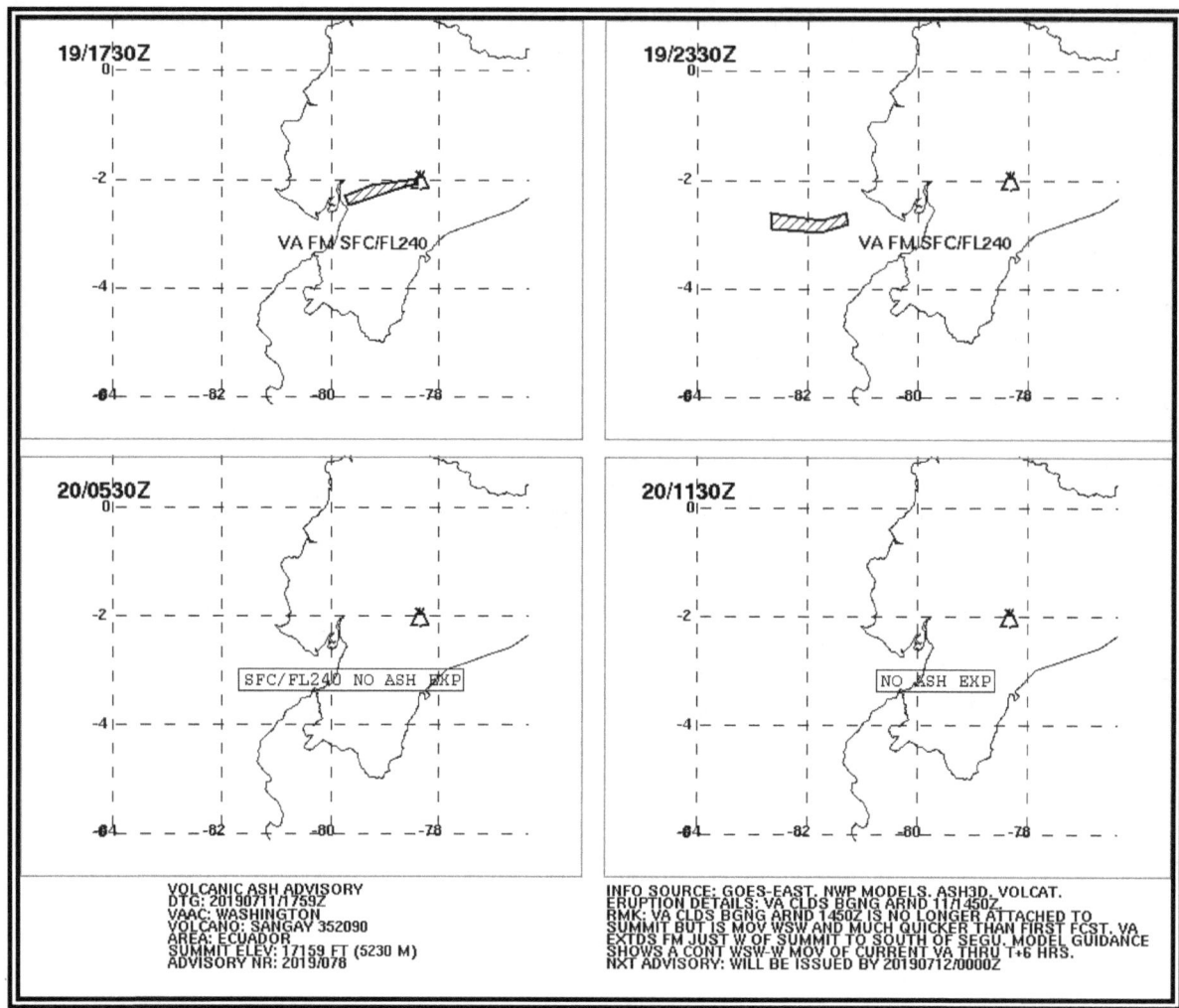

Figure 26-12. VAA in Graphical Format

26.6 Aviation Tropical Cyclone Advisory (TCA)

The aviation TCA is intended to provide tropical cyclone forecast guidance, through 24 hours, for international aviation safety and routing purposes. TCAs are issued by the WMO's TCAC for all ongoing tropical cyclone activity in their respective areas of responsibility. Two of the world's TCACs are in the United States: the NWS NHC in Miami, FL, and the NWS CPHC in Honolulu, HI.

The NHC is responsible for all ongoing tropical cyclone activity in the Atlantic and eastern Pacific, north of the Equator, while the CPHC is responsible for the central Pacific, north of the Equator. TCAs are available on their respective websites.

26.6.1 TCA Issuance

The NHC and CPHC issue TCAs at 0300, 0900, 1500, and 2100 UTC and are valid from the time of issuance until the next scheduled issuance or update. The forecast position information in the TCA is interpolated from the official forecast data, which is valid at 0000, 0600, 1200, and 1800 UTC.

26.6.2 TCA Content

TCAs list the current tropical cyclone position, motion, and intensity, and 3-, 9-, 15-, 21-, and 27-hour forecast positions and intensities. It is an alphanumeric text product produced by hurricane forecasters and consists of information extracted from the official forecasts.

This forecast is produced from subjective evaluation of current meteorological and oceanographic data as well as output from Numerical Weather Prediction (NWP) models, and is coordinated with affected NWS offices, the NWS national centers, and the DOD.

26.6.3 TCA Example

```
FKNT22 KNHC 151436
TCANT2
TROPICAL STORM BILL ICAO ADVISORY NUMBER   5
NWS NATIONAL HURRICANE CENTER MIAMI FL
1500 UTC TUE JUN 15 2021

AL022021

TC ADVISORY
DTG: 20210615/1500Z
TCAC: KNHC
TC: BILL
ADVISORY NR: 2021/005
OBS PSN: 15/1500Z N4030 W06200
MOV: NE 33KT
INTST CHANGE: NC
C: 0998HPA
MAX WIND: 050KT
FCST PSN +3 HR: 15/1800Z N4225 W05939
FCST MAX WIND +3 HR: 050KT
FCST PSN +9 HR: 16/0000Z N4425 W05722
FCST MAX WIND +9 HR: 050KT
FCST PSN +15 HR: 16/0600Z N4628 W05507
FCST MAX WIND +15 HR: 045KT
FCST PSN +21 HR: 16/1200Z N//// W/////
FCST MAX WIND +21 HR: ///KT
FCST PSN +27 HR: 16/1800Z N//// W/////
FCST MAX WIND +27 HR: ///KT
RMK: SOME FORECAST INFORMATION IN THIS PRODUCT IS INTERPOLATED FROM OFFICIAL
FORECAST DATA.
NXT MSG: 20210615/2100Z
$$
```

26.6.4 Additional Tropical Cyclone Information Products

The NHC, CPHC, and select WFOs issue many public tropical storm and hurricane/typhoon information products. Many are in web-based graphical format. For more information on these public forecasts, see the websites of the NHC, CPHC, WFO Guam, and WFOs along the Gulf Coast and the East Coast.

26.7 Space Weather Advisory

ICAO implemented a Space Weather Advisory program in late 2019. Under this program, ICAO designated four global space weather service providers:

- The United States, which is done by the NOAA SWPC.

- The Australia, Canada, France, and Japan (ACFJ) consortium, comprising of space weather agencies from Australia, Canada, France, and Japan.

- The Pan-European Consortium for Aviation Space Weather User Services (PECASUS), comprising of space weather agencies from Finland (lead), Belgium, the United Kingdom, Poland, Germany, Netherlands, Italy, Austria, and Cyprus.

- The China-Russian Federation Consortium (CRC) comprising of space weather agencies from China and the Russian Federation.

26.7.1 Space Weather Advisory Issuance

The SWPC, ACFJ consortium, PECASUS consortium, and CRC consortium serve as four global space weather centers that share the responsibility to issue global Space Weather Advisories, on a rotating basis, when there are impacts to HF COM, SATCOM, satellite-based navigation and surveillance systems (GNSS), or when heightened radiation occurs above FL250.

A Space Weather Advisory is issued whenever space weather conditions exceed predefined ICAO thresholds for both moderate impacts (MOD) and severe impacts (SEV) as given in Table 26-6.

Table 26-6. Space Weather Advisory Issuance Criteria

Effect	Sub-Effect	Parameter Used	Thresholds		Impact within Advisory Area	
			MOD	SEV	MOD	SEV
GNSS	Amplitude Scintillation	S4 (dimensionless)	0.5	0.8	Possible degraded service	Possible unreliable service
GNSS	Phase Scintillation	Sigma-phi (radians)	0.4	0.7		
GNSS	Vertical Total Electron Content (TEC)	TEC units	125	175		
RADIATION		Effective dose (micro-Sieverts/hour)*	30	80	Possible increased dose rates above normal levels	
HF COM	Auroral Absorption (AA)	Kp index	8	9	Possible degraded service	Possible unreliable service
HF COM	Polar Cap Absorption (PCA)	dB from 30 MHz riometer data	2	5		
HF COM	Shortwave Fadeout (SWF)	Solar X rays (0.0–0.8 NM) (W-m^{-2})	1×10^{-4} (X1)	1×10^{-3} (X10)		
HF COM	Post-Storm Depression	Maximum Usable Frequency (MUF)	30%	50%		
SATCOM	No threshold has been set for this effect				Possible degraded service	Possible unreliable service

MOD advisories will only be issued when the MOD threshold is reached between FL250 and FL460. SEV advisories will be issued when the SEV threshold is reached at any FL above FL250. For context, the background effective dose rate at FL370 at very high latitudes is approximately 9 micro-Sieverts/hour during solar minimum and 6 micro-Sieverts/hour during solar maximum. These rates decrease progressively toward the equatorial regions to values approximately one quarter of what is observed at very high latitudes.

Note: SEV radiation is a rare event with only a few short-lived events occurring during an 11-year solar cycle.

26.7.2 Space Weather Advisory Format

The Space Weather Advisory provides an observed or expected location for the impact and 6-, 12-, 18-, and 24-hour forecasts. The advisory describes the affected part of the globe in one of three ways:

- Six pre-defined latitude bands of width 30° shown in Table 26-7 (multiple bands may be given in one advisory), followed by a longitude range in 15° increments;*
- The term "DAYLIGHT SIDE," meaning the extent of the planet that is in daylight; or
- A polygon using latitude and longitude coordinates.

Note: *E18000-W18000 (or E180-W180) is used when the entire band is affected.

Table 26-7. Latitude Bands Used in Space Weather Advisories

Latitude Bands Used in Space Weather Advisories	
High latitudes northern hemisphere (HNH)	N90 to N60
Middle latitudes northern hemisphere (MNH)	N60 to N30
Equatorial latitudes northern hemisphere (EQN)	N30 to equator
Equatorial latitudes southern hemisphere (EQS)	Equator to S30
Middle latitudes southern hemisphere (MSH)	S30 to S60
High latitudes southern hemisphere (HSH)	S60 to S90

By design, the vertical and temporal resolutions of the advisory are very coarse. The use of 30° latitude bands, 15° longitude increments, 1,000-ft vertical increments (for radiation), and 6-hour time intervals will at times result in over-forecasting the affected airspace. In addition, while an entire latitude band may be forecast to have MOD or SEV space weather, there will often be times that the effect does not cover the entire width of the band or is intermittent or temporary. Users should refer to the remarks section of the advisory for additional information. Users can also go to the center's website, where a graphical depiction of the space weather event may be provided along with additional information.

Changes to the Space Weather Advisory content and format are possible in the coming years as experience is gained with the use of this product.

The Space Weather Advisory is not a replacement for the SWPC's other products or the NOAA Space Weather Scales, which continue to be provided by the SWPC. Refer to the SWPC's website for information on these products and the scales.

Table 26-8. Format of the Space Weather Advisory

Format	Explanation	Examples

Communication header	Product's coded identification for the issuing centers. KWNP is the SWPC, LFPW and YMMC are the ACFJ, EFKL and EGRR are the PECASUS, and ZBBB and UUAG are the CRC.
	
	FNXX01 is for GNSS, FNXX02 is for HF COM, FNXX03 is for Radiation, and FNXX04 is for SATCOM.
		`FNXX01 KWNP`
		`FNXX01 LFPW`
		`FNXX01 YMMC`
		`FNXX01 EFKL`
		`FNXX02 EGRR`
		`FNXX03 UUAG`
		`FNXX04 ZBBB`

SWX ADVISORY	Space Weather (SWX) Advisory.	`SWX ADVISORY`
STATUS:	Status indicator (optional) for test or exercise.	`TEST` `EXER`
DTG:	Date and time of origin, in YYYYMMDD/HHMMZ.	`20190418/0100Z`

SWXC:	Name of the Space Weather Advisory Center (SWXC).	`ACFJ`
		`PECASUS`
		`SWPC`
		`CRC`

ADVISORY NR:	Advisory number (NR).	`2019/9`
NR RPLC:	Advisory number being replaced by this advisory (optional).	`2019/8`

SWX EFFECT:	Space weather effect.	`HF COM MOD`
		`HF COM SEV`
		`SATCOM MOD`
		`SATCOM SEV`
		`GNSS MOD`
		`GNSS SEV`
		`RADIATION MOD`
		`RADIATION SEV`

OBS (or FCST) SWX:	Observed (OBS) or expected (FCST) space weather effect date/time, location, and altitudes (altitudes are only used in the radiation advisory).	`18/0100Z EQN W18000-W12000`
		`18/0100Z HNH HSH E180-W180 ABV FL370`
		`18/0100Z DAYLIGHT SIDE`
		`18/0100Z NO SWX EXP`

FCST SWX +6 HR:	6-hour forecast. Date/time, location, and altitudes.	Same as above
FCST SWX +12 HR:	12-hour forecast. Date/time, location, and altitudes.	Same as above
FCST SWX +18 HR:	18-hour forecast. Date/time, location, and altitudes.	Same as above
FCST SWX +24 HR:	24-hour forecast. Date/time, location, and altitudes.	Same as above

Format	Explanation	Examples
RMK:	Remarks (RMK).	Additional information
NXT ADVISORY:	Date/time when the next (NXT) scheduled advisory will be issued.	2010418/0700Z

26.7.3 Space Weather Advisory Examples

26.7.3.1 Example Space Weather Advisory—GNSS

Note: "GNSS" is the acronym for Global Navigation Satellite System, which is the term for all the world's navigation satellites, which includes the United States' Global Position Satellites.

```
FNXX03 KWNP 020100
SWX ADVISORY
DTG:                20190502/0100Z
SWXC:               SWPC
ADVISORY NR:        2019/59
NR RPLC:            2019/58
SWX EFFECT:         GNSS MOD
OBS SWX:            02/0100Z HNH HSH E18000-W18000
FCST SWX +  6 HR:   02/0700Z HNH HSH E18000-W18000
FCST SWX + 12 HR:   02/1300Z HNH HSH E18000-W18000
FCST SWX + 18 HR:   02/1900Z NO SWX EXP
FCST SWX + 24 HR:   03/0100Z NO SWX EXP
RMK:                IONOSPHERIC STORM CONTINUES TO CAUSE LOSS-OF-LOCK
                    OF GNSS IN AURORA ZONE. THIS ACTIVITY IS
                    EXPECTED TO SUBSIDE IN THE FORECAST PERIOD
NXT ADVISORY:       20190502/0700Z=
```

26.7.3.2 Example Space Weather Advisory—RADIATION

```
FNXX02 EFKL 190300
SWX ADVISORY
DTG:                20190219/0300Z
SWXC:               PECASUS
ADVISORY NR:        2019/20
SWX EFFECT:         RADIATION MOD
OBS SWX:            19/0300Z HNH HSH E18000-W18000 ABV FL370
FCST SWX +  6 HR:   19/0900Z NO SWX EXP
FCST SWX + 12 HR:   19/1500Z NO SWX EXP
FCST SWX + 18 HR:   19/2100Z NO SWX EXP
FCST SWX + 24 HR:   20/0300Z NO SWX EXP
RMK:                RADIATION AT AIRCRAFT ALTITUDES ELEVATED
                    BY SMALL ENHANCEMENT JUST ABOVE PRESCRIBED
                    THRESHOLD. DURATION TO BE SHORT-LIVED
NXT ADVISORY:       NO FURTHER ADVISORIES=
```

26.7.3.3 Example Space Weather Advisory—HF COM

```
FNXX01 YMMC 020100
SWX ADVISORY
DTG:                20190202/0100Z
SWXC:               ACFJ
ADVISORY NR:        2019/10
SWX EFFECT:         HF COM MOD
OBS SWX:            02/0100Z DAYLIGHT SIDE
FCST SWX +  6 HR:   02/0700Z DAYLIGHT SIDE
FCST SWX + 12 HR:   02/1300Z DAYLIGHT SIDE
FCST SWX + 18 HR:   02/1900Z NO SWX EXP
FCST SWX + 24 HR:   03/0100Z NO SWX EXP
RMK:                LOW END OF BAND HF COM DEGRADED
                    ON SUNLIT ROUTES. NEXT 12 HOURS
                    MOST POSSIBLE, DECLINING THEREAFTER.
NXT ADVISORY:       20190202/0700Z=
```

26.8 Low-Level Wind Shear (LLWS)/Microburst Advisories

When LLWS/microburst is reported by pilots or the integrated terminal weather system (ITWS), or detected on wind shear detection systems, such as the LLWS Alert System, the weather systems processor (WSP), or the TDWR, ATC issues the alert to all arriving and departing aircraft. ATC will continue providing the alert to aircraft until it is broadcast on the ATIS and pilots indicate they have received the appropriate ATIS code. A statement is included on the ATIS for 20 minutes following the last report or indication of the wind shear/microburst.

At facilities without the ATIS, ATC broadcasts wind shear/microburst information to all arriving and departing aircraft for 20 minutes following the last report or indication of the wind shear/microburst.

26.9 Airport Weather Warning (AWW)

The AWW addresses weather phenomena that can adversely impact airport ground operations. Information contained in this product may be useful to airport managers, fixed-based operators, airline ground personnel, and others responsible for the safety of ground operations. Ground decisions supported by the AWW information may include: fueling delays during thunderstorms, deicing frequency, and other similar ground operations. AWWs are not intended for use by in-flight operations.

Note: The AWW is not to be confused with the Aviation Watch Notification Message (see Section 27.16.4.1), which the NWS refers to as "SAW," but is also known by "AWW" on the websites of some weather provider services. The Aviation Watch Notification Message was formerly known as the Alert Severe Weather Watch Bulletin as well as the Severe Weather Forecast Alert.

26.9.1 AWW Issuance

The AWW is issued for select larger airports and is available on NWS WFOs' websites who have responsibility for issuing the TAFs for that airport. The AWW is *not* available on the AWC's website.

Issuance criteria are established according to local airport requirements. Some examples of criteria are: strong surface winds, freezing rain, heavy snow, and lightning within 5 miles of the airport.

26.9.2 AWW Format and Example

The AWW is written in a plain language, free-text format.

```
WWUS82 KJAX 112212
AWWJAX
FLZ025-112315-
AIRPORT WEATHER WARNING FOR JACKSONVILLE INTERNATIONAL AIRPORT
NATIONAL WEATHER SERVICE JACKSONVILLE FL
612 PM EDT WED MAR 11 2015
...AN AIRPORT WEATHER WARNING HAS BEEN ISSUED FOR JACKSONVILLE
INTERNATIONAL AIRPORT FOR STRONG WINDS AND POSSIBLE WINDSHIFT
VALID UNTIL 700 PM...
A CLUSTER OF THUNDERSTORMS WILL MOVE EASTWARD AT 12 KNOTS AND
AFFECT JACKSONVILLE INTERNATIONAL AIRPORT BY 620 PM. WIND GUSTS
OVER 35 KNOTS ARE LIKELY WITH THESE STORMS...AS WELL AS CLOUD-
TO-GROUND LIGHTNING AND REDUCED VISIBILITIES FROM HEAVY RAIN.
WINDSHIFTS ACCOMPANYING THE THUNDERSTORMS MAY ALTER THE RUNWAY
LANDING PATTERN WITH SOME STRONG WESTERLY WIND GUSTS.
$$
```

27 Forecasts

27.1 Introduction

The fourth of five types of aviation weather information discussed in this handbook is forecasts.

This chapter will discuss many forecast products produced primarily by the NWS that are either specific to aviation or are public products of interest to aviation users.

As with other products discussed in this handbook, the visualization of these products has evolved over the past decade with the use of internet websites. The use of static black and white depictions of aviation forecasts is almost a thing of the past. Today's websites provide the forecast products in color and offer options to select and overlay multiple products.

Today's aviation weather websites, including those of the NWS, continue to improve the content and visualization of forecast products. Unfortunately, it is not practical to update this handbook with every change to a weather product.

Examples of weather products in this handbook represent one way of how they can be visualized on a user's viewing device (e.g., computer, tablet, mobile phone, or cockpit display). The examples shown in this handbook are from the NWS' websites.

For this handbook, forecasts include the following:
- Winds and Temperatures Aloft Forecasts.
- Terminal Aerodrome Forecasts (TAF).
- Aviation Surface Forecasts.
- Aviation Clouds Forecasts.
- Area Forecasts (FA):
  - Alaska FA.
  - Hawaii FA.
  - Gulf of Mexico FA.
  - Caribbean FA.
- Alaska Graphical Forecasts:
  - Flying Weather.
  - Surface Forecast.
  - Icing Forecast.
  - Turbulence Forecast.
  - Convective Outlook.
- World Area Forecast System (WAFS) Forecasts.
- Significant Weather (SIGWX) Forecasts:
  - Low-Level.
  - Mid-Level.
  - High-Level.
- Surface Prognostic Forecasts.
- Upper-Air Forecasts.
- Freezing Level Forecasts.
- Forecast Icing Product (FIP).
- Graphical Turbulence Guidance (GTG) Forecasts.
- Cloud Tops Forecasts.
- Localized Aviation Model Output Statistics (MOS) Program (LAMP) Forecasts.
  - Alaska Aviation Guidance (AAG) Weather Product.
- Additional Convection Products:
  - Convective Outlook.
  - Traffic Flow Management (TFM) Convective Forecast (TCF).
  - Extended Convective Forecast Product (ECFP).
  - Watch Notification Messages.

- Route Forecasts (ROFOR).
- Aviation Forecast Discussions (AFD).
- Meteorological Impact Statements (MIS).
- Soaring Forecasts.
- Balloon Forecasts.

27.2 Winds and Temperatures Aloft

There are many wind and temperature aloft forecasts and products produced by the NWS. Each NWP model (i.e., sometimes referred to as computer models) outputs wind and temperature at multiple levels. The primary output of these forecasts is a gridded binary code format intended for use in flight planning software.

There is no official wind and temperature aloft model for flight planning. Depending on the computer model, the validity times, time intervals, and altitude levels will vary. Some models produce wind and temperature forecasts at hourly time-steps while others produce forecasts at 6-hour time steps. A few models provide wind forecasts at 1,000-ft altitude levels while others are mostly at 3,000-ft altitude levels. The data points (location) will also vary depending on the source.

Because each computer model is based on different algorithms and physics, the wind and temperature forecasts will vary from model to model. These differences are due, in part, to the model's forecast pressure patterns on the surface and aloft. In addition, some models have more detailed terrain as well as finer spacing between data points.

For many years there was just one set of wind and temperature forecasts, known as the FD Winds (then later as the FB Winds), which were presented in a coded text table format (see Section 27.2.1.1.2).

Today's flight planning software directly imports wind and temperature data from various computer models, which has effectively made the coded text table format obsolete.

Although FB winds are still produced today, they are archaic compared with model output available to the pilot and flight planner. FB winds:

- Are updated only four times daily, so winds that differ from the forecast are not updated for up to 6 hours.
- Provide a single value for each of three broad periods of time: from issuance time through 7 hours, 7 hours through 16 hours, and 16 hours through 28 hours.
- Provide a value at scattered locations in the country separated by about 100 to 150 mi.

In contrast, for example, the NWS' Rapid Refresh model winds (available on the AWC GFA Tool):

- Are updated every hour based partly on automated wind reports from airliners.
- Provide a wind forecast for each hour into the future.
- Provide values at grid points separated by as low as 9 mi (at full zoom on the GFA Tool).

This section describes the details of the FB Wind and Temperature Aloft Forecast product. Other sections within this handbook provide additional wind and temperature aloft forecasts (i.e., constant pressure level forecasts (see Section 27.10.1) and the global wind and temperature forecasts provided under the WAFS (see Section 27.7)).

27.2.1 FB Wind and Temperature Aloft Forecast

FB Wind and Temperature Aloft Forecasts are computer-prepared forecasts of wind direction, wind speed, and temperature at specified times, altitudes, and locations.

27.2.1.1 FB Wind and Temperature Aloft Forecast Issuance

The NWS NCEP produces scheduled FB Wind and Temperature Aloft Forecasts four times daily for specified locations in the CONUS, the Hawaiian Islands, Alaska and coastal waters, and the western Pacific Ocean. Specified locations are documented on the AWC's website under "Winds/Temps" forecasts.

Amendments are not issued to the forecasts. Wind forecasts are not issued for altitudes within 1,500 ft of a location's elevation. Temperature forecasts are not issued for altitudes within 2,500 ft of a location's elevation.

27.2.1.1.1 FB Wind and Temperature Aloft Forecast Text Format

The text format for the FB Wind and Temperature Aloft Forecast uses the symbolic form **DDff+TT** in which **DD** is the wind direction (true), **ff** is the wind speed, and **TT** is the temperature.

Wind direction is indicated in tens of degrees (two digits) with reference to true north and wind speed is given in knots (two digits). Light and variable wind or wind speeds of less than 5 kt are expressed by **9900**. Forecast wind speeds of 100 through 199 kt are indicated by adding 50 to the first two digits of the wind direction and subtracting 100 from the speed. For example, a forecast of 250°, 145 kt, is encoded as **7545**. Forecast wind speeds of 200 kt or greater are indicated as a forecast speed of 199 kt. For example, **7799** is decoded as 270° at 199 kt or greater.

Temperature is indicated in degrees Celsius (two digits) and is preceded by the appropriate algebraic sign for the levels from 6,000 through 24,000 ft. Above 24,000 ft, the sign is omitted since temperatures are always negative at those altitudes.

The product header includes the date and time observations were collected, the forecast valid date and time, and the time period during which the forecast is to be used.

27.2.1.1.2 FB Wind and Temperature Aloft Forecast Coding Example

Sample winds aloft text message:

```
DATA BASED ON 010000Z
VALID 010600Z FOR USE 0500-0900Z. TEMPS NEG ABV 24000
FT  3000 6000    9000    12000   18000   24000   30000  34000  39000
MKC 9900 1709+06 2018+00 2130-06 2242-18 2361-30 247242 258848 550252
```

Sample message decoded:

```
DATA BASED ON 010000Z
```

Forecast data is based on computer forecasts generated the 1st day of the month at 0000 UTC.

```
VALID 010600Z FOR USE 0500-0900Z. TEMPS NEG ABV 24000
```

The valid time of the forecast is the 1st day of the month at 0600 UTC. The forecast winds and temperatures are to be used between 0500 and 0900 UTC. Temperatures are negative above 24,000 ft.

```
FT  3000 6000 9000 12000 18000 24000 30000 34000 39000
```

FT indicates the altitude of the forecast.

MKC 9900 1709+06 2018+00 2130-06 2242-18 2361-30 247242 258848 550252

MKC indicates the location of the forecast. The rest of the data is the wind and temperature aloft forecast for the respective altitudes.

Table 27-1 shows data for MKC (Kansas City, MO). Table 27-2 provides the time periods for the use of FB Wind and Temperature Forecasts.

Table 27-1. Wind and Temperature Aloft Forecast Decoding Examples

FT 3000 6000 9000 12000 18000 24000 30000 34000 39000 MKC 9900 1709+06 2018+00 2130-06 2242-18 2361-30 247242 258848 550252			
Altitude (ft)	**Coded**	**Wind**	**Temperature (°C)**
3,000 ft	9900	Light and variable	Not forecast
6,000 ft	1709+06	170° at 9 kt	+06 °C
9,000 ft	2018+00	200° at 18 kt	Zero °C
12,000 ft	2130-06	210° at 30 kt	-06 °C
18,000 ft	2242-18	220° at 42 kt	-18 °C
24,000 ft	2361-30	230° at 61 kt	-30 °C
30,000 ft	247242	240° at 72 kt	-42 °C
34,000 ft	258848	250° at 88 kt	-48 °C
39,000 ft	750252	250° at 102 kt	-52 °C

Table 27-2. Wind and Temperature Aloft Forecast Periods

Model Run	Product Available	6-Hour Forecast		12-Hour Forecast		24-Hour Forecast	
		Valid	**For Use**	**Valid**	**For Use**	**Valid**	**For Use**
0000Z	~0200Z	0600Z	0200-0900Z	1200Z	0900-1800Z	0000Z	1800-0600Z
0600Z	~0800Z	1200Z	0800-1500Z	1800Z	1500-0000Z	0600Z	0000-1200Z
1200Z	~1400Z	1800Z	1400-2100Z	0000Z	2100-0600Z	1200Z	0600-1800Z
1800Z	~2000Z	0000Z	2000-0300Z	0600Z	0300-1200Z	1800Z	1200-0000Z

27.2.1.1.3 Graphical FB Wind and Temperature Aloft Forecast

Graphical depictions of FB Wind and Temperature Aloft Forecasts vary depending on the website. Details and information on these graphical depictions can usually be found on the website's help or information page.

The AWC's website provides an interactive display of the FB Wind and Temperature Aloft Forecasts (see Figure 27-1). As mentioned earlier, AWC's GFA Tool provides very high-resolution wind forecasts from the NWS' Rapid Refresh model (see Figure 27-2).

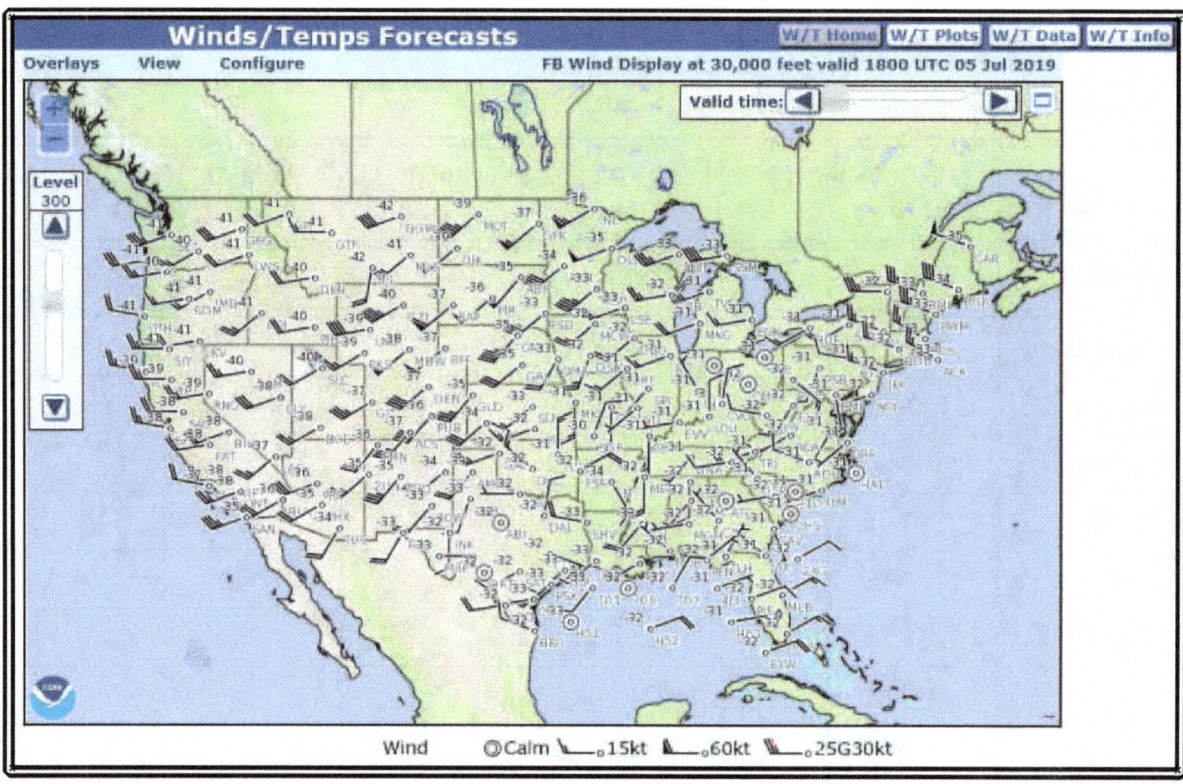

Figure 27-1. FB Wind and Temperature Aloft Interactive Display Example

The interactive graphic in Figure 27-1 depicts wind speed (knots) and direction (referenced to true north) using standard windbarb display. Temperature (Celsius) is placed to the upper left of the station circle.

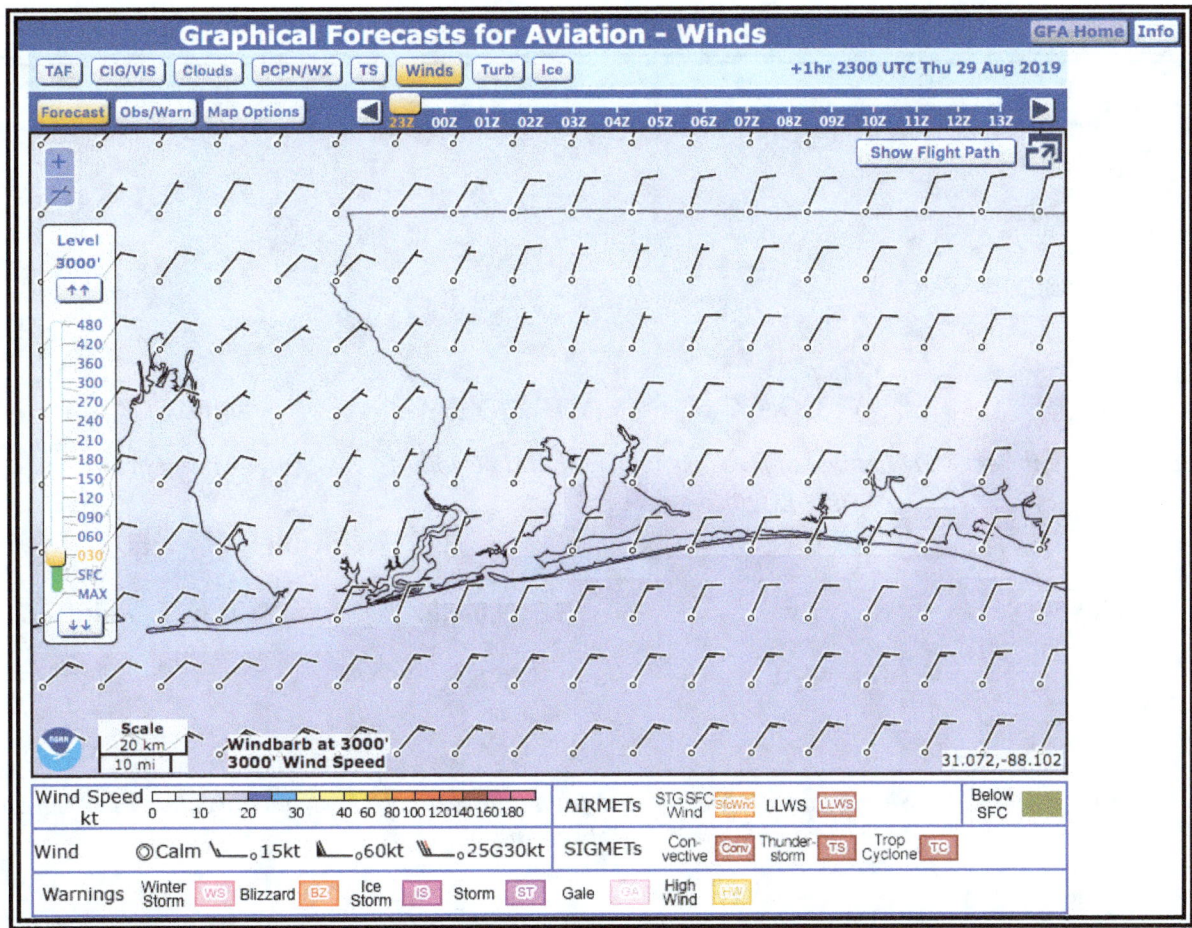

Figure 27-2. Example of the GFA Tool Wind Aloft Forecast Display at 3,000 ft Zoomed in Over Pensacola, FL

27.3 Terminal Aerodrome Forecast (TAF)

A TAF is a concise statement of the expected meteorological conditions significant to aviation for a specified time period within 5 sm of the center of the airport's runway complex (terminal). TAFs use the same weather codes found in METARs (see Section 24.3).

27.3.1 TAF Responsibility

TAFs are issued by NWS WFOs for nearly 700 U.S. airports. The majority of TAFs provide a 24-hour forecast for the airport, while TAFs for some major airports provide a 30-hour forecast.

27.3.2 Generic Format of the Forecast Text of an NWS-Prepared TAF

Refer to Table 27-3 for the generic format of the NWS' TAFs.

Table 27-3. Generic Format of the National Weather Service's TAFs

TAF or **TAF AMD** or **TAF COR** *Type of report*			
CCCC *Location identifier*	**YYGGggZ** *Date/time of forecast origin group*	**$Y_1Y_1G_1G_1/Y_2Y_2G_2G_2$** *Valid period*	**$dddffGf_mf_mKT$** *Wind group*
VVVV *Visibility group*	**w'w'** or **NSW** *Significant weather group*	**$N_sN_sN_sh_sh_sh_s$** or **$VVh_sh_sh_s$** or **SKC** *Cloud and vertical obscuration groups*	**$WSh_{ws}h_{ws}h_{ws}/dddftKT$** *Nonconvective LLWS group*
TTGGgg *Forecast change indicator groups*			
FMY_1Y_1GGgg *From group*	**$TEMPO\ Y_1Y_1GG/Y_eY_eG_eG_e$** *Temporary group*	**$PROB30\ Y_1Y_1GG/Y_eY_eG_eG_e$** *Probability group*	

27.3.2.1 Type of Report (TAF, TAF AMD, or TAF COR)

The report-type header always appears as the first element in the TAF and is produced in three forms: a routine forecast (**TAF**), an amended forecast (**TAF AMD**), or a corrected forecast (**TAF COR**).

TAFs are amended whenever they become, in the forecaster's judgment, unrepresentative of existing or expected conditions, particularly regarding those elements and events significant to aircraft and airports. An amended forecast is identified by **TAF AMD** (in place of **TAF**) on the first line of the forecast text.

Select airports have amendments routinely issued. See Table 27-7.

27.3.2.2 Location Identifier (CCCC)

After the line containing either **TAF**, **TAF AMD**, or **TAF COR**, each TAF begins with its four-letter ICAO location identifier.

Examples:

KDFW   Dallas-Fort Worth

PANC   Anchorage, Alaska

PHNL   Honolulu, Hawaii

27.3.2.3 Date/Time of Forecast Origin Group (YYGGggZ)

The date/time of the forecast origin group (**YYGGggZ**) follows the terminal's location identifier. It contains the day of the month in two digits (**YY**) and the time in four digits (**GGgg** in hours and minutes) in which the forecast is completed and ready for transmission, with a **Z** appended to denote UTC. This time is entered by the forecaster. A routine forecast, TAF, is issued 20 to 40 minutes before the beginning of its valid period.

Examples:

061737Z

The TAF was issued on the 6th day of the month at 1737 UTC.

121123Z

The TAF was issued on the 12th day of the month at 1123 UTC.

27.3.2.4 Valid Period ($Y_1Y_1G_1G_1/Y_2Y_2G_2G_2$)

The TAF valid period ($Y_1Y_1G_1G_1/Y_2Y_2G_2G_2$) follows the date/time of the forecast origin group. Scheduled 24- and 30-hour TAFs are issued four times per day, at 0000, 0600, 1200, and 1800Z. The first two digits (Y_1Y_1) are the day of the month for the start of the TAF. The next two digits (G_1G_1) are the starting hour (UTC). Y_2Y_2 is the day of the month for the end of the TAF, and the last two digits (G_2G_2) are the ending hour (UTC) of the valid period. A forecast period that begins at midnight UTC is annotated as **00**. If the end time of a valid period is at midnight UTC, it is annotated as **24**. For example, a 00Z TAF issued on the 9th of the month and valid for 24 hours would have a valid period of **0900/0924**.

Whenever an amended TAF (**TAF AMD**) is issued, it supersedes and cancels the previous TAF. That is, users should not wait until the start of the valid period indicated within the **TAF AMD** to begin using it.

Examples

1512/1612

The TAF is valid from the 15th day of the month at 1200 UTC until the 16th day of the month at 1200 UTC.

2306/2412

This is a 30-hour TAF valid from the 23rd day of the month at 0600 UTC until the 24th day of the month at 1200 UTC.

0121/0218

This is an amended TAF valid from the 1st day of the month at 2100 UTC until the 2nd day of the month at 1800 UTC.

```
0600/0624
```

This TAF is valid from the 6th day of the month at 0000 UTC until the 6th day of the month at 2400 UTC (or 7th day of the month at 0000 UTC).

27.3.2.5 Wind Group (dddffGf$_m$f$_m$KT)

The initial time period and any subsequent "from" (**FM**) groups begin with a mean surface wind forecast (**dddffGf$_m$f$_m$KT**) for that period. Wind forecasts are expressed as the mean three-digit direction (**ddd**, relative to true north) from which the wind is blowing, rounded to the nearest 10°, and the mean wind speed in knots (**ff**) for the time period. If wind gusts are forecast (gusts are defined as rapid fluctuations in wind speeds with a variation of 10 kt or more between peaks and lulls), they are indicated immediately after the mean wind speed by the letter **G**, followed by the peak gust speed expected. **KT** is appended to the end of the wind forecast group. Any wind speed of 100 kt or more will be encoded in three digits. Calm winds are encoded as **00000KT**.

The prevailing wind direction is forecast for any speed greater than or equal to 7 kt. When the prevailing surface wind direction is variable (variations in wind direction of 30° or more), the forecast wind direction is encoded as **VRBffKT**. Two conditions where this can occur are very light winds and convective activity. Variable wind direction for very light winds must have a wind speed of 1 to 6 kt inclusive. For convective activity, the wind group may be encoded as **VRBffGf$_m$f$_m$KT**, where **Gf$_m$f$_m$** is the maximum expected wind gusts. **VRB** is not used in the nonconvective LLWS group.

Squalls are forecast in the wind group as gusts (**G**) but must be identified in the significant weather group with the code **SQ**.

Examples:

```
23010KT
```

Wind from 230° "true" (southwest) at 10 kt.

```
28020G35KT
```

Wind from 280° "true" (west) at 20 kt gusting to 35 kt.

```
VRB05KT
```

Wind variable at 5 kt.

```
VRB15G30KT
```

Wind variable at 15 kt gusting to 30 kt due to forecast convective activity.

```
00000KT
```

Wind calm.

27.3.2.6 Visibility Group (VVVV)

The initial time period and any subsequent **FM** groups include a visibility forecast (**VVVV**) in statute miles appended by the contraction **SM**.

When the prevailing visibility is forecast to be less than or equal to 6 sm, one or more significant weather groups are included in the TAF. However, drifting dust (**DRDU**), drifting sand (**DRSA**), drifting snow (**DRSN**), shallow fog (**MIFG**), partial fog (**PRFG**), and patchy fog (**BCFG**) may be forecast with prevailing visibility greater than or equal to 7 sm.

When a whole number and a fraction are used to forecast visibility, a space is included between them (e.g., **1 1/2SM**). Visibility greater than 6 sm is encoded as **P6SM**.

If the visibility is not expected to be the same in different directions, prevailing visibility is used.

When volcanic ash (**VA**) is forecast in the significant weather group, visibility is included in the forecast, even if it is unrestricted (**P6SM**). For example, an expected reduction of visibility to 10 sm by volcanic ash is encoded in the forecast as **P6SM VA**.

Although *not* used by the NWS in U.S. domestic TAFs, the contraction **CAVOK** (ceiling and visibility OK) may replace the visibility, weather, and sky condition groups if all of the following conditions are forecast: visibility of 10 km (6 sm) or more, no clouds below 1500 m (5,000 ft) or below the highest minimum sector altitude (whichever is greater), no cumulonimbus, and no significant weather phenomena.

Examples:

```
P6SM
```

Visibility unrestricted.

```
1 1/2SM
```

Visibility 1 and ½ sm.

```
4SM
```

Visibility 4 sm.

27.3.2.7 Significant Weather Group (w'w' or NSW)

The significant weather group (**w'w'** or **NSW**) consists of the appropriate qualifier(s) and weather phenomenon contraction(s) or **NSW** (no significant weather).

If the initial forecast period and subsequent **FM** groups are not forecast to have explicit significant weather, the significant weather group is omitted. **NSW** is *not* used in the initial forecast time period or **FM** groups.

One or more significant weather group(s) is (are) included when the visibility is forecast to be 6 sm or less. The exceptions are: volcanic ash (**VA**), low drifting dust (**DRDU**), low drifting sand (**DRSA**), low drifting snow (**DRSN**), shallow fog (**MIFG**), partial fog (**PRFG**), and patchy fog (**BCFG**). Obstructions to vision are only forecast when the prevailing visibility is less than 7 sm or, in the opinion of the forecaster, is considered operationally significant.

Volcanic ash (**VA**) is always forecast when expected. When **VA** is included in the significant weather group, visibility is included in the forecast as well, even if the visibility is unrestricted (**P6SM**).

NSW is used in place of significant weather only in a temporary (**TEMPO**) group to indicate when significant weather (including in the vicinity (**VC**)) included in a previous subdivided group is expected to end.

Multiple precipitation elements are encoded in a single group (e.g., -**TSRASN**). If more than one type of precipitation is forecast, up to three appropriate precipitation contractions can be combined in a single group (with no spaces) with the predominant type of precipitation being first. In this single group, the intensity refers to the total precipitation and can be used with either one or no intensity qualifier, as appropriate. In TAFs, the intensity qualifiers (light, moderate, and heavy) refer to the intensity of the precipitation and not to the intensity of any thunderstorms associated with the precipitation.

Intensity is coded with precipitation types (except ice crystals and hail), including those associated with thunderstorms and those of a showery nature (**SH**). No intensity is ascribed to blowing dust (**BLDU**), blowing sand (**BLSA**), or blowing snow (**BLSN**). Only moderate or heavy intensity is ascribed to a sandstorm (**SS**) and dust storm (**DS**).

27.3.2.7.1 Exception for Encoding Multiple Precipitation Types

When more than one type of precipitation is forecast in a time period, any precipitation type associated with a descriptor (e.g., **FZRA**) is encoded first in the precipitation group, regardless of the predominance or intensity of the other precipitation types. Descriptors are not encoded with the second or third precipitation type in the group. The intensity is associated with the first precipitation type of a multiple precipitation type group. For example, a forecast of moderate snow and light freezing rain is coded as **-FZRASN**, although the intensity of the snow is greater than the freezing rain.

Examples:

Combinations of one precipitation and one nonprecipitation weather phenomenon:

- ```-DZ FG```

    Light drizzle and fog (obstruction that reduces visibility to less than 5/8 sm).

- ```RA BR```

    Moderate rain and mist.

- ```-SHRA FG```

    Light rain showers and fog.

- ```+SN FG```

    Heavy snow and fog.

Combinations of more than one type of precipitation:

- ```-RASN FG HZ```

    Light rain and snow (light rain predominant), fog, and haze.

- ```TSSNRA```

    Thunderstorm with moderate snow and rain (moderate snow predominant).

- ```FZRASNPL```

    Moderate freezing rain, snow, and ice pellets (freezing rain mentioned first due to the descriptor, followed by other precipitation types in order of predominance).

- ```SHSNPL```

    Moderate snow showers and ice pellets.

27.3.2.7.2 Thunderstorm Descriptor

The **TS** descriptor is treated differently than other descriptors in the following cases:

- When nonprecipitating thunderstorms are forecast, **TS** may be encoded as the sole significant weather phenomenon; and
- When forecasting thunderstorms with freezing precipitation (**FZRA** or **FZDZ**), the **TS** descriptor is included first, followed by the intensity and weather phenomena.

Example:

```TS -FZRA```

When a thunderstorm is included in the significant weather group (even using vicinity, **VCTS**), the cloud group ($N_sN_sN_sh_sh_sh_s$) includes a forecast cloud type of **CB**. See the following example for encoding **VCTS**.
```

Example:

```
-FZRA VCTS BKN010CB
```

27.3.2.7.3 Fog Forecast

A visibility threshold must be met before a forecast for fog (**FG**) is included in the TAF. When forecasting a fog-restricted visibility from 5/8 to 6 sm, the phenomenon is coded as **BR** (mist). When a fog-restricted visibility is forecast to result in a visibility of less than 5/8 sm, the code **FG** is used. The forecaster never encodes weather obstruction as mist (**BR**) when the forecast visibility is greater than 6 sm (**P6SM**).

Fog-related terms are used as described in Table 27-4.

Table 27-4. TAF Fog Terms

Term	Description
Freezing Fog (**FZFG**)	Any fog (visibility less than 5/8 sm) consisting predominantly of water droplets at temperatures less than or equal to 32 °F (0 °C), whether or not rime ice is expected to be deposited. **FZBR** is not a valid significant weather combination and will not be used in TAFs.
Shallow Fog (**MIFG**)	The visibility at 6 ft AGL is greater than or equal to 5/8 sm and the apparent visibility in the fog layer is less than 5/8 sm.
Patchy Fog (**BCFG**)	Fog patches covering part of the airport. The apparent visibility in the fog patch or bank is less than 5/8 sm, with the foggy patches extending to at least 6 ft AGL.
Partial Fog (**PRFG**)	A substantial part of the airport is expected to be covered by fog while the remainder is expected to be clear of fog (e.g., a fog bank). *Note: MIFG, PRFG, and BCFG may be forecast with prevailing visibility of P6SM.*

Examples:

```
1/2SM FG
```

Fog is reducing visibilities to less than 5/8 sm, therefore **FG** is used to encode the fog.

```
3SM BR
```

Fog is reducing visibilities to between 5/8 and 6 sm, therefore **BR** is used to encode the fog.

27.3.2.8 Vicinity (VC)

In the United States, vicinity (**VC**) is defined as a donut-shaped area between 5 and 10 sm from the center of the airport's runway complex. NWS TAFs may include a prevailing condition forecast of fog, showers, and thunderstorms in the airport's vicinity. A prevailing condition is defined as a greater than or equal to 50 percent probability of occurrence for more than half of the subdivided forecast time period. **VC** is not included in temporary (**TEMPO**) or probability (**PROB**) groups.

The significant weather phenomena in Table 27-5 are valid for use in prevailing portions of NWS' TAFs in combination with **VC**.

Table 27-5. TAF Use of Vicinity (VC)

Phenomenon	Coded
Fog*	VCFG
Shower(s)**	VCSH
Thunderstorm	VCTS

*Always coded as **VCFG** regardless of visibility in the obstruction, and without qualification as to intensity or type (frozen or liquid).
The **VC group, if used, should be the last entry in any significant weather group.

27.3.2.9 Cloud and Vertical Obscuration Groups ($N_sN_sN_sh_sh_sh_s$ or $VVh_sh_sh_s$ or SKC)

The initial time period and any subsequent **FM** groups include a cloud or obscuration group ($N_sN_sN_sh_sh_sh_s$ or $VVh_sh_sh_s$ or **SKC**), used as appropriate to indicate the cumulative amount ($N_sN_sN_s$) of all cloud layers in ascending order and height ($h_sh_sh_s$), to indicate vertical visibility ($VVh_sh_sh_s$) into a surface-based obstructing medium, or to indicate a clear sky (**SKC**). All cloud layers and obscurations are considered opaque.

27.3.2.9.1 Cloud Group ($N_sN_sN_sh_sh_sh_s$)

The cloud group ($N_sN_sN_sh_sh_sh_s$) is used to forecast cloud amount as indicated in Table 27-6.

Table 27-6. TAF Sky Cover

Sky Cover Contraction	Sky Coverage
SKC	0 oktas
FEW	0 to 2 oktas
SCT	3 to 4 oktas
BKN	5 to 7 oktas
OVC	8 oktas

When 0 oktas of sky coverage is forecast, the cloud group is replaced by **SKC**. The contraction **CLR**, which is used in the METAR code, is not used in TAFs. TAFs for sites with an ASOS or AWOS contain the cloud amount and/or obscurations, which the forecaster expects, not what is expected to be reported by an ASOS/AWOS.

Heights of clouds ($h_sh_sh_s$) are forecast in hundreds of feet AGL.

The lowest level at which the cumulative cloud cover equals 5/8 or more of the celestial dome is understood to be the forecast ceiling. For example, **VV008**, **BKN008**, or **OVC008** all indicate an 800-ft ceiling.

27.3.2.9.2 Vertical Obscuration Group (VVh$_s$h$_s$h$_s$)

The vertical obscuration group (**VVh$_s$h$_s$h$_s$**) is used to forecast, in hundreds of feet AGL, the vertical visibility (**VV**) into a surface-based total obscuration. **VVh$_s$h$_s$h$_s$** is this ceiling at the height indicated in the forecast. TAFs do not include forecasts of partial obscurations (i.e., **FEW000**, **SCT000**, or **BKN000**).

Example:

```
1SM BR VV008
```

Ceiling is 800 ft due to vertical visibility into fog.

27.3.2.9.3 Cloud Type (CB)

The only cloud type included in the TAF is **CB**. **CB** follows cloud or obscuration height (**h$_s$h$_s$h$_s$**) without a space whenever thunderstorms are included in the significant weather group (**w'w'**), even if thunderstorms are only forecast in the vicinity (**VCTS**). **CB** can be included in the cloud group (**N$_s$N$_s$N$_s$h$_s$h$_s$h$_s$**) or the vertical obscuration group (**VVh$_s$h$_s$h$_s$**) without mentioning a thunderstorm in the significant weather group (**w'w'**). Therefore, situations may occur where nearly identical **N$_s$N$_s$N$_s$h$_s$h$_s$h$_s$** or **VVh$_s$h$_s$h$_s$** appear in consecutive time periods, with the only change being the addition or elimination of **CB** in the forecast cloud type.

Examples:

```
1/2SM TSRA OVC010CB
```

Thunderstorms are forecast at the airport.

27.3.2.9.4 Nonconvective Low-Level Wind Shear (LLWS) Group (WSh$_{ws}$h$_{ws}$h$_{ws}$/dddffKT)

Wind shear (**WS**) is defined as a rapid change in horizontal wind speed and/or direction, with distance and/or a change in vertical wind speed and/or direction with height. A sufficient difference in wind speed, wind direction, or both can severely impact airplanes, especially within 2,000 ft AGL because of limited vertical airspace for recovery.

Forecasts of LLWS in the TAF refer only to nonconvective LLWS from the surface up to and including 2,000 ft AGL. LLWS is always assumed to be present in convective activity. LLWS is included in TAFs on an "as-needed" basis to focus the aircrew's attention on LLWS problems that currently exist or are expected. Nonconvective LLWS may be associated with the following: frontal passage, inversion, low-level jet, lee-side mountain effect, sea breeze front, Santa Ana winds, etc.

When LLWS conditions are expected, the nonconvective LLWS code **WS** is included in the TAF as the last group (after cloud forecast). Once in the TAF, the **WS** group remains the prevailing condition until the next **FM** change group or the end of the TAF valid period if there are no subsequent **FM** groups. Forecasts of nonconvective LLWS are not included in **TEMPO** or **PROB** groups.

The format of the nonconvective LLWS group is:

```
WShwshwshws/dddffKT
```

WS	Indicator for nonconvective LLWS.
Hwshwshws	Height of the top of the wind shear layer in hundreds of feet AGL.
Ddd	True direction in 10° increments at the indicated height.
	(**VRB** is not used for direction in the nonconvective LLWS forecast.)

Ff Speed in knots of the forecast wind at the indicated height.

KT Unit indicator for wind.

Example:

TAF…13012KT…WS020/27055KT

Wind shear from the surface to 2,000 ft. Surface winds from 130° (southeast) at 12 kt changes to 270° (west) at 55 kt at 2,000 ft.

In this example, the indicator **WS** is followed by a three-digit number that is the top of the wind shear layer. LLWS is forecast to be present from the surface to this level. After the solidus (/), the five-digit wind group is the wind direction and speed at the top of the wind shear layer. It is not a value for the amount of shear.

A nonconvective LLWS forecast is included in the initial time period or an **FM** group in a TAF whenever:

- One or more PIREPs are received of nonconvective LLWS within 2,000 ft of the surface, at or in the vicinity of the TAF airport, causing an indicated air speed loss or gain of 20 kt or more, and the forecaster determines the report(s) reflect a valid nonconvective LLWS event rather than mechanical turbulence; or
- Nonconvective vertical wind shear of 10 kt or more per 100 ft in a layer more than 200 ft thick is expected or reliably reported within 2,000 ft of the surface at, or in the vicinity of, the airport.

27.3.2.10 Forecast Change Indicator Groups

Forecast change indicator groups are contractions that are used to subdivide the forecast period (24 hours for scheduled TAFs; less for amended or delayed forecasts) according to significant changes in the weather.

The forecast change indicators **FM**, **TEMPO**, and **PROB** are used when a change in any or all of the forecast elements is expected.

27.3.2.10.1 From (FM) Group (FMYYGGgg)

The change group **FMYYGGgg** (voiced as "from") is used to indicate when prevailing conditions are expected to change significantly over a period of less than 1 hour. In these instances, the forecast is subdivided into time periods using the contraction **FM**, followed, without a space, by six digits, the first two of which indicate the day of the month and the final four indicate the time (in hours and minutes Z) the change is expected to occur. While the use of a four-digit time in whole hours (e.g., **2100Z**) is acceptable, if a forecaster can predict changes and/or events with higher resolution, then more precise timing of the change to the minute will be indicated. All forecast elements following **FMYYGGgg** relate to the period of time from the indicated date and time (**YYGGgg**) to the end of the valid period of the terminal forecast, or to the next **FM** if the terminal forecast valid period is divided into additional periods.

The **FM** group will be followed by a complete description of the weather (i.e., self-contained), and all forecast conditions given before the **FM** group are superseded by those following the group. All elements of the TAF (e.g., surface wind, visibility, significant weather, clouds, obscurations, and when expected, nonconvective LLWS) will be included in each **FM** group, regardless if they are forecast to change or not. For example, if forecast cloud and visibility changes warrant a new **FM** group but the wind does not, the new **FM** group will include a wind forecast, even if it is the same as the most recently forecast wind.

The only exception to this involves the significant weather group. If no significant weather is expected in the **FM** time period group, then significant weather group is omitted. A TAF may include one or more **FM** groups, depending on the prevailing weather conditions expected. In the interest of clarity, each **FM** group starts on a new line of forecast text, indented five spaces.

Examples:

```
TAF
KDSM 022336Z 0300/0324 20015KT P6SM BKN015
FM030230 29020G35KT 1SM +SHRA OVC005
TEMPO 0303/0304 30030G45KT 3/4SM -SHSN
FM030500 31010G20KT P6SM SCT025...
```

A change in the prevailing weather is expected on the 3rd day of the month at 0230 UTC and the 3rd day of the month at 0500 UTC.

```
TAF
KAPN 312330Z 0100/0124 13008KT P6SM SCT030
FM010320 31010KT 3SM -SHSN BKN015
FM010500 31010KT 1/4SM +SHSN VV007...
```

Note that the wind in the **FM010500** group is the same as the previous **FM** group, but is repeated since all elements are to be included in a **FM** group.

27.3.2.10.2   Temporary (TEMPO) Group (TEMPO YYGG/$Y_eY_eG_eG_e$)

The change-indicator group **TEMPO YYGG/$Y_eY_eG_eG_e$** is used to indicate temporary fluctuations to forecast meteorological conditions that are expected to:

- Have a high percentage (greater than 50 percent) probability of occurrence;
- Last for one hour or less in each instance; and
- In the aggregate, cover less than half of the period **YYGG** to **$Y_eY_eG_eG_e$**.

The first two digits (**YY**) are the day of the month for the start of the **TEMPO**. The next two digits (**GG**) are the starting hour (UTC). After the solidus (/), the next two digits (**Y_eY_e**) are the ending day of the month, while the last two digits (**G_eG_e**) are the ending hour (UTC) of the **TEMPO** period.

Each **TEMPO** group is placed on a new line in the TAF. The **TEMPO** identifier is followed by a description of all the elements in which a temporary change is forecast. A previously forecast element that has not changed during the **TEMPO** period is understood to remain the same and will not be included in the **TEMPO** group. Only those weather elements forecast to temporarily change are included in the **TEMPO** group.

TEMPO groups will not include forecasts of either significant weather in the vicinity (**VC**) or nonconvective LLWS.

Examples:

```
TAF
KDDC 221130Z 2212/2312 29010G25KT P6SM SCT025
TEMPO 2215/2217 30025G35KT 1 1/2SM SHRA BKN010...
```

In the example, all forecast elements in the **TEMPO** group are expected to be different than the prevailing conditions. The **TEMPO** group is valid on the 17th day of the month from 1500 UTC to 1700 UTC.

```
TAF
KSEA 091125Z 0912/1012 19008KT P6SM SCT010 BKN020 OVC090
TEMPO 0912/0915 -RA SCT010 BKN015 OVC040...
```

In this example the visibility is *not* forecast in the **TEMPO** group. Therefore, the visibility is expected to remain the same (**P6SM**) as forecast in the prevailing conditions group. Also, note that in the **TEMPO 0912/0915** group, all three cloud layers are included, although the lowest layer is not forecast to change from the initial time period.

27.3.2.10.3 Probability (PROB) Group (PROB30 YYGG/Y$_e$Y$_e$G$_e$G$_e$)

The probability group (**PROB30 YYGG/Y$_e$Y$_e$G$_e$G$_e$**) is only used by NWS forecasters to forecast a low-probability occurrence (30 percent chance) of a thunderstorm or precipitation event and its associated weather and obscuration elements (e.g., wind, visibility, and/or sky condition) at an airport.

The **PROB30** group is the forecaster's assessment of probability of occurrence of the weather event that follows it. The first two digits (**YY**) are the day of the month for the start of the **PROB30**. The next two digits (**GG**) are the starting hour (UTC). After the solidus (/), the next two digits (**Y$_e$Y$_e$**) are the ending day of the month, while the last two digits (**G$_e$G$_e$**) are the ending hour (UTC) of the **PROB30** period. **PROB30** is the only **PROB** group used in NWS' TAFs.

Note: The U.S. military and international TAFs may use the **PROB40** (40 percent chance) group as well.

The **PROB30** group is located within the same line of the prevailing condition group, continuing on the line below if necessary.

The NWS' TAFs do not use the **PROB30** group in the first 9 hours of the TAF's valid period, including amendments. Also, only one **PROB30** group may be used in the initial forecast period and in any subsequent **FM** groups. Note that the U.S. military and international TAFs do not have these restrictions.

PROB30 groups do not include forecasts of significant weather in the vicinity (**VC**) or nonconvective LLWS.

Example:

```
FM012100 18015KT P6SM SCT050 PROB30 0123/0201 2SM TSRA OVC020CB
```

In this example, the **PROB30** group is valid on the 1st day of the month at 2300 UTC to the 2nd day of the month at 0100 UTC.

27.3.2.10.4 TAFs for Joint-Use (Joint Civilian/Military) Airports

The TAF format at some joint-use airports is different from the NWS TAF format as follows:

- Visibility is in meters instead of statute miles.
    - Example: BECMG 0504/0505 21006KT **9000** BR SKC QNH3005INS.
- Includes a forecast of the lowest barometric altimeter setting (QNH) during the forecast period in inches of mercury.
    - Example: BECMG 0504/0505 21006KT 9000 BR SKC **QNH3005INS**.
- Includes a forecast of the maximum temperature (in whole degrees Celsius) and expected time of occurrence.
    - Example: BECMG 0511/0512 16005KT 9999 NSW SKC QNH3006INS **TX28/0420Z** TN22/0410Z.
- Includes a forecast of the minimum temperature (in whole degrees Celsius) and expected time of occurrence.
    - Example: TN22/0410Z BECMG 0511/0512 16005KT 9999 NSW SKC QNH3006INS TX28/0420Z **TN22/0410Z**.

27.3.3 TAF Examples

```
TAF
KPIR 111140Z 1112/1212 13012KT P6SM BKN100 WS020/35035KT
TEMPO 1112/1114 5SM BR
```

```
FM111500 16015G25KT P6SM SCT040 BKN250
FM120000 14012KT P6SM BKN080 OVC150 PROB30 1200/1204 3SM TSRA BKN030CB
FM120400 14008KT P6SM SCT040 OVC080 TEMPO 1204/1208 3SM TSRA OVC030CB
```

TAF	Terminal Aerodrome Forecast.
KPIR	Pierre, South Dakota.
111140	Prepared on the 11th day of the month at 1140 UTC.
1112/1212	Valid from the 11th day of the month at 1200 UTC until the 12th day of the month at 1200 UTC.
13012KT	Wind 130° true at 12 kt.
P6SM	Visibility greater than 6 sm.
BKN100	Ceiling 10,000 ft broken.
WS020/35035KT	Wind shear at 2,000 ft, wind from 350° true at 35 kt.
TEMPO 1112/1114	Temporary conditions between the 11th day of the month at 1200 UTC and the 11th day of the month at 1400 UTC.
5SM	Visibility 5 sm.
BR	Mist.
FM111500	From the 11th day of the month at 1500 UTC.
16015G25KT	Wind 160° true at 15 kt gusting to 25 kt.
P6SM	Visibility greater than 6 sm.
SCT040 BKN250	4,000 ft scattered, ceiling 25,000 ft broken.
FM120000	From the 12th day of the month at 0000Z.
14012KT	Wind 140° true at 12 kt.
P6SM	Visibility greater than 6 sm.
BKN080 OVC150	Ceiling 8,000 ft broken, 15,000 ft overcast.
PROB30 1200/1204	30 percent probability between the 12th day of the month at 0000 UTC and the 12th day of the month at 0400 UTC.
3SM	Visibility 3 sm.
TSRA	Thunderstorm with moderate rain showers.
BKN030CB	Ceiling 3,000 ft broken with cumulonimbus.
FM120400	From the 12th day of the month at 0400 UTC.
14008KT	Wind 140° true at 8 kt.
P6SM	Visibility greater than 6 sm.
SCT040 OVC080	4,000 ft scattered, ceiling 8,000 ft overcast.
TEMPO 1204/1208	Temporary conditions between the 12th day of the month at 0400 UTC and the 12th day of the month at 0800 UTC.

3SM	Visibility 3 sm.
TSRA	Thunderstorms with moderate rain showers.
OVC030CB	Ceiling 3,000 ft overcast with cumulonimbus.

```
TAF AMD
KEYW 131555Z 1316/1412 VRB03KT P6SM VCTS SCT025CB BKN250
TEMPO 1316/1318 2SM TSRA BKN020CB
FM131800 VRB03KT P6SM SCT025 BKN250 TEMPO 1320/1324 1SM TSRA OVC010CB
FM140000 VRB03KT P6SM VCTS SCT020CB BKN120 TEMPO 1408/1412 BKN020CB
```

TAF AMD	Amended Terminal Aerodrome Forecast.
KEYW	Key West, Florida.
131555Z	Prepared on the 13th day of the month at 1555 UTC.
1316/1412	Valid from the 13th day of the month at 1600 UTC until the 14th day of the month at 1200 UTC
VRB03KT	Wind variable at 3 kt.
P6SM	Visibility greater than 6 sm.
VCTS	Thunderstorms in the vicinity.
SCT025CB BKN250	2,500 ft scattered with cumulonimbus, ceiling 25,000 ft broken.
TEMPO 1316/1318	Temporary conditions between the 13th day of the month at 1600 UTC and the 13th day of the month at 1800 UTC.
2SM	Visibility 2 sm.
TSRA	Thunderstorms with moderate rain showers.
BKN020CB	Ceiling 2,000 ft broken with cumulonimbus.
FM131800	From the 13th day of the month at 1800 UTC.
VRB03KT	Wind variable at 3 kt.
P6SM	Visibility greater than 6 sm.
SCT025 BKN250	2,500 ft scattered, ceiling 25,000 ft broken.
TEMPO 1320/1324	Temporary conditions between the 13th day of the month at 2000 UTC and the 14th day of the month at 0000 UTC.
1SM	Visibility 1 sm.
TSRA	Thunderstorms with moderate rain showers.
OVC010CB	Ceiling 1,000 ft overcast with cumulonimbus.
FM140000	From the 14th day of the month at 0000 UTC.
VRB03KT	Variable wind at 3 kt.
P6SM	Visibility greater than 6 sm.
VCTS	Thunderstorms in the vicinity.

`SCT020CB BKN120`	2,000 ft scattered with cumulonimbus, ceiling 12,000 ft broken.
`TEMPO 1408/1412`	Temporary conditions between the 14th day of the month at 0800 UTC and the 14th day of the month at 1200 UTC.
`BKN020CB`	Ceiling 2,000 ft broken with cumulonimbus.

27.3.4 Issuance

Scheduled TAFs prepared by NWS offices are issued at least four times a day, every 6 hours. Some locations have amendments routinely issued 3 hours after the initial issuance. The issuance schedule is shown in Table 27-7.

The issuance of a new TAF cancels any previous TAF for the same location.

Table 27-7. TAF Issuance Schedule

Scheduled Issuance	Valid Period	End Time for 30 Hour	Issuance Window
0000 UTC	0000 to 0000	0600 UTC	2320 to 2340 UTC
0300 UTC (AMD)	0300 to 0000 UTC	0600 UTC	
0600 UTC	0600 to 0600	1200 UTC	0520 to 0540 UTC
0900 UTC (AMD)	0900 to 0600 UTC	1200 UTC	
1200 UTC	1200 to 1200	1800 UTC	1120 to 1140 UTC
1500 UTC (AMD)	1500 to 1200 UTC	1800 UTC	
1800 UTC	1800 to 1800	0000 UTC	1720 to 1740 UTC
2100 UTC (AMD)	2100 to 1800 UTC	0000 UTC	

27.3.4.1 Minimum Observational Criteria for Routine TAF Issuance and Continuation

The NWS forecaster must have certain information for the preparation and scheduled issuance of each individual TAF. Although integral to the TAF writing process, a complete surface (METAR/SPECI) observation is not needed. Forecasters use the "total observation concept" to write TAFs with data including nearby surface observations, radar, satellite, radiosonde, model data, aircraft, and other sources.

If information sources, such as surface observations, are missing, unreliable, or not complete, forecasters will append **AMD NOT SKED** to the end of a TAF. The use of **AMD NOT SKED** indicates the forecaster has enough data, using the total observation concept, to issue a forecast, but will not provide updates. This allows airport operations to continue using a valid TAF.

In rare situations where observations have been missing for extended periods of time (i.e., more than one TAF cycle of 6 hours) and the total observation concept cannot provide sufficient information, the TAF may be suspended by the use of **NIL TAF**.

27.3.4.2 Sites with Scheduled Part-Time Observations

For TAFs with less than 24-hour observational coverage, the TAF will be valid to the end of the routine scheduled forecast period even if observations cease prior to that time. The time observations are scheduled to end and/or resume will be indicated by expanding the **AMD NOT SKED** statement. Expanded statements will include the observation ending time (**AFT Y_1Y_1HHmm**, e.g., AFT 120200), the scheduled observation resumption time (**TIL Y_1Y_1HHmm**, e.g., TIL 171200Z) or the period of observation unavailability (Y_1Y_1HH/$Y_eY_e$hh, e.g., 2502-2512). **TIL** will be used only when the beginning of the scheduled TAF valid period coincides with the time of the last observation or when observations are scheduled to resume prior to the next scheduled issuance time. When used, these remarks will immediately follow the last forecast group. If a routine TAF issuance is scheduled to be made after observations have ceased, but before they resume, the remark **AMD NOT SKED** will immediately follow the valid period group of the scheduled issuance. After sufficient data using the total observation concept has been received, the **AMD NOT SKED** remark will be removed.

Examples of Scheduled Part-Time Observations TAFs:

```
TAF AMD
KRWF 150202Z 1502/1524 {TAF text}
AMD NOT SKED 1505Z-1518Z=
```

No amendments will be available between the 15th day of the month at 0500 UTC and the 15th day of the month at 1800 UTC due to lack of a complete observational set between those times.

```
TAF AMD
KPSP 190230Z 1903/1924 {TAF text}
AMD NOT SKED=
```

Amendments are not scheduled.

27.3.4.3 Automated Observing Sites Requiring Part-Time Augmentation

TAFs for automated stations without present weather and obstruction to vision information and have no augmentation or only part-time augmentation are prepared using the procedures for part-time manual observation sites detailed in the previous section, with one exception. This exception is the remark used when the automated system is unattended. Specifically, the time an augmented automated system is scheduled to go into unattended operation and/or the time augmentation resumes is included in a remark unique to automated observing sites: **AMD LTD TO CLD VIS AND WIND (AFT YYHHmm**, or **TIL YYhhmm**, or **YYHH-YYhh**), where **YY** is the date, **HHmm** is the time, in hours and minutes, of last augmented observation, and **hhmm** is the time, in hours and minutes, the second complete observation is expected to be received. This remark, which does not preclude amendments for other forecast elements, is appended to the last scheduled TAF issued prior to the last augmented observation. It will also be appended to all subsequent amendments until augmentation resumes.

The **AMD LTD TO** (elements specified) remark is a flag for users and differs from the **AMD NOT SKED AFT Z** remark for part-time manual observation sites. **AMD LTD TO** means users should expect amendments only for those elements and the times specified.

Example:

```
TAF AMD
KCOE 150202Z 1502/1524 text
AMD LTD TO CLD VIS AND WIND 1505-1518=
```

The amended forecast indicates that amendments will only be issued for wind, visibility, and clouds, between the 15th day of the month at 0500Z and the 15th day of the month at 1800Z.

An amendment includes forecasts for all appropriate TAF elements, even those not reported when the automated site is not augmented. If unreported elements are judged to be crucial to the TAF and cannot be adequately determined (e.g., fog versus moderate snow), the TAF will be suspended (i.e., an amended TAF stating "AMD NOT SKED").

AWOS systems with part-time augmentation, which the forecaster suspects are providing unreliable information when not augmented, will be reported for maintenance and treated the same as part-time manual observation sites. In such cases, the **AMD NOT SKED AFT YY/aaZ** remark will be used.

27.3.4.4 Non-Augmented Automated Observing Sites

The TAF issued for automated observing stations with no augmentation may be suspended in the event the forecaster is notified of, or strongly suspects, an outage or unrepresentative data. Forecasters may also suspend TAF amendments when an element the forecaster judges to be critical is missing from the observation and cannot be obtained using the total observation concept. The term **AMD NOT SKED** will be appended, on a separate line and indented five spaces, to the end of an amendment to the existing TAF when appropriate.

27.4 Aviation Surface Forecast and Aviation Clouds Forecast

The Aviation Surface Forecast and Aviation Clouds Forecast graphics are snapshot images derived from a subset of the aviation weather forecasts within the GFA Tool (see Chapter 28, Aviation Weather Tools). A CONUS view is available as well as several regional views. Forecasts are provided for 3, 6, 9, 12, 15, and 18 hours.

The Aviation Surface Forecast (see Figure 27-3) provides obscurations, visibility, weather phenomena, and winds (including wind gusts) with AIRMET Sierra for IFR conditions and AIRMET Tango for sustained surface winds of 30 kt or more overlaid. The Aviation Clouds Forecast (see Figure 27-4) provides cloud coverage, bases, layers, and tops with AIRMET Sierra for mountain obscuration and AIRMET Zulu for icing overlaid.

The Aviation Clouds Forecast graphic provides a forecast of cloud coverage and height (in hundreds of feet MSL). Tops of the highest broken (**BKN**) or overcast (**OVC**) layer are shown when bases are below FL180. Overlays of AIRMETs for icing and mountain obscuration are included when applicable.

Both of these products are updated every 3 hours and provide forecast snapshots for 3, 6, 9, 12, 15, and 18 hours in the future.

These forecasts are presented on regional maps as well as a CONUS map. The regional maps provide more detail than the CONUS map. The NWS plans to expand the coverage beyond the CONUS.

Complete product information can be found on the AWC's website.

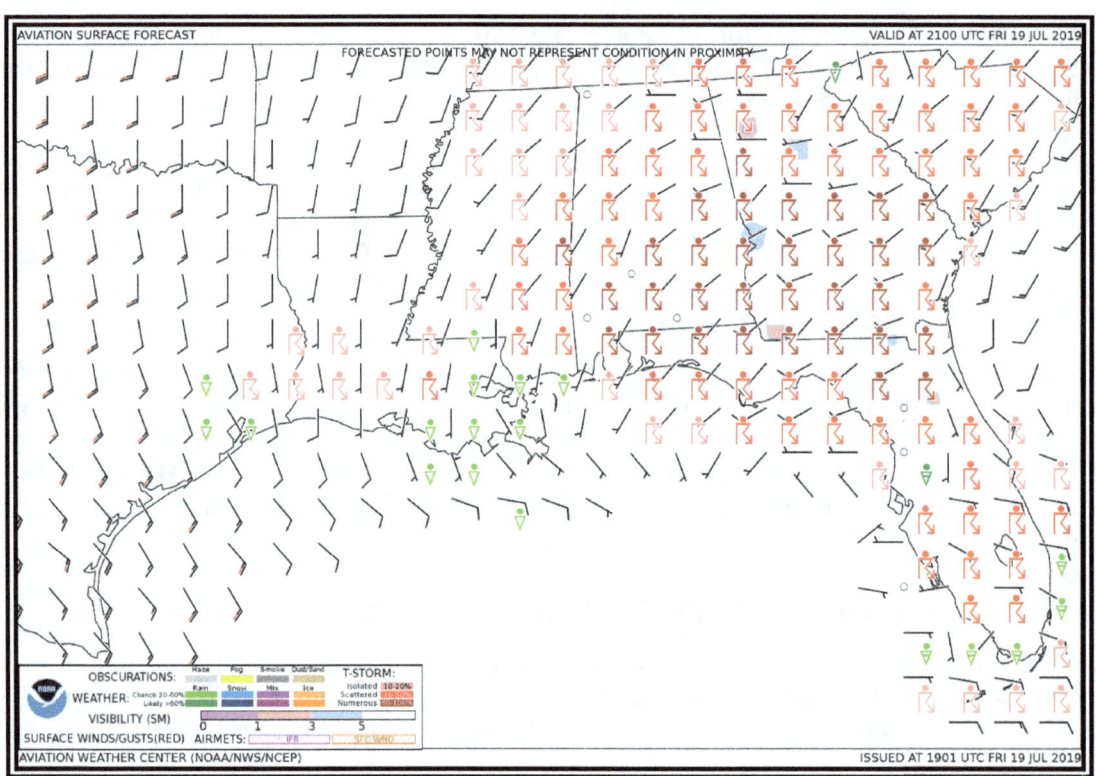

Figure 27-3. Aviation Surface Forecast Example

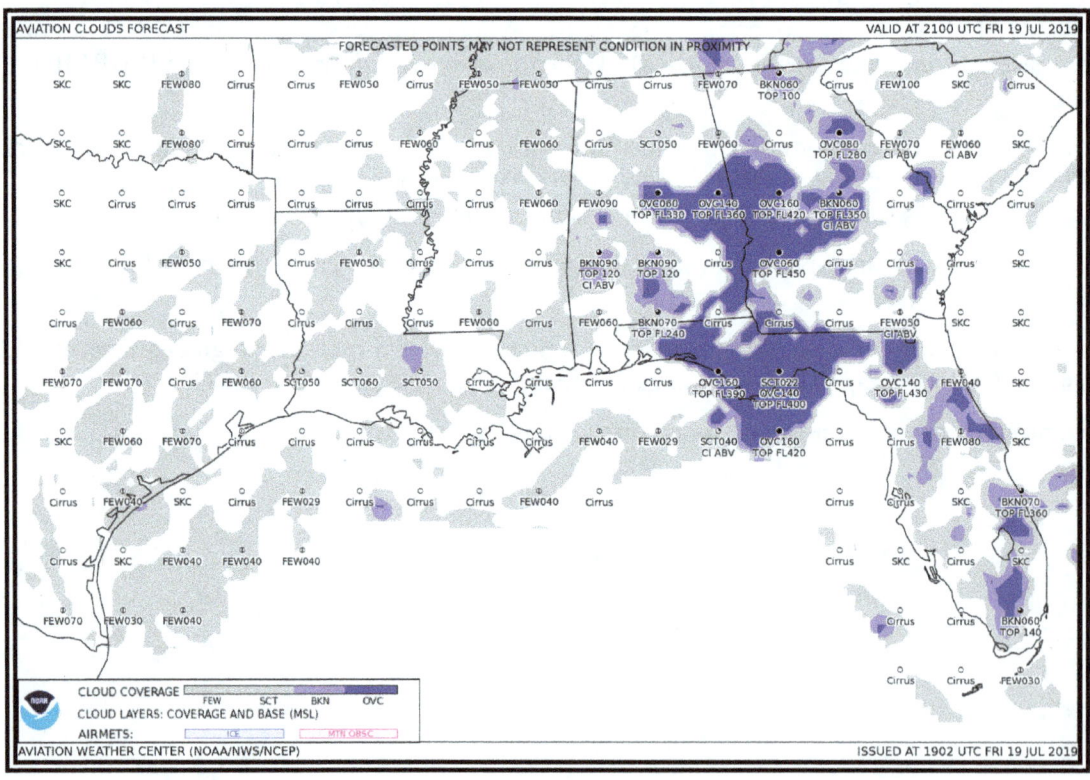

Figure 27-4. Aviation Clouds Forecast Example

Chapter 27, Forecasts

27.5 Area Forecasts (FA)

An FA is an abbreviated plain language (text) forecast concerning the occurrence or expected occurrence of specified en route weather phenomena. FAs cover an 18 to 24-hour period, depending on the region, are issued three to four times daily, depending on the region, and are updated as needed. The exact phenomenon contained in FAs also varies by region.

Text FAs are produced by the NWS for Alaska, Hawaii, the Gulf of Mexico, and the Caribbean. Text FAs for the CONUS were retired in late 2017 and replaced by the NWS' GFA Tool (see Chapter 28) and the aforementioned Aviation Surface Forecast and Aviation Clouds Forecast graphics (see Section 27.4). The NWS plans to replace the Alaska, Hawaii, Gulf of Mexico, and Caribbean FAs with GFAs in the coming years. Text FAs for Hawaii, the Gulf of Mexico, and the Caribbean are expected to be retired in 2024. These text products will be replaced by the GFA Tool.

27.5.1 FA Standardization

All FAs follow these standards:

- All heights or altitudes are referenced to MSL, unless otherwise noted (i.e., prefaced by AGL or CIG), and annotated using the height in hundreds of feet, consisting of three digits (e.g., 040). For heights at or above 18,000 ft, the level is preceded by "FL" to represent flight levels (e.g., FL180). Tops are always referenced to MSL.

- References to latitude and longitude are in whole degrees and minutes following the model: Nnn[nn] or Snn[nn], Wnnn[nn], or Ennn[nn] with a space between latitude and longitude and a hyphen between successive points (e.g., N3106 W07118 – N3011 W7209).

- Messages are prepared in abbreviated plain language using contractions from FAA Order 7340.2, Contractions, for domestic products, and from ICAO Doc 8400, ICAO Abbreviations and Codes, for products issued for Oceanic FIRs. A limited number of nonabbreviated words, geographical names, and numerical values of a self-explanatory nature may also be used.

- Weather and obstructions to visibility are described using the weather abbreviations for surface weather observations (METAR/SPECI) (see Section 24.3).

27.5.2 FA Issuance Schedule

FAs are scheduled products issued at the times listed in Table 27-8 below.

Table 27-8. FA Issuance Schedule

	Gulf of Mexico (UTC)	Caribbean (UTC)	Hawaii (UTC)	Alaska (UTC)
1st Issuance	0130	0330	0340	0415 (DT)/0515 (ST)
2nd Issuance	1030	0930	0940	1215 (DT)/1315 (ST)
3rd Issuance	1830	1530	1540	2015 (DT)/2115 (ST)
4th Issuance	None	2130	2140	None

Note: DT—During Alaska Daylight Time; ST—During Alaska Standard Time; UTC—Coordinated Universal Time.

27.5.3 FA Amendments and Corrections

Amendments are issued whenever the weather significantly improves or deteriorates based upon the judgment of the forecaster. **AMD** is included after the date/time group. The date/time group on the WMO and FAA lines is updated to indicate the time of the correction. The ending valid time remains unchanged.

FAs containing errors will be corrected. **COR** is included after the date/time group.

27.5.4 FA Issuance

See the following sections for the NWS offices that issue FAs and for which areas.

27.5.4.1 Alaska

27.5.4.1.1 FA Issuance

Issued by the AAWU. There are seven FAs containing a total of 25 zones (see Table 27-9), covering separate geographical areas of Alaska and the adjacent coastal waters, including the Pribilof Islands and Southeast Bering Sea (see Figure 27-5).

FAs issued for Alaska cover the airspace between the surface and 45,000 ft MSL and include the following elements:

1. Synopsis: A brief description of the significant synoptic weather affecting the FA area during the first 18 hours of the forecast period.

2. Significant Clouds and Weather: A description of the significant clouds and weather for each geographical zone during the first 12 hours of the forecast period, including the following elements:

   - AIRMET information for IFR ceiling and visibility, mountain obscuration, and strong surface winds.
   - Cloud amount (**SCT**, **BKN**, or **OVC**) with bases and tops.
   - Visibility below 7 sm and obstruction(s) to visibility.
   - Precipitation and thunderstorms.
   - Surface wind greater than 20 kt.
   - Mountain pass conditions using categorical terms (for selected zones only).
   - Categorical outlook (VFR, MVFR, and IFR) for 12 to 18 hours.

3. Icing and Freezing Level: A description of expected icing conditions, including the following elements:

   - AIRMET information for icing and freezing precipitation.
   - Icing not meeting SIGMET or AIRMET criteria during the 6- to 12-hour period.
   - Freezing level.
   - If no significant icing is forecast, **NIL SIG** will be entered, followed by the freezing level.

4. Turbulence: A description of expected turbulence conditions, including the following elements:

- AIRMET information for turbulence or LLWS.
- Turbulence not meeting SIGMET or AIRMET criteria during the 6- to 12-hour period.
- If no significant turbulence is forecast, **NIL SIG** will be entered.

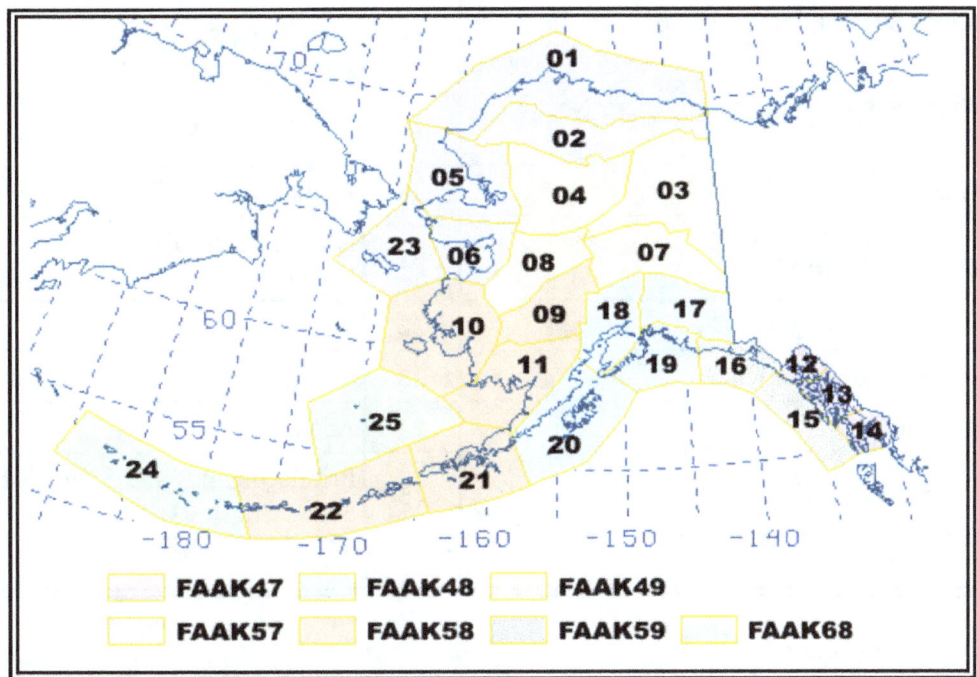

Figure 27-5. AAWU Flight Advisory and FA Zones—Alaska

Table 27-9. AAWU Area Forecast (FA) Zones—Alaska

1	Arctic Coast Coastal	14	Southern Southeast Alaska
2	North Slopes of the Brooks Range	15	Coastal Southeast Alaska
3	Upper Yukon Valley	16	Eastern Gulf Coast
4	Koyukuk and Upper Kobuk Valley	17	Copper River Basin
5	Northern Seward Peninsula-Lower Kobuk Valley	18	Cook Inlet-Susitna Valley
6	Southern Seward Peninsula-Eastern Norton Sound	19	Central Gulf Coast
7	Tanana Valley	20	Kodiak Island
8	Lower Yukon Valley	21	Alaska Peninsula-Port Heiden to Unimak Pass
9	Kuskokwim Valley	22	Unimak Pass to Adak
10	Yukon-Kuskokwim Delta	23	St. Lawrence Island-Bering Sea Coast
11	Bristol Bay	24	Adak to Attu
12	Lynn Canal and Glacier Bay	25	Pribilof Islands and Southeast Bering Sea
13	Central Southeast Alaska		

27.5.4.1.2 FA—Alaska Example

```
FAAK47 PAWU 222010 (ICAO product header)
FA7H (NWS AWIPS Communication header)
JNUH FA 222015 (Area Forecast region, product type, issuance date/time)
EASTERN GULF COAST AND SE AK...
.
AIRMETS VALID UNTIL 230415
CB IMPLY POSSIBLE SEV OR GREATER TURB SEV ICE LLWS AND IFR CONDS.
NON MSL HEIGHTS NOTED BY AGL OR CIG.
.
SYNOPSIS VALID UNTIL 231400
989 MB LOW 275 NM SE KODIAK IS WILL MOV SE WARD TO ABOUT 350 NM S
PASI BY 14Z WHILE FILLING TO 998 MB. ASSOCD OCFNT ARCING E AND SE FM
LOW WILL MOV ONSHR SE AK AND DSIPT BY END OF PD.
.
LYNN CANAL AND GLACIER BAY JB...VALID UNTIL 230800
...CLOUDS/WX...
FEW025 SCT050 BKN100 TOP 120.
OTLK VALID 230800-231400...VFR.
PASSES...WHITE...CHILKOOT...VFR.
...TURB...
NIL SIG.
...ICE AND FZLVL...
NIL SIG. FZLVL 020.
.
CNTRL SE AK JC...VALID UNTIL 230800
```

```
...CLOUDS/WX...
FEW025 SCT050 BKN100 TOP 120.
AFT 03Z ISOL BKN050 -SHRA.
OTLK VALID 230800-231400...VFR.
...TURB...
AFT 05Z SW PAFE ISOL MOD TURB BLW 040.
...ICE AND FZLVL...
NIL SIG. FZLVL 025.
.
SRN SE AK JD...VALID UNTIL 230800
...CLOUDS/WX...
FEW025 SCT050 BKN100 TOP 120.
AFT 00Z OCNL BKN050 -RA. ISOL BKN025 -RA.
AFT 03Z SFC WND SE 25G35KT.
OTLK VALID 230800-231400...VFR.
...TURB...
AFT 02Z CLARENCE STRAIT SW ISOL MOD TURB BLW 040.
...ICE AND FZLVL...
AFT 02Z OUTER CST PAHY S ISOL MOD ICEIC 040-100. FZLVL 025.
```

27.5.4.2 Hawaii

27.5.4.2.1 FA Issuance

Issued by the WFO Honolulu for the main Hawaiian Islands and adjacent coastal waters extending out 40 NM from the coastlines (see Figure 27-6).

FAs issued for Hawaii cover the airspace between the surface and 45,000 ft MSL and include the following elements:

1. Synopsis: A brief discussion of the significant synoptic weather affecting the FA area during the 18-hour valid period.

2. Significant Clouds and Weather: A description of the significant clouds and weather for the first 12 hours of the forecast period, including the following elements:

    - Cloud amount (**SCT**, **BKN**, or **OVC**) with bases and tops.
    - Visibilities of 6 sm or less with obstruction(s) to visibility.
    - Precipitation and thunderstorms.
    - Sustained surface winds of 20 kt or greater.
    - Categorical outlook (IFR, MVFR, or VFR) for 12 to 18 hours, including expected precipitation and/or obstructions to visibility.

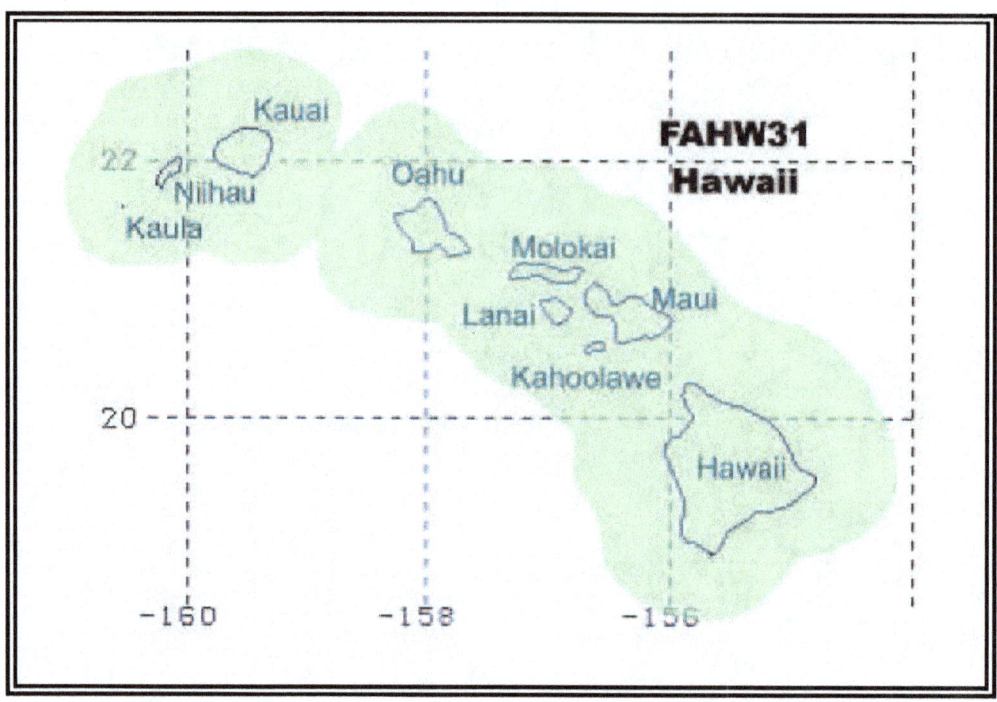

Figure 27-6. WFO Honolulu FA Region and WMO Header—Hawaii

27.5.4.2.2 FA—Hawaii Example

```
FAHW31 PHFO 080940 (ICAO product header)
FA0HI (NWS AWIPS Communication header)

HNLC FA 080940 (Area Forecast region, product type, issuance date/time)
SYNOPSIS AND VFR CLD/WX
SYNOPSIS VALID UNTIL 090400
CLD/WX VALID UNTIL 082200...OUTLOOK VALID 082200-090400
.
SEE AIRMET SIERRA FOR IFR CLD AND MT OBSC.
TS IMPLY SEV OR GREATER TURB SEV ICE LOW LEVEL WS AND IFR COND.
NON MSL HGT DENOTED BY AGL OR CIG.
.
SYNOPSIS...SFC HIGH FAR N PHNL NEARLY STNR.
.
BIG ISLAND ABOVE 060.
SKC. 20Z SCT090. OUTLOOK...VFR.
.
BIG ISLAND LOWER SLOPES...COAST AND ADJ WATERS FROM UPOLU POINT TO CAPE KUMUKAHI
TO APUA POINT.
SCT030 BKN050 TOPS 080 ISOL BKN030 VIS 3-5SM -SHRA BR. 21Z SCT030 SCT BKN050
TOPS 080 ISOL BKN030 5SM -SHRA. OUTLOOK...VFR.
.
BIG ISLAND LOWER SLOPES…COAST AND ADJ WATERS FROM APUA POINT TO SOUTH CAPE TO
UPOLU POINT. SKC. 21Z SCT-BKN060 TOPS 080. 23Z SCT030 SCT-BKN060 TOPS 080 ISOL
BKN030 -SHRA. OUTLOOK...VFR.
.
BIG ISLAND LOWER SLOPES...COAST AND ADJ WATERS FROM SOUTH CAPE TO PHKO TO UPOLU
POINT.
SCT050 ISOL BKN050 TOPS 080. 18Z FEW050. 23Z SCT-BKN050 TOPS 080. OUTLOOK...VFR.
.
```

```
N AND E FACING SLOPES...COAST AND ADJ WATERS OF THE REMAINING ISLANDS.
SCT020 BKN045 TOPS 070 TEMPO BKN020 VIS 3-5SM -SHRA...FM OAHU EASTWARD ISOL CIG
BLW 010 AND VIS BLW 3SM SHRA BR WITH TOPS 120. 22Z SCT025
SCT-BKN050 TOPS 070 ISOL BKN025 3-5SM -SHRA. OUTLOOK...VFR.
.
REST OF AREA.
SCT035 SCT-BKN050 TOPS 070 ISOL BKN030 -SHRA. 20Z SCT050 ISOL SCT030 BKN045
TOPS 070 -SHRA. OUTLOOK...VFR
```

27.5.4.3 Gulf of Mexico

27.5.4.3.1 FA Issuance

Issued by the AWC for the northern Gulf of Mexico, including the Houston Oceanic FIR, the Gulf of Mexico portion of the Miami Oceanic FIR, and the coastal waters west of 85° W longitude (see Figure 27-7).

FAs issued for the Gulf of Mexico (FAGX) cover the airspace between the surface and 45,000 ft MSL and include the following elements with each geographical section having an entry, even if it is negative:

1. Synopsis: A brief discussion of the significant synoptic weather affecting the FAGX area during the entire 24-hour valid period.

2. Significant Clouds and Weather: A description of the significant clouds and weather for the first 12 hours of the forecast period, including the following elements:

   - Cloud amount (**SCT**, **BKN**, or **OVC**) for clouds with bases below FL180.
   - Cloud bases and tops associated with the cloud amount listed above.
   - Precipitation and thunderstorms.
   - Visibility below 7 sm and obstruction(s) to visibility.
   - Sustained surface winds greater than or equal to 20 kt.
   - Categorical outlook (LIFR, IFR, MVFR, or VFR) for 12 to 24 hours.

3. Icing and Freezing Level: Moderate or severe icing and freezing level.

4. Turbulence: Moderate or greater turbulence.

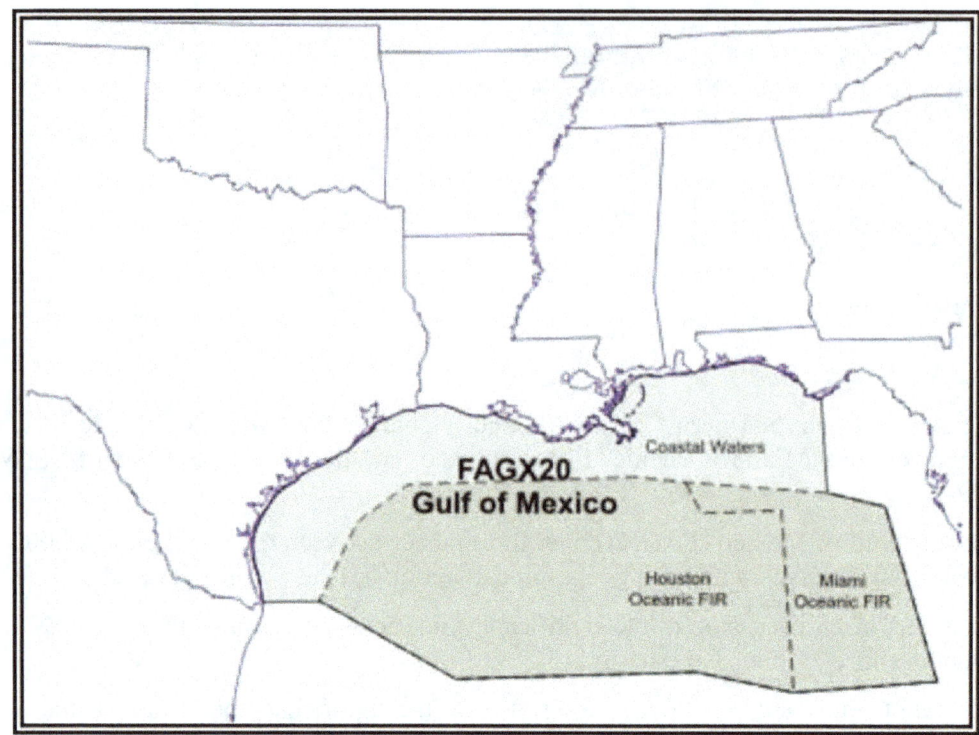

Figure 27-7. AWC FA Region and WMO Header—Gulf of Mexico

27.5.4.3.2 FA—Gulf of Mexico (FAGX) Example

```
FAGX20 KKCI 091812 (ICAO product header)
OFAGX (NWS AWIPS Communication header)
SYNOPSIS VALID TIL 101900Z
FCST...091900Z-100700Z
OTLK...100700Z-101900Z
INTERNATIONAL OPERATIONS BRANCH
AVIATION WEATHER CENTER KANSAS CITY MISSOURI
.
CSTL WATERS FROM COASTLINE OUT TO HOUSTON OCEANIC FIR AND GLFMEX MIAMI OCEANIC
FIR AND W OF 85W. HOUSTON OCEANIC FIR AND GLFMEX MIAMI OCEANIC FIR.
.
TS IMPLY SEV OR GTR TURB SEV ICE LLWS AND IFR CONDS. HGTS MSL.
.
01 SYNOPSIS...HIGH PRES OVR NRN GLFMEX.
.
02 SIGNIFICANT CLD/WX...
.
CSTL WATERS...
SCT020. OTLK...VFR.
.
HOUSTON OCEANIC FIR...
SCT020. OTLK...VFR.
.
GLFMEX MIAMI OCEANIC FIR...
SCT020. OTLK...VFR.
.
03 ICE AND FRZLVL...
CSTL WATERS...SEE AIRMETS ZULU WAUS44 KKCI AND WAUS42 KKCI.
```

```
HOUSTON OCEANIC FIR... NO SGFNT ICE EXP OUTSIDE CNVTV ACT.
GLFMEX MIAMI OCEANIC FIR...NO SGFNT ICE EXP OUTSIDE CNVTV ACT.
FRZLVL...140 THRUT.
.
04 TURB...
CSTL WATERS...SEE AIRMETS TANGO WAUS44 KKCI AND WAUS42 KKCI.
HOUSTON OCEANIC FIR... NO SGFNT TURB EXP OUTSIDE CNVTV ACT.
GLFMEX MIAMI OCEANIC FIR...NO SGFNT TURB EXP OUTSIDE CNVTV ACT.
```

27.5.4.4 Caribbean

27.5.4.4.1 FA Issuance

Issued by the AWC for portions of the Gulf of Mexico (south of the Houston Oceanic FIR to approximately 22° N latitude), the Caribbean Sea, and adjacent portions of the North Atlantic (see Figure 27-8).

FAs issued for the Caribbean (FACA) cover the airspace between the surface and 24,000 ft MSL and include the following elements with each geographical section having an entry, even if it is negative:

1. Synopsis: A brief discussion of the significant synoptic weather affecting the FACA area during the 24-hour valid period.

2. Significant Clouds and Weather: A description of the significant clouds and weather for the first 12 hours of the forecast period, including the following elements:

   - Cloud amount (**SCT**, **BKN**, or **OVC**) for cloud bases below FL180.
   - Cloud bases and tops associated with the cloud amount listed above.
   - Precipitation and thunderstorms.
   - Visibility below 7 sm and obstruction(s) to visibility.
   - Sustained surface winds greater than or equal to 20 kt.
   - Categorical outlook (IFR, MVFR, or VFR) for 12 to 24 hours.

3. Icing and Freezing Level: Moderate or greater icing and freezing level.

4. Turbulence: Moderate or greater turbulence.

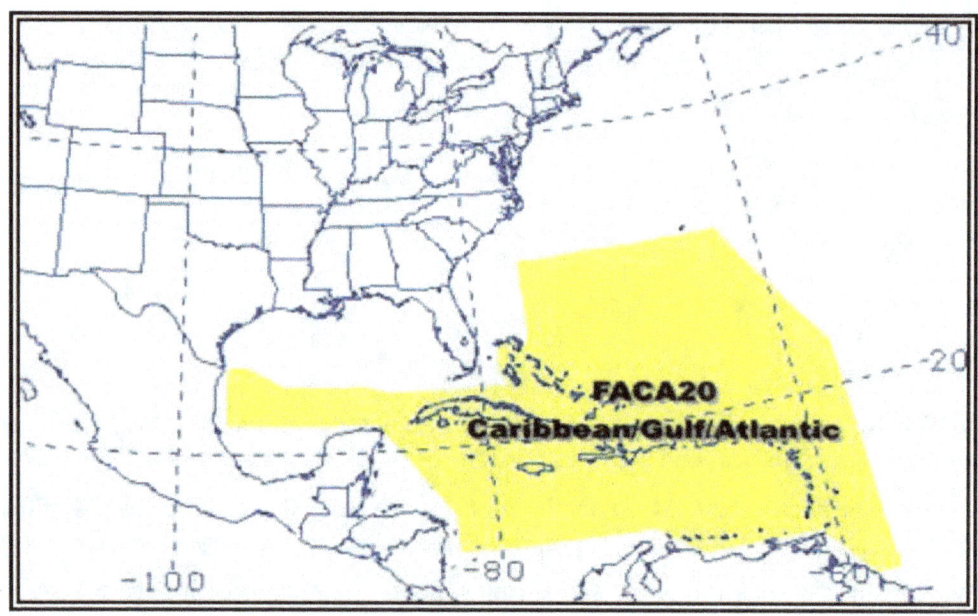

Figure 27-8. AWC FA Region and WMO Header—Caribbean

27.5.4.4.2 FA—Caribbean (FACA) Example

```
FACA20 KKCI 121530 (ICAO product header)
OFAMKC (NWS AWIPS Communication header)
INTERNATIONAL OPERATIONS BRANCH
AVIATION WEATHER CENTER KANSAS CITY MISSOURI
VALID 121600-130400
OUTLOOK...130400-131600
.
ATLANTIC S OF 32N W OF 57W...CARIBBEAN...GULF OF MEXICO BTN 22N AND
24N.
.
TS IMPLY SEV OR GTR TURB SEV ICE LLWS AND IFR CONDS. SFC TO 400 MB.
.
SYNOPSIS...WK CDFNT EXTDS FM NR 28N60W TO 23N63W TO THE MONA
PASSAGE. CDFNT WL MOV EWD AND WKN TODAY. EXP NARROW BAND OF
CLDS WITH ISOL SHRA INVOF CDFNT.

SIGNIFICANT CLD/WX...
ERN MONTERREY FIR...NRN MERIDA FIR
SCT025 SCT060. OTLK...VFR.
.
ATLC SWRN NEW YORK FIR...SAN JUAN FIR
NW OF CDFNT...SCT025 SCT060. LYR OCNL BKN. TOP 120. ISOL SHRA.
OTLK...VFR.
VCNTY CDFNT...SCT025 BKN060. OCNL BKN025. TOP 120. WDLY SCT
SHRA. ISOL TSRA TIL 20Z. OTLK...VFR SHRA.
SE OF CDFNT...SCT025 SCT060. ISOL SHRA. OTLK...VFR.
.
ATLC MIAMI FIR
SCT025 SCT060. LYR OCNL BKN. TOP 120. ISOL SHRA. OTLK...VFR.
.
WRN PIARCO FIR...NRN MAIQUETIA FIR...CURACAO FIR
BTN 61W-63W...SCT025 BKN060. OCNL BKN025. TOP 120. WDLY SCT
```

```
SHRA. OTLK...VFR SHRA.
RMNDR...SCT025 SCT060. ISOL SHRA. OTLK...VFR.
.
SANTO DOMINGO FIR...PORT-AU-PRINCE FIR
SCT025 SCT060. LYR OCNL BKN. TOP 120. ISOL SHRA. OTLK...VFR.
.
NRN BARRANQUILLA FIR...NRN PANAMA FIR
SCT025 SCT060. ISOL SHRA. SFC WND NE 20-25KT. OTLK...VFR.
.
KINGSTON FIR...NERN CNTRL AMERICAN FIR...HABANA FIR
SCT025 SCT060. ISOL SHRA. OTLK...VFR.
.
ICE AND FRZLVL...
NO SGFNT ICE EXP OUTSIDE CNVTV ACT.
FRZLVL... 145-170.
.
TURB...
NO SGFNT TURB EXP OUTSIDE CNVTV ACT.
```

27.6 Alaska Graphical Forecasts

The NWS AAWU produces a series of graphical forecasts to complement the text-based FA for Alaska (see Section 27.5.4.1). These forecasts are available from the AAWU's website. Forecasts include:

- Flying Weather,
- Surface Forecast,
- Icing Forecast,
- Turbulence Forecast, and
- Convective Outlook (seasonal product and only issued from May 1 through September 30).

Additional products may be available. Some of these may be labeled experimental; thus, the contents and format are subject to change.

27.6.1 AAWU Flying Weather

The Flying Weather graphic (see Figure 27-9) includes flying weather conditions and any active volcanoes in Alaska. This product consists of two 6-hour forecasts valid for a total of 12 hours. Each forecast specifies where such conditions can be expected within the 6-hour valid time.

Areas of occasional or continuous MVFR/IFR are represented by shaded regions (red for IFR, blue for MVFR), whereas areas of predominately VFR weather are not shaded. MVFR/IFR conditions are possible outside these shaded regions, but only isolated in coverage. Strong surface winds are shown in a circle hatch overlay. Active volcanoes are denoted by a volcano symbol at the location of the volcano.

Note: This forecast is also referred to as the "IFR/MVFR" graphic on their website.

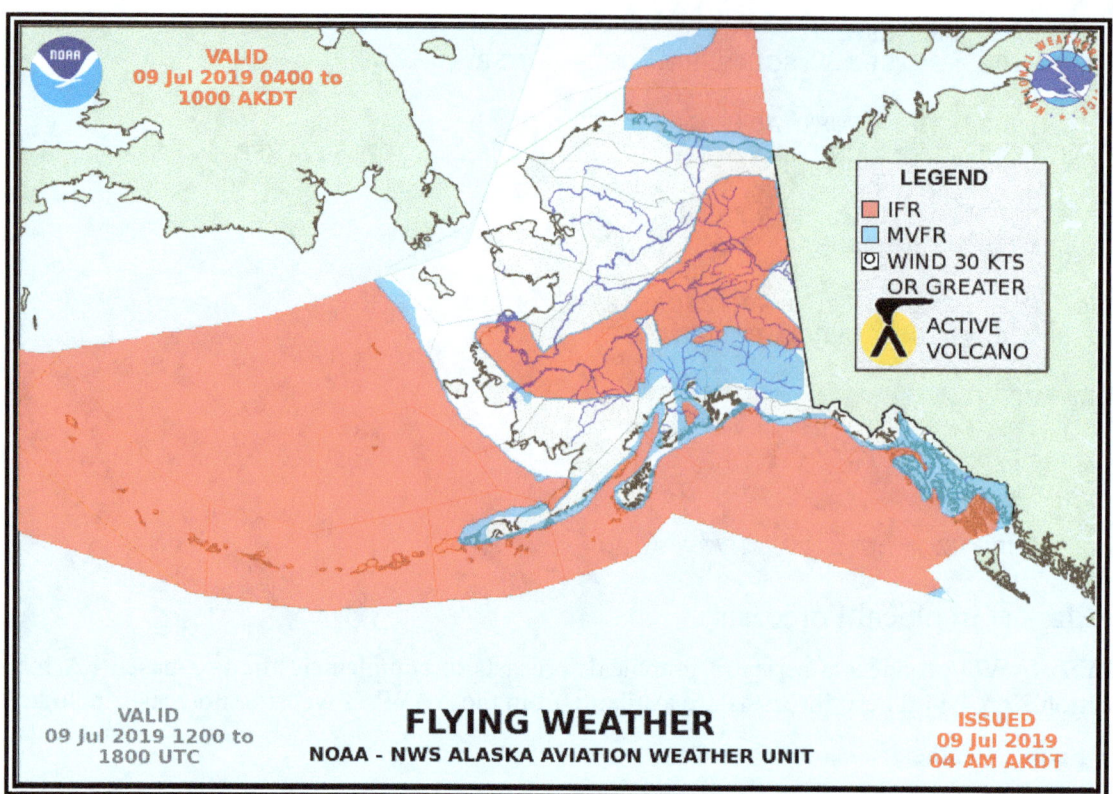

Figure 27-9. Alaska Flying Weather Example

27.6.2 Alaska Surface Forecast

The Surface Forecast graphic (see Figure 27-10) illustrates prominent surface features, including sea level pressure, areas of high and low pressure, fronts and troughs, and precipitation. Each forecast shows the surface weather that can be expected within 1 hour of the designated time.

Areas of high pressure are depicted along with the maximum sea level pressure. Areas of low pressure are depicted with the minimum sea level pressure. The mean 12-hour motion of low pressure systems is also shown. Areas of occasional or continuous precipitation and/or fog are represented by shaded regions (green for precipitation, yellow for fog), whereas isolated or scattered precipitation is not shaded. This product is issued every 6 hours with forecasts valid for 00Z, 06Z, 12Z, and 18Z.

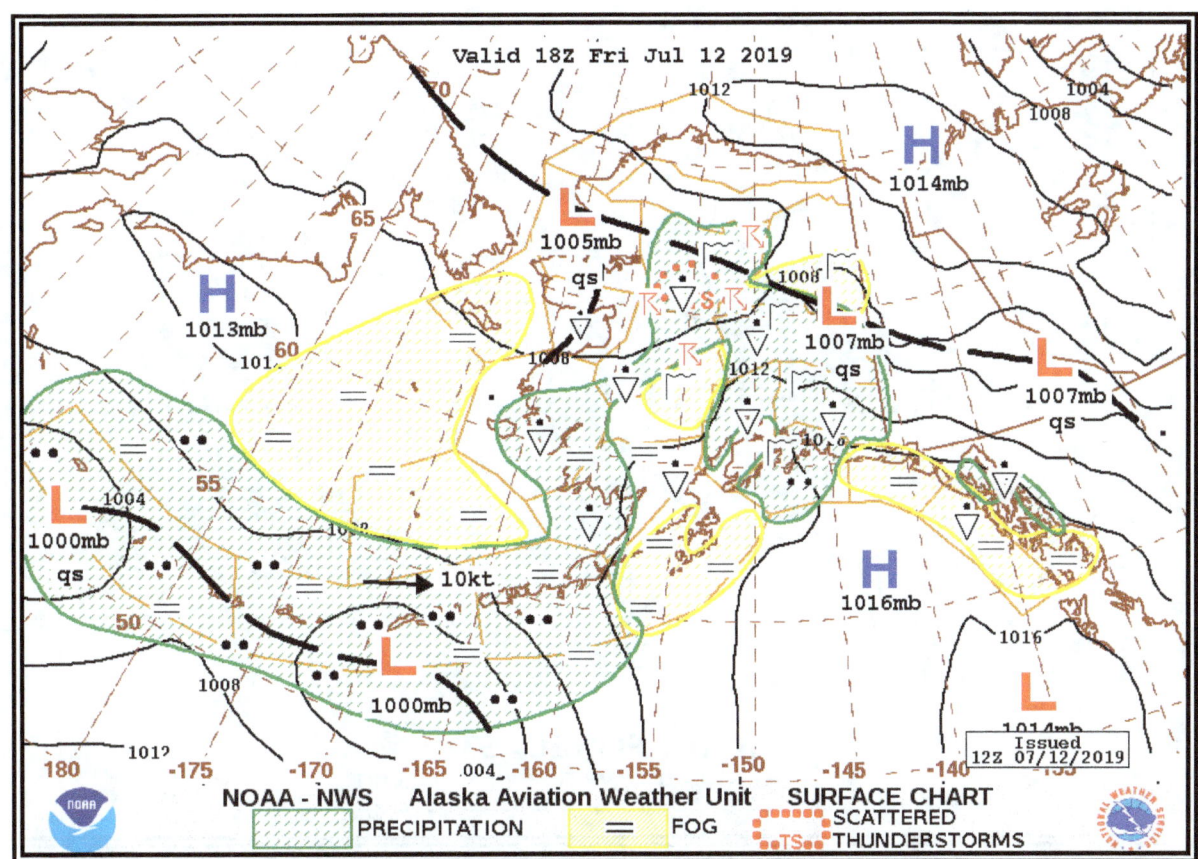

Figure 27-10. Alaska Surface Forecast Example

27.6.3 Alaska Icing Forecast

The Icing Forecast graphic (see Figure 27-11) provides information about freezing levels and the potential for significant icing at specified valid times.

Freezing level heights are blue-filled contours (every 2,000 ft). Areas of isolated (**ISOL**) moderate (**MOD**) icing are shaded yellow, areas of occasional (**OCNL**) or continuous (**CONS**) moderate icing are shaded orange, and red is used for moderate with isolated severe (**SEV**) icing (refer to SIGMETs for occasional or greater severe icing). These forecasts are issued every 8 hours and amended as needed.

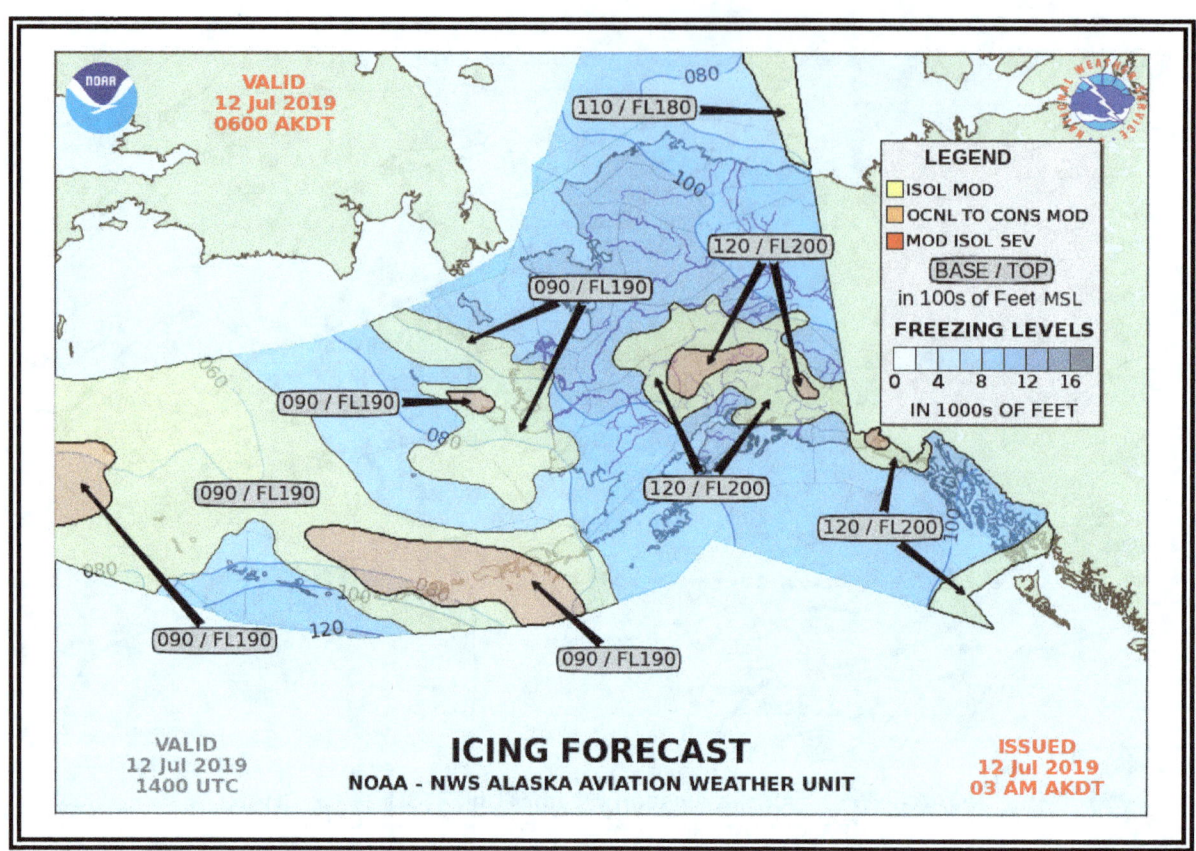

Figure 27-11. Alaska Icing Forecast Example

27.6.4 Alaska Turbulence Forecast

The Turbulence Forecast graphic (see Figure 27-12) depicts areas of significant turbulence at specified valid times.

Areas of isolated (**ISOL**) moderate (**MOD**) turbulence are shaded yellow, areas of occasional (**OCNL**) or continuous (**CONS**) moderate turbulence are shaded orange, and red is used for moderate with isolated severe (**SEV**) turbulence (refer to SIGMETs for occasional or greater severe turbulence).

Separate graphics are provided for low-level (defined for this product as FL180 and below) and high-level (defined for this product as above FL180) turbulence.

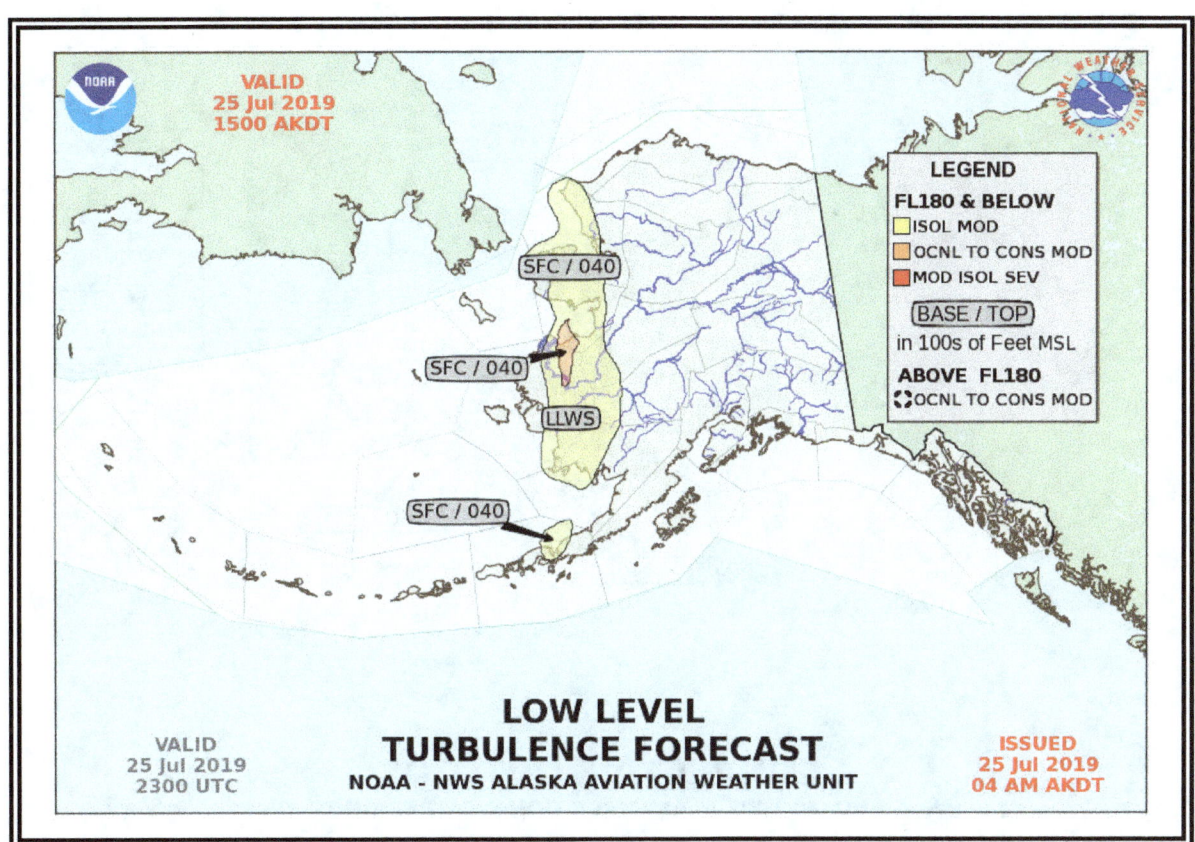

Figure 27-12. Alaska Turbulence Forecast Example

27.6.5 Alaska Convective Outlook

The Convective Outlook graphic (see Figure 27-13) is a seasonal product that provides information about convective activity at specific valid times. Each forecast indicates where conditions are favorable for the development of towering cumulus and thunderstorms.

Locations of towering cumulus are depicted in yellow. Locations of isolated (**ISOL**), scattered (**SCT**), and widespread (**WDSPRD**) thunderstorms (**TS**) are depicted in orange, red, and dark red, respectively. Cloud bases and tops are also depicted.

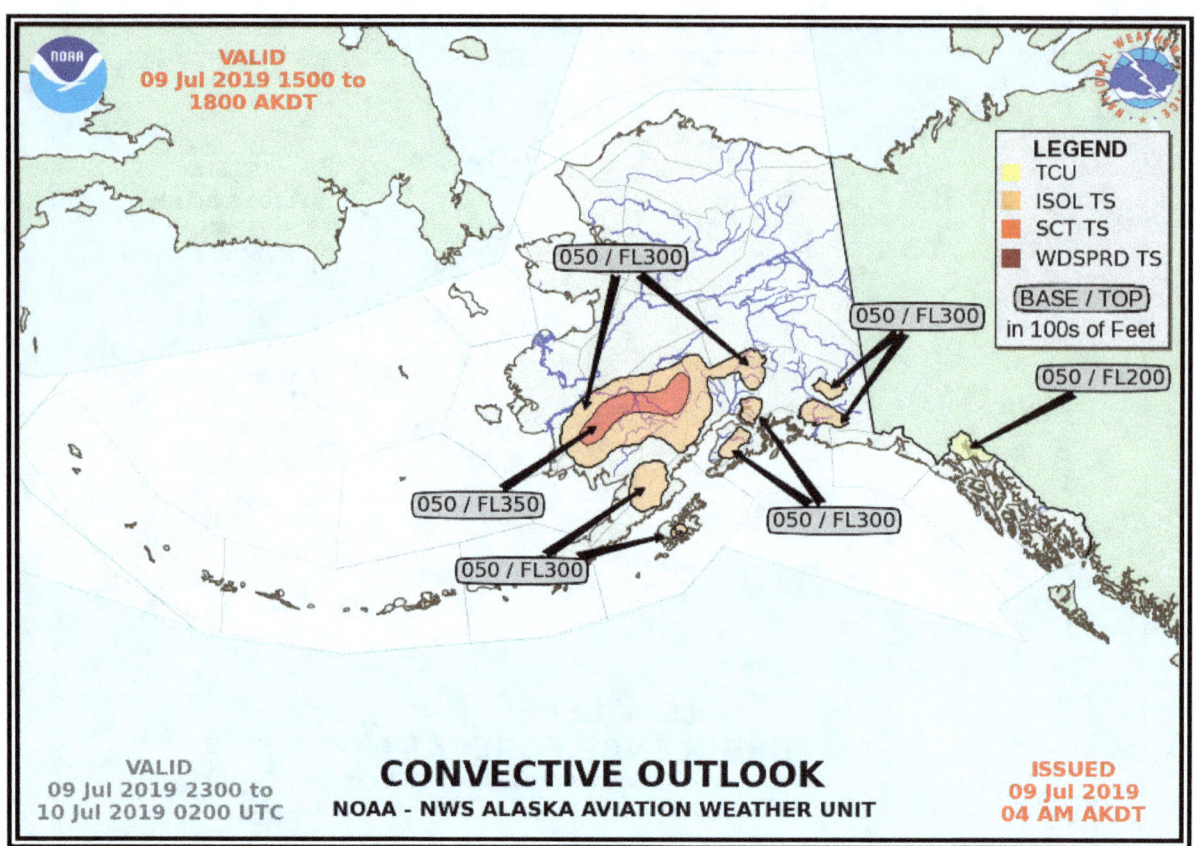

Figure 27-13. Alaska Convective Outlook

27.7 World Area Forecast System (WAFS)

ICAO's WAFS supplies aviation users with global aeronautical meteorological en route forecasts suitable for use in flight-planning systems and flight documentation.

Two WAFCs, WAFC Washington and WAFC London, have the responsibility to issue the WAFS forecasts. WAFC Washington is operated by the NWS NCO in College Park, MD, and the NWS AWC in Kansas City, MO. WAFC London is operated by the United Kingdom's Meteorological Office in Exeter, United Kingdom.

27.7.1 WAFS Forecasts

Both WAFC Washington and WAFC London issue the following WAFS forecasts in accordance with ICAO Annex 3, Meteorological Service for International Air Navigation.

- Global forecasts of:
  - Upper wind and temperature (i.e., wind and temperature aloft, which is also issued in chart form for select areas);
  - Upper-air humidity;
  - Geopotential altitude of FLs;
  - FL and temperature of tropopause (i.e., tropopause forecast);
  - Direction, speed, and FL of maximum wind;

- o Cumulonimbus clouds;
- o Icing; and
- o Turbulence.
- Global forecasts of SIGWX (i.e., High-Level SIGWX forecasts (see Section 27.8.3).
- Select regional areas of Mid-Level SIGWX forecasts (see Section 27.8.2).

27.7.1.1 Issuance

The WAFS forecasts of upper wind, temperature, and humidity; direction, speed, and FL of maximum wind; FL and temperature of tropopause; areas of cumulonimbus clouds; icing; turbulence; and geopotential altitude of FLs are issued four times a day by both WAFC Washington and WAFC London.

These forecasts are produced from weather computer models and are not modified by WAFC forecasters. WAFC Washington's forecast is from the Global Forecast System (GFS) model. These forecasts are issued in grid-point format (i.e., WMO GRIdded Binary, Edition 2 (GRIB2) format).

These forecasts are valid for fixed valid times at 6, 9, 12, 15, 18, 21, 24, 27, 30, 33, and 36 hours after the time (0000, 0600, 1200, and 1800 UTC) on which the forecasts were based. Additional valid times are planned to be implemented in 2024.

27.7.1.2 WAFS Wind and Temperature Forecasts

Wind and temperature forecasts are issued for FLs 050 (850 mb), 100 (700 mb), 140 (600 mb), 180 (500 mb), 240 (400 mb), 270 (350 mb), 300 (300 mb), 320 (275 mb), 340 (250 mb), 360 (225 mb), 390 (200 mb), 410 (175 mb), 450 (150 mb), and 530 (100 mb). Additional levels are planned to be implemented in 2024.

Note: ICAO uses FLs below 18,000 MSL for global weather products.

WAFC wind and temperature forecasts use a plotting model where the air temperature (degrees Celsius) is the center of the data point and the wind direction and speed follows the standard model (see Figure 27-14) with the exception that wind speed for points in the Southern Hemisphere is flipped. Note the data points do not correspond to any airports or reference points with names or identifiers.

WAFS global wind and temperature forecasts are provided in grid point format (e.g., computer format) for use in flight-planning systems. Chart format is also provided on the AWC's website in the User Tools under "Flight Folder."

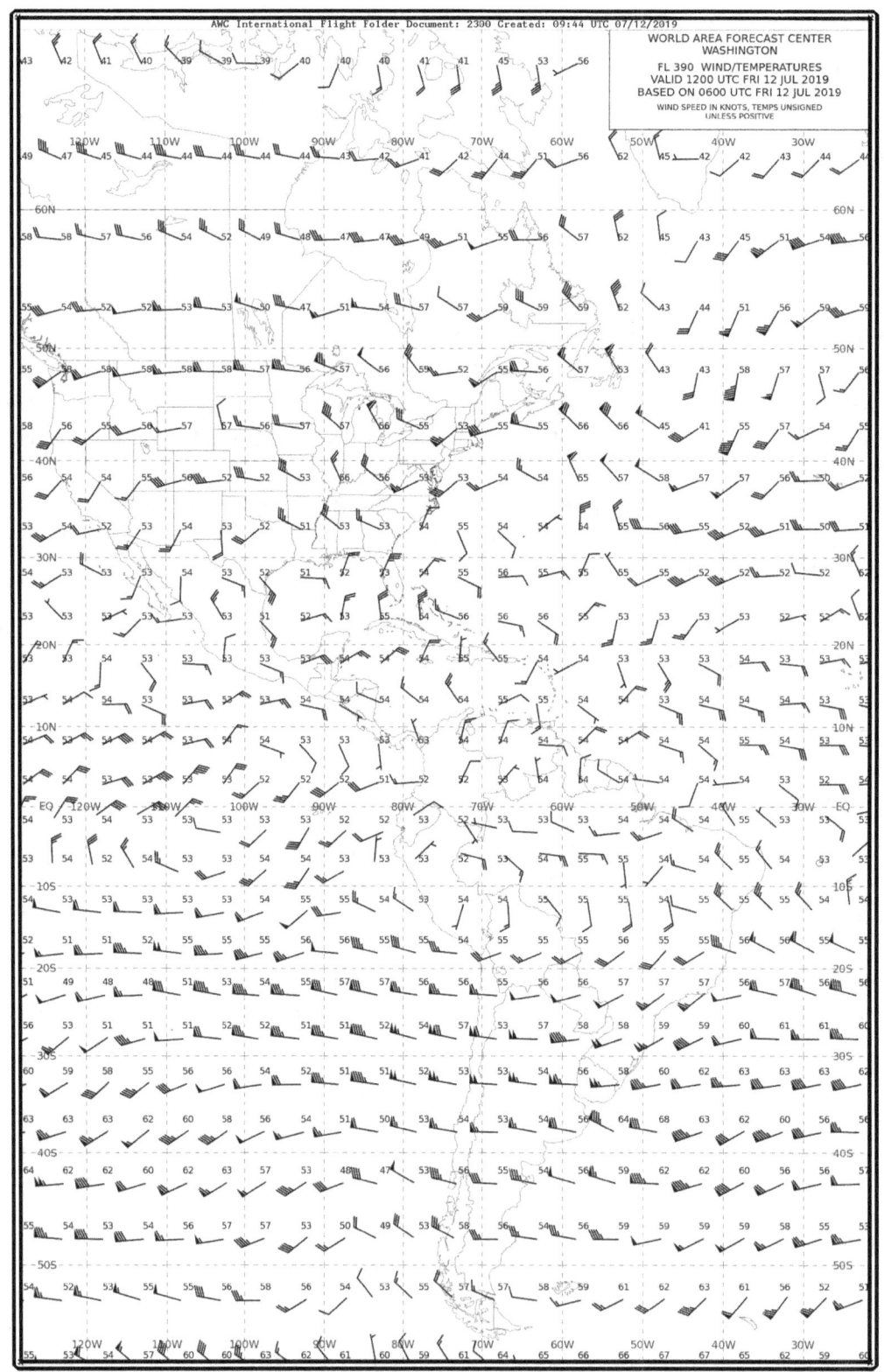

Figure 27-14. WAFS Wind and Temperature 6-Hour Forecast at FL390—Example

27.7.1.3 Humidity, Maximum Wind, and Tropopause Forecasts

No specific charts are issued for global upper-air humidity, maximum wind, height of tropopause, and altitude of FLs. These products are provided in grid point format (e.g., computer format) for use in flight-planning systems. Data from these forecasts are used by the WAFC forecasters to produce the High-Level and Medium-Level SIGWX forecasts, which contain tropopause and jet stream forecasts.

Humidity data is produced for FLs 50 (850 mb), 100 (700 mb), 140 (600 mb), and 180 (500 mb). Additional valid times are planned to be implemented in 2024.

27.7.1.4 WAFS Turbulence, Icing, and Cumulonimbus Cloud Forecasts

WAFS global turbulence, icing, and cumulonimbus cloud forecasts are provided in grid point format (e.g., computer format) for use in flight-planning systems, but the AWC does make these available on their website in web display format and not chart format. The web display allows the user to select various products and FLs and view the forecasts as single time steps or in a movie-loop sequence. More detailed information is provided on the AWC's website under "WAFS Forecasts."

The WAFS global turbulence, icing, and cumulonimbus cloud forecasts are actually a blend of the WAFC Washington global turbulence, icing, and cumulonimbus cloud forecasts and the WAFC London global turbulence, icing, and cumulonimbus cloud forecasts. In other words, each WAFC produces their own global turbulence, icing, and cumulonimbus cloud forecasts using their own global computer models (WAFC Washington uses the NCEP's GFS model). The two WAFCs' forecasts, for turbulence, icing, and cumulonimbus cloud only, are then merged together to eliminate any differences between the two sets of forecasts.

27.8 Significant Weather (SIGWX)

SIGWX forecasts may be depicted in monochrome or color display. The colors used for symbols as well as the color and style of lines are not standard. The colors of jet streams, turbulence, cloud cover, and other elements may vary depending on the website and service provider. The examples shown in this section are from the NWS AWC's and the AAWU's websites. Refer to any legend, Help Page, or user information on the websites for details on the content and display of the weather information.

27.8.1 Low-Level Significant Weather (SIGWX) Charts

The Low-Level SIGWX Charts (see Figure 27-15) provide an overview of selected aviation weather hazards up to FL240 at 12 and 24 hours into the future.

The forecast domain covers the CONUS and the coastal waters. Each depicts a "snapshot" of weather expected at the specified valid time.

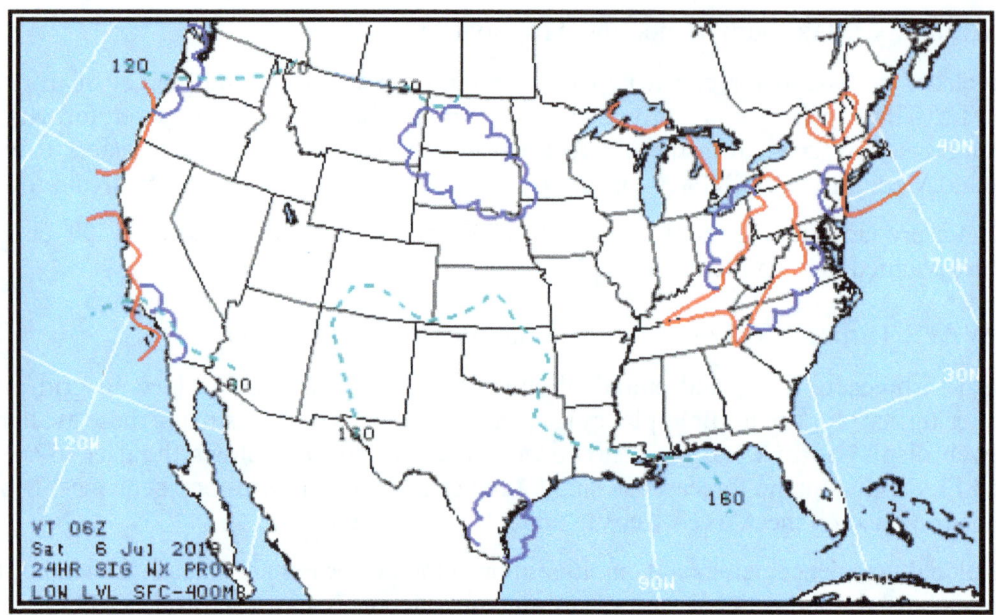

Figure 27-15. 12-Hour Low-Level SIGWX Chart—Example

27.8.1.1 Issuance

Low-Level SIGWX Charts are issued four times per day by the NWS AWC (see Table 27-10). Two charts are issued: a 12-hour and a 24-hour prognostic (prog) chart.

Table 27-10. Low-Level SIGWX Chart Issuance Schedule

	Issuance Time			
	~1720Z	~2310Z	~0530Z	~0935Z
Chart	Valid Time			
12-Hour Prog	00Z	06Z	12Z	18Z
24-Hour Prog	12Z	18Z	00Z	06Z

27.8.1.2 Content

Low-Level SIGWX Charts depict weather flying categories, turbulence, and freezing levels (see Figure 27-16). In-flight icing is not depicted on the Low-Level SIGWX Chart.

Depending on the website or service provider, the Low-Level SIGWX Charts may be combined with Surface Prog Charts to create a four-panel presentation. For example, the left two panels represent the 12-hour forecast interval and the right two panels represent the 24-hour forecast interval. The upper two panels depict the SIGWX Charts and the lower two panels the Surface Prog.

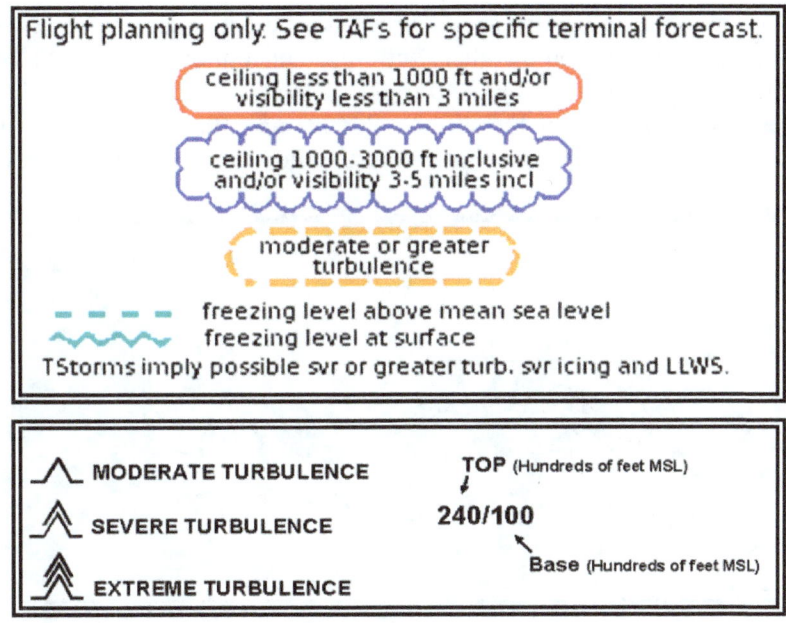

Note: The colors used in the Low-Level SIGWX Charts may vary depending on the website or service provider.

Figure 27-16. Low-Level SIGWX Chart Symbols

27.8.1.2.1 Flying Categories

IFR areas are outlined with a solid red line, MVFR areas are outlined with a scalloped blue line, and VFR areas are not depicted (see Figure 27-17).

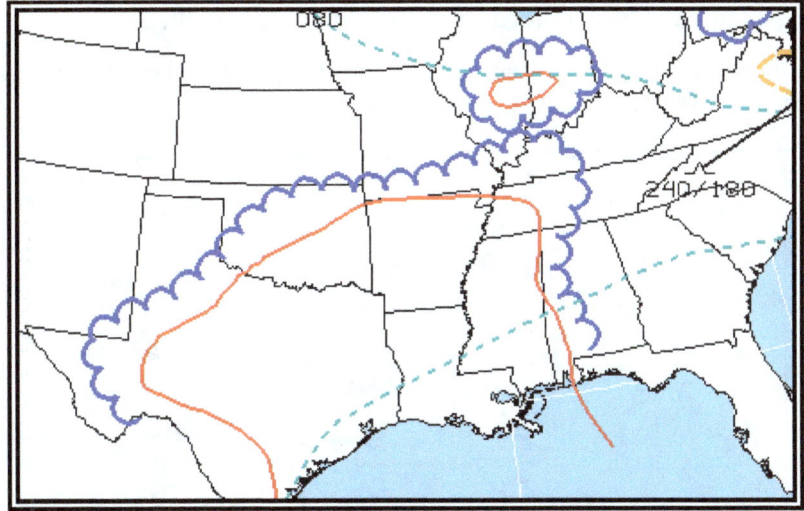

Figure 27-17. Low-Level SIGWX Chart Flying Categories—Example

27.8.1.2.2 Turbulence

Areas of moderate or greater turbulence are enclosed by bold, dashed, brown lines (see Figure 27-18). Turbulence intensities are identified by standard symbols (see Figure 27-16). The vertical extent of turbulence layers is specified by top and base heights separated by a slant. The intensity symbols and height information may be located within or adjacent to the forecasted areas of turbulence. If located adjacent to an area, an arrow will point to the associated area. Turbulence height is depicted by two numbers separated by a solidus (/). For example, an area on the chart with turbulence indicated as **240/100** indicates that the turbulence can be expected from the top at FL240 to the base at 10,000 ft MSL. When the base height is omitted, the turbulence is forecast to reach the surface. For example, **080/** identifies a turbulence layer from the surface to 8,000 ft MSL. Turbulence associated with thunderstorms is not depicted on the chart.

Figure 27-18. Low-Level SIGWX Chart Turbulence Forecast—Example

27.8.1.2.3 Freezing Levels

The freezing level at the surface is depicted by a blue, saw-toothed symbol (see Figure 27-19). The surface freezing level separates above-freezing from below-freezing temperatures at the Earth's surface.

Freezing levels above the surface are depicted by blue dashed lines labeled in hundreds of feet MSL beginning at 4,000 ft using 4,000-ft intervals (see Figure 27-19). If multiple freezing levels exist, these lines are drawn to the highest freezing level. For example, **80** identifies the 8,000-ft freezing level contour (see Figure 27-19). The lines are discontinued where they intersect the surface.

The freezing level for locations between lines is determined by interpolation. For example, the freezing level midway between the 4,000 and 8,000-ft lines is 6,000 ft.

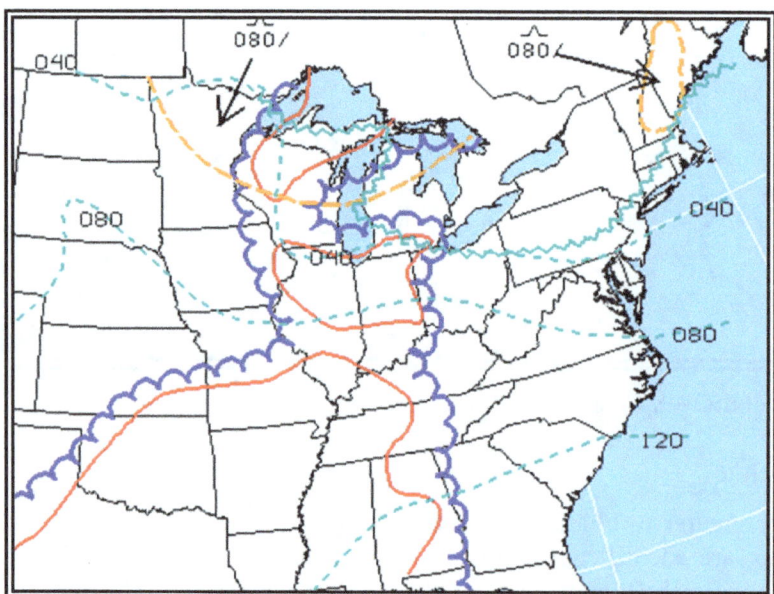

Figure 27-19. Low-Level SIGWX Chart Freezing Level Forecast—Example

Multiple freezing levels occur when the temperature is 0 °C at more than one altitude aloft (see Figure 27-20). Multiple freezing levels can be forecasted on the Low-Level SIGWX Prog Charts in situations where the temperature is below freezing (negative) at the surface with multiple freezing levels aloft.

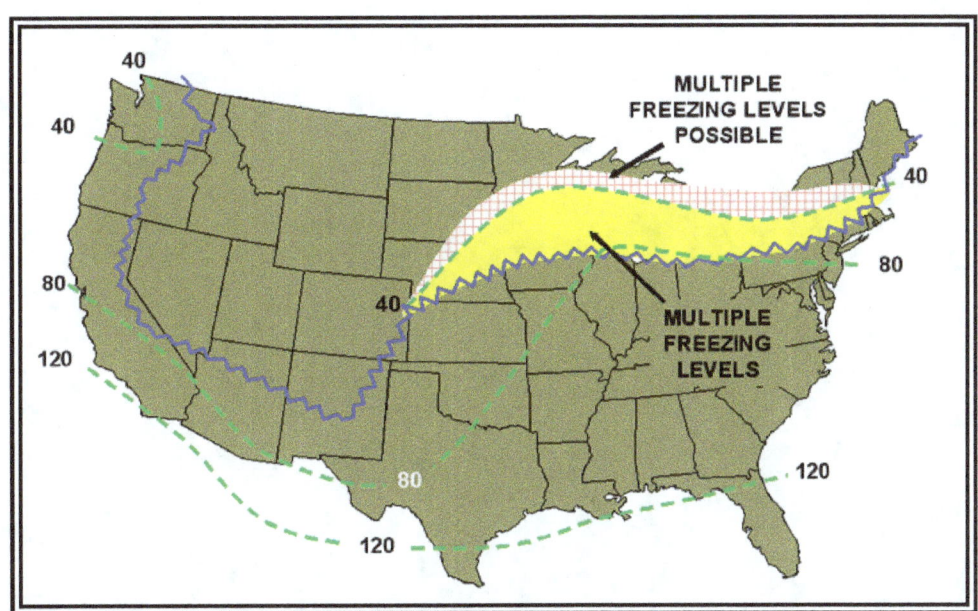

The words "Multiple Freezing Levels Possible" and/or "Multiple Freezing Levels," and/or associated shading and hatched area do not appear on the chart.
Note: The colors used in this example are different from those used in other examples.

Figure 27-20. Low-Level SIGWX Chart Multiple Freezing Levels—Example

In Figure 27-20, areas with multiple freezing levels are located on the below-freezing side of the surface freezing level contour and bounded by the 4,000-ft freezing level. Multiple freezing levels are possible beyond the 4,000-ft freezing level (i.e., below 4,000 ft MSL), but the exact cutoff cannot be determined.

27.8.2 Mid-Level Significant Weather (SIGWX) Chart

The Mid-Level SIGWX Chart (see Figure 27-21) is a product of ICAO's WAFS. The Mid-Level SIGWX Chart is also known as the Medium-Level SIGWX Chart. The Mid-Level SIGWX Chart is planned to be phased out in 2024 and replaced by new WAFS SIGWX forecasts.

The Mid-Level SIGWX Chart provides a forecast of significant en route weather phenomena over a range of FLs from 10,000 ft MSL to FL450. The chart depicts a "snapshot" of weather expected at the specified valid time. It can be used by airline dispatchers for flight planning and weather briefings before departure and by flightcrew members during flight.

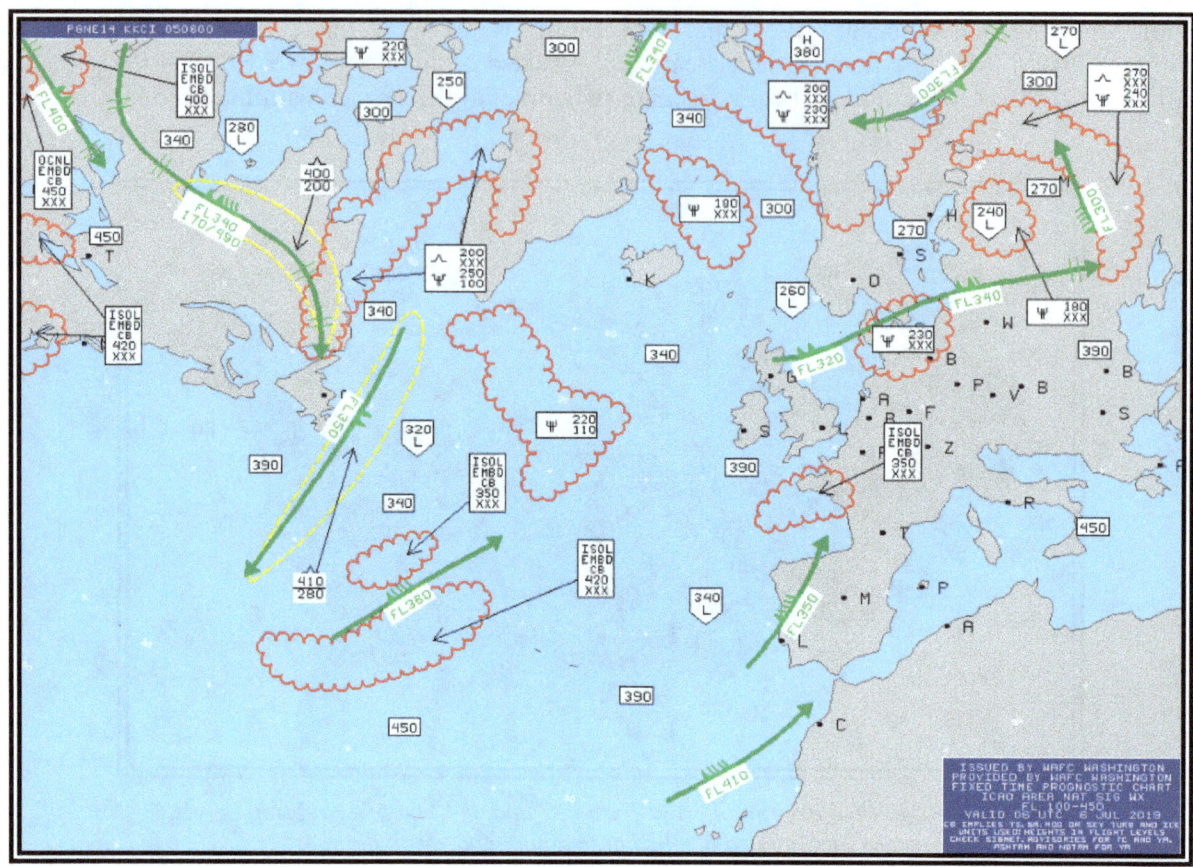

Figure 27-21. Mid-Level SIGWX Chart—Example

27.8.2.1 Issuance

The AWC in Kansas City, MO, has the responsibility, as part of the WAFC Washington, to provide global weather forecasts of SIGWX phenomena. The AWC issues a 24-hour Mid-Level SIGWX Chart, four times daily, for the North Atlantic Ocean Region (see Table 27-11).

Table 27-11. Mid-Level SIGWX Chart Issuance Schedule

Issued (UTC)	Valid (UTC)
0800	0000 (next day)
1400	0600 (next day)
2000	1200 (next day)
0200	1800

27.8.2.2 Content

The Mid-Level SIGWX Chart depicts numerous weather elements that can be hazardous to aviation. The weather elements and their presentation are the same as in the High-Level SIGWX Charts (see Section 27.8.3) except for the addition of nonconvective clouds with moderate or severe icing and/or moderate or severe turbulence. See Section 27.8.3.2 for details on these other weather elements.

27.8.2.2.1 Nonconvective Clouds with Moderate or Severe Icing and/or Moderate or Severe Turbulence

Areas of nonconvective clouds with moderate or severe icing and/or moderate or severe turbulence are depicted by enclosed (red) scalloped lines (see Figure 27-21). The type of icing (i.e., rime, clear, or mixed) is not forecast.

Note: Cumulonimbus clouds are also depicted by enclosed (red) scalloped lines.

The identification and characterization of each area appears within or adjacent to the outlined area. If the identification and characterization is adjacent to an outlined area, an arrow points to the appropriate area.

The identification box uses the standard icing symbol (see Table 27-12). The vertical extent of the icing layer is specified by top and base heights. When the bases extend below 10,000 ft MSL, they are identified with **XXX**.

Table 27-12. Icing and Turbulence Intensity Symbols

Intensity	Icing Symbol	Turbulence Symbol
Moderate	⩗⊢⊢⩘	⌃
Severe	⩗⊢⊢⊢⩘	⌃

27.8.3 High-Level Significant Weather (SIGWX) Charts

High-Level SIGWX Charts (see Figure 27-22) provide a forecast of significant en route weather phenomena over a range of FLs from FL250 to FL630. Each chart depicts a "snapshot" of weather expected at the specified valid time. The vertical range is planned to change in 2024 to FL100 to FL600 with the planned implementation of new WAFS SIGWX forecasts, which will have 3-hour time steps (valid times) from 0 to 48 hours.

High-Level SIGWX forecasts are provided for the en route portion of international flights. These products are used by airline dispatchers for flight planning and weather briefings before departure and by flightcrew members during flight.

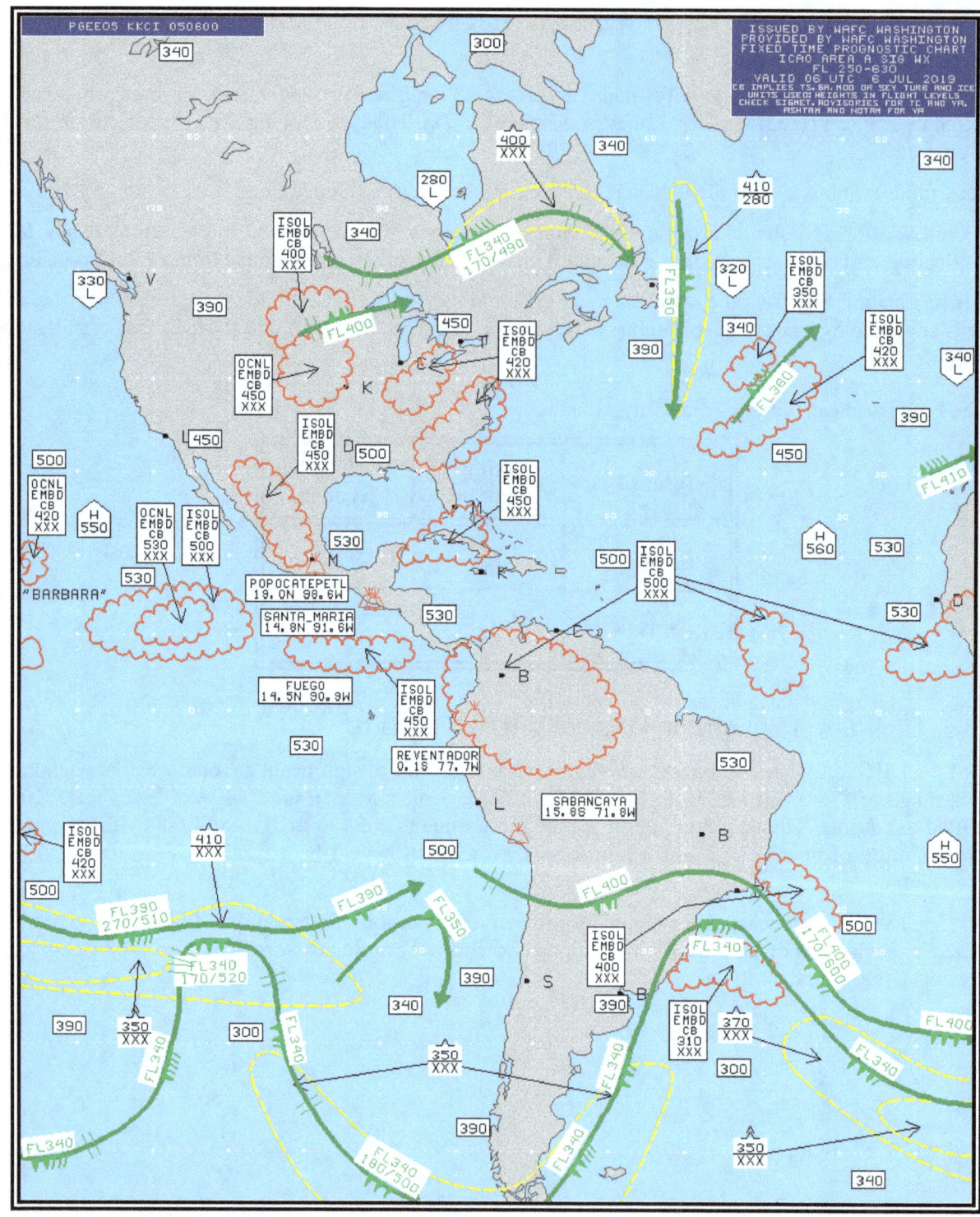

Figure 27-22. High-Level SIGWX Chart—Example

27.8.3.1 Issuance

In accordance with the WAFS of ICAO, High-Level SIGWX forecasts are provided for the en route portion of international flights.

High-Level SIGWX forecasts are issued as a global data set in digital format by two WAFCs, one at the NWS AWC and the other at the United Kingdom's Meteorological Office. Each center produces a global data set of SIGWX that is then made available (displayed) in chart form for different areas of the globe. These charts are available on the AWC's website.

Corrections are issued for format errors or missing information. These charts are not amended.

Table 27-13 provides the issuance schedule.

Table 27-13. High-Level SIGWX Forecast Issuance Schedule

Issued (UTC)	Valid (UTC)
0800	0000 (next day)
1400	0600 (next day)
2000	1200 (next day)
0200	1800

27.8.3.2 Content

27.8.3.2.1 Thunderstorms and Cumulonimbus Clouds

The abbreviation **CB** is only included where it refers to the expected occurrence of an area of widespread cumulonimbus clouds, cumulonimbus along a line with little or no space between individual clouds, cumulonimbus embedded in cloud layers, or cumulonimbus concealed by haze. It does not refer to isolated cumulonimbus not embedded in cloud layers or concealed by haze.

Each cumulonimbus area is identified with **CB** and characterized by coverage, bases, and tops.

Coverage is identified as isolated (**ISOL**) meaning less than 4/8, occasional (**OCNL**) meaning 4/8 to 6/8, and frequent (**FRQ**) meaning more than 6/8 coverage. Isolated CBs can only be depicted when they are embedded (**EMBD**) in clouds or concealed by haze. Occasional cumulonimbus can be depicted with or without **EMBD**.

The vertical extent of cumulonimbus layer is specified by top and base heights. Bases that extend below FL250 (the lowest altitude limit of the chart) are encoded **XXX**.

Cumulonimbus are depicted by enclosed (red) scalloped lines. The identification and characterization of each cumulonimbus area will appear within or adjacent to the outlined area. If the identification and characterization is adjacent to an outlined area, an arrow will point to the associated cumulonimbus area.

On SIGWX charts, the inclusion of **CB** shall be understood to include all weather phenomena normally associated with cumulonimbus (e.g., thunderstorm, moderate or severe icing, moderate or severe turbulence, and hail).

27.8.3.2.2 Moderate or Severe Turbulence

Forecast areas of moderate or severe turbulence (see Figure 27-23) associated with wind shear zones and/or mountain waves are enclosed by bold yellow dashed lines. Intensities are identified by standard symbols (see Table 27-12).

The vertical extent of turbulence layers is specified by top and base heights, separated by a horizontal line. Turbulence bases that extend below FL250 are identified with **XXX**.

Convective or thunderstorm turbulence is not identified.

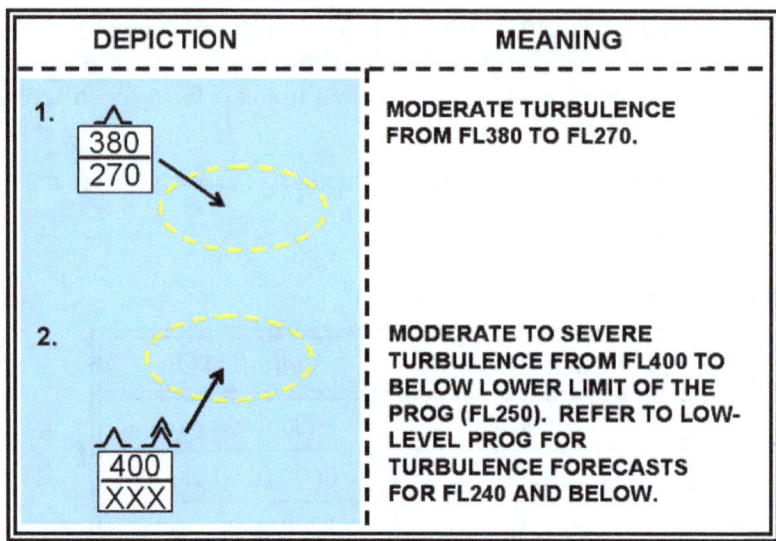

Figure 27-23. High-Level SIGWX Chart Turbulence—Examples

27.8.3.2.3 Moderate or Severe Icing

Moderate or severe icing (outside of thunderstorms) above FL240 is rare and is not generally forecasted on High-Level SIGWX Charts.

27.8.3.2.4 Jet Streams

A jet stream axis with a wind speed of more than 80 kt is identified by a bold (green) line. An arrowhead is used to indicate wind direction. Wind change bars (double-hatched, light green lines) positioned along a jet stream axis identify 20-kt wind speed changes (see Figure 27-24).

Symbols and altitudes are used to further characterize a jet stream axis. A standard wind symbol (light green) is placed at each pertinent position to identify wind velocity. The FL is placed adjacent to each wind symbol to identify the altitude of the jet stream core or axis.

Jet stream vertical depth forecasts are included when the maximum speed is 120 kt or more. Jet depth is defined as the vertical depths to the 80-kt wind field above and below the jet stream axis using FLs. Jet depth information is placed at the maximum speed point only, normally at one point on each jet stream. When the jet stream is very long and there are several wind maxima, then each maximum should include forecasts of the vertical depth.

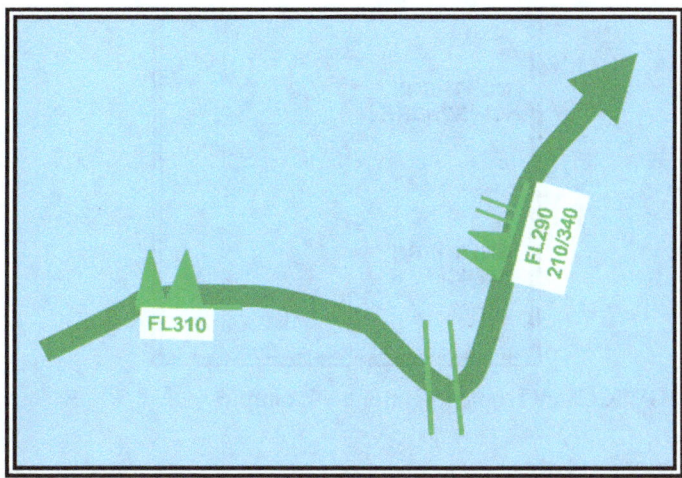

Forecast maximum speeds of 100 kt at FL310 at one location and 120 kt at FL290 at another location. At the latter location, the base of the 80-kt wind field is FL210, and the top of the 80-kt wind field is FL340.

Figure 27-24. High-Level SIGWX Chart Jet Stream—Example

27.8.3.2.5 Tropopause Heights

Tropopause heights are plotted at selected locations on the chart. They are enclosed by rectangles and plotted in hundreds of feet MSL (see Figure 27-25). Centers of high (**H**) and low (**L**) tropopause heights are enclosed by polygons and plotted in hundreds of feet MSL.

Figure 27-25. High-Level SIGWX Chart Tropopause Height—Examples

27.8.3.2.6 Tropical Cyclones

Tropical cyclones (i.e., with surface wind speed 34 kt or greater) are depicted by the symbol in Figure 27-26 with the storm's name positioned adjacent to the symbol. Cumulonimbus clouds meeting chart criteria are identified and characterized relative to each storm.

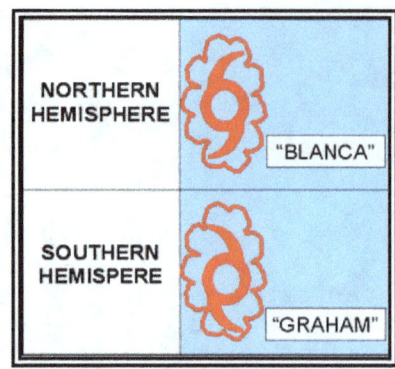

Figure 27-26. High-Level SIGWX Chart Tropical Cyclone—Examples

27.8.3.2.7 Volcanic Eruption Sites

Volcanic eruption sites are identified by a trapezoidal symbol depicted in Figure 27-27. The dot on the base of the trapezoid identifies the location of the volcano. The name of the volcano, its latitude, and its longitude are noted adjacent to the symbol.

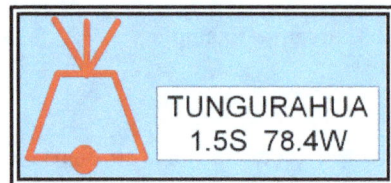

Figure 27-27. High-Level SIGWX Chart Volcanic Eruption Site—Example

27.8.4 Alaska Significant Weather (SIGWX) Charts

The Alaska SIGWX Charts (see Figure 27-28) are a series of four forecasts (24-hour, 36-hour, 48-hour, and 60-hour) valid at specified times. These charts provide a graphical overview of the specified forecast weather primarily for lower flight altitudes.

27.8.4.1 Issuance

The AAWU issues the Alaska SIGWX Charts (see Figure 27-28). These charts are issued twice a day at 0530 and 1330 UTC during Alaska Standard Time, and 0430 and 1230 UTC during Alaska Daylight Time. The 1330/1230 UTC-issued 24-hour SIGWX chart may be updated around 2145/2045 UTC valid at 1200 UTC the next day.

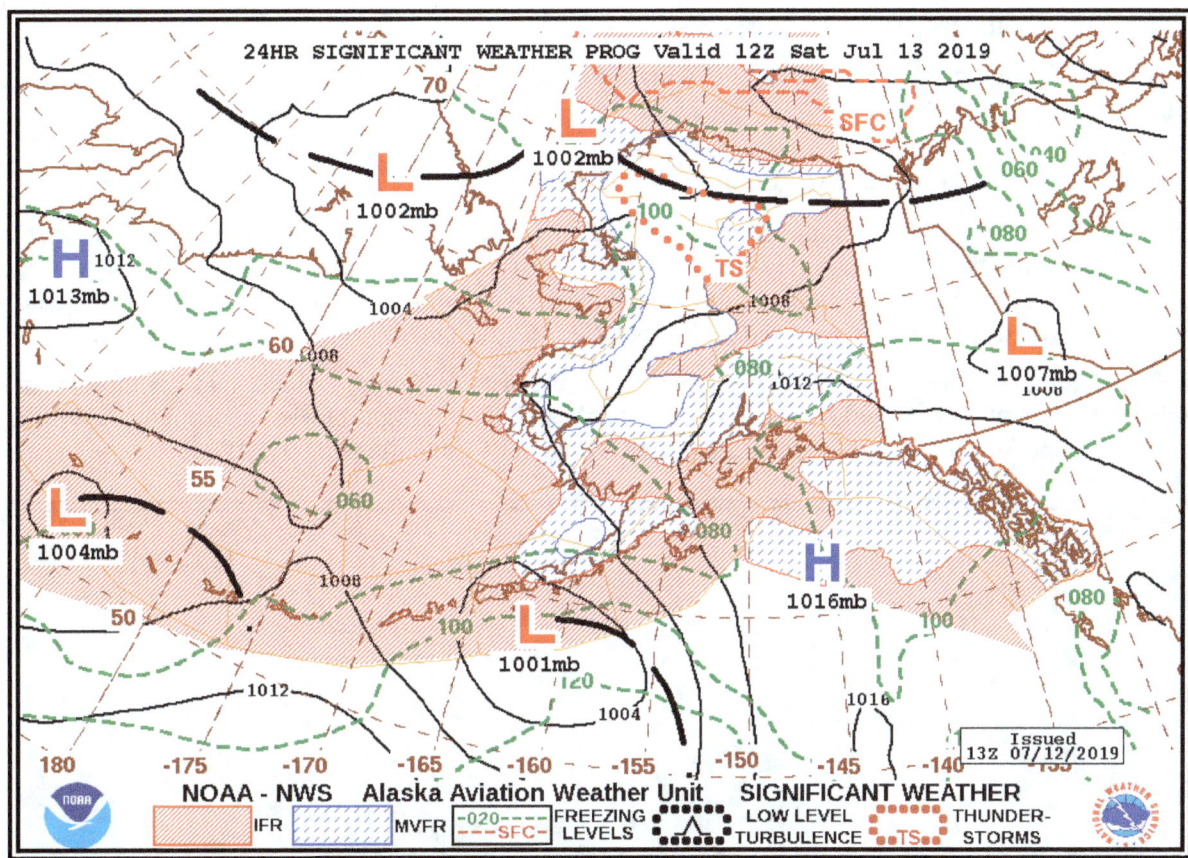

Figure 27-28. Alaska SIGWX Chart—Example

27.8.4.1.1 Content

27.8.4.1.1.1 Surface Pressure Systems and Fronts

Pressure systems and fronts are depicted using standard symbols. Isobars are denoted by solid, thin black lines and labeled with the appropriate pressure in millibars. The central pressure is plotted near the respective pressure center.

27.8.4.1.1.2 Areas of IFR and MVFR Weather Conditions

Areas of forecast IFR and MVFR conditions are shown in red and blue hatching, respectively.

27.8.4.1.1.3 Freezing Levels

Forecast freezing levels are depicted for the surface (dashed red line) and at 2,000-ft intervals (dashed green lines).

Note: Areas of in-flight icing forecasts are not included in the Alaska SIGWX forecasts.

27.8.4.1.1.4 Low-Level Turbulence

Areas of forecast moderate or greater nonconvective low-level turbulence are depicted with black dots. Turbulence altitudes are not included but can be considered as turbulence that is near the surface as a result of wind interactions with the terrain. In most cases, it would be within 6,000 ft above the terrain.

27.8.4.1.1.5 Thunderstorms

Areas of forecast thunderstorms are depicted with red dots. Thunderstorm areal coverage, cloud bases, and tops are not included.

27.9 Short-Range Surface Prognostic (Prog) Charts

The NWS WPC provides Short-Range Surface Prog Charts (see Figure 27-29) of surface pressure systems, fronts, and precipitation for a multiday period. The forecast area covers the CONUS and coastal waters. The forecasted conditions are divided into selected forecast valid time periods. Each chart depicts a "snapshot" of weather elements expected at the specified valid time.

The Short-Range Surface Prog Charts combine WPC forecasts of fronts, isobars, and high/low pressure systems with the NWS' National Digital Forecast Database (NDFD) digital forecasts from the NWS WFO. The Short-Range Surface Prog Forecasts are issued by the WPC in College Park, MD.

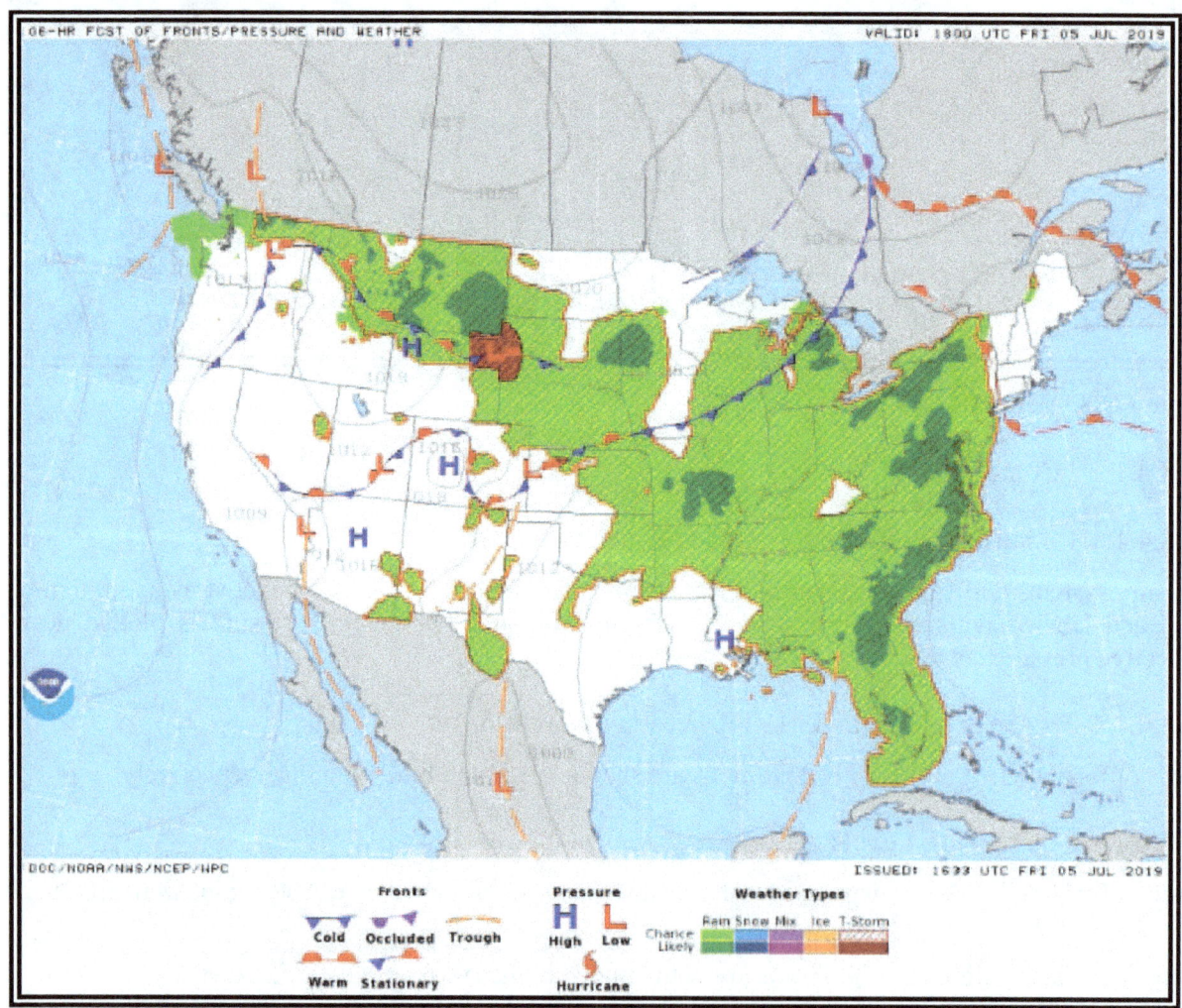

Figure 27-29. NDFD Short-Range Surface Prog Forecast—Example

27.9.1 Content

27.9.1.1 Precipitation

The Short-Range Surface Prog Forecast provides precipitation forecasts in the following depiction:

- NDFD Rain (Chance—light green):–There is a 25 to less than 55 percent probability of measurable rain (≥0.01 in) at the valid time.
- NDFD Rain (Likely—dark green): There is a greater than or equal to 55 percent probability for measurable rain (≥0.01 in) at the valid time.
- NDFD Snow (Chance—light blue): There is a 25 to less than 55 percent probability of measurable snowfall (≥0.01 in liquid equivalent) at the valid time.
- NDFD Snow (Likely—dark blue): There is a greater than or equal to 55 percent probability of measurable snow (≥0.01 in liquid equivalent) at the valid time.
- NDFD Mix (Chance—light purple): There is a 25 to less than 55 percent probability of measurable mixed precipitation (≥0.01 in liquid equivalent) at the valid time. "Mix" can refer to precipitation where a combination of rain and snow, rain and sleet, or snow and sleet are forecast.
- NDFD Mix (Likely—dark purple): There is a greater than or equal to 55 percent probability of measurable mixed precipitation (≥0.01 in liquid equivalent) at the valid time. "Mix" can refer to precipitation where a combination of rain and snow, rain and sleet, or snow and sleet are forecast.
- NDFD Ice (Chance—light brown): There is a 25 to less than 55 percent probability of measurable freezing rain (≥0.01 in) at the valid time.
- NDFD Ice (Likely—brown): There is a greater than or equal to 55 percent probability of measurable freezing rain (≥0.01 in) at the valid time.
- NDFD T-Storm (Chance—red hatching): There is a 25 to less than 55 percent probability of thunderstorms at the valid time. Areas are displayed with diagonal red hatching enclosed in a red border.
- NDFD T-Storm (Likely and/or Severe—dark red): There is a greater than or equal to 55 percent probability of thunderstorms, and/or the potential exists for some storms to reach severe levels at the valid time.

27.9.1.2 Symbols

Figure 27-30 shows the Surface Prog Forecast symbols.

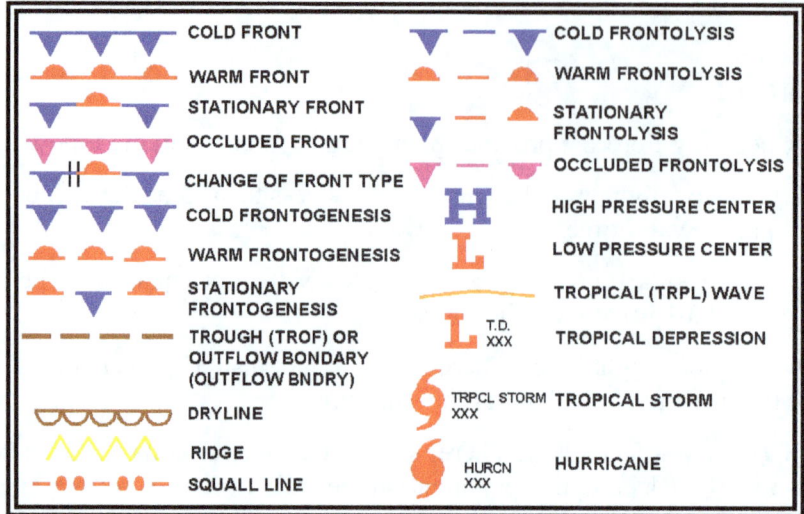

Figure 27-30. Surface Prog Forecast Symbols

27.9.1.3 Pressure Systems

Pressure systems are depicted by pressure centers, troughs, isobars, drylines, tropical waves, tropical storms, and hurricanes using standard symbols (see Figure 27-30). Isobars are denoted by solid, thin black lines and labeled with the appropriate pressure in millibars. The central pressure is plotted near the respective pressure center.

27.9.1.4 Fronts

Fronts are depicted using the standard symbols in Figure 27-30.

27.9.1.5 Squall Lines

Squall lines are denoted using the standard symbol in Figure 27-30.

27.10 Upper-Air Forecasts

NWP models, run on supercomputers, generate surface and upper-air forecasts, known as "Model Guidance" to meteorologists. The NWS NCO runs several models daily and produces hundreds of surface and upper-air guidance products, valid from model run time (i.e., 00-hour) out to several days or weeks (e.g., 340 hours after model run time) depending on the model. Their "Model Analyses and Guidance" website (see Figure 27-31) contains a user's guide as well as a product description link that provides details on the various products.

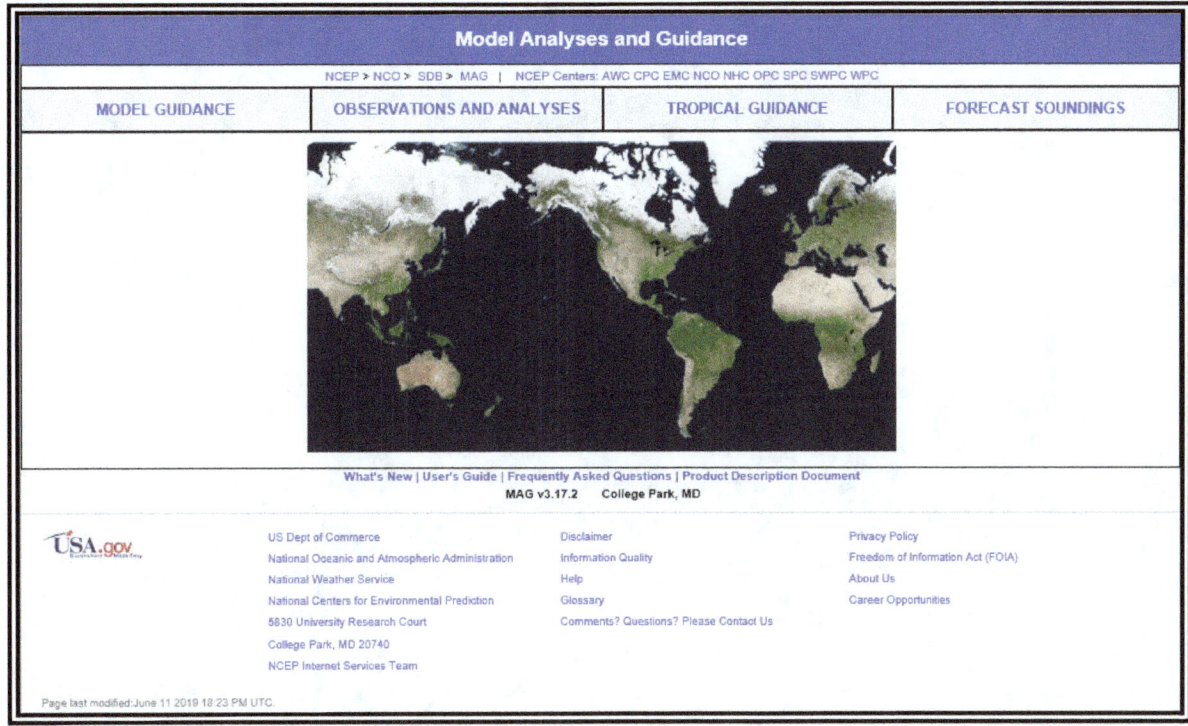

A user's guide, located below the image (lower left), provides descriptions, details, and examples of the various products.

Figure 27-31. NWS NCO Model Analyses and Guidance Website

27.10.1 Constant Pressure Level Forecasts

Constant pressure level forecasts (see Figure 27-32) are just one of the many products produced by NWP models. Constant pressure level forecasts are the computer model's depiction of select weather (e.g., wind) at a specified constant pressure level (e.g., 300 mb), along with the altitudes (in meters) of the specified constant pressure level. When considered together, constant pressure level forecasts describe the 3D aspect of pressure systems. Each product provides a plan-projection view of a specified pressure altitude at a given forecast time.

Constant pressure level forecasts are used to provide an overview of weather patterns at specified times and pressure altitudes and are the source for wind and temperature aloft forecasts.

Pressure patterns cause and characterize much of the weather. Typically, lows and troughs are associated with clouds and precipitation while highs and ridges are associated with fair weather, except in winter when valley fog may occur. The location and strength of the jet stream can be viewed at 300 mb, 250 mb, and 200 mb levels.

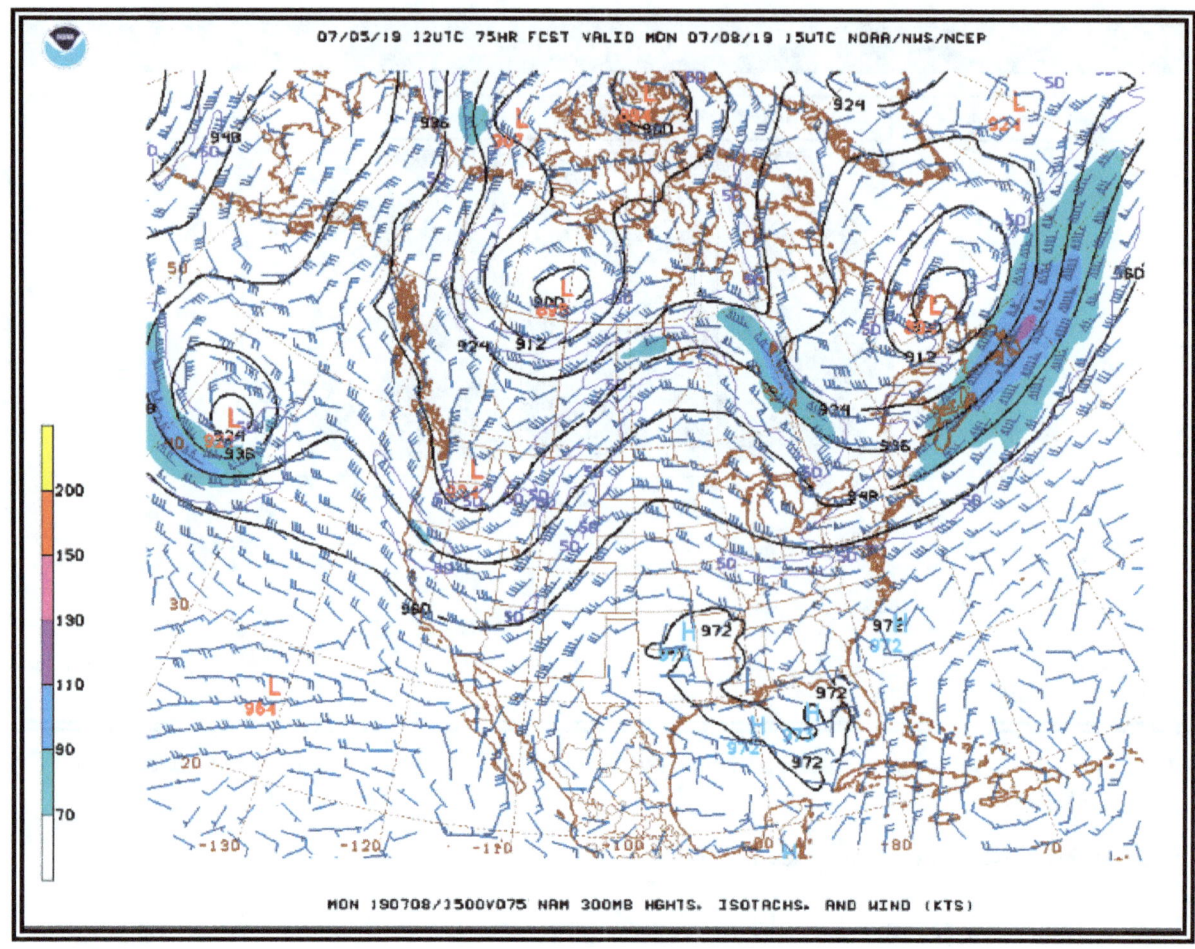

Contours of the height of the 300 mb surface are presented as solid lines. Windbarbs are used to show the direction and speed of the wind. Shading is done for wind speeds greater than 70 kt and generally represents the jet stream.

Figure 27-32. 300 mb Constant Pressure Forecast—Example

27.10.1.1 Issuance

Constant pressure level forecasts are produced several times a day depending on the model. The NCEP's GFS model and North American Model (NAM) produce forecasts four times per day, with initial times of 00, 06, 12, and 18 UTC. Other higher resolution models such as the High Resolution Rapid Refresh (HRRR) produce forecasts at hourly intervals.

27.10.1.2 Content

Constant pressure level forecasts vary in content depending on the selected model and product. Most provide a wind forecast that may be combined with temperature, relative humidity, or certain derived parameters (e.g., vorticity).

Many constant pressure levels are available for display, depending on the model. For example, the NCEP's "Model Analyses and Guidance" website (see Figure 27-31) provides displays of the NCEP's GFS model constant pressure levels contained in Table 27-14. It should be noted that the levels provided on the website are only a subset of the levels available from the model that are routinely made available to NWS meteorologists and others (e.g., 400 mb, 600 mb).

Table 27-14. Select Constant Pressure Levels from the GFS Model

Constant Pressure Level	Approximate Altitude (MSL)
925 mb	2,500
850 mb	5,000
700 mb	10,000
500 mb	18,000
300 mb	30,000
250 mb	34,000
200 mb	39,000

27.11 Freezing Level Forecast Graphics

The freezing level forecast graphics provide an initial analysis and forecasts at specified times into the future. The forecasts are based on output from NWS computer models. They supplement the forecast freezing level information contained in the icing AIRMETs.

The freezing level is the lowest altitude in the atmosphere over a given location at which the air temperature reaches 0 °C. This altitude is also known as the height of the 0 °C constant-temperature surface. A freezing level forecast graphic (see Figure 27-33) shows the height of the 0 °C constant-temperature surface.

The initial analysis and forecast graphics are updated hourly. The colors represent the height in hundreds of feet above MSL of the lowest freezing level. Regions with white indicate the surface and the entire depth of the atmosphere are below freezing. Hatched or spotted regions (if present) represent areas where the surface temperature is below freezing with multiple freezing levels aloft.

More information on the freezing level forecast graphics is available on the AWC's website.

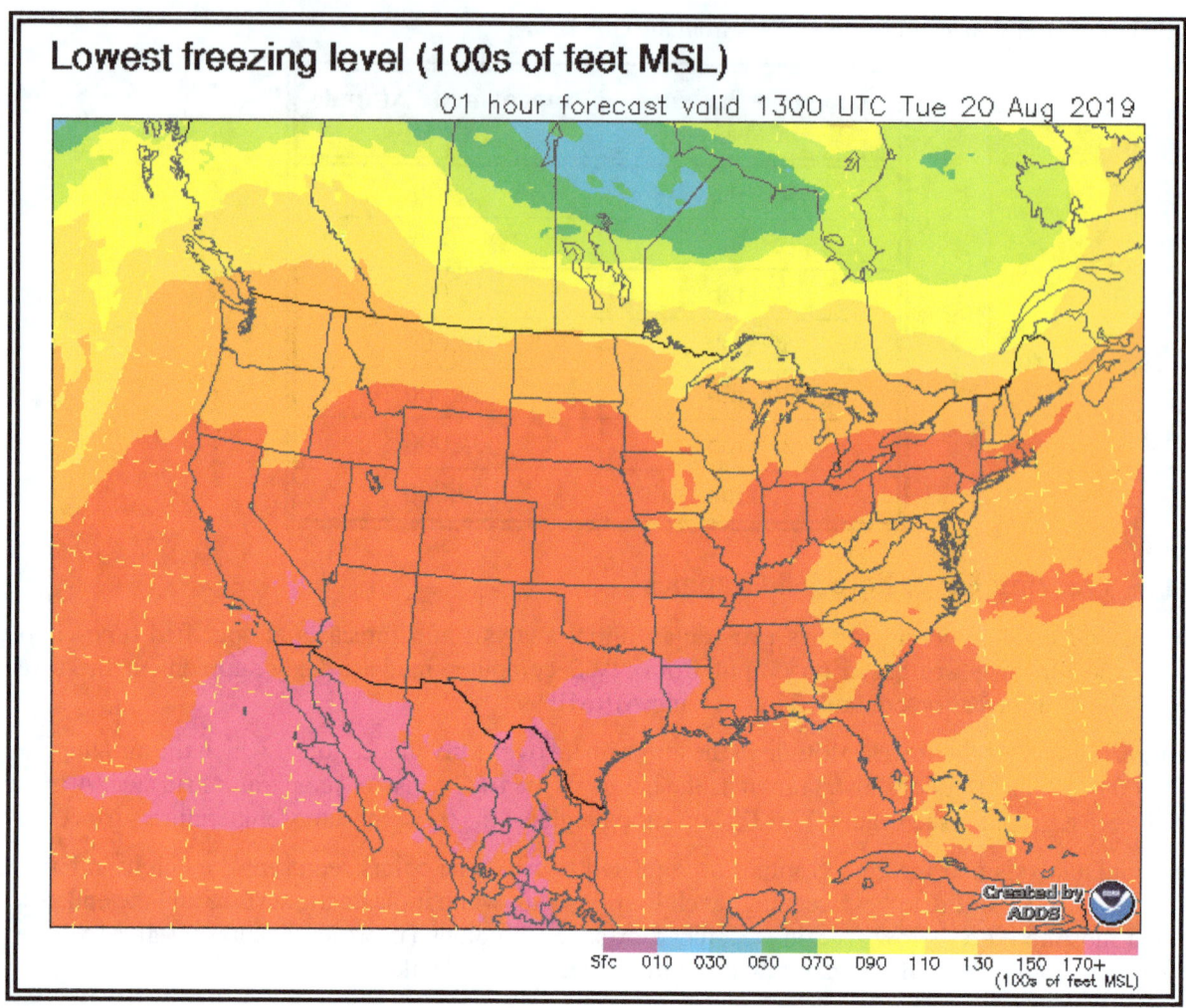

Figure 27-33. Freezing Level Forecast Graphic—Example

27.12 Forecast Icing Product (FIP)

The NWS produces icing products that are derived from NWS computer model data and PIREPs with no forecaster modifications. This includes the following products:

- The CIP, and
- The FIP.

The FIP provides the same suite of products as the CIP (see Section 25.5), describing the icing environment in the future and being solely NWP model based. Information on the graphics is determined from NWP model output; observational data, including WSR-88D; satellite, PIREPs, and surface weather reports; and lightning network data.

FIPs contain a heavy intensity level. Heavy icing has been defined, in part, as the rate of ice accumulation that requires maximum use of ice protection systems to minimize ice accretions on the airframe. Heavy icing is not an intensity level used in PIREPs.

FIPs will continue to evolve over the coming years with increased model resolutions, additional horizontal layers, and improvements to the algorithms and/or data sets used to produce the products. Along with these improvements may come a change in references to the product update version. Users can find additional information on these products and any changes on the AWC's "Icing" web page.

The FIP suite as it appears on the AWC's website consists of five graphics, including:

- Icing Probability;
- Icing Severity;
- Icing Severity—Probability > 25 percent;
- Icing Severity—Probability > 50 percent; and
- Icing Severity plus SLD.

The CIPs/FIPs are generated for select altitudes from 1,000 ft MSL to FL300. FIPs are available at select forecast times through 18 hours.

The CIPs/FIPs can be viewed at single altitudes and FLs or as a composite of all altitudes from 1,000 ft MSL to FL300, which is referred to as the "maximum" or "max."

The FIP should be used in conjunction with the report and forecast information contained in an AIRMET and SIGMET.

27.12.1 Icing Probability

The Icing Probability product displays the probability of icing at *any* level of intensity. Probabilities range from 0 percent (no icing expected) to 85 percent or greater (nearly certain icing). The product is available in single altitudes (e.g., 3,000 ft MSL) or a composite of all altitudes from 1,000 ft MSL to FL300.

27.12.2 Icing Severity

The Icing Severity product depicts the icing intensity likelihood at locations where the Icing Probability product depicts possible icing. Icing intensity is displayed using icing intensity categories: trace, light, moderate, and heavy. The product is available in single altitudes (e.g., 17,000 ft MSL) or a composite of all altitudes from 1,000 ft MSL to FL300 (i.e., max level).

27.12.3 Icing Severity—Probability > 25 Percent

The Icing Severity—Probability > 25 percent product depicts icing intensity where at least a 26 percent probability exists for any icing intensity. Icing intensity is displayed using icing intensity categories: trace, light, moderate, and heavy. The product is available in single altitudes (e.g., 3,000 ft MSL) or a composite of all altitudes from 1,000 ft MSL to FL300 (i.e., max level).

27.12.4 Icing Severity—Probability > 50 Percent

The Icing Severity—Probability > 50 percent product depicts icing intensity where at least a 51 percent probability exists for any icing intensity. Icing intensity is displayed using icing intensity categories: trace, light, moderate, and heavy. The product is available in single altitudes (e.g., 3,000 ft MSL) or a composite of all altitudes from 1,000 ft MSL to FL300 (i.e., max level).

27.12.5 Icing Severity Plus SLD

The Icing Severity plus SLD product depicts the intensity of icing expected as well as locations where a threat for SLD exists. The product is available in single altitudes (e.g., 3,000 ft MSL) or a composite of all altitudes from 1,000 ft MSL to FL300 (i.e., max level (see Figure 27-34)).

SLD is defined as supercooled water droplets larger than 50 micrometers in diameter. These size droplets include freezing drizzle and/or freezing rain aloft.

Icing intensity is displayed using icing intensity categories: trace, light, moderate, and heavy.

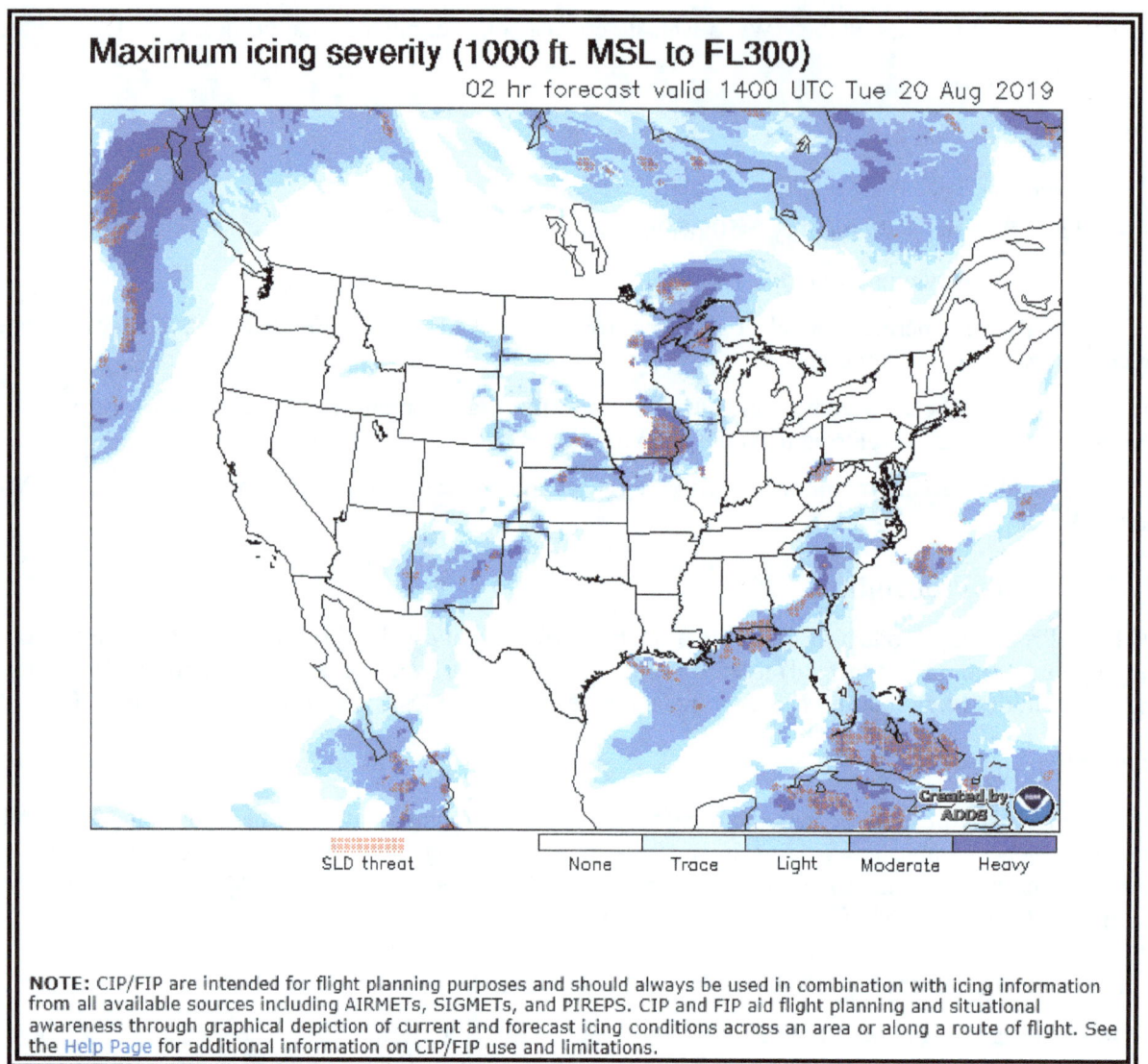

Figure 27-34. FIP Icing Severity Plus SLD—Max Example

27.13 Graphical Turbulence Guidance (GTG)

The NWS produces a turbulence product that is derived from airborne turbulence observations and NWS model data with no forecaster modifications. This product is GTG.

GTG computes the results from more than 10 turbulence algorithms, then compares the results of each algorithm with turbulence observations from both PIREPs and AMDAR data to determine how well each algorithm matches reported turbulence conditions from these sources. GTG then weighs the results of this comparison to produce a single turbulence forecast. Note that the success of GTG is proportional to the number of PIREPs and AMDAR reports available to verify the algorithms. This means the accuracy of GTG improves during daylight hours and where there is more traffic making PIREPs and sending of AMDAR data. GTG produces its forecasts every hour. Currently, GTG has separate forecasts for each hour through the first 3 hours, followed by forecasts at 3-hour intervals through 18 hours. GTG forecasts are available at select altitudes from 1,000 ft MSL through FL450. GTG forecasts are also scaled to three ICAO weight class sizes for aircraft, with light-sized aircraft being less than 15,500 lb, heavy-sized aircraft being more than 300,000 lb, and medium-sized in between.

See Figure 27-35 for an example of a GTG forecast for CAT and Mountain Wave Turbulence (MWT).

GTG does not specifically predict turbulence associated with convective clouds or small-scale local terrain features, but it does predict turbulence associated with upper-level clear and mountain wave sources.

GTG provides three depictions of turbulence:

- CAT,
- MWT, and
- Combined Turbulence (the Combined GTG product depicts the higher of CAT values and MWT values at any given point).

This turbulence product will continue to evolve over the coming years with increased model resolutions, additional horizontal layers, and improvements to the algorithms and/or data sets used to produce the product. Users can find additional information on these products and any changes on the AWC's "Turbulence" web page.

The GTG product suite is issued and updated every hour by the AWC and is available on the AWC's website and other sources.

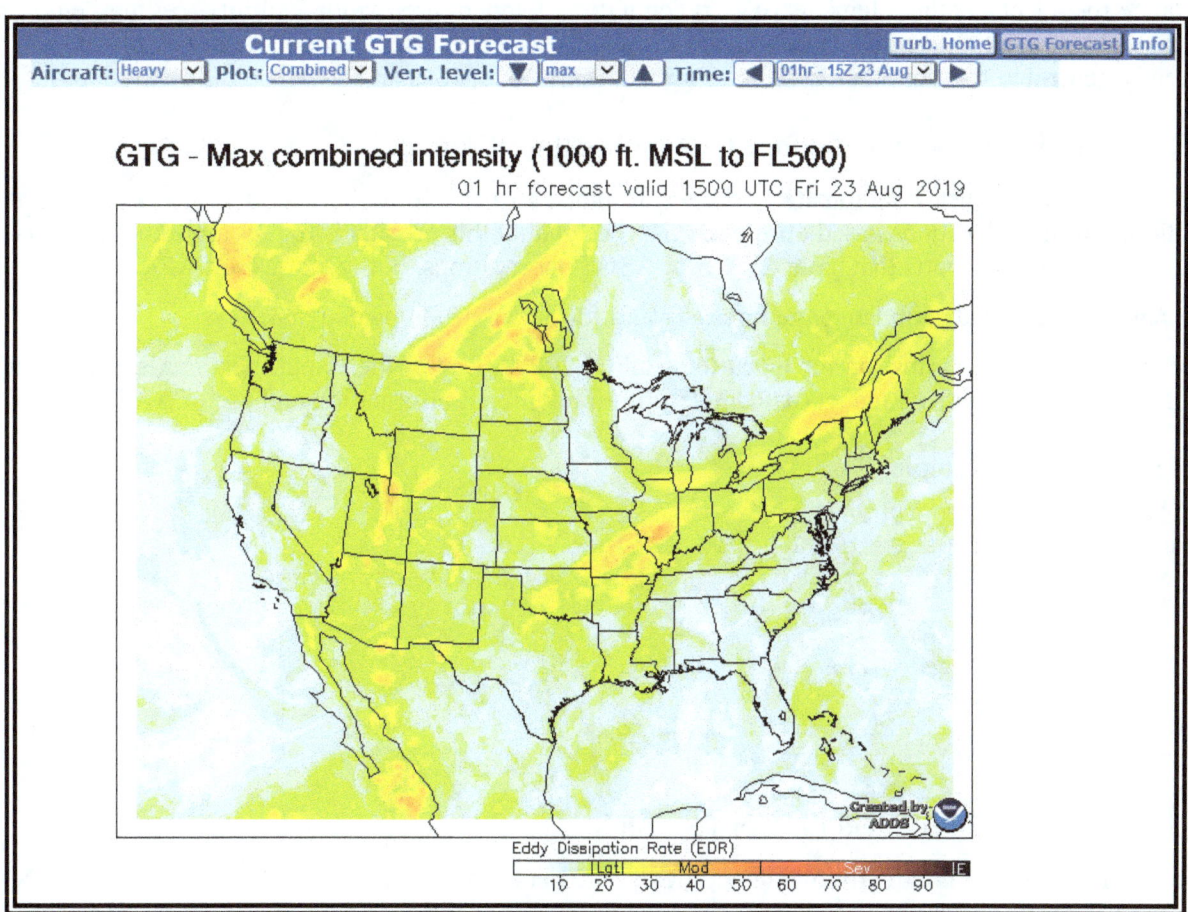

Figure 27-35. GTG Forecast—Max Intensity (CAT+MWT All Levels) Example

27.14 Cloud Tops

The Cloud Tops product is one of the products transmitted through the FIS-B. This product uses HRRR model data, which currently provides a 1- and 2-hour forecast of the altitude of cloud tops and the cloud amounts. The FIS-B data source receives the cloud tops data from this model. The HRRR model data is updated hourly and the transmission interval occurs every 15 minutes.

This product is only currently available for the CONUS.

27.15 Localized Aviation Model Output Statistics (MOS) Program (LAMP)

The NWS has a long history of developing and using statistical analysis of historical and model weather data to produce forecast guidance for forecasters, which is known as MOS.

The LAMP system was developed to provide aviation forecast guidance. LAMP is designed to frequently update the central MOS product suite primarily by incorporating the most recent observational data. The guidance is available at over 2,000 stations in the CONUS, Alaska, Hawaii, and Puerto Rico. The products are updated hourly and valid over a 25-hour period.

The LAMP products are derived from a statistical model program that provides specific-point forecast guidance for select weather elements (e.g., precipitation, temperature, wind, visibility, ceiling height, sky cover). LAMP aviation weather products are provided in both graphical and coded text format, and are currently generated for more than 2,000 airports in the CONUS, Alaska, Hawaii, and Puerto Rico.

27.15.1 Alaska Aviation Guidance (AAG) Weather Product

The AAG is a completely automated product designed to provide a short-term projection of weather conditions at select locations based off the LAMP. The goal of this product is to provide additional aviation guidance to Alaska airports that have AWOS or ASOS observations, but do not have TAFs.

The AAG is a decoded plain language forecast valid for 6 hours and updated hourly.

Refer to the FAA's Information for Operators (InFO) 20002, Use of the Experimental Alaskan Aviation Guidance (AAG) Weather Product (dated 3/25/20) for additional information and use of AAG.

27.15.1.1 AAG Example

```
Guidance for: PXXX (Someplace, AK) issued at 0900 UTC 12 Jun 2019
Forecast period: 0900 to 1000 UTC 12 June 2019
Forecast type: FROM: standard forecast or significant change
Winds: from the E (90 degrees) at 21 MPH (18 knots; 9.3 m/s) gusting to 28 MPH
(24 knots; 12.3 m/s)
Visibility: 2.00 sm (3.22 km)
Ceiling: 1500 feet AGL
Clouds: overcast cloud deck at 1500 feet AGL
Weather: -RA BR (light rain, mist)
```

27.16 Additional Products for Convection

This section will describe the following additional thunderstorm forecast products produced by the NWS that are of interest to aviation users:

- Convective Outlook,
- TCF, and
- ECFP.

Note: The National Convective Weather Forecast (NCWF) was retired in 2018.

27.16.1 Convective Outlook (AC)

The NWS SPC issues narrative and graphical Convective Outlooks (**AC**) to provide the CONUS NWS WFOs, the public, the media, and emergency managers with the potential for severe (tornado, wind gusts 50 kt or greater, or hail with diameter 1 in or greater) and non-severe (general) convection and specific severe weather threats during the following 8 days. The Convective Outlook defines areas of marginal risk (**MRGL**), slight risk (**SLGT**), enhanced risk (**ENH**), moderate risk (**MDT**), or high risk (**HIGH**) of severe weather based on a percentage probability, which varies for time periods from 1 day to 3 days, and then two probabilistic thresholds for days 4 through 8. The day 1, day 2, and day 3 Convective Outlooks also depict areas of general thunderstorms (**TSTM**). The outlooks in graphical (see Figure 27-36) and text formats are available on the SPC's website. See Figure 27-37 for the legend.

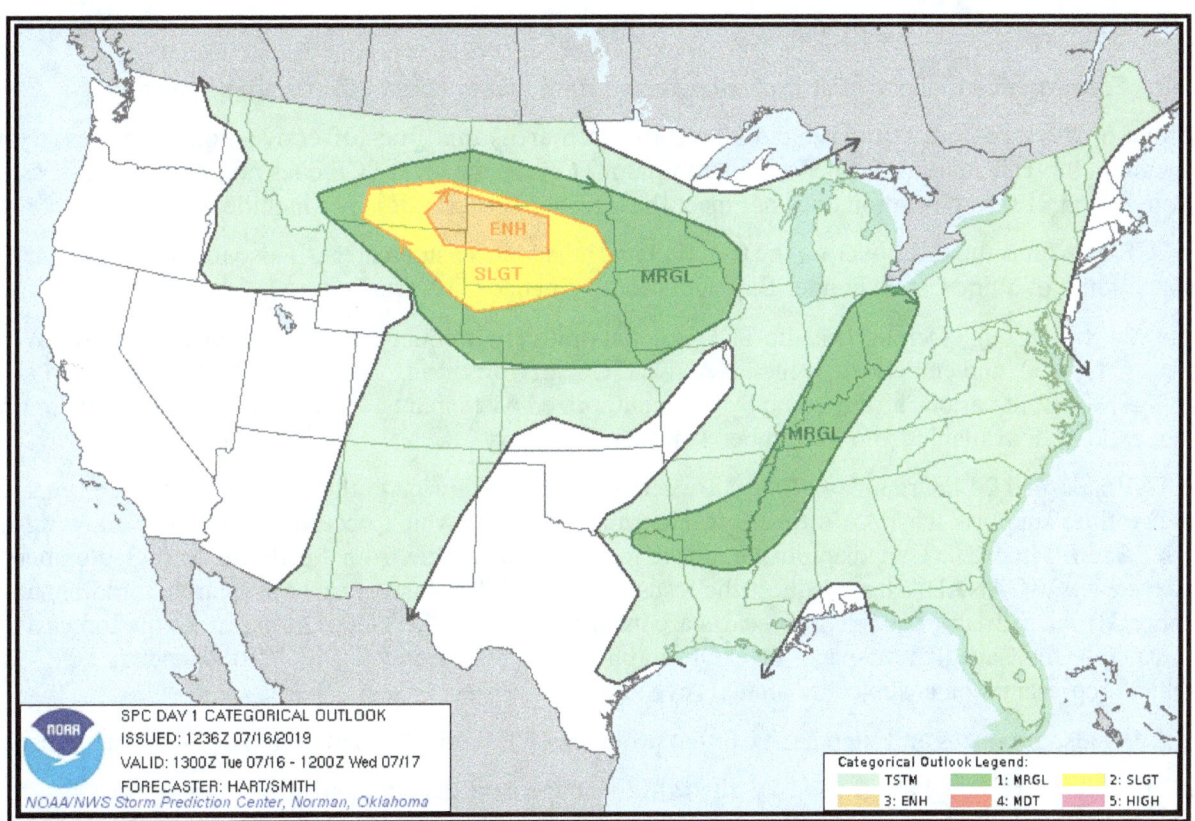

Figure 27-36. Day 1 Categorical Convective Outlook Graphic Example

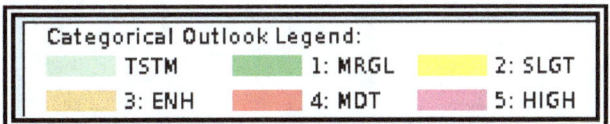

Figure 27-37. Categorical Outlook Legend for Days 1-3 Convective Outlook Graphic Example

27.16.2 Traffic Flow Management (TFM) Convective Forecast (TCF)

The TCF is a high-confidence graphical representation of forecasted convection meeting specific criteria of coverage, intensity, and echo top height. The TCF graphics are produced every 2 hours and are valid at 4, 6, and 8 hours after issuance time.

Areas of convection in the TCF include any area of convective cells meeting the following criteria (at a minimum):

1. Composite radar reflectivity of at least 40 dBZ;
2. Echo tops at or above FL250;
3. Coverage (criteria 1 and 2) of at least 25 percent of the polygon area; and
4. Forecaster confidence of at least 50 percent (high) that criteria 1, 2, and 3 will be met.

Lines of convection in the TCF include any lines of convective cells meeting the following criteria (at a minimum):

1. Composite radar reflectivity of at least 40 dBZ having a length of at least 100 NM;
2. Linear coverage of 75 percent or greater;
3. Echo tops at or above FL250; and
4. Forecaster confidence of at least 50 percent (high) that criteria 1, 2, and 3 will be met.

All four of the threshold criteria listed above for both areas and lines of convection are necessary for inclusion in the TCF. This is defined as the minimum TCF criteria. The TCF does not include a forecast for all convection. If the convection does not meet the threshold criteria, it is not included in the TCF.

The TCF domain is the FIR covering the CONUS and adjacent coastal waters. It also includes the Canadian airspace south of a line from Thunder Bay, Ontario, to Quebec City, Quebec.

From March 1 through October 31, the TCF is collaboratively produced by meteorologists at the AWC in Kansas City, MO, and embedded at the FAA ATCSCC in Warrenton, VA; at the CWSU embedded at the FAA's ARTCC; at various airlines; and by other authorized participants. Automated routines will continue to make the TCF available from November 1 through February 28.

The TCF is issued 24 hours a day, 7 days a week at 30 minutes prior to the indicated issuance time. The issuance time supports the FAA's Strategic Planning Webinar, which occurs 15 minutes following odd hours Eastern Time. The Canadian portion of the forecast is available from April 1 through September 30. However, NAV CANADA may request the issuance of each forecast as early as March 1 and as late as October 31. All available Canadian forecasts are incorporated into the TCF. During times the forecasts are not available for Canadian airspace, the TCF graphics will be annotated with "No Canadian TCF." The graphical representation is subject to annual revision.

The AWC also produces an Extended TCF that provides TCFs from 10 to 30 hours at 2-hour increments.

The TCF and Extended TCF is used by air traffic management decisionmakers in support of convective weather mitigation strategies within the NAS. It is designed to meet the needs of TFM decisionmakers at the FAA's ATCSCC, the FAA's ARTCC TMUs, and airline and corporate flight operations centers (FOC).

Figure 27-38 shows an example of a TCF.

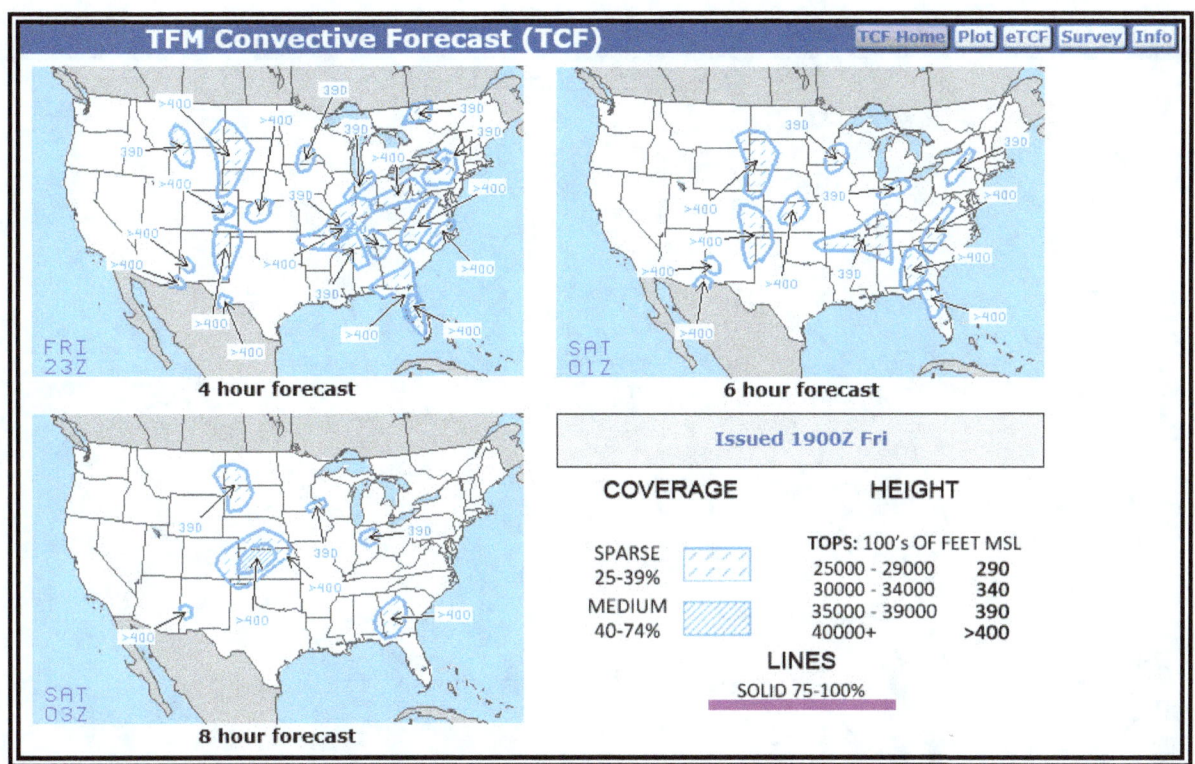

Figure 27-38. TCF Example

27.16.3 Extended Convective Forecast Product (ECFP)

The ECFP planning tool (see Figure 27-39) is a graphical representation of the forecast probability of thunderstorms and is intended to support the long-range planning for convective constraints in the NAS. The product identifies graphically where thunderstorms are expected through the next 72 hours over the CONUS. Although the ECFP uses TCF-style graphics to facilitate ease of interpretation, the ECFP does not forecast TCF criteria.

The development of the ECFP planning tool was a response to FAA and industry needs in planning for weather hazards, specifically convection, 1 to 3 days in advance. To meet these planning needs, the ECFP provides traffic planners and collaborators a quick look at where the probability of convection is greatest. By utilizing TCF-style graphics, users familiar with the TCF can easily determine where traffic constraints are most likely to occur over the next 3 days.

The ECFP is an automated forecast product issued by the AWC. It is issued four times a day at approximately 0100, 0700, 1300, and 1900 UTC.

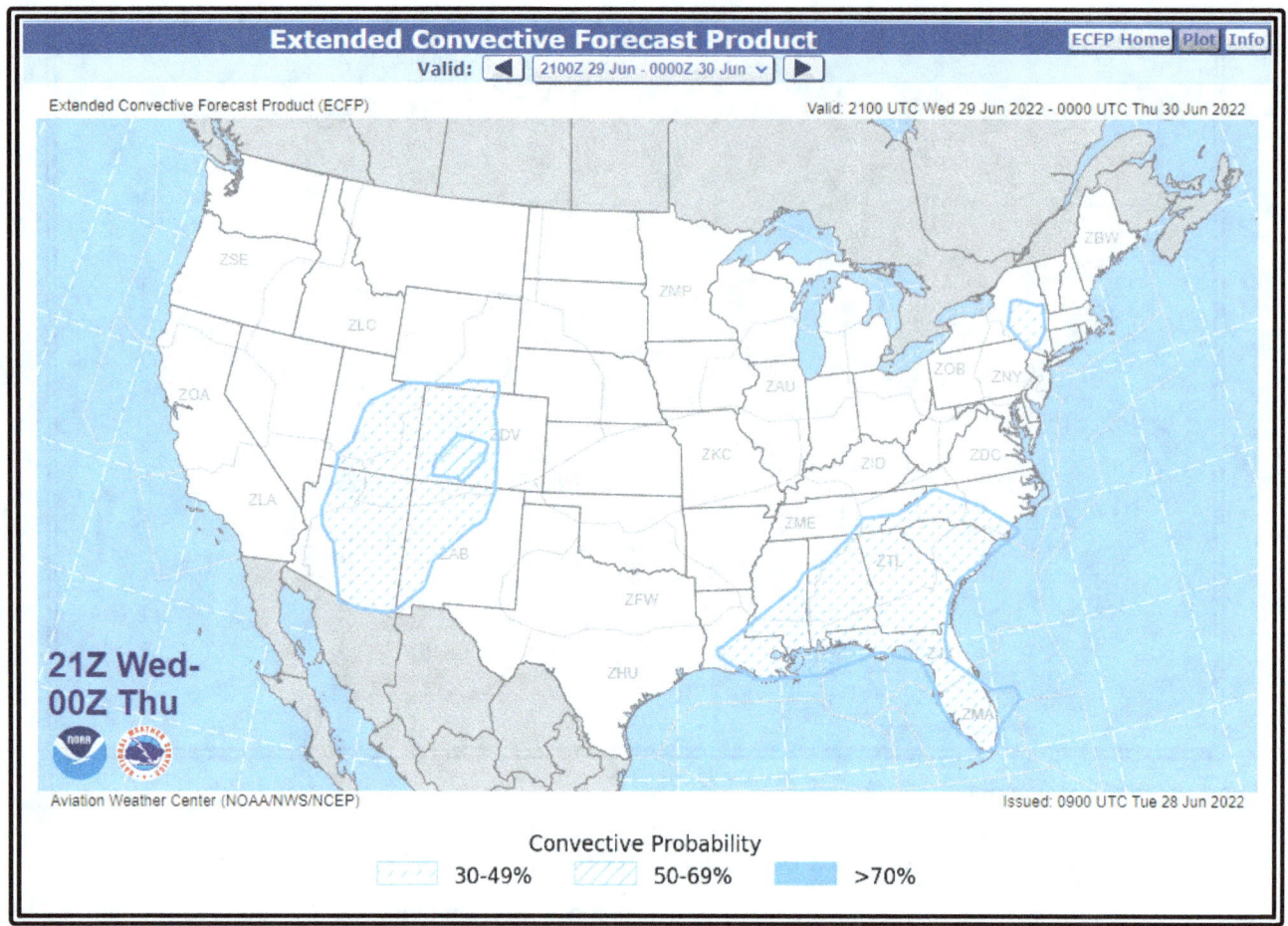

Figure 27-39. ECFP Example

27.16.4 Watch Notification Messages

The NWS SPC issues severe weather Watch Notification Messages to provide an area threat alert for the aviation meteorology community to forecast organized severe thunderstorms that may produce tornadoes, large hail, and/or convective damaging winds within the CONUS.

The SPC issues three types of Watch Notification Messages:

- Aviation Watch Notification Message (SAW),
- Public Severe Thunderstorm Watch Notification Message, and
- Public Tornado Watch Notification Message.

The SAW was formerly known as the Alert Severe Weather Watch Bulletin (AWW), as well as the Severe Weather Forecast Alert. The NWS no longer uses these titles or acronym for this product. The NWS uses the acronym SAW for the Aviation Watch Notification Message, but retains "AWW" in the product header for processing by weather data systems. The NWS uses the acronym AWW for their Airport Weather Warning product, which is a completely different product from the SAW (see Section 27.16.4.1).

The Severe Thunderstorm and Tornado Watch Notification Messages were formerly known as the Severe Weather Watch Bulletins (WW). The NWS no longer uses that title or acronym for this product but retains "WW" in the product header for processing by weather data systems.

It is important to note the difference between a Severe Thunderstorm (or Tornado) Watch and a Severe Thunderstorm (or Tornado) Warning. A "watch" means severe weather is possible during the watch valid time, while a "warning" means that severe weather has been observed or is expected within the hour. Only the SPC issues Severe Thunderstorm and Tornado Watches, while only NWS WFOs issue Severe Thunderstorm and Tornado Warnings.

27.16.4.1 Aviation Watch Notification Message (SAW)

The SPC issues the SAW to provide an area threat alert for the aviation meteorology community to forecast organized severe thunderstorms that may produce tornadoes, large hail, and/or convective damaging winds as indicated in Public Watch Notification Messages.

The SAW product is an *approximation* of the area in a watch. For the official area covered by a watch, see the corresponding Public Watch product. To illustrate, Figure 27-40 is an example of the Aviation Watch (polygon) compared to the Public Watch (shaded). Also, the SAW is easier to communicate verbally over the radio and telephone than reciting the entire Public Watch product.

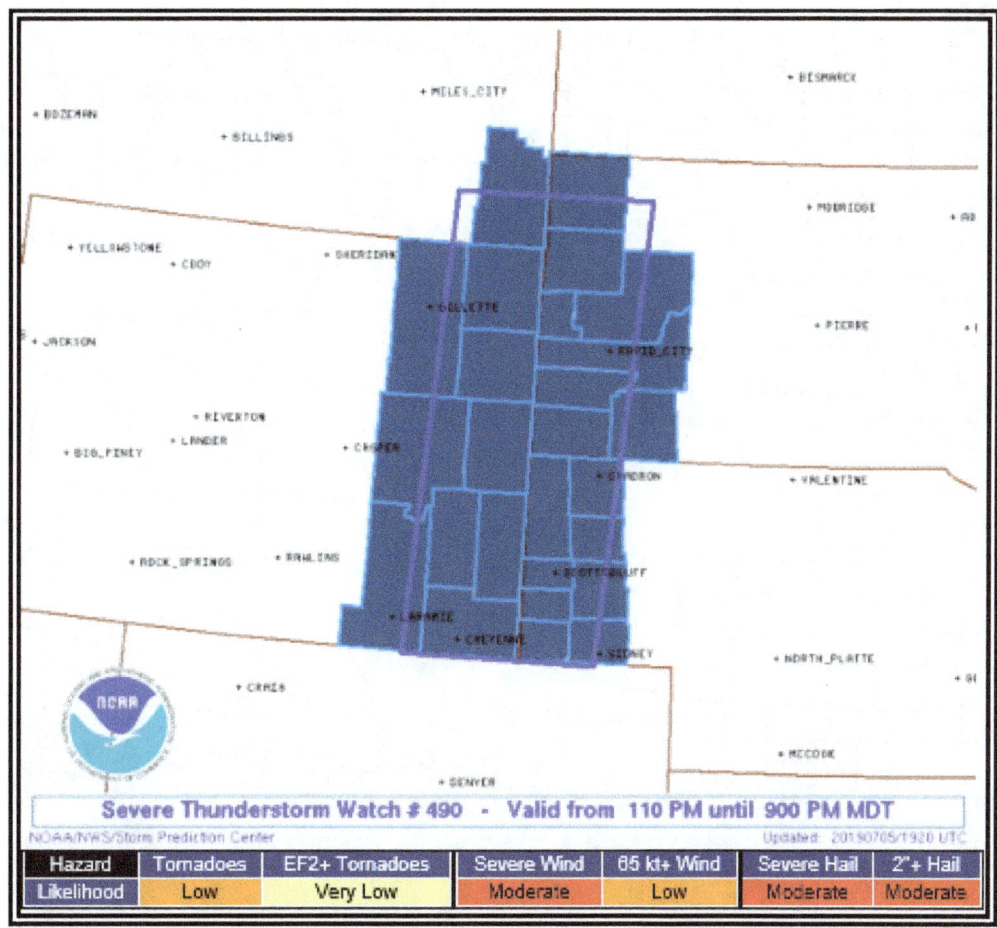

Figure 27-40. Aviation Watch (polygon) Compared to Public Watch (shaded) Example

The SPC will issue the SAW after the proposed convective watch area has been collaborated with the impacted NWS WFOs defining the approximate areal outline of the watch.

SAWs are nonscheduled, event-driven products valid from the time of issuance to expiration or cancellation time. Valid times are in UTC. SPC will correct watches for formatting and grammatical errors.

When tornadoes or severe thunderstorms have developed, the local NWS WFO will issue the warnings for the storms.

27.16.4.1.1 Format and Example of a SAW

SPC forecasters may define the area as a rectangle or parallelogram (X miles either side of the line from point A to point B, or X miles north and south or east and west of the line from point A to point B). Distances of the axis coordinates should be in statute miles. The aviation coordinates referencing VOR locations and state distances will be in nautical miles. Valid times will be in UTC. The watch half width will be in statute miles. The SAW will contain hail size in inches or half inches (forecaster discretion for tornado watches associated with hurricanes) surface and aloft, surface convective wind gusts in knots, maximum cloud tops and the Mean Storm Motion Vector, and replacement information, if necessary.

```
WWUS30 KWNS 271559
SAW2
SPC AWW 271559
WW 568 TORNADO AR LA MS 271605Z - 280000Z
AXIS..65 STATUTE MILES EAST AND WEST OF LINE..
45ESE HEZ/NATCHEZ MS/ - 50N TUP/TUPELO MS/
..AVIATION COORDS.. 55NM E/W /18WNW MCB - 60E MEM/
HAIL SURFACE AND ALOFT..3 INCHES. WIND GUSTS..70 KNOTS. MAX TOPS TO 550. MEAN
STORM MOTION VECTOR 26030.
LAT...LON 31369169 34998991 34998762 31368948
THIS IS AN APPROXIMATION TO THE WATCH AREA. FOR A
COMPLETE DEPICTION OF THE WATCH SEE WOUS64 KWNS
FOR WOU2.
```

Table 27-15. Decoding an Aviation Weather Watch Notification Message

Line(s)	Content	Description
1	WWUS30 KWNS 271559	Communication header with issuance date/time
2	SAW2	NWS product type (SAW) and issuance number (2)
3	SPC AWW 271559	Issuing office Product type Issuance date/time
4	WW 568 TORNADO AR LA MS 271605Z - 280000Z	Watch number Watch type States affected Valid date/time period
5	AXIS..65 STATUTE MILES EAST AND WEST OF A LINE...	Watch axis
6	45ESE HEZ/NATCHEZ MS/ - 50N TUP/TUPELO MS/	Anchor points
7	...AVIATION COORDS..55NM E/W /18WNW MCB - 60E MEM/	Aviation coordinates
8–9	HAIL SURFACE AND ALOFT...3 INCHES. WIND GUSTS..70 KNOTS. MAX TOPS TO 550. MEAN STORM MOTION VECTOR 26030.	Type, intensity, max tops and mean storm motion using standard contractions.
10	(blank line)	
11	LAT...LON 31369169 34998991 4998762 31368948	Latitude and longitude coordinates
12	(blank line)	
13–15	THIS IS AN APPROXIMATION TO THE WATCH AREA. FOR A COMPLETE DEPICTION OF THE WATCH SEE WOUS64 KWNS FOR WOU2.	Notice that this is an approximation of the watch area and for users to refer to the referenced product for the actual area

27.16.4.2 Public Severe Thunderstorm Watch Notification Message

The SPC issues a Public Severe Thunderstorm Watch Notification Message when forecasting six or more hail events of 1 in (quarter-sized) diameter or greater or damaging winds of 50 kt (58 mph) or greater. The forecast event minimum threshold is at least 2 hours over an area of at least 8,000 mi^2. Below these thresholds, the SPC, in collaboration with affected NWS offices, may issue a watch for smaller areas and for shorter periods of time when conditions warrant, and for convective watches along coastlines, near the Canadian border, and near the Mexican border.

A Public Severe Thunderstorm Watch Notification Message contains three bulleted blocks of information:

- The geographic area of the watch,
- The valid time of the watch, and
- A description of the primary threats anticipated within the watch.

A plain text watch summary is included beneath the bulleted information followed by a more detailed description of the area and axis of the watch.

The SPC includes the term "adjacent coastal waters" when the watch affects coastal waters adjacent to the Pacific/Atlantic coast, the Gulf of Mexico, or the Great Lakes. Adjacent coastal waters refers to a WFO's near-shore responsibility (out to 20 NM for oceans), except for convective watches.

The SPC issues a watch cancellation message when no counties, parishes, independent cities, and/or marine zones remaining are in the watch area prior to the expiration time. The text of the message will specify the number and area of the cancelled watch.

27.16.4.3 Public Tornado Watch Notification Message

The SPC issues a Public Tornado Watch Notification Message when forecasting two or more tornadoes or any tornado that could produce EF-2 or greater damage. The forecast event minimum thresholds are at least 2 hours over an area at least 8,000 mi^2. Below these thresholds, the SPC, in collaboration with affected NWS offices, may issue a watch for smaller areas and for shorter periods of time when conditions warrant, and for convective watches along coastlines, near the Canadian border, and near the Mexican border.

A Public Tornado Watch Notification Message contains the following:

- The area description and axis,
- The watch expiration time,
- The term "damaging tornadoes,"
- A description of the largest hail size and strongest thunderstorm wind gusts expected,
- The definition of the watch,
- A call-to-action statement,
- A list of other valid watches,
- A brief discussion of meteorological reasoning, and
- Technical information for the aviation community.

The SPC may enhance a Public Tornado Watch Notification Message by using the words "THIS IS A PARTICULARLY DANGEROUS SITUATION" when there is a likelihood of multiple strong (damage of EF-2 or EF-3) or violent (damage of EF-4 or EF-5) tornadoes.

The SPC includes the term "adjacent coastal waters" when the watch affects coastal waters adjacent to the Pacific/Atlantic coast or the Gulf of Mexico. Adjacent coastal waters refers to a WFO's near-shore responsibility (out to 20 NM for oceans), which includes portions of the Great Lakes.

The SPC issues a watch cancellation message whenever it cancels a watch prior to the expiration time. The text of the message will specify the number and area of the cancelled watch.

27.17 Route Forecast (ROFOR)

The ROFOR product is no longer issued by the NWS and has been replaced by the GFA in the Pacific. See Section 28.2 for information on the GFA.

27.18 Aviation Forecast Discussion (AFD)

AFDs describe the weather conditions within a multistate or substate-sized area. They also may:

- Describe the weather conditions as they relate to a specific TAF or group of TAFs; and
- Provide additional aviation weather-related issues that cannot be encoded into the TAF, such as the reasoning behind the forecast.

AFDs are a free-form plain language text product. Common or well-known aviation weather contractions are used as well as local or regional geographic names, such as valleys, mountain ranges, and bodies of water.

Technically, the AFD is not a discrete product; it is the aviation section in the NWS WFO's AFD. The NWS AWC extracts the aviation section from the WFO's AFD and makes it available on the AWC's website under the Forecasts tab, titled "Avn. Forecast Disc." The aviation section of the AFD can also be found on the WFO's website under "Forecaster Discussion."

All WFOs in the CONUS, and most outside the CONUS, produce the aviation section of the AFD for their area of responsibility (see Figure 27-41). They are issued roughly every 6 hours and correspond to the issuance of TAFs from the respective NWS WFO. Each NWS office may tailor the format to meet the needs of their local aviation users.

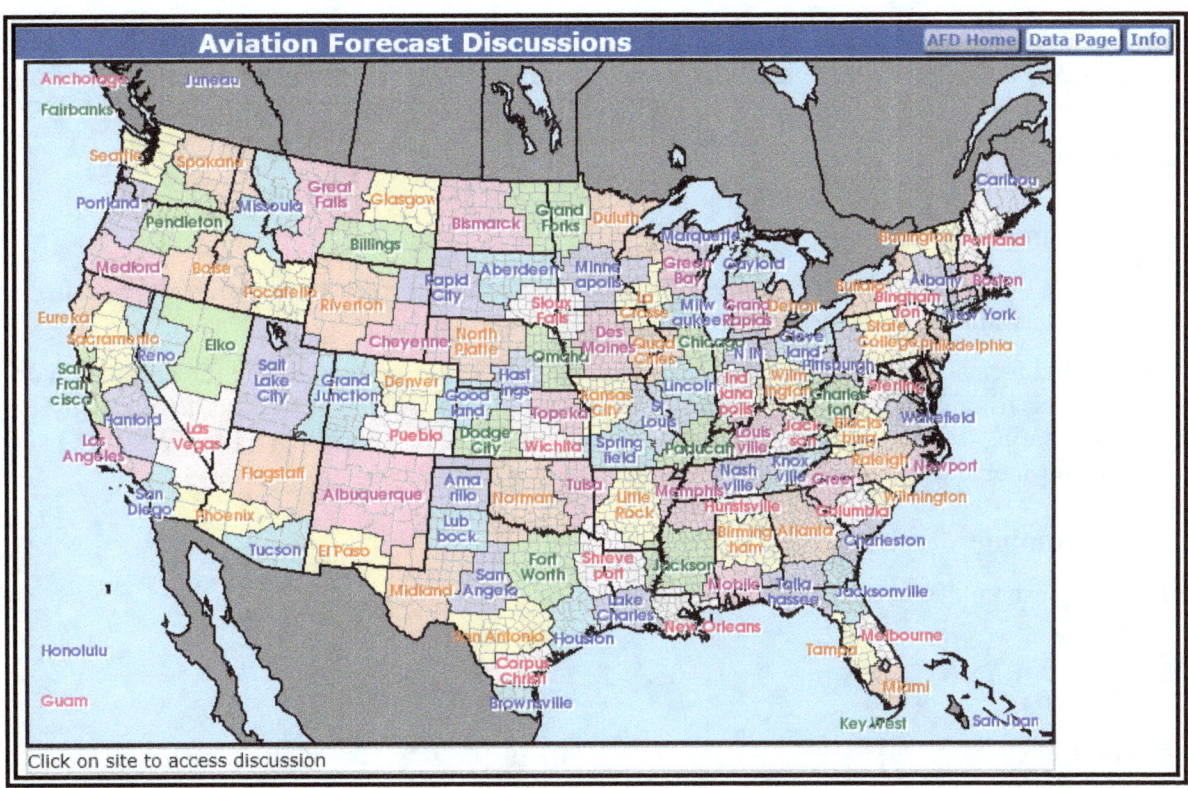

Figure 27-41. Map of NWS WFO's Area of Responsibility

27.18.1 Example

```
NWS Boise, ID
COLD FRONT CURRENTLY OVER SW IDAHO WEST OF THE MAGIC VALLEY. IFR IN HEAVIER
RAIN/SNOW SHOWERS BEHIND THE FRONT MOSTLY IN THE MTNS UNTIL THIS EVENING.
OTHERWISE LOW VFR THROUGH TOMORROW WITH ISOLATED SHOWERS INTO THE EVENING.
SURFACE WINDS...W TO NW WITH GUSTS 20-30 KTS...BECOMING 35-45 KTS IN THE UPPER
TREASURE AND MAGIC VALLEYS FOR A FEW HOURS THIS AFTERNOON...DROPPING DOWN TO
20-30 KTS OVERNIGHT INTO THE MORNING IN THE MAGIC VALLEY AND 5-10 KTS ELSEWHERE.
WINDS ALOFT NEAR 10K FT MSL...NW 30-40 KTS...40-50 KTS OVER THE UPPER TREASURE
AND MAGIC VALLEYS OVERNIGHT...BECOMING 20-30 KTS BY 15/12Z.
```

27.19 Meteorological Impact Statement (MIS)

The MIS is a nontechnical plain language product intended primarily for FAA traffic managers and those involved in planning aircraft routing. MISs are issued by the NWS CWSU.

MISs are available on the AWC's website as well as CWSU's websites.

The MIS valid times are determined according to local policy. The MIS is limited to a 48-hour valid period.

27.19.1 Example

```
ZAB MIS 02 VALID 281300-290300
...FOR ATC PLANNING PURPOSES ONLY...
AN UPPER-LVL DISTURBANCE OVER COLORADO COMBINED WITH A STRONG JET STREAM MOVING
ACROSS THE SWRN U.S. IS FCST TO PRODUCE AREAS OF TURBULENCE ACROSS PORTIONS OF
ZAB. THE TURBULENCE IS FCST TO SUBSIDE AFT 00Z AS THE DISTURBANCE AND JETSTREAM
MOVE FURTHER EAST.
```

27.20 Soaring Forecast

Select NWS WFOs issue soaring forecasts. These are automated forecasts primarily derived from the radiosonde observation or model-generated soundings.

The content and format of soaring forecasts vary with the NWS WFO providing the forecast, based on the needs of their soaring community. It is beyond the scope of this handbook to describe all of the many variations of soaring forecasts and their content. Soaring pilots should consult with the NWS WFO in their soaring area for more information.

27.20.1 Example

The following example is for Salt Lake City, Utah:

```
UXUS97 KSLC 091233
SRGSLC

Soaring Forecast
National Weather Service Salt Lake City, Utah
0633 MDT Tuesday, July 9, 2019

This forecast is for Tuesday, July 9, 2019:

If the trigger temperature of 81.4 F/27.4 C is reached...then
    Thermal Soaring Index...................... Excellent
    Maximum rate of lift....................... 1239 ft/min (6.3 m/s)
    Maximum height of thermals................. 17411 ft MSL (13185 ft AGL)
```

```
Forecast maximum temperature................... 88.0 F/31.6 C
Time of trigger temperature.................... 1200 MDT
Time of overdevelopment........................ None
Middle/high clouds during soaring window....... None
Surface winds during soaring window............ 20 mph or less
Height of the -3 thermal index................. 10097 ft MSL (5872 ft AGL)
Thermal soaring outlook for Wednesday 07/10.... Excellent

Wave Soaring Index............................. Not available

Remarks...

Sunrise/Sunset.................... 06:05:02 / 21:01:07 MDT
Total possible sunshine........... 14 hr 56 min 5 sec (896 min 5 sec)
Altitude of sun at 13:33:04 MDT... 70.27 degrees

Upper air data from rawinsonde observation taken on 07/09/2019 at 0600 MDT

Freezing level.................. 13975 ft MSL (9749 ft AGL)
Convective condensation level... 15400 ft MSL (11174 ft AGL)
Lifted condensation level....... 16064 ft MSL (11838 ft AGL)
Lifted index.................... -0.7
K index......................... +19.1
```

This product is issued once per day by approximately 0600 MST/0700 MDT (1300 UTC). This product is not continuously monitored nor updated after the initial issuance.

The information contained herein is based on the 1200 UTC rawinsonde observation at the Salt Lake City, Utah International Airport and/or numerical weather prediction model data representative of the airport. These data may not be representative of other areas along the Wasatch Front. Erroneous data such as these should not be used.

The content and format of this report as well as the issuance times are subject to change without prior notice.

27.21 Balloon Forecast

Select NWS WFOs issue balloon forecasts. These are automated forecasts primarily derived from the radiosonde observation or model-generated soundings.

The content and format of balloon forecasts vary with the NWS WFO providing the forecast, based on the needs of their ballooning community. It is beyond the scope of this handbook to describe all of the many variations of balloon forecasts and their content. Balloon pilots should consult with the NWS WFO in their area for more information.

27.21.1 Example

The following example is for St. Louis, Missouri:

```
SXUS43 KLSX 090850
RECSTL

Morning Hot Air Balloon Forecast
National Weather Service Saint Louis MO
350 AM CDT Tue Jul 9 2019
```

```
...HERE IS THE MORNING HOT AIR BALLOON FORECAST
FOR THE SAINT LOUIS METROPOLITAN AREA...

.THIS MORNING...
    Sunrise:                            545 AM.

    Surface Wind Forecast:              6 AM...110/05 mph.
                                        8 AM...120/05 mph.
    Latest Geostrophic Wind:            1 AM...160/15 mph.
    Boundary Layer Wind:                7 AM...120/06 mph.
                                       10 AM...160/12 mph.

    NWS Doppler Winds (at 0334 AM):  1000ft...130/15 mph.
                                     2000ft...150/22 mph.
                                     3000ft...160/20 mph.

    Surface Lifted Index:               7 AM...+3.
                                       10 AM...-1.

    Density Altitude:                   6 AM...1708 FT.
                                        8 AM...1948 FT.

    Additional Weather Info:         Slight chance of thunderstorms.

.OUTLOOK FOR THIS EVENING...
    Sunset:         829 PM.
    Weather:        Chance of thunderstorms.
    Surface Wind:   6 PM...150/07 mph.
                    8 PM...150/06 mph.
    Boundary Wind:  4 PM...160/14 mph.
                    7 PM...160/08 mph.
```

This forecast is not routinely updated or amended outside of scheduled issuances.

&&

This forecast is also available at phone number 636-441-8467 ext 5.

$$

TES

28 Aviation Weather Tools

28.1 Introduction

The fifth of five types of aviation weather information discussed in this handbook is aviation weather tools. Aviation weather tools are web-based applications that incorporate multiple weather products into a web-based interactive display.

This chapter will describe three web-based interactive displays: the GFA Tool, the Flight Service's Interactive Map, and the Helicopter Emergency Medical Services (HEMS) Tool.

This chapter describes these web-based displays as they are configured in 2022. Enhancements to these displays will likely occur in the coming years as technology improves. Users will find information on any enhancements on the providers' websites.

Note: The AWC Flight Path Tool is not included in this handbook since the NWS plans to retire this web-based tool.

28.2 Graphical Forecasts for Aviation (GFA) Tool

The GFA Tool is a set of web-based displays intended to provide the necessary aviation weather information to give users a complete picture of the weather that may impact their flight operations. It is a one-stop shop for multiple data fields. The AWC website includes observational data, forecasts, and advisories including thunderstorms, clouds, flight category, precipitation, icing, turbulence, and wind. Hourly model data and forecasts, including information on clouds, flight category, precipitation, icing, turbulence, wind, and graphical output from the NWS' NDFD, are available. Built with modern geospatial information tools, users can pan and zoom to focus on areas of greatest interest. An example is seen in Figure 28-1.

Descriptions of the observations, forecasts, and advisories that are used in the GFA Tool have been presented in other chapters within this handbook. This section does not show or describe the many possible overlays for the GFA Tool, since the AWC has produced a very thorough and complete description under the "Info" tab on the AWC's GFA Tool website, which also includes a tutorial video. In addition, the GFA Tool will continue to be enhanced; therefore, users will be assured of being informed of the latest GFA Tool information by referring to the "Info" tab on AWC's GFA Tool website.

Caution should be applied as users configure the GFA Tool for use. Users can turn on and turn off certain functions (e.g., AIRMET and SIGMET overlays for a given geographical area). This can lead to hidden areas of hazardous weather for a given flightpath.

Currently, the GFA Tool covers the CONUS, the Gulf of Mexico, the Caribbean, portions of the Atlantic Ocean, and portions of the Pacific Ocean, including the Hawaiian Islands and Alaska. The experimental GFA Tool became available for Alaska in December 2020 and transitioned into operational status in February 2022.

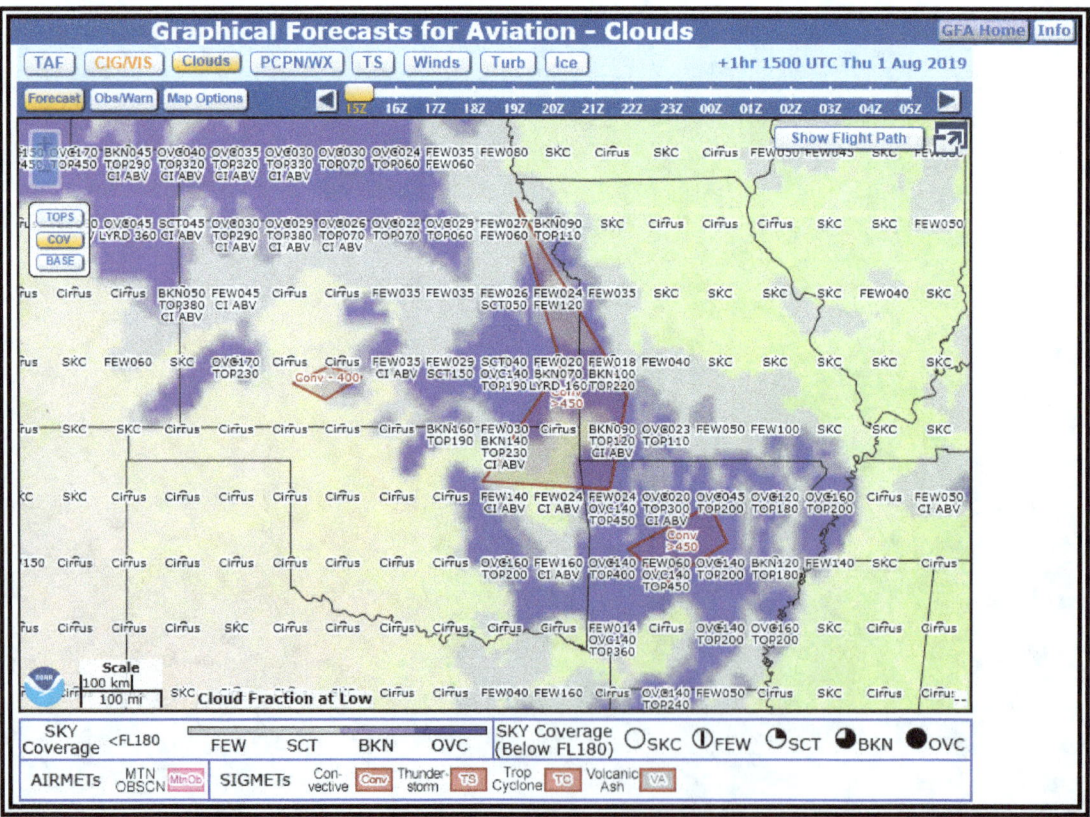

Figure 28-1. GFA Aviation Forecast for Clouds—Example

28.2.1 GFA Static Images

Some users with limited internet connectivity may access static images via the AWC's website as well as from Flight Service's website (see Appendix G, Websites). There are two static graphical images available titled, "Aviation Clouds Forecast" and "Aviation Surface Forecast." See Section 27.4 for more information.

28.3 FAA Flight Service's Interactive Map

Similar in presentation and functionality to the GFA Tool for weather data and products is the interactive map on Flight Service's website. It provides users with interactive graphical capabilities to view a variety of weather and aeronautical information. Users logged in to the website are able to enter basic route information to display the flightpath on the map. Two primary types of weather information can be displayed on the map: overlay data and weather imagery.

Overlay data includes the following and can be displayed simultaneously:

- Winds aloft,
- METARs,
- TAFs,
- PIREPs,
- SIGMETs,
- AIRMETs,
- CWAs, and
- NWS' severe weather watches and warnings.

Weather imagery includes the following and can only be displayed one product at a time:

- Radar (NEXRAD precipitation),
- Satellite,
- CIP/FIP, and
- GTG.

Certain products will enable additional controls, such as an FL slider, a time slider, and animation controls.

Information on the above products and imagery is presented in other chapters of this handbook, as well as in the User Guide provided on Flight Service's website.

28.4 Helicopter Emergency Medical Services (HEMS) Tool

The HEMS Tool overlays multiple fields of interest: ceiling, visibility, flight category, winds, relative humidity, temperature, icing, satellite, radar (Base and Composite Reflectivity), AIRMETs, SIGMETs, METARs, TAFs, and PIREPs. Data including temperature, relative humidity, winds, and icing potential is interpolated to AGL altitudes and can be sliced horizontally on 1,000-ft intervals up to 5,000 ft.

The display includes a rolling 2-hour archive of observed data (with 15-minute intervals) and hourly forecasts out to 6 hours. Overlays include roads, NAVAIDs, airports, and heliports for the entire United States. More detail is revealed as you zoom in, and individual layers can be turned on or off independently.

Additional information, including a tutorial, can be found on the AWC's website.

Note: The NWS plans to replace the HEMS Tool with the next upgrade of the GFA Tool. This upgrade will have a Low Altitude mode that focuses on the 1- to 5-hour timeframe and altitudes up to 5,000 ft AGL.

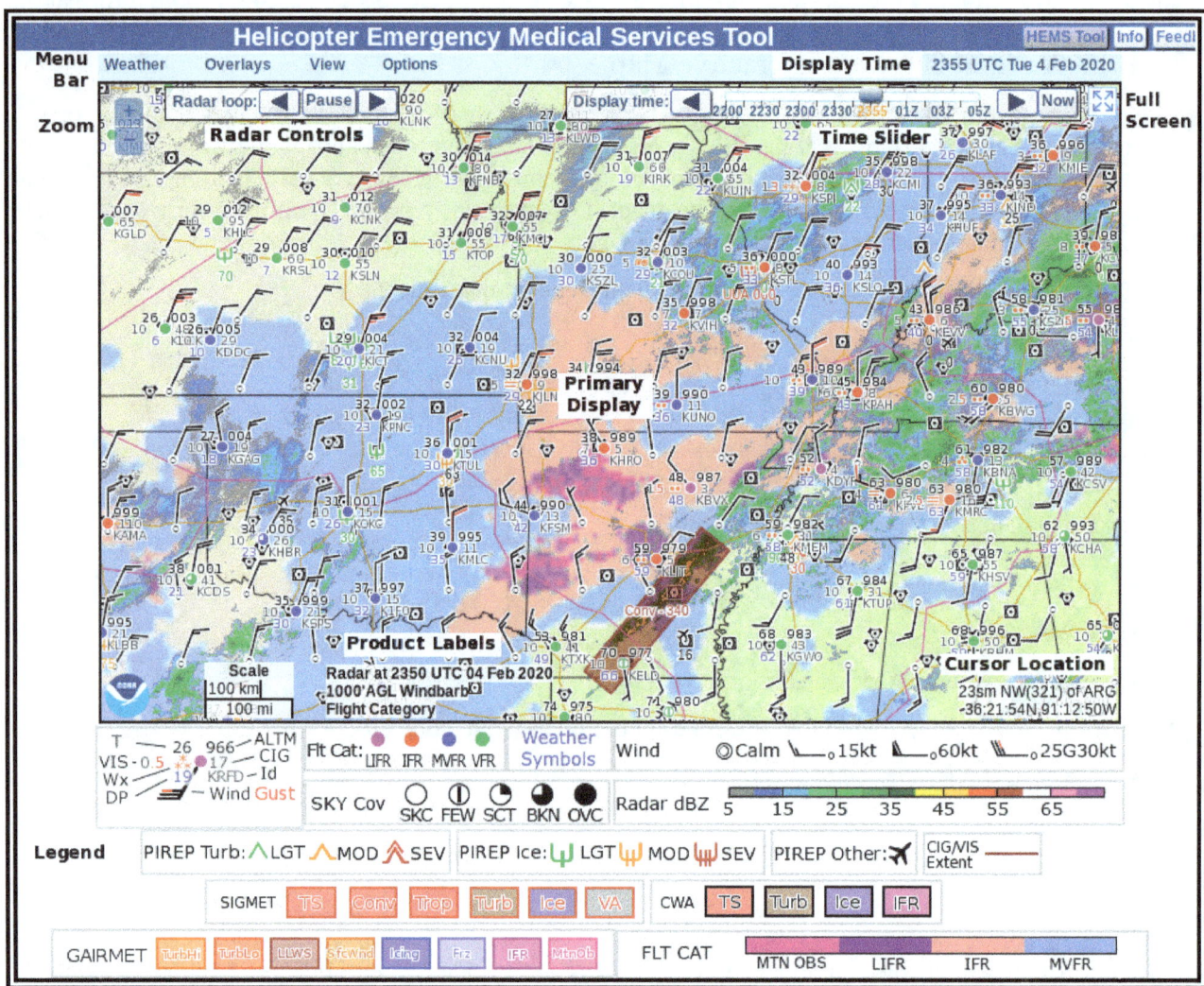

Figure 28-2. HEMS Tool—Example

Appendix A Cloud Types

In each level, the clouds may be divided by type (genera). The definitions of the cloud types (given below) do not cover all possible aspects, but are limited to a description of the main types and of the essential characteristics necessary to distinguish a given type from another type having a somewhat similar appearance.

Clouds are identified based upon one's observation point *at a particular elevation*. From sea level, one might observe stratus clouds enveloping the top of a mountain. However, on the mountaintop, one would observe and report that same cloud as fog.

The exception to this is cumulus or cumulonimbus clouds over a mountain. Even though these low-level clouds typically have bases under 6,500 ft (2,000 m), if one were to observe them over a mountaintop, their base might be 12,000 ft (3,600 m) or higher relative to a particular location closer to sea level. However, due to their location over the mountain, one would still call them cumulus or cumulonimbus clouds, as appropriate.

A.1 High Clouds

Cirrus (Ci), cirrocumulus (Cc), and cirrostratus (Cs) are high-level clouds. They are typically thin and white in appearance, but can appear in a magnificent array of colors when the Sun is low on the horizon. They are composed almost entirely of ice crystals.

A.1.1 Cirrus (Ci)

Cirrus (Ci) (see Figure A-1) is a cloud type composed of detached cirriform elements in the form of white, delicate filaments of white (or mostly white) patches, or of narrow bands. These clouds have a fibrous (hairlike) appearance and/or a silky sheen. Many of the ice crystal particles of cirrus are sufficiently large to acquire an appreciable speed of fall; therefore, the cloud elements often trail downward in well-defined wisps called "mares' tails." Cirrus clouds in themselves have little effect on aircraft and contain no significant icing or turbulence.

Figure A-1. Cirrus (Ci)

A.1.2 Cirrocumulus (Cc)

Cirrocumulus (Cc) is a cirriform cloud type appearing as a thin white patch, sheet, or layer of cloud without shading, and is composed of very small elements in the form of grains, ripples, etc. The elements may be merged or separate, and more or less regularly arranged; they subtend an angle of less than 1° when observed at an angle of more than 30° above the horizon.

Cirrocumulus (see Figure A-2) may be composed of highly supercooled water droplets, as well as small ice crystals, or a mixture of both; usually, the droplets are rapidly replaced by ice crystals. Pilots can expect some turbulence and icing.

Cirrocumulus is not very common. It is composed of very small elements, which never show shading. The cloud is frequently associated with cirrus or cirrostratus.

Figure A-2. Cirrocumulus (Cc)

A.1.3 Cirrostratus (Cs)

Cirrostratus (Cs) (see Figure A-3) is a cloud type appearing as a whitish veil, usually fibrous (hairlike) but sometimes smooth, that may totally cover the sky, and that often produces halo phenomena, either partial or complete. Cirrostratus occasionally may be so thin and transparent as to render it nearly indiscernible, especially through haze or at night. At such times, the existence of a halo around the Sun or Moon may be the only revealing feature.

The angle of incidence of illumination upon a cirrostratus layer is an important consideration in evaluating the identifying characteristics. When the Sun is high (generally above 50° elevation), cirrostratus never prevents the casting of shadows by terrestrial objects, and a halo might be completely circular. At progressively lower angles of the Sun, halos become fragmentary, and light intensity noticeably decreases. When near the horizon, cirrostratus may be impossible to distinguish from cirrus.

Cirrostratus clouds are composed primarily of ice crystals and contain little, if any, icing and no turbulence.

Figure A-3. Cirrostratus (Cs)

A.2 Middle Clouds

Altocumulus (Ac), altostratus (As), and nimbostratus (Ns) are mid-level clouds. They are composed primarily of water droplets; however, they can also be composed of supercooled liquid water droplets and/or ice crystals when temperatures are below freezing. Altostratus is usually found in the middle level, but it often extends higher. Nimbostratus is almost invariably found in the middle level, but it usually extends into the other levels.

A.2.1 Altocumulus (Ac)

Altocumulus (Ac) (see Figure A-4) is a cloud type, white and/or gray in color, that occurs as a layer or patch with a waved aspect, the elements of which appear as laminae, rounded masses, rolls, etc. These elements usually are sharply outlined, but they may become partly fibrous or diffuse; they may or may not be merged.

Small liquid water droplets invariably compose the major part of the composition of altocumulus. This results in a sharp outline and small internal visibility. At very low temperatures, however, ice crystals may form. Pilots flying through altocumulus can expect some turbulence and small amounts of icing.

The altocumulus in Figure A-4 is a single level, the greater part of which is sufficiently transparent to reveal the position of the Sun and Moon. The clouds do not progressively invade the sky, and the individual elements change very little. These thin altocumulus clouds usually produce a corona.

Figure A-4. Altocumulus (Ac)

A.2.1.1 Altocumulus Lenticularis

Altocumulus Lenticularis, commonly known as "Altocumulus Standing Lenticular (ACSL)" (see Figure A-5), are an orographic type of cloud. They often appear to be dissolving in some places and forming in others. They also often form in patches in the shape of almonds or wave clouds. These formations are caused by wave motions in the atmosphere and are frequently seen in mountainous or hilly areas. They may be triggered off by hills only a few thousand feet high and may extend downwind for more than 60 mi (100 km). The cloud elements form at the windward edge of the cloud and are carried to the downwind edge where they evaporate. The cloud as a whole is usually stationary or slow moving. These clouds often have very smooth outlines and show definite shading.

ACSL clouds indicate the position of the wave crests, but they do not necessarily give an indication on the intensity of turbulence or strength of updrafts and downdrafts. This is because the clouds depend on both lifting and moisture. A well-defined wave may be visible (i.e., ACSL cloud) in weak updrafts where there is an adequate supply of moisture, but may not be visible when the environment is very dry, even if the wave is intense.

Figure A-5. Altocumulus Standing Lenticular (ACSL)

A.2.2 Altostratus (As)

Altostratus (As) (see Figure A-6) is a cloud type in the form of a gray or bluish (never white) sheet or layer of striated, fibrous, or uniform appearance. Altostratus very often totally covers the sky and may, in fact, cover an area of several thousand square miles. The layer has parts thin enough to reveal the position of the Sun, and if gaps and rifts appear, they are irregularly shaped and spaced.

Within the rather large vertical extent of altostratus (from several hundred to thousands of feet), a very heterogeneous particulate composition may exist. In this most complete case, there may be distinguished:

1. An upper part, mostly or entirely ice crystals;

2. A middle part, a mixture of ice crystals and/or snowflakes and supercooled water droplets; and

3. A lower part, mostly or entirely supercooled or ordinary water droplets.

Note: A number of partial combinations of these composition types may occur, but never an entire cloud like item 3 above.

The particles are widely dispersed enough so as not to obscure the Sun except by its thickest parts, but rather to impose a ground-glass effect upon the Sun's image, and to prevent sharply outlined shadows from being cast by terrestrial objects. Halo phenomena do not occur. Pilots can expect little or no turbulence, but light to moderate icing in the supercooled water regions.

Thin altostratus usually evolves from the gradual thickening of a veil of cirrostratus. This means that unlike most clouds, which increase in height as they grow, altostratus (and nimbostratus) grow as the base of the cloud *lowers*. Altostratus is grayish or bluish color, never white, and the greater part is always translucent enough to reveal the Sun (or Moon) as through ground glass. Objects on the ground do not cast shadows and halo phenomena are never seen.

Figure A-6. Thin Altostratus (As)

A.2.3 Nimbostratus (Ns)

Nimbostratus (Ns) (see Figure A-7) is a gray cloud layer, often dark, rendered diffuse by more or less continuously falling rain, snow, ice pellets, etc., which in most cases reaches the ground. It is not accompanied by lightning, thunder, or hail.

Nimbostratus is composed of suspended water droplets, sometimes supercooled, and of falling raindrops and/or snow crystals or snowflakes. It occupies a layer of large horizontal and vertical extent. The great density and thickness (usually many thousands of feet) of this cloud prevent observation of the Sun. This, plus the absence of small droplets in its lower portion, gives nimbostratus the appearance of dim and uniform lighting from within. It also follows that nimbostratus has no well-defined base, but rather a deep zone of visibility attenuation. Frequently, a false base may appear at the level where snow melts into rain. It is officially classified as a middle cloud, although it may merge into very low stratus or stratocumulus. Other cloud classification systems may identify it as a low-level cloud. Nimbostratus produces very little turbulence but can pose a serious icing problem if temperatures are near or below freezing.

Thick altostratus is denser and of a darker gray or bluish gray than thin altostratus with the greater part sufficiently dense to completely mask the Sun or Moon. With further thickening of the altostratus and a lowering of its base, the cloud may begin to produce precipitation, at which point it is called nimbostratus. Some cloud charts will depict nimbostratus as a low-level cloud. This is because oftentimes, during continuously falling precipitation, the base of nimbostratus clouds decrease into the low level. But officially and historically, nimbostratus is classified as a mid-level cloud.

Figure A-7. Thick Altostratus (As) or Nimbostratus (Ns)

A.3 Low Clouds

Cumulus (Cu), towering cumulus (TCu), stratocumulus (Sc), stratus (St), and cumulonimbus (Cb) are low clouds composed of water droplets. However, they can also be composed of supercooled liquid water droplets and/or ice crystals when temperatures are below freezing. Cumulus and cumulonimbus usually have bases in the low level, but their vertical extent is often so great that their tops may reach into the middle and high levels.

A.3.1 Cumulus (Cu) and Towering Cumulus (TCu)

Cumulus is a cloud type in the form of individual, detached elements that are generally dense and possess sharp, nonfibrous outlines. These elements develop vertically, appearing as rising mounds, domes, or towers, the upper parts of which often resemble a cauliflower. The sunlit parts of these clouds are mostly brilliant white; their bases are relatively dark and nearly horizontal. Near the horizon, the vertical development of cumulus often causes the individual clouds to appear merged. If precipitation occurs, it is usually of a showery nature. Various effects of wind, illumination, etc., may modify many of the aforementioned characteristics.

Cumulus is composed of a great density of small water droplets, frequently supercooled. Within the cloud, larger water drops are formed that may, as the cloud develops, fall from the base as rain or virga.[13] Ice crystal formation will occur within the cloud at sufficiently low temperatures, particularly in upper portions as the cloud grows vertically.

For cumulus with little vertical development (see Figure A-8), pilots can expect some turbulence and no significant icing. However, for towering cumulus (i.e., cumulus of moderate/strong development), pilots can expect very strong turbulence and some clear icing above the freezing level (where temperatures are negative). Towering cumulus (see Figure A-9) is also referred to as the first stage of a thunderstorm.

---

[13] Virga are wisps or streaks of water or ice particles falling out of a cloud, but vaporizing before reaching the Earth's surface as precipitation.

Figure A-8. Cumulus (Cu) with Little Vertical Development

Cumulus formation is often preceded by hazy spots out of which the clouds evolve. When completely formed, the clouds have clear-cut horizontal bases and flattened or slightly rounded tops. At this stage of development, they are known as fair weather cumulus. Over land, on clear mornings, cumulus may form as the Sun rapidly heats the ground. Near coasts, cumulus may form over the land by day in a sea breeze and over the sea during the night in a land breeze.

Figure A-9. Towering Cumulus (TCu)

A.3.2 Stratocumulus (Sc)

Stratocumulus (Sc) (see Figure A-10) is a cloud type, predominantly stratiform, in the form of a gray and/or whitish layer or patch, which nearly always has dark parts and is nonfibrous (except for virga). Its elements are tessellated, rounded, roll-shaped, etc.; they may or may not be merged, and usually are arranged in orderly groups, lines, or undulations, giving the appearance of a simple (or occasionally a cross-pattern) wave system. These elements are generally flat-topped, smooth, and large; observed at an angle of more than 30° above the horizon, the individual stratocumulus element subtends an angle of greater than 5°. When a layer is continuous, the elemental structure is revealed in true relief on its undersurface.

Stratocumulus is composed of small water droplets, sometimes accompanied by larger droplets, soft hail, and (rarely) snowflakes. Under ordinary conditions, ice crystals are too sparse even to give the cloud a fibrous aspect; however, in extremely cold weather, ice crystals may be numerous enough to produce abundant virga and sometimes even halo phenomena. The highest liquid water contents are in the tops of these clouds where the icing threat is the greatest, if cold enough. Virga may form under the cloud, particularly at very low temperatures. Precipitation rarely occurs with stratocumulus.

Pilots can expect some turbulence and possible icing at subfreezing temperatures. Ceiling and visibility are usually better than with low stratus.

Figure A-10. Stratocumulus (Sc)

The stratocumulus in Figure A-10 occurs in patches or layers, composed of rounded masses or rolls, at one or more levels. The clouds are gray or whitish and always have dark parts. Sometimes the elements lie in parallel bands. Due to perspective, these may appear to converge towards the horizon. It may also occur in the shape of lenses or almonds, although this is fairly rare.

A.3.3 Stratus (St)

Stratus (St) (see Figure A-11 and Figure A-12) is a cloud type in the form of a gray layer with a fairly uniform base. Stratus does not usually produce precipitation, but when it does occur, it is in the form of minute particles, such as drizzle, ice crystals, or snow grains. Stratus often occurs in the form of ragged patches or cloud fragments, in which case rapid transformation is a common characteristic. When the Sun is seen through the cloud, its outline is clearly discernible. In the immediate area of the solar disk, stratus may appear very white. Away from the Sun, and at times when the cloud is sufficiently thick to obscure it, stratus gives off a weak, uniform luminance.

The particulate composition of stratus is quite uniform, usually of fairly widely dispersed water droplets and, at lower temperatures, of ice crystals (although this is much less common). Halo phenomena may occur with this latter composition.

Stratus produces little or no turbulence, but temperatures near or below freezing can create hazardous icing conditions. When stratus is associated with fog or precipitation, the combination can become troublesome for visual flying.

Stratus most commonly occurs as a gray, fairly uniform, and featureless single layer of low cloud. Occasionally, it can be dark or threatening, although at most it can only produce weak precipitation. This feature makes it fairly easy to distinguish it from nimbostratus (Ns), which nearly always produces rain, snow, or ice pellets. Fog will often lift into a layer of stratus by an increase in wind or a rise in temperature. Stratus is sometimes comparatively thin, and the disk of the Sun or Moon may be seen with a clear outline.

Figure A-11. Stratus (St)

Stratus fractus (StFra) and/or cumulus fractus (CuFra) (see Figure A-12) are ragged shreds of low clouds that always appear in association with other clouds for a short time before, during, and a short time after precipitation. They often form beneath lowering altostratus (As) or nimbostratus (Ns). They also occur beneath cumulonimbus (Cb) and precipitating cumulus (Cu) and are collectively known as "scud clouds."

Figure A-12. Stratus Fractus (StFra) and/or Cumulus Fractus (CuFra) of Bad Weather

A.3.4 Cumulonimbus (Cb)

Cumulonimbus (Cb) (see Figure A-13 and Figure A-14) is a cloud type, exceptionally dense and vertically developed, occurring either as isolated clouds or as a line or wall of clouds with separated upper portions. These clouds appear as mountains or huge towers, at least a part of the upper portions of which are usually smooth, fibrous, or striated, and almost flattened as it approaches the tropopause. This part often spreads out in the form of an anvil or vast plume. Under the base of cumulonimbus, which is often very dark, there frequently exist virga, precipitation, and low, ragged clouds, either merged with it or not. Its precipitation is often heavy and always of a showery nature. The usual occurrence of lightning and thunder within or from this cloud leads to its popular appellations: thundercloud, thunderhead (the latter usually refers only to the upper portion of the cloud), and thunderstorm.

Cumulonimbus is composed of water droplets and ice crystals, the latter almost entirely in its upper portions. It also contains large water drops, snowflakes, snow pellets, and sometimes hail. The liquid water forms may be substantially supercooled. Cumulonimbus contains nearly the entire spectrum of flying hazards, including extreme turbulence.

In Figure A-13, no part of the cloud top has acquired a fibrous appearance or any anvil development. The protuberances tend to form a whitish mass without striations. Showers or thunderstorms may occur. The presence of lightning/thunder differentiate between this cloud and towering cumulus. Cumulus, stratocumulus, or stratus may also be present.

Figure A-13. Cumulonimbus (Cb) Without Anvil

In Figure A-14, the characteristic shape of these clouds can only be seen as a whole when viewed from a distance. The tops of these massive clouds show a fibrous or striated structure that frequently resembles an anvil, plume, or huge mass of hair. They may occur as an isolated cloud or an extensive wall and squalls, hail, and/or thunder often accompany them. Underneath the base, which is often very dark, stratus fractus frequently form and, in storms, these may be only a few hundred feet above the Earth's surface, and they can merge to form a continuous layer. Mammatus may form, especially on the underside of the projecting anvil, and may appear particularly prominent when the Sun is low in the sky. A whole variety of other clouds, such as dense cirrus, altocumulus, altostratus, stratocumulus, cumulus, and stratus may also be present.

Figure A-14. Cumulonimbus (Cb) with Anvil

Appendix A, Cloud Types

Appendix B Standard Conversion Chart

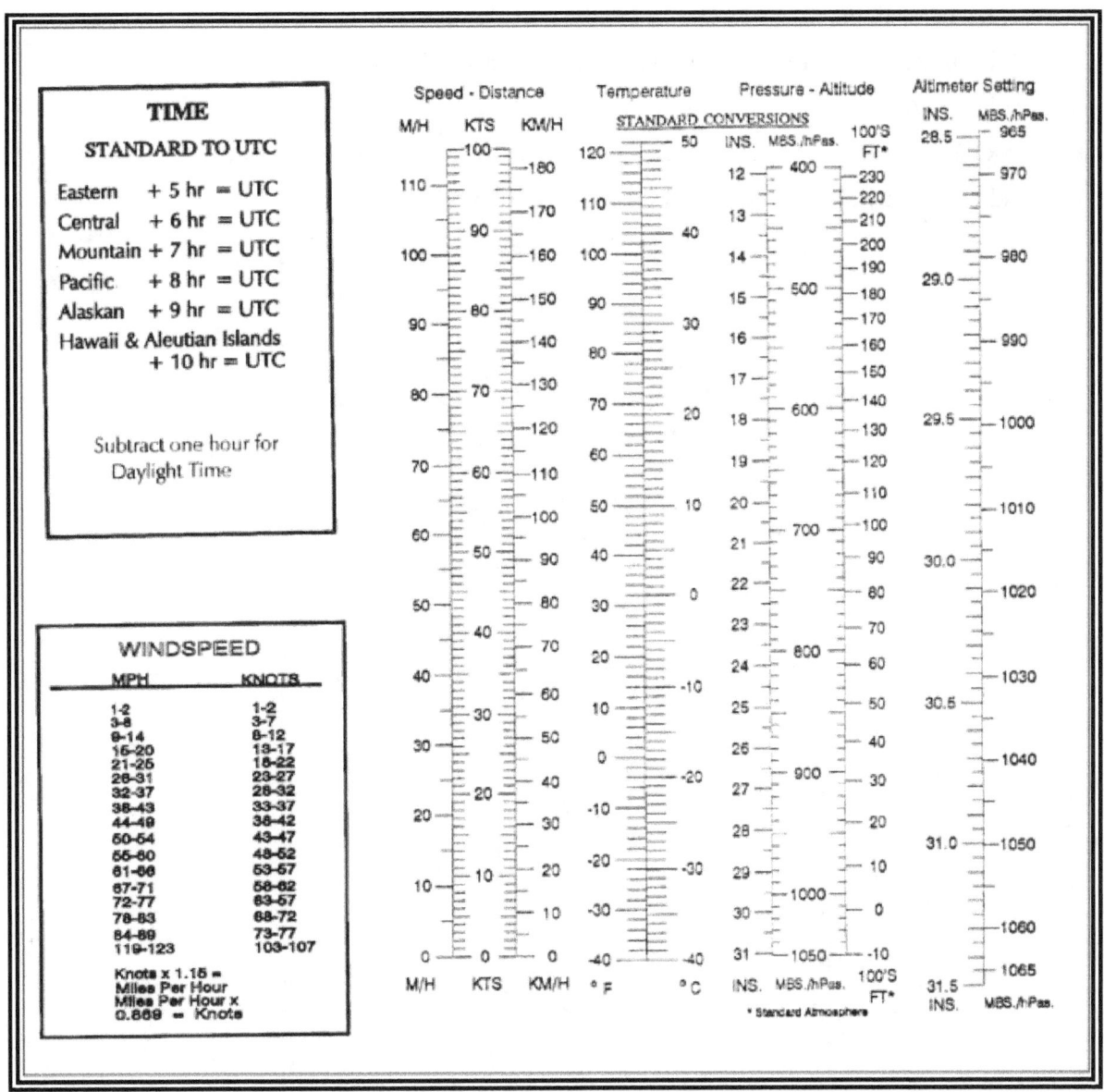

Figure B-1. Standard Conversion Chart

Appendix C Density Altitude Calculation

To determine density altitude:

1. Set the aircraft's altimeter to 29.92 inHg. The altimeter will indicate pressure altitude.
2. Read the OAT.
3. Mark the intersection of pressure altitude (diagonal) and temperature (vertical) lines on Figure C-1.
4. Read the density altitude from the horizontal lines on Figure C-1.

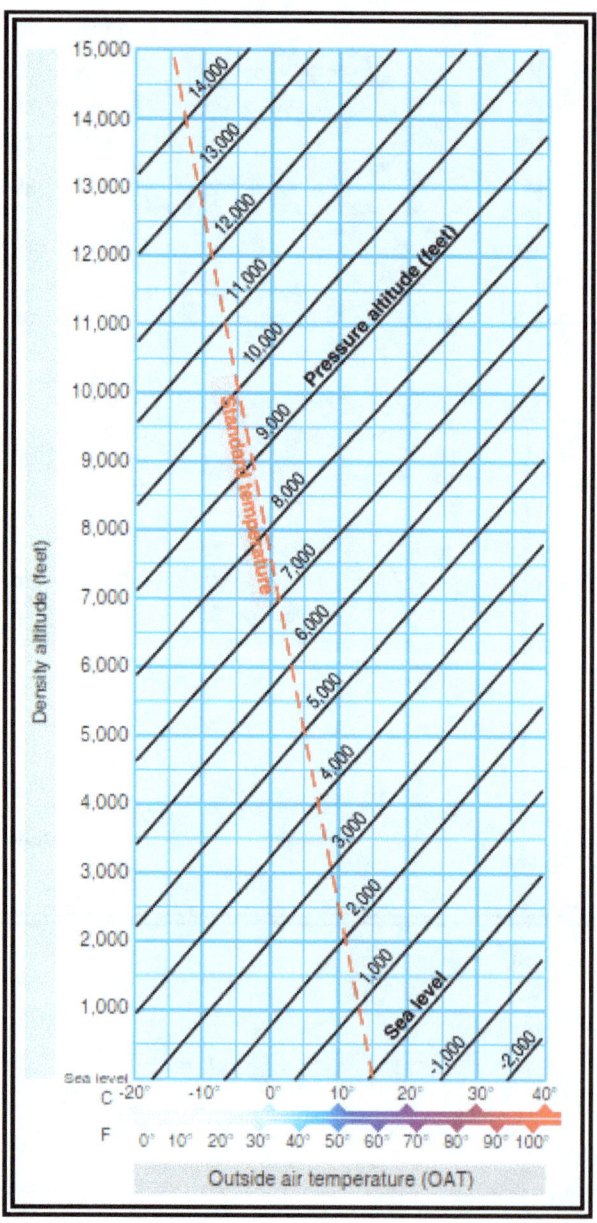

Figure C-1. Density Altitude Computation Chart

Appendix C, Density Altitude Calculation

Appendix D Special Terminal Operation Procedures (STOP) for Operations in a Noncontiguous State

D.1 Introduction

Air carriers often need to release flights to locations with missing weather reports (e.g., surface observations such as METAR), missing elements of weather reports, or locations that do not have an ASOS or AWOS facility or human weather observers.

The STOP program is designed for 14 CFR part 121 domestic and supplemental operations at airports in a noncontiguous state (e.g., Alaska) without a complete surface observation and/or forecast.

This guidance provides acceptable methods and guidelines for the certificate holder (CH) to use to develop the STOP program. The Aviation Safety Inspector—Aircraft Dispatcher (ASI-AD) and the POI may refer to this guidance as they review part 121 domestic or supplemental operations conducted in a noncontiguous state to airports with no METARs, missing METAR elements, or no weather reporting facilities.

D.2 Weather Information

D.2.1 General

In accordance with part 121, § 121.599(a), no aircraft dispatcher may release a flight unless thoroughly familiar with reported and forecast weather conditions on the route to be flown. Section 121.601 requires the aircraft dispatcher to provide the PIC with all available current reports or information on airport conditions and irregularities of navigation facilities that may affect the safety of the flight. Section 121.601 applies to every phase of flight. In accordance with § 121.599(b), no PIC may begin a flight unless thoroughly familiar with reported and forecast weather conditions on the route to be flown. Section 121.597 refers to the individuals authorized by the CH to exercise operational control over the flight and to execute a flight release setting forth the conditions under which the flights will be conducted.

IFR operations under part 121 require current weather reports and forecasts. Those approved sources are official surface observation (e.g., METAR) and a TAF. If no NWS TAF is available, an approved EWINS TAF may be used. Refer to FAA Order 8900.1 Volume 3, Chapter 26, Section 4, Enhanced Weather Information Systems, for additional information about EWINS.

Landing limitations and aircraft performance require a CH to consider forecast (anticipated/probable) winds at the destination airport. Therefore, weather forecasts that include wind information for the destination airport are always required.

D.2.2 Supplements to NWS or EWINS TAFs

The AAG forecast product provides TAF-like information for locations where an NWS or EWINS TAF is not available at the intended destination in Alaska. AAG is intended for use under VMC. Operators conducting operations under part 121 cannot use AAG at alternate airports. The AAG is also limited for flight times of 2 hours or less in duration. The AAG, in conjunction with METARs, PIREPs, the FA, SIGMETs, and AIRMETs, helps inform operators of the weather conditions at the ETA and whether those conditions will be at or above the minimums, which determines whether operations can be conducted within the performance limitations of the aircraft per § 121.101. In addition, the AAG does not indicate blowing snow or convective activity. If precipitation is forecasted, other products, including SIGMETs and NEXRAD weather radar, where available, can determine if hazardous convective weather will be present when the aircraft arrives. InFO 20002 provides more information on the AAG.

Table D-1 lists weather products that supplement NWS or EWINS TAFs.

Table D-1. Weather Products to Supplement NWS or EWINS TAFS

Weather Product		Location in Handbook (Section)	Website
AIRMET		26.3	https://www.weather.gov/aawu/airmets https://www.weather.gov/aawu/
Alaska Aviation Guidance (AAG)		27.15.1	https://www.weather.gov/arh/aag
Alaska Graphical Forecasts	AAWU Flying Weather	27.6.1	https://www.weather.gov/aawu/#map
	Alaska Surface Forecast	27.6.2	
	Alaska Icing Forecast	27.6.3	
	Alaska Turbulence Forecast	27.6.4	
	Alaska Convective Outlook	27.6.5	
Alaska Significant Weather Forecasts		27.8.4	https://www.weather.gov/aawu/4panelsigwx
Area Forecast (FA)[1]		27.5	https://weather.gov/aawu/AreaForecasts
Center Weather Advisory (CWA)		26.4	https://aviationweather.gov/cwamis
FAA Aviation Weather Cameras		24.9	https://weathercams.faa.gov/
Graphical Forecasts for Aviation (GFA)		28.2	https://aviationweather.gov/gfa
Localized Aviation Model-Output Statistic (MOS) Product (LAMP)		27.15	https://www.nws.noaa.gov/mdl/gfslamp/stnplots.php?&region=Alaska&elm=flight
Pilot Weather Reports (PIREP)		24.5.1	https://aviationweather.gov/airep
Real-Time Mesoscale Analysis (RTMA)		25.7	https://nomads.ncep.noaa.gov/pub/data/nccf/com/rtma/prod/airport_temps/
SIGMET		26.2	https://weather.gov/aawu/sigmets https://aviationweather.gov/sigmet

Weather Product		Location in Handbook (Section)	Website
Surface Observations[2]	METAR	24.4.1	https://www.weather.gov/aawu/obs https://www.weather.gov/aawu/
	SPECI	24.4.2	
	ASOS	24.3.1	
	AWOS	24.3.2	

Notes:

1. *An FA alone, without a valid TAF, is not an acceptable standalone forecast to satisfy the relevant part 121 weather reporting requirements. The FA is not intended for use in operational decision making or as a substitute for a TAF. However, the FA may provide vital weather information for the entire en route portion of flight, from point of origin to destination and all applicable alternates.*
2. *Aircraft performance is predicated on aircraft weight, ambient temperature, wind direction, wind speed, and altimeter setting. In order to ensure aircraft takeoff and landing limitations are met and obstacles are avoided, a surface observation contains at least the following elements:*
   - *The station identifier (e.g., airport code),*
   - *Date and time of observation (to establish relevance of the report),*
   - *Ambient air temperature at the station (e.g., airport or seaport),*
   - *Wind speed and direction,*
   - *Altimeter setting at the station (unless the current published approach plate lists an alternate source),*
   - *Visibility (for obstacle avoidance and IAPs), and*
   - *Ceiling.*

D.2.3 Manual Surface Observations by Flightcrew Using a Portable Weather Observation Device at STOP Airports

D.2.3.1 Portable Weather Observation Device

A portable weather observation device is issued to the flightcrews who conduct STOP flights as part of the preflight preparation. An acceptable portable weather observation device provides the following:

1. Date and time.
2. Wind direction (degrees magnetic north).
3. Wind speed (knots).
4. Temperature (°C).
5. Dewpoint temperature (°C).
6. Altimeter or pressure (inHg).

Note: Apply the appropriate variation to get the wind direction in true north, which is needed for documentation and reporting.

Calibration of the device is done in accordance with the manufacturer's procedures. A log with the date of last battery change, latest calibration date, the result (satisfactory or unsatisfactory), and the name of the individual who performed the calibration should be kept with the device.

D.2.3.2 Visibility Estimation by the Pilot

Estimates of visibility can be made using the runway length (e.g., a 5,300-ft runway is 1 sm, geographical features or prominent obstructions at a known distance, markers placed for the purpose of measuring distance visually or with FAA cameras).

D.2.3.3 Ceiling Estimation by the Pilot

Ceilings can be estimated using geographical features or prominent obstructions of a known elevation of height. Ceilings can be estimated using:

- Known geographic points;
- The conditions on arrival, provided they have not radically changed; and
- Temperature/dewpoint spread, which is the difference between the temperature and dewpoint, divided by 2.5, multiplied by 1,000.
    - If the temperature is 18 °C and the dewpoint 9 °C, a difference of 9 °C exists.
    - This result indicates that the lowest cloud layer should be 3,600 ft AGL.
    - Comparing this to a visual assessment, the forecast (TAF (NWS or EWINS)), and PIREPs (your own on arrival or provided by other aircraft) should provide reasonable estimated ceiling information.

D.2.4 Documentation and Reporting

Prior to departure, the PIC records all required elements on the weather record log, then communicates with the responsible aircraft dispatcher or flight follower. A mutual agreement is achieved on these conditions for a flight release. In this process, the PIC advises, and the aircraft dispatcher records the following information:

- Station identifier.
- Date and time.
- Wind direction (degrees true north).
- Wind speed (knots).
- Visibility (sm).
- Ceiling (BKN, OVC).
- Temperature (°C).
- Dewpoint temperature (°C).
- Altimeter (inHg).

This information is part of the dispatch and flight release documents and is retained in the company's 3-month retention file.

D.3 Airport Data

D.3.1 Data Maintenance

All airports and runways on which STOP operations are conducted are documented in the CH's Safety Assurance System (SAS), which is maintained and updated by active surveillance.

D.3.2 Airport Analysis Program

For airports that do not appear in the Alaska Chart Supplement, the CH uses an airport analysis program to ensure the facility is adequate for the proposed STOP operation. This may require a full airport risk assessment, strip check, or airport condition report to be completed. If the airport appears in the Alaska Chart Supplement but runway conditions are not monitored, only the runway contaminants section of the strip check or an airport condition report is completed. The CH determines how this will be accomplished and requests the POI's approval.

For airports not listed in the Alaska Chart Supplement, or non-part 139 in the State of Alaska only, the airport risk assessment, strip check, or airport condition report includes airport pictorials to familiarize the flightcrew with the area. Suitable pictorials include pictures taken from the air or from the ground, satellite pictures, and excerpts from aeronautical charts, etc.

Flight operations to airports with weather reporting deficiencies are thoroughly evaluated to ensure the facility is adequate for the proposed operation. For all STOP flight operations to temporary runways (frozen bodies of water or flat open terrain), the CH may need to accomplish a physical, onsite inspection of the runway and facilities and complete the company's documented report.

The dispatcher, flight follower, and flightcrew member use the completed airport analysis, airport risk assessment, strip check, or airport condition report during the crew briefing. If during the crew briefing the PIC, dispatcher, or flight follower determines the conditions at the destination airport are not adequate for safe operations, the flight is not released.

D.3.3 Civil Twilight Considerations

Flight operations to airports with weather reporting deficiencies are initiated unless the ETA and estimated time of departure falls within the hours of the start of morning civil twilight and the end of evening civil twilight. Dispatchers or flight followers note these hours in the "Remarks" section of the dispatch or flight release.

Civil twilight is determined by the following U.S. Military database: https://aa.usno.navy.mil/data/RS_OneYear.

Night operations may be authorized if the airport is equipped with lighting at least equivalent to medium intensity edge lights (easily identifiable as lighting during daylight conditions). No operations outside the start of morning civil twilight and the end of evening civil twilight will be conducted where this lighting standard is not met.

The flightcrew determines the wind direction from an illuminated wind direction indicator or ground communications before landing.

If temporary lighting (e.g., flare pots) are used, they are evenly spaced on each side of the runway, to replace at least 50 percent of the normal lights. If permanent lighting is not installed at the airport, the temporary light spacing is 200 ft apart. Four lights are placed at each runway end to clearly mark the threshold and runway end. Lights are turned on 30 minutes prior to arrival and for 30 minutes after departure.

The Director of Operations (DO), PIC, dispatcher, or flight follower should agree that the lighting is adequate for the proposed operation, with the approval by the DO or designee annotated on the flight release. Temporary lighting may be placed and illuminated by qualified company personnel who are familiar with these procedures.

D.4 Flight Planning

D.4.1 General

Flight operations to airports with weather reporting deficiencies are operated under IFR with the exception of operations in the terminal area. In the terminal area of the STOP airport, flightcrews maintain the basic cloud clearance as specified in § 91.155, and maintain the minimum altitudes prescribed in § 91.126, § 91.129, § 91.130, or § 91.131, as applicable for the airspace class in which the flight is operated.

D.4.2 IFR Flight Plan

Both domestic and supplemental flights under STOP are planned as an IFR flight plan when the intended flight is 50 NM or more from origin to destination.

D.4.3 Composite IFR/VFR Flight Plan

When the flight is to be greater than 50 NM but will exit the IFR structure within 50 NM of the destination under VFR, a composite flight plan is filed. The IFR portion will change to a VFR transition point, at which the PIC will activate the VFR flight plan.

D.4.4 VFR Flight Plan

Only to be used when the intended STOP flight is less than 50 NM from origin to destination. Only Group I airplanes (propeller-driven, including reciprocating and turbopropeller-powered) operate under a part 121 VFR flight plan.

Note: For a VFR flight, this distance will consider the route most likely to be flown. For example, a 50 NM direct route would not be appropriate to consider if high terrain would make point-to-point direct impracticable. In this case, the flight would be released on an IFR flight plan.

D.4.5 Fuel Requirements

Domestic and supplemental STOP flights, regardless of aircraft type, are not released without considering fuel to meet the requirements of § 121.647. This includes a landing area visual inspection (flyover) of 15 additional minutes of the intended landing strip. This fuel is noted on the release in a manner which is readily identifiable to the flightcrew. The CH determines according to aircraft type if an additional 15 minutes of fuel is sufficient, or adjust accordingly to accommodate a possible flyover pass of the intended airstrip or airport.

D.4.6 Weather Minimums at STOP Destination Airports

The Administrator considers the ETA at the destination and the arrival minimums that are listed below, to contribute to mitigating risks and hazards to STOP destinations, and will allow the CH to operate at the highest level of safety. These weather minimums apply regardless if the whole weather report is missing or only missing a required element.

D.4.6.1 Domestic Operations

Forecast weather conditions 1 hour before and 1 hour after ETA at the STOP destination airport to be at least a cloud ceiling of 3,500 ft above the airport elevation and visibility of 5 sm, and appropriate wind velocity and direction.

D.4.6.2 Supplemental Operations

Forecast weather conditions 1 hour before and 1 hour after ETA at the STOP destination airport to be at least a cloud ceiling of 3,000 ft AGL and visibility of 3 sm, and appropriate wind velocity and direction. If the STOP destination airport has an approved IAP the required release minimums are a ceiling of 1,500 ft above the airport elevation and visibility of 3 sm 1 hour before and 1 hour after ETA.

D.4.7 Required Destination Primary Alternate

A destination's primary alternate is always included for both domestic and supplemental operations when the STOP process is applied. It always includes the appropriate fuel requirements for the alternate. The destination primary alternate must have a TAF and all required reporting elements of a METAR, and meet the air carrier's alternate minimum requirements.

D.4.8 Designation of Secondary Destination Alternate

When weather conditions at the destination airport and the first alternate airport are marginal, § 121.619(a) requires that at least one additional alternate airport be designated. The term "marginal," as applied to domestic alternate minimums, is not defined by regulation. To some extent this is because the definition of what constitutes marginal depends on the nature of the weather phenomena, the operation conducted, and the equipment used.

If marginal conditions at the primary alternate are forecast, the CH must, per §§ 121.619, 121.623, and 121.625, list a second alternate in the dispatch or flight release the flightcrew briefed prior to release. When marginal conditions are present at destination and primary alternates, the secondary alternate may not be less than 2,000 ft above the airport elevation and the visibility must be at least 3 sm. Both the primary and the secondary alternates must have a complete METAR and TAF at ETA.

D.4.9 Minimum Equipment List (MEL) Restrictions

When required by an MEL, flight operations to airports with weather reporting deficiencies are not to be initiated with, or continue with, the following items deferred or inoperative:

- Any long-range navigation systems (LRNS) (e.g., GPS) that may be installed on the aircraft. In a dual installation, one must be operational.
- OAT reporting system installed on the aircraft.
- One or both flight management system(s) (FMS).
- One or both radio altimeter system(s).
- Ground proximity system.
- Satellite voice communication available over the entire route and at any destination airport.

D.5 Preflight Planning and Briefing to STOP Airport

D.5.1 Preflight Planning

See Sections D.2, D.3, and D.4 for weather, airport data, and flight planning information.

D.5.2 Preflight Briefing

Before releasing any flight, both the aircraft dispatcher or flight follower and the PIC should become thoroughly familiar with the reported and forecast weather conditions, including adverse weather phenomena, the status of communications, navigation, and airport facilities. This requires the dispatcher or flight follower to provide the PIC with information on each of these items. It also requires a preflight briefing between the dispatcher or flight follower and the PIC.

The preflight briefing is either delivered verbally or in writing. Communication facilities must be available for the aircraft dispatcher or flight follower and the PIC to communicate directly by voice if direct communication is required or desired. If communications are conducted verbally, the preflight briefing communication is documented.

Prior to conducting a STOP flight operation on a flight to an unfamiliar destination, over an unfamiliar route, or a route they have not flown over in the last 90 days, the PIC and second in command (SIC) update themselves using all information pertinent to the safe conduct of the flight. This may include, but is not limited to, one or more of the following resources:

- Personal briefing (route qualification required by § 121.463).
- Aeronautical charts, with regard to prominent geographical features such as mountain ranges and passes, rising terrain, locally hazardous areas, lakes, rivers, and coastlines.
- FAA webcams (https://weathercams.faa.gov/).
- The Alaska Airport/Facilities Directory website (https://dot.alaska.gov/airport-portal.shtml).
- Appropriate company wind chart(s) for airport runway reports (if available).
- NOTAMs.

D.6 En Route, Flight Monitoring, and Communications

D.6.1 General

An aircraft dispatcher or fight follower monitors the progress of each flight under that individual's control until the flight has landed or passed beyond the area of control, or until the individual is properly relieved by another dispatcher or flight follower. Flight monitoring, at a minimum, consists of the monitoring of each aircraft's fuel state, flight time remaining, destination and alternate airport weather trends, en route winds and weather (including PIREPs), ATC constraints, and the status of airport and navigational facilities. Section 121.99 requires that reliable and rapid two-way communications between each flight and the aircraft dispatcher be available at any point in the flight, including overwater portions of international flights.

D.6.2 Domestic Operations

Section 121.601(c) requires the aircraft dispatcher to report to the PIC any additional information that could affect the safety of the flight. This information may be delivered by voice message or by other means, such as the ACARS.

D.6.3 Supplemental Operations

Section 121.603(b) requires that during a flight, the PIC shall obtain any additional available information of meteorological conditions and irregularities of facilities and services that may affect the safety of the flight.

D.6.4 Flight Monitoring

All flights released under STOP are monitored, tracked, and controlled by a dispatcher or flight follower. For flights to remote sites without phone service, flightcrews are provided with appropriate satellite communications. During the flight:

- The dispatcher or flight follower:
  - Provides the PIC with any additional available information of meteorological conditions (including any changes in primary weather information or adverse weather phenomena, such as clear air turbulence, thunderstorms, or low-altitude wind shear) that may affect the safety of the flight.
  - Provides the PIC with any additional available information regarding irregularities of facilities and services that may affect the safety of the flight, including any changes in NOTAMs (refer to § 121.601).
  - Monitors any and all information available for weather at the destination and alternates, as well as the progress of the flight. All flights on an IFR flight plan are tracked using aircraft situation display software.
- The PIC:
  - Ensures contact is established with the dispatcher or flight follower as soon as practical after takeoff, reaching cruise altitude in nonturbine airplanes or leaving 10,000 MSL in turbine airplanes.
  - Requests any available updated information for the destination and alternate airports at approximately 30 minutes before ETA when a flight has a cruise time of more than 1 hour. If the report does not indicate conditions suitable for a visual approach, the flight proceeds to the alternate airport. Ensures that the dispatcher or flight follower is informed of the intent, either directly or via relay.
  - When contacting the dispatcher or flight follower, or forwards progress reports to the dispatcher or flight follower, includes ETA at the destination (or alternate), fuel remaining, and aircraft maintenance status.

D.6.5 En Route Terrain Clearance

Part 121 subpart I contains limitations on weights at which aircraft may be dispatched due to terrain clearance requirements. These limitations apply to all types of aircraft operated under part 121.

D.6.6 En Route IFR to VFR Transition Point

Filing to the destination will normally be IFR to a point where VFR can be expected to be achieved. The IFR to VFR transition point should be no more than 50 NM from the intended destination. If upon arrival at the transition point the crew is unable to establish and maintain VFR through landing, then the flight proceeds to the primary alternate. If the weather unexpectedly begins to deteriorate while en route to the destination, but after transitioning to VFR, a course change may need to be made to maintain VFR. If VFR cannot be maintained to destination, coordinate with ATC for an IFR clearance to the alternate airport.

Maintain terrain clearance visually until at or above minimum safe altitudes and an IFR clearance has been obtained.

D.7 Destination Approach and Arrival

D.7.1 General

Flight operations to destination airports with weather reporting deficiencies proceed no further than the final approach fix on an instrument approach to the airport, or descend no lower than the minimum en route altitude, unless the flightcrew has visual contact with the runway and can maintain VFR in VMC.

D.7.2 Visual Inspection

In VMC conditions, the flightcrew conduct a visual inspection of the airport prior to landing. This inspection will allow the flightcrew to assess the winds using a windsock and/or environmental factors (e.g., blowing/drifting snow, wind over water, or leaf and tree movement). When available, these visual cues are compared to the onboard FMS-derived wind calculations, ensuring optimum situational awareness. This wind check prior to landing ensures an approach and landing into the wind and compliance with the company's crosswind limitation. On final approach, the flightcrew verify the indicated air speed is greater than the aircraft's ground speed.

D.7.3 Barometric Altimeter Validation

In addition to verifying wind conditions, the visual inspection of the airport prior to landing allows the flightcrew to validate or correct the barometric altimeter using the aircraft's radio altimeter system.

D.7.4 Radio Altimeter Setting

Prior to the descent for approach to the airport, the flightcrew obtain and set the nearest reporting station in accordance with § 91.121(a)(1) and the AIM Chapter 7, Section 2, Paragraph 7-2-2, Barometric Pressure Altimeter Errors, subparagraph a. To ensure an equivalent level of safety, the STOP requires the nearest reporting station to be within 75 NM of the destination airport. During the visual inspection, the flightcrew note the aircraft's radio altimeter reading. This above-ground height is added to the airport's elevation, which then gives an MSL height. When required, the flightcrew makes an adjustment of the barometric altimeter to synchronize and ascertain the correct MSL altitude.

D.8 Departure Procedures From a STOP Airport

D.8.1 General

Airports with weather deficiencies are likely to not have instrument departure procedures (i.e., a Standard Instrument Departure (SID) or Obstacle Departure Procedure (ODP)).

D.8.2 Departures

Departures from airports with weather deficiencies are done in VFR until reaching the minimum IFR altitude. Where an ODP is published, it is briefed immediately before departure and adhered to as published until:

- A safe altitude is reached,
- The flight has joined the IFR structure, or
- The flight is in VMC where the terrain and obstacles are clearly identified and avoided by a minimum of 1,000 ft.

D.8.3 Departure Data

Departure data is determined using reported METAR elements. When elements are missing, weather data elements are obtained from portable weather observation device or estimation techniques (see Section D.2.3).

D.9 Training for Dispatchers, Flight Followers, and Flightcrew

D.9.1 General

The training curriculum ensures that the dispatcher, flight follower, and flightcrew learn and are able to demonstrate understanding of the following policies and procedures associated with STOP:

- When the program applies.
- The requirements for releasing flights under the program.
- A proven understanding about the pilot and management notification requirements when a flight is released under the program.

The STOP training is required for both initial and annual recurrent training events. Documentation of both initial and annual recurrent training is maintained for each individual.

D.9.2 Introduction for Dispatchers and Flight Followers

Introduction to STOP training for dispatchers and flight followers covers the following topics:

- Regulatory basis for the approved program.
- Minimum weather requirements.
- Dispatch and flight following procedures in accordance with STOP.
- Verification of need for STOP.
- Notification and approval requirements.
- Contents of the weather package.
- Civil twilight or nighttime operational requirements.
- Fuel requirements.
- Briefing requirements.
- Airport risk assessment, strip check, or airport condition report.

- Flight operation in accordance with STOP.
- Airport and runway requirements.
- Operations in Class G Airspace.
- Portable weather observation devices.
- Usage.
- Storage.
- Weather Device Observation Log.

D.9.3 Introduction for Flightcrews

Introduction to STOP training for flightcrews covers the following topics:

- Regulatory basis for the approved program.
- Flightcrew member procedures in accordance with STOP.
- Verification of a need for STOP.
- Notification and approval requirements.
- Weather and daylight requirements (or lighting).
- Fuel requirements—inclusion of 15 minutes additional fuel for airport survey.
- Briefing requirements, including pictorials as available and risk assessment factors, in accordance with § 121.463 route qualification.
- Flight operation in accordance with STOP.
- Flight planning (IFR and/or VFR).
- Airport and runway requirements.
- Operations in Class G Airspace.
- Operations within 50 NM of the departure point.
- Operations 50 NM from the IFR en route system.
- Operations when an approach is not possible (diversion).
- Required communications during STOP.
- Use of portable weather observation devices.
- Storage of devices.
- Weather Device Observation Log.
- Classroom training with devices.
- Calibration training.
- Taking a weather observation and recording that information.

D.10 Approval

STOP programs are approved by the CH's certificate management office (CMO).

The approved STOP program can be a standalone approved manual, or it may be incorporated into the CH's accepted manual system.

D.11 Compliance Monitoring, Quality Assurance (QA), and Auditing

D.11.1 Compliance Monitoring

Compliance monitoring is intended to ensure, through a department audit, the following:

- The dispatcher, flight follower, and flightcrew comply with the procedures, provisions, and limitations of the STOP program.
- They all perform the process and procedures required to support the STOP,
- Compliance monitoring is documented, including the individual who conducted the audit and the date it was completed.

D.11.2 Quality Assurance (QA)

The QA process is an audit of the CH's compliance in the use of STOP process and procedures. STOP procedure audits are conducted at the conclusion of each STOP flight by the CH's designated individual. QA audits are documented with the individual who conducted the audit and the date it was completed.

D.11.3 STOP Audit

The contents of each STOP flight envelope are audited for compliance; to include procedures, weather reports, NWS and EWINS TAFs, manager approver signatures, and the portable weather observation log, as well as a complete audit of all flightplan paperwork. The STOP audit is documented with the individual who conducted the audit and the date it was completed.

All or any deficiencies identified as a result of the monitoring and QA audits are reported to the DO and followed through with the CH's Safety Management System (SMS) program followup. If systemic deficiencies with the STOP procedures are identified, a plan for corrective action is developed. The evaluation of these discrepancies is conducted in accordance with the CH's SMS program. For any identified deficiencies, an analysis is conducted to ascertain whether the deficiencies are addressed through Safety Risk Management (SRM) (per 14 CFR part 5, § 5.73(b)) or a correction (per § 5.75). If the root cause of the deficiencies is a shortfall in compliance with the designed operational process (e.g., required training was not completed), the deficiency is addressed through a correction (per § 5.75). However, if the deficiency is associated with the basic design of the control (e.g., inadequate training curriculum), the SRM process would be repeated to revise the procedure in the areas where it was found deficient (per § 5.73(b)). The performance, results, and corrective actions of these audits are measured at least annually through the company's quality audit program.

Appendix E Abbreviations, Acronyms, and Initialisms

14 CFR	Title 14 of the Code of Federal Regulations
3D	Three Dimensional
A	Arctic
AAG	Alaska Aviation Guidance
AAWU	Alaska Aviation Weather Unit
AC	Advisory Circular
AC	Convective Outlook
ACARS	Aircraft Communications Addressing and Reporting System
ACFJ	Australia, Canada, France, and Japan
ACSL	Altocumulus Standing Lenticular
ADS-B	Automatic Dependent Surveillance-Broadcast
AFD	Aviation Forecast Discussion
AFIS	Automatic Flight Information Service
AGL	Above Ground Level
AIM	Aeronautical Information Manual
AIREP	Aircraft Report
AIRMET	Airmen's Meteorological Information
ALP	Airport Location Point
AMDAR	Aircraft Meteorological Data Relay
AP	Anomalous Propagation
app	Application
ARTCC	Air Route Traffic Control Center
ASI-AD	Aviation Safety Inspector—Aircraft Dispatcher
ASOS	Automated Surface Observing System
ATC	Air Traffic Control
ATCSCC	Air Traffic Control Systems Command Center

ATCT	Air Traffic Control Tower	
ATIS	Automatic Terminal Information Service	
AWC	Aviation Weather Center	
AWIPS	Advanced Weather Interactive Processing System	
AWOS	Automated Weather Observing System	
AWRP	Aviation Weather Research Program	
AWW	Airport Weather Warning	

Note: AWW formerly stood for the Alert Severe Weather Watch Bulletin as well as the Severe Weather Forecast Alert.

°C or C	degrees Celsius or Celsius
c	Continental
cA	Continental Arctic
CAPE	Convective Available Potential Energy
CAT	Clear Air Turbulence
Cb	Cumulonimbus
CCSL	Cirrocumulus Standing Lenticular
CH	Certificate Holder
CIP	Current Icing Product
CIT	Convectively Induced Turbulence
cm	centimeter
C-MAN	Coastal-Marine Automated Network
CME	Coronal Mass Ejection
CMO	Certificate Management Office
CONUS	Contiguous United States
cP	Continental Polar
CPHC	Central Pacific Hurricane Center
CRC	China-Russian Federation Consortium
cT	Continental Tropical
CWA	Center Weather Advisory
CWSU	Center Weather Service Unit

dBZ	decibels of Z
DO	Director of Operations
DOC	Department of Commerce
DOD	Department of Defense
DOT	Department of Transportation
E	East
ECFP	Extended Convective Forecast Product
EDR	Eddy Dissipation Rate
EF	Enhanced Fujita
EFB	Electronic Flight Bag
ENE	East-Northeast
ESE	East-Southeast
ETA	Estimated Time of Arrival
EUV	Extreme Ultraviolet
EWINS	Enhanced Weather Information System
°F or F	degrees Fahrenheit or Fahrenheit
FA	Area Forecast
FAA	Federal Aviation Administration
FACA	Area Forecast for Caribbean
FAGX	Area Forecast for Gulf of Mexico
FB	Wind and Temperature Aloft Forecast
FIP	Forecast Icing Product
FIR	Flight Information Region
FIS	Flight Information Service
FIS-B	Flight Information Service-Broadcast
FL	Flight Level
FOC	Flight Operations Center
fpm	feet per minute
FSS	Flight Service Station
ft	foot

g	gram
GA	General Aviation
GCR	Galactic Cosmic Rays
GFA	Graphical Forecasts for Aviation
GFS	Global Forecast System
GNSS	Global Navigation Satellite System
GOES	Geostationary Operational Environmental Satellite
GPS	Global Positioning System
GTG	Graphical Turbulence Guidance
HEMS	Helicopter Emergency Medical Services
HF	High Frequency
HF COM	High Frequency Communications
Hg	Mercury
HIWC	High Ice Water Content
hPa	hectopascal
hr	hour
HRRR	High Resolution Rapid Refresh
IAP	Instrument Approach Procedure
IATA	International Air Transport Association
ICAO	International Civil Aviation Organization
IFR	Instrument Flight Rules
IMC	Instrument Meteorological Conditions
in	inch
inHg	inches of mercury
IR	Infrared
ITCZ	Intertropical Convergence Zone
ITWS	Integrated Terminal Weather System
J	joule
J/g	joules per gram
$J\ g^{-1}\ K^{-1}$	joules per gram Kelvin

J/kg	joules per kilogram
K	Kelvin
kg	kilogram
kg/m	kilogram per meter
kg/m³	kilogram per cubic meter
K-H	Kelvin-Helmholtz
km	kilometer
km/h	kilometers per hour
kt	knot
kW	kilowatt
LAA	Local Airport Advisory
LAMP	Localized Aviation MOS Program
lb	pound
LCL	Lifted Condensation Level
LFC	Level of Free Convection
LI	Lifted Index
LIFR	Low Instrument Flight Rules
LLWAS	Low-Level Wind Shear Alert System
LLWS	Low-Level Wind Shear
LRNS	Long-Range Navigation System
m	Maritime
m	meter
m³	cubic meter
mA	Maritime Arctic
mb or mbar	millibar
MCS	Mesoscale Convective System
MEL	Minimum Equipment List
METAR	Aviation Routine Weather Report
mg/m³	milligrams per cubic meter
MHz	megahertz

mi	mile
min	minute
MIS	Meteorological Impact Statement
MOS	Model Output Statistics
mP	Maritime Polar
mph	miles per hour
m/s	meters per second
MSL	Mean Sea Level
mT	Maritime Tropical
MVFR	Marginal Visual Flight Rules
MWO	Meteorological Watch Office
MWT	Mountain Wave Turbulence
N	North
N/A	Not Applicable
NAM	North American Model
NAS	National Airspace System
NASA	National Aeronautics and Space Administration
NAVAID	Navigational Aid
NCEP	National Centers for Environmental Prediction
NCO	NCEP Central Operations
NCWF	National Convective Weather Forecast
NDFD	National Digital Forecast Database
NE	Northeast
NESDIS	National Environmental Satellite, Data, and Information Service
NEXRAD	Next Generation Weather Radar
NextGen	Next Generation Air Transportation System
NHC	National Hurricane Center
nm	nanometer
NM	nautical mile
NNE	North-Northeast

NNW	North-Northwest
NOAA	National Oceanic and Atmospheric Administration
NOTAM	Notice to Air Missions
NW	Northwest
NWP	Numerical Weather Prediction
NWS	National Weather Service
OAT	Outside Air Temperature
ODP	Obstacle Departure Procedure
OMO	One-Minute Observation
OPC	Ocean Prediction Center
OPMET	Operational Meteorological
P	Polar
PECASUS	Pan-European Consortium for Aviation Space Weather User Services
PGF	Pressure Gradient Force
PIC	Pilot in Command
PIREP	Pilot Weather Report
POES	Polar Operational Environment Satellite
POI	Principal Operations Inspector
Prog	Prognostic
psi	pounds per square inch
QA	Quality Assurance
Radar	Radio Detection and Ranging
RCM	Radar Coded Message
RCO	Remote Communications Outlet
RH	Relative Humidity
ROFOR	Route Forecast
RTMA	Real-Time Mesoscale Analysis
RVR	Runway Visual Range
S	South
s or sec	second

SAA	Special Activity Airspace
SAB	Satellite Analysis Branch
SAS	Safety Assurance System
SATCOM	Satellite Communications
SAW	Aviation Watch Notification Message
SE	Southeast
SFRA	Special Flight Rules Area
SI	International System of Units
SIC	Second in Command
SID	Standard Instrument Departure
SIGMET	Significant Meteorological Information
SIGWX	Significant Weather
SLD	Supercooled Large Drop
SLWC	Supercooled Liquid Water Content
sm	statute mile
SMS	Safety Management System
SO_2	Sulfur Dioxide
SPC	Storm Prediction Center
SPECI	Aviation Selected Special Weather Report
SRM	Safety Risk Management
SSE	South-Southeast
SSW	South-Southwest
STOP	Special Terminal Operation Procedures
SUA	Special Use Airspace
SW	Southwest
SWPC	Space Weather Prediction Center
T	Tropical
TAF	Terminal Aerodrome Forecast
TCA	Tropical Cyclone Advisory
TCAC	Tropical Cyclone Advisory Center

TCF	TFM Convective Forecast
TCu	Towering Cumulus
TDWR	Terminal Doppler Weather Radar
TDZ	Touchdown Zone
TFM	Traffic Flow Management
TFR	Temporary Flight Restriction
TMS	Traffic Management System
TMU	Traffic Management Unit
TRACON	Terminal Radar Approach Control
TUTT	Tropical Upper Tropospheric Trough
U.S.	United States
UAT	Universal Access Transceiver
UCAR	University Corporation for Atmospheric Research
UTC	Coordinated Universal Time
VAA	Volcanic Ash Advisory
VAAC	Volcanic Ash Advisory Center
VAD	Velocity Azimuth Display
VAG	Graphical Version of the Volcanic Ash Advisory
VAR	Volcanic Activity Report
VCP	Volume Coverage Pattern
VEIA	Visibility Estimation through Image Analytics
VFR	Visual Flight Rules
VHF	Very High Frequency
VMC	Visual Meteorological Conditions
VNR	VFR Flight Not Recommended
VOR	Very High Frequency Omni-Directional Range Station
V_R	Rotation Speed
W	watt
W	West
$W\ m^{-1}\ K^{-1}$	watt per meter-Kelvin

WADL	West African Disturbance Line
WAFC	World Area Forecast Center
WAFS	World Area Forecast System
WARP	Weather and Radar Processor
WFO	Weather Forecast Office
WIFS	WAFS Internet File Service
WMO	World Meteorological Organization
WNW	West-Northwest
WPC	Weather Prediction Center
WSP	Weather Systems Processor
WSR-88D	Weather Surveillance Radar—1988 Doppler
WSW	West-Southwest
Wx	Weather
yd	yard
Z or z	Zulu

Appendix F Units of Measurement

°C or C	degrees Celsius or Celsius
cm	centimeter
dBZ	decibels of Z
°F or F	degrees Fahrenheit or Fahrenheit
fpm	feet per minute
ft	foot
g	gram
hPa	hectopascal
hr	hour
in	inch
inHg	inches of mercury
J	joule
J/g	joules per gram
J g^{-1} K^{-1}	joules per gram-Kelvin
J/kg	joules per kilogram
K	Kelvin
kg	kilogram
kg/m	kilogram per meter
kg/m^3	kilogram per cubic meter
km	kilometer
km/h	kilometers per hour
kt	knot
kW	kilowatt
lb	pound
m	meter
m^3	cubic meter

mb or mbar	millibar
mg/m^3	milligrams per cubic meter
MHz	megahertz
mi	mile
min	minutes
mph	miles per hour
m/s	meters per second
nm	nanometer
NM	nautical mile
psi	pounds per square inch
s or sec	second
sm	statute mile
W	watt
W m^{-1} K^{-1}	watt per meter-Kelvin

Note: The units of measurement shown above are occasionally listed in uppercase and/or lowercase in figures, tables, and examples throughout the document.

Appendix G Websites

- Federal Aviation Administration (FAA) Aviation Weather Cameras: https://weathercams.faa.gov/.
- FAA Air Traffic Notices to Air Missions (NOTAM): https://www.faa.gov/pilots/safety/notams_tfr.
- FAA Flight Service: https://www.1800wxbrief.com/.
- FAA home page: https://www.faa.gov/.
- Flight Information Service-Broadcast (FIS-B) malfunctions not attributed to aircraft system failures or covered by active Notices to Air Missions (NOTAM) via the ADS-B/TIS-B/FIS-B Problem Report: https://www.faa.gov/exit/?pageName=this%20form&pgLnk=http%3A%2F%2Fgoo%2Egl%2Fforms%2FisWDKYpYYv/.
- International Civil Aviation Organization (ICAO) home page: https://www.icao.int/.
- National Weather Service (NWS) Alaska Aviation Guidance (AAG): https://weather.gov/arh/aag.
- NWS Alaska Aviation Weather Unit (AAWU): https://www.weather.gov/aawu/.
- NWS Aviation Weather Center (AWC): https://www.aviationweather.gov/.
- NWS Center Weather Service Units (CWSU): https://www.weather.gov/aviation/cwsu.
- NWS Central Pacific Hurricane Center (CPHC): https://www.nhc.noaa.gov/?cpac.
- NWS Graphical Forecasts for Aviation (GFA) code charts: https://www.aviationweather.gov/gfa/help?page=products.
- NWS Localized Aviation Model Output Statistics (MOS) Program (LAMP) airports: https://www.nws.noaa.gov/mdl/gfslamp/gfslamp.shtml.
- NWS LAMP description: https://www.nws.noaa.gov/mdl/gfslamp/docs/LAMP_description.shtml.
- NWAS LAMP home page: https://www.nws.noaa.gov/mdl/lamp/index.shtml.
- NWS National Centers for Environmental Prediction (NCEP) Central Operations (NCO) Model Analyses and Guidance: https://mag.ncep.noaa.gov/.
- NWS National Hurricane Center (NHC): https://www.nhc.noaa.gov/.
- NWS Ocean Prediction Center (OPC): https://ocean.weather.gov/.
- NWS Real-Time Mesoscale Analysis (RTMA) Alternative Report of Surface Temperature: https://nomads.ncep.noaa.gov/pub/data/nccf/com/rtma/prod/airport_temps/.
- National Oceanic and Atmospheric Administration (NOAA) Washington Volcanic Ash Advisory Centers (VAAC): https://www.ssd.noaa.gov/VAAC/vaac.html.
- NWS Weather Forecast Office (WFO) Honolulu: https://www.weather.gov/HFO.
- NWS Weather Prediction Center (WPC): https://www.wpc.ncep.noaa.gov/.
- Weather radars (Department of Defense (DOD), FAA, and NWS' next generation weather radar (NEXRAD) and FAA's terminal Doppler weather radar (TDWR)): https://radar.weather.gov/.